Martin Schlichenmaier

Krichever–Novikov Type Algebras

De Gruyter Studies in Mathematics

Volume 53

Martin Schlichenmaier

Krichever–Novikov Type Algebras

Theory and Applications

DE GRUYTER

Mathematics Subject Classification 2010
Primary: 17B65, 17B66, 17B67, 17B68, 17B81; Secondary: 14H10, 14H15, 4H55, 17B56, 30F30, 32G15, 81R10, 81T40

Author
Prof. Dr. Martin Schlichenmaier
University of Luxembourg
Mathematics Research Unit
Rue Richard Coudenhove-Kalergi 6
1359 Luxembourg
martin.schlichenmaier@uni.lu

ISBN 978-3-11-026517-0
e-ISBN 978-3-11-027964-1
Set-ISBN 978-3-11-028025-8
ISSN 0179-0986

Library of Congress Cataloging-in-Publication Data
A CIP catalog record for this book has been applied for at the Library of Congress.

Bibliographic information published by the Deutsche Nationalbibliothek
The Deutsche Nationalbibliothek lists this publication in the Deutsche Nationalbibliografie; detailed bibliographic data are available on the Internet at http://dnb.dnb.de.

Typesetting: PTP-Berlin, Protago-TEX-Production GmbH, Berlin
Printing and binding: CPI books GmbH, Leck
♾Printed on acid-free paper
Printed in Germany

MIX
Papier aus verantwortungsvollen Quellen
FSC® C003147

www.degruyter.com

Preface

Krichever–Novikov type algebras are examples of infinite dimensional Lie algebras. It is the goal of this monograph to introduce them, to study their properties and to give some exemplary applications.

The elements of the Krichever–Novikov type algebras are meromorphic objects on a compact Riemann surface which are holomorphic outside a fixed set of points. Objects could mean functions, vector fields, forms of certain weights, matrix valued functions and more. Algebras could mean Lie algebras, but also associative algebras, super algebras, Clifford algebras, and more.

The theory of infinite dimensional Lie algebras (and of course algebras of other types) is still a very lively research field. Without additional structures there is no hope of really understanding them and possibly obtaining classification results as in the finite-dimensional Lie case.

From the purely mathematical point of view, the Krichever–Novikov type algebras are of interest as they come as geometric objects. As such, their geometric interpretation will give additional information for algebraic understanding. Vice versa, from their algebraic structures, in particular from their representations, we expect results to be useful in the geometric context. An example of such an application will be given in the context of the moduli space of marked curves in this book.

Another important approach to introduce additional structures for infinite dimensional Lie algebras is to study only graded Lie algebras. In particular, those for which the homogeneous subspaces are finite-dimensional. With such a grading one can introduce triangular decompositions, highest weight representations, semi-infinite wedge forms, and many more things. Our Krichever–Novikov type algebras are in general not graded. It was the important observation of Krichever and Novikov that a weaker concept, an *almost-grading*, will be enough to do most of the above-mentioned constructions. A graded structure means that the result of an operation with homogeneous elements will again be homogeneous of degree given by the sum of the degree of the individual elements. If an algebra is only almost-graded, then the result will not necessarily be homogeneous again. But the degree of its homogeneous components will lie in a fixed range around the ideal value given by the sum of the degree. See Definition 3.1 for a precise statement. In the case where our set of points A where poles are allowed consists of more than one point but is still finite, then the Krichever–Novikov type algebras are almost-graded. The almost-grading will depend on the splitting of A into two disjoint non-empty subsets I and O. Often such a splitting is naturally given by the applications. Hence, our Krichever–Novikov type algebras have two additional properties: they are given by geometric objects, and they admit an almost-grading, which in some important aspects is nearly as good as a grading.

Krichever–Novikov type algebras did not come out of nowhere. They generalize in a natural manner the Virasoro algebra and its relatives, like the affine Lie algebras and many more, in the sense explained in the following. The Virasoro algebra without central extension, also sometimes called Witt algebra, has two different realizations. One realization is as a complexification of the Lie algebra of polynomial vector fields on the circle, i.e., of vector fields which have only a finite number of Fourier modes. By analytic extensions of these vector fields to the punctured complex plane $\mathbb{C}^* = \mathbb{C} \setminus \{0\}$ we obtain the second realization. In this realization the Virasoro algebra is the Lie algebra of those meromorphic vector fields on the Riemann sphere which are holomorphic outside of $\{0, \infty\}$. The grading is given by the order of the vector field evaluated at $z = 0$ shifted by one.

Without any doubt the Virasoro algebra is an extremely important algebra. It appears in mathematics, mathematical physics, theoretical physics, and in other fields. A huge amount of literature has appeared about it and its representations. See [96, 118, 122] as references to some books which contain more references to the original work. It turned out to be fundamental e.g. in Conformal Field Theory (CFT), integrable systems, Korteweg-de Vries (KDV) equations, partial differential equations, and Schramm–Loewner evolution equation (SLE).

In two-dimensional CFT the application of Virasoro algebra is naturally restricted to either genus zero field theory, or alternatively to a local description yielding local operators.

Hence, from this point of view it is natural to ask for a global operator approach to arbitrary genus CFT. This was the crucial point of view of Krichever and Novikov in 1987 [140–142]. They replaced the genus zero Riemann sphere by an arbitrary compact Riemann surface without any restriction of the genus, and the set of points $\{0, \infty\}$ in the classical case by an arbitrary set of two points $\{P_-, P_+\}$. The first step, to introduce the meromorphic objects of the generalization, is rather straightforward. The second step, to introduce a replacement for the graded structure, existing for Virasoro algebra and its representations, is much harder. This structure is the above-mentioned *almost-grading*. Next they gave an explicit method of constructing certain types of central extensions and certain representations for the vector field algebra and the function algebra. Central extensions are indispensable in the context of quantization of field theories. There, some naive constructions, which work well in finite dimensions, will become ill-defined in the infinite dimensional case. Here dimension refers to the dimension of the symmetry algebra and not to the dimension of space time. To make the constructions well-defined again they have to be *regularized*, as physicists call it. Regularization is a sound mathematical concept. In many cases regularization is related to passing to a projective representation of the symmetry algebra. It can be made into an honest Lie action again by enlarging the symmetry algebra and taking a central extension of it.

Quite a number of mathematicians and physicists contributed fundamentally to the development of the theory of the two-point Krichever–Novikov type algebras. For example, from the mathematical side, Sheinman studied the representation of the algebras of affine type intensively [233–236]. Different group of physicists adopted the algebras and carried out further constructions in the context of quantum field theories. See Section 14.5 for an incomplete but nevertheless quite long list of contributions and references.

From the applications the need for a multi-point generalization, including an almost-grading, was clearly evident. The two-point situation (meaning that poles are only allowed at two points) was too restrictive. As an example, in string theory (a two-dimensional CFT), where the Riemann surface corresponds to the world sheet of the string, we have to deal with K free incoming strings and M free outgoing strings. The numbers K and M should be finite but arbitrary. The only condition is that the "sum of the momenta of the incoming strings" has to be equal to the "sum of the momenta of the outgoing strings". Hence, the natural geometric data is a Riemann surface Σ and a finite set of points A. If $A = \emptyset$, then our objects will be globally holomorphic. In particular for a genus greater than or equal to two we will not have nontrivial vector fields. If $A \neq \emptyset$ it should be divided into two disjoint non-empty subsets I and O. The points in I correspond to the incoming free strings and the points in O to the outgoing free strings.

The author of this monograph developed systematically the theory of the Krichever–Novikov type algebras (as they are nowadays called) in this more general context [204–207, 210, 214, 215]. Again, the first step of introducing the meromorphic objects is straightforward. However, introducing the almost-grading turns out to be more difficult. Indeed, only after fixing a splitting of A into I and O is such an almost-grading essentially fixed. Also, the understanding of central extensions is much more complicated. The author succeeded, after fixing an arbitrary almost-grading, in the classification of equivalence classes of central extensions which are compatible with the almost-grading [214, 215]. In particular, for the vector field algebra, after fixing an almost-grading by a splitting of A, there is up to equivalence and rescaling a unique nontrivial central extension. Hence, the situation is similar to the Virasoro case. Without the almost-grading this would not in general be true. Also, we have to stress that an essentially different almost-grading will yield a non-equivalent central extension.

Other researchers also worked on such generalizations, see e.g. Dick [53, 55]. But the crucial point to introduce an almost-grading was not done. Only Sadov [199] obtained some results which are related.

In this book I will present the theory of the Krichever–Novikov type algebras and give some of their applications. We will do this in the frame of the multi-point higher genus approach of the author. Both the two-point Krichever and Novikov results and the classical genus zero, the Virasoro case, are obtained as special results. No prior knowledge about them is necessary.

We have attempted to separate (and explain) which chapters and parts are necessary in order to follow the chain of arguments of the book, and which are special results, proofs etc., to be studied when needed. In particular, as some of the classification results are rather technical, I try to formulate the statements and their consequences first, and postpone the technical proofs to a separate section or chapter, so that the reader in a hurry could skip them.

We assume that the reader has some background knowledge of Lie algebras and the theory of compact Riemann surfaces. Nevertheless, for his or her convenience we will recall the basic concepts needed. For example, in Chapter 1 we present them for the Lie algebras. The experienced reader might safely skip this chapter. We do not expect that all readers are aware of Lie algebra cohomology and its relation to central extensions. As the construction and classification of central extensions for these Krichever–Novikov type algebras is one of the core topics of this monograph, we explain this theory in detail in Chapter 6. Modulo the above remarks the main part of the monograph is rather self-contained and shall be accessible for new-comers to the field. For some of the applications towards the end of the book we use some more knowledge from other fields.

As a kind of *Leitfaden*, or guide, let me first give the chapters which are the main body of the book.

The basic geometric structure and the basic algebraic objects are given in Chapter 2. Everything is constructed with the help of meromorphic forms of integer or half-integer weights which are holomorphic outside the set A of points.

In Chapter 3 we introduce the concept of an almost-grading. We will construct in our context an almost-grading for our algebras and modules which is associated with the splitting of A into two non-empty disjoint subsets.

Chapter 6 on central extensions is rather voluminous. After an introduction to Lie algebra cohomology, we discuss the relation to central extensions. We introduce central extensions for our algebras which are given by geometric means. The classification of almost-graded central extensions of the introduced algebras is one of the core results presented in this book. In particular, it turns out that all these central extensions can be given in geometric terms. For the vector field algebra with fixed almost-grading we essentially obtain a unique nontrivial central extension which is compatible with its almost-grading.

Chapter 9 deals with the current and the affine Lie algebras of Krichever–Novikov type. Higher genus current algebras of Krichever–Novikov type are Lie algebras of $\mathfrak{g}$-valued meromorphic functions, where $\mathfrak{g}$ is any finite-dimensional Lie algebra. In the two-point case these were already studied by Sheinman. The multi-point cases were given by the author. In this chapter we will show the existence and give the classification of almost-graded central extensions of the current algebras. The centrally extended algebras are called affine Lie algebras of Krichever–Novikov type.

As already mentioned, the above chapters contain the basics of the Krichever–Novikov type algebras. The intention was that they should stay rather self-contained. Chapter 4 and Chapter 5 are of more technical nature; they supply proofs and contain additional material. They could be skipped for a first reading. In Chapter 3 only the fact that there is an almost-grading and its fundamental properties have been formulated. The proof of the existence of an almost-grading is postponed to Chapter 4. Roughly speaking, we exhibit a set of basis elements by order considerations which gives the "homogeneous" subspaces. For this we apply Riemann–Roch type arguments. In Chapter 5 we give explicit expressions for the homogeneous basis elements. Depending on the genus this is done with the help of rational functions, theta functions, and σ-functions. This chapter is rather independent of the others. It is of importance in the context of explicit calculations for integrable systems and also in Chapter 11 in this book.

The next chapters deal with representations of the algebras. In Chapter 7 we study the fermionic Fock space representations of our algebras given by semi-infinite wedge forms. In the interpretation of Quantum CFT they give a representation of the global operators. In fact, this is the first occasion in the book where we are forced to use a regularization procedure yielding central extensions of the involved algebras.

In Chapter 8 we show that the semi-infinite wedge form also comes with the representation of an algebra which has a Clifford algebra-like structure. The corresponding field theoretical system is called $b - c$ system. We will show that for such $b - c$ systems an energy momentum tensor is associated whose modes will give a representation of a centrally extended Krichever–Novikov vector field algebra.

From a field-theoretical point of view, the current, respectively affine algebras, correspond to gauge symmetries. In this field-theoretical interpretation, the Sugawara construction relates the gauge symmetry to conformal symmetry which is realized by the vector field algebras, and respectively their central extensions. In Chapter 10 we present the arbitrary genus and arbitrary number of points version of the Sugawara construction. These results were jointly obtained with Oleg Sheinman [225].

Now we come to the first major application of the algebras. In Chapter 11 we present our global operator approach to Wess–Zumino–Novikov–Witten (WZNW) models and the Knizhnik–Zamolodchikov (KZ) connection. These models live over the moduli space of Riemann surfaces with marked points. The Krichever–Novikov objects give global objects in contrast to the semi-local approach of Tsuchiya, Ueno, and Yamada [254]. Again, the results were jointly obtained with Oleg Sheinman [226, 227]. The crucial observation is that a certain finite-dimensional subspace of the Krichever–Novikov vector field algebra (the so called "critical strip") operates as infinitesimal deformation of the moduli space. The corresponding action of the vector field on the representation space is given via the Sugawara construction.

Another application of the Krichever–Novikov algebra, in this case related to deformations of algebras, is given in Chapter 12. There, families of algebras on the torus and on the Riemann sphere are given in terms of generators and structure equations which deform the Witt, Virasoro and classical current algebra, despite the fact that they are formally rigid. This is partly joint work with Alice Fialowski [76, 77]. The reader might also be interested in having a look at them to get a feeling which type of algebras will appear as Krichever–Novikov type algebras. The chapter can be studied with the knowledge of the basic chapters.

The same is true for Chapter 13 on Lax operator algebras, where we report on a new class of current type algebras introduced by Krichever and Sheinman. More details about them can be found in the recent book by Sheinman called *Current Algebras on Riemann Surfaces*, [250].

In the final chapter, Chapter 14, I will mention some related subjects and further developments, mainly by giving references.

I would like to thank my former academic teacher and friend, the late Julius Wess. Besides other things, one of his influences was getting me interested in Krichever–Novikov algebras. Also important was my contact with other theoretical physicists. I would especially like to mention Loriano Bonora and the other members of his group at SISSA, in particular Marco Matone.

I would like also to thank Rainer Weissauer, for his willingness to accept [207] as a mathematical dissertation at the University of Mannheim.

My acquaintance with Oleg K. Sheinman from the Moscow Independent University and the Steklov Mathematical Institute in Moscow was very influential. On several research stays, e.g., at the Mathematisches Forschungsinstitut in Oberwolfach (MFO), Steklov Mathematical Institute, University of Mannheim and now University of Luxembourg, we further developed the theory and its applications. The following publications appeared as the fruit of this interaction [225–227], and [228].

I gratefully acknowledge the partial support during the process of obtaining the presented results and writing this book of the Volkswagenstiftung (RIP Programme at the MFO), the DFG Forschergruppe Mannheim - Heidelberg, the ESF Research Networking Programmes *Harmonic and Complex Analysis with Applications (HCAA)*, and *Interaction of low-dimensional Topology and Geometry with Mathematical Physics (ITGP)*, the Mittag-Leffler Institute, several internal Research Projects of the University of Luxembourg (the most recent one is GEOMQ11), and the support in the frame of the OPEN scheme of the Fonds National de la Recherche Luxembourg (FNR) with the project QUANTMOD O13/5707106.

Luxembourg, May 2014 Martin Schlichenmaier

Contents

1 Some background on Lie algebras

In the book we expect the reader to have some mathematical background on the theory of Lie algebras and the theory of Riemann surfaces. For the convenience of the reader we will collect in this chapter the necessary notions on Lie algebras. Furthermore, we will show certain special properties which we will use later. We suggest that the reader browse through the material in a first reading and return to it if a concept needs to be refreshed. The background on Riemann surfaces will be recalled in the relevant sections.

1.1 Basic definitions on Lie algebras

Let L be a vector space over a field $\mathbb{K}$ of characteristics different from two. In fact, in this book we will be dealing exclusively with the field $\mathbb{K} = \mathbb{C}$ of complex numbers.

Definition 1.1. A bilinear map

$$[\,.\,,\,.\,] : L \times L \to L \tag{1.1}$$

is called a *Lie product* for L if and only if we have

(1) (antisymmetry)

$$\forall x, y \in L : \quad [x, y] = -[y, x], \tag{1.2}$$

(2) (Jacobi identity)

$$\forall x, y, z \in L : \quad [[x, y], z] + [[y, z], x] + [[z, x], y] = 0. \tag{1.3}$$

The pair $(L, [\,.\,,\,.\,])$ is called a *Lie algebra* .

If the Lie product is known, we will drop it mostly in the notation.

Example 1.2. If $(A, \cdot)$ is an associative algebra, then $(A, [\,.\,,\,.\,])$ with the commutator

$$[x, y] := x \cdot y - y \cdot x \tag{1.4}$$

as Lie product is a Lie algebra.

Example 1.3. Let $\mathfrak{gl}(n, \mathbb{K})$ be the algebra of $n \times n$ K-valued matrices under the matrix product, then $\mathfrak{gl}(n, \mathbb{K})$ becomes a Lie algebra under the commutator.

Example 1.4. As a generalisation of Example 1.3, but still a special case of Example 1.2, we consider a vector space of arbitrary dimension. Denote by $\mathrm{End}(V)$ the algebra of linear endomorphism with product given by the composition of maps. Hence, $\mathrm{End}(V)$ with the commutator defines a Lie algebra. Quite often it is denoted by $gl(V)$.

Example 1.5. A differently constructed example is given by the following procedure: let A be an algebra (neither necessarily commutative, nor associative, nor with unit). Considered as vector space as in Example 1.4 the vector space of $\mathbb{K}$-linear maps. Then $\mathrm{End}_{\mathbb{K}}(A)$ is an associative algebra and hence a Lie algebra under the commutator.

Definition 1.6. A linear endomorphism $\phi \in \mathrm{End}_{\mathbb{K}}$ is called a *derivation* if and only if

$$\forall a, b \in A : \ \phi(a \cdot b) = \phi(a) \cdot b + a \cdot \phi(b). \tag{1.5}$$

Let $\mathrm{Der}_{\mathbb{K}}(A, A)$ be the space of all derivations, then it is a Lie algebra under the commutator in $\mathrm{End}_{\mathbb{K}}(A)$.

Many more infinite dimensional Lie algebras will appear in the main text. Furthermore, we will introduce Lie superalgebras in Section 2.8.

1.2 Subalgebras and ideals

Definition 1.7. Given a Lie algebra $(L, [.,.])$.
(a) A vector subspace $M \leq L$ of L is called a Lie subalgebra of L if

$$[M, M] \subseteq M. \tag{1.6}$$

(b) A subspace $M \leq L$ of L is called a (Lie) ideal of L if

$$[L, M] \subseteq M. \tag{1.7}$$

(c) The Lie algebra L is called abelian if

$$[L, L] = \{0\}. \tag{1.8}$$

Obviously, a Lie subalgebra M is itself a Lie algebra by restricting the Lie product of L to M. Also every Lie ideal is a Lie subalgebra. Given an element $x \in L$, then $\mathbb{C}x$ is a Lie subalgebra of L which is abelian.

Important examples of Lie subalgebras of $\mathfrak{gl}(n, \mathbb{C})$, (of $\mathfrak{gl}(2n, \mathbb{C})$ respectively) are given by the following classical Lie algebras

$$\begin{aligned} \mathfrak{sl}(n, \mathbb{C}) &:= \{X \in \mathfrak{gl}(n, \mathbb{C}) \mid \mathrm{tr}(X) = 0\}, \\ \mathfrak{so}(n, \mathbb{C}) &:= \{X \in \mathfrak{gl}(n, \mathbb{C}) \mid {}^tX = -X\}, \\ \mathfrak{sp}(2n, \mathbb{C}) &:= \{X \in \mathfrak{gl}(2n, \mathbb{C}) \mid {}^tX = -\sigma X \sigma^{-1}\}, \end{aligned} \tag{1.9}$$

where $\sigma = \begin{pmatrix} 0 & I_n \\ -I_n & 0 \end{pmatrix}$ is the standard symplectic form.

For every Lie algebra L we have the following ideals. The *center* is defined as

$$C(L) := \{x \in L \mid \forall y \in L : [x, y] = 0\}. \tag{1.10}$$

and the *derived subalgebra* defined as

$$L' := [L, L] := \langle [x, y] \mid x, y \in L \rangle_{\mathbb{K}}. \tag{1.11}$$

Note that $C(L)$ and L' are indeed ideals.

1.3 Lie homomorphism

Given two Lie algebras $(L, [..,]^{(1)})$ and $(M, [.,.]^{(2)}$. A linear map $\Phi : L \to M$ is called a *Lie homomorphism* if it respects the Lie algebra structure:

$$\forall x, y \in L: \; [\phi(x), \phi(y)]^{(2)} = \phi([x, y]^{(1)}). \tag{1.12}$$

The kernel $\ker \phi$ is a Lie ideal of L, the image $\operatorname{im} \phi$ a Lie subalgebra of M. We have the canonical short exact sequence

$$0 \longrightarrow \ker \phi \xrightarrow{\;i\;} L \xrightarrow{\;\phi\;} \operatorname{im} \phi \longrightarrow 0. \tag{1.13}$$

Given a Lie ideal J of L, then the vector space quotient L/J carries a natural Lie product given by

$$[x \bmod J, y \bmod J] := [x, y] \bmod J. \tag{1.14}$$

One obtains the short exact sequence

$$0 \longrightarrow J \xrightarrow{\;i\;} L \xrightarrow{\;\nu\;} L/J \longrightarrow 0. \tag{1.15}$$

1.4 Representations and modules

Given a Lie algebra L and a vector space V (over the same field $\mathbb{K}$). A *Lie action* of L on V (or for short an L-action on V) is given by a bilinear map

$$L \times V \to V, \quad (x, v) \mapsto x \,.\, v, \tag{1.16}$$

such that $\forall x, y \in L$, $\forall v \in V$ we have

$$[x, y] \,.\, v = x \,.\, (y \,.\, v) - y \,.\, (x \,.\, v). \tag{1.17}$$

Given such an action, we call V also a *Lie module* over L (or L-module) with module structure given by the action.

Equivalently, we can describe a Lie action in the following way: if we fix $x \in L$, then $v \mapsto \phi_x(v) := x \,.\, v$ defines a linear map $\phi_x \in \operatorname{End}(V)$. The condition (1.17) is equivalent to the fact that the map

$$\phi : L \to gl(V), \quad x \mapsto \phi_x \tag{1.18}$$

is a Lie homomorphism. In other words, Lie module structures on V correspond to Lie homomorphism $\phi : L \to gl(V)$. Such a Lie homomorphism is also called a *Lie representation* and V is called a representation space.

Strictly speaking, our requirement (1.17) defines a left Lie module structure. In a similar manner, we can describe right Lie modules by

$$v \,.\, [x, y] = (v \,.\, x) \,.\, y - (v \,.\, y) \,.\, x. \tag{1.19}$$

Given a right module $v \,.\, x$, it gives via $x \,.\, v := -v \,.\, x$, a left module structure and vice versa.

Example 1.8. Let $\mathbb{K}$ be the base field, then the *trivial module* is defined by

$$x \,.\, v = 0, \quad \forall x \in L, \forall v \in \mathbb{K}. \tag{1.20}$$

Example 1.9. Every Lie algebra operates on itself via

$$\mathrm{ad} : x \mapsto \begin{cases} L \to gl(L) \\ y \mapsto \mathrm{ad}_x(y) := [x, y]. \end{cases} \tag{1.21}$$

From the Jacobi identity it follows that it is a Lie homomorphism. This representation is called the *adjoint representation*. The kernel of the adjoint representation is the center of the Lie algebra L, i.e.,

$$\ker \mathrm{ad} = C(L) = \{x \in L \mid \forall y \in L : [x, y] = 0\}. \tag{1.22}$$

1.5 Simple Lie algebras

A Lie algebra L is called *simple* if and only if L is not abelian and does not have nontrivial ideals. In other words, if J is an ideal of L, then $J = \{0\}$ or $J = L$. Hence, for a simple Lie algebra L, the center and the derived subalgebra are trivial ideals. In this case this means

$$C(L) = \{0\} \quad \text{and} \quad L' = [L, L] = L. \tag{1.23}$$

A Lie algebra L is called *perfect* if $L' = [L, L] = L$. In particular, simple Lie algebras are perfect.

Over the field $\mathbb{C}$ the finite-dimensional simple Lie algebras are completely classified. This is done via the (irreducible) root systems ($n \in \mathbb{N}$)

$$\begin{gathered} A_n, n \geq 1, \quad B_n, n \geq 2, \quad C_n, n \geq 2, \quad D_n, n \geq 4, \\ E_6, \quad E_7, \quad E_8, \quad F_4, \quad G_2. \end{gathered} \tag{1.24}$$

For example, A_n corresponds to $\mathfrak{sl}(n+1, \mathbb{C})$.

Definition 1.10. A symmetric bilinear form $\beta : L \times L \to \mathbb{K}$ is called invariant if

$$\beta([x, y], z) = \beta(x, [y, z]) \quad \forall x, y, z \in L. \tag{1.25}$$

For a finite-dimensional Lie algebra we define the *Cartan–Killing form* to be

$$\beta(x, y) := \mathrm{tr}(\mathrm{ad}_y \circ \mathrm{ad}_y), \quad x, y \in L. \tag{1.26}$$

One verifies directly that it is an invariant symmetric bilinear form. This makes perfect sense for arbitrary finite-dimensional Lie algebras. But for example, if L is abelian, the adjoint representation will be zero and the Cartan–Killing form will vanish.

If L is a simple Lie algebra, then the Cartan–Killing form is non-degenerate. Moreover, in this case every invariant symmetric bilinear form will be a multiple of the Cartan–Killing form. For the Lie algebra $\mathfrak{sl}(n, \mathbb{C})$ (and the other matrix groups introduced above) up to multiplication with a non-zero constant, the Cartan–Killing form is given by $\beta(A, B) = \mathrm{tr}(A \cdot B)$.

1.6 Direct sum and semidirect sum

Given two Lie algebras $(L, [.,.]_1)$ and $(M, [.,.]_2)$, we introduce the *direct sum Lie algebra* on the vector space direct sum $L \oplus M$ by defining the Lie product on the summands independently, i.e.,

$$[(l_1, m_1), (l_2, m_2)] := ([l_1, l_2], [m_1, m_2]). \tag{1.27}$$

In a direct sum, both L and M are ideals.

A finite-dimensional Lie algebra L is called *semisimple* if L is the direct sum of simple Lie algebras L_i, $i = 1, \ldots, M$

$$L = L_1 \oplus L_2 \oplus \cdots \oplus L_M. \tag{1.28}$$

If L is semisimple, then L equals its derived algebra $L = L' = [L, L]$. In particular, semisimple Lie algebras are perfect Lie algebras. The Cartan–Killing form for a semisimple Lie algebra is also non-degenerate. In fact, the decomposition (1.28) is an orthogonal decomposition with respect to the Cartan–Killing form (1.26). But in the semisimple and not simple case an arbitrary invariant symmetric bilinear form will not necessarily be a multiple of the Cartan–Killing form. We will follow up on this in Chapter 9.

A finite-dimensional Lie algebra is called *reductive* if L is the direct sum of an abelian Lie algebra with a semisimple Lie algebra. The Lie algebra $\mathfrak{gl}(n, \mathbb{C})$ is a typical example of a reductive Lie algebra

$$\mathfrak{gl}(n) = \mathfrak{s}(n) \oplus \mathfrak{sl}(n), \tag{1.29}$$

where $\mathfrak{s}(n)$ denotes the abelian algebra of scalar matrices.

Later we will need a more general construction, the *semidirect sum* $L \rtimes M$ of Lie algebras . Here L and M are again Lie algebras. Furthermore, L is required to be an M-module, and M operates on L as a derivation, i.e.,

$$m\,.\,[l_1, l_2] = [m\,.\,l_1, l_2] + [l_1, m\,.\,l_2], \quad m \in M,\ l_1, l_2 \in L. \tag{1.30}$$

We define

$$[(l_1, m_1), (l_2, m_2)] := ([[l_1, l_2] + m_1 . l_2 - m_2 . l_1, [m_1, m_2]). \tag{1.31}$$

Proposition 1.11. *The prescription* (1.31) *defines a Lie algebra structure on the vector space direct sum $L \oplus M$. This Lie structure is called the semidirect sum of L by M and denoted by $L \rtimes M$.*

Proof. The antisymmetry and bilinearity is clear. It remains to check the Jacobi identity. For the triple product we obtain

$$\begin{aligned}[][(l_1, m_2), (l_2, m_2)], (l_3, m_3)] = ([[l_1, l_2], l_3] &+ [m_1 . l_2, l_3] - [m_2 . l_1, l_3] \\ &+ [m_1, m_2] . l_3 - m_3 . [l_1, l_2] - m_3 . (m_1 . l_2) \\ &+ m_3 . (m_2 . l_1), [[m_1, m_2], m_3])\end{aligned} \tag{1.32}$$

If we cyclically permute and add this up, then the sum of the terms of both types $[[l_1, l_2], l_3]$ and $[[m_1, m_2], m_3]$ is zero, as L and M are Lie algebras. Adding up the terms

$$[m_1, m_2] . l_3 - m_3 . (m_1 . l_2) + m_3 . (m_2 . l_1) \tag{1.33}$$

gives zero, as L is an M-module. Also adding up

$$[m_1 . l_2, l_3] - [m_2 . l_1, l_3] - m_3 . [l_1, l_2] \tag{1.34}$$

gives zero as M operates as derivation on L. Hence, in total the Jacobi identity is fulfilled. □

Both L and M are Lie subalgebras of $L \rtimes M$. Moreover, L is an ideal. We have the short exact sequence of Lie algebras

$$0 \longrightarrow L \xrightarrow{i_1} L \rtimes M \xrightarrow{p_2} M \longrightarrow 0, \tag{1.35}$$

and $L \rtimes M/L \cong M$. Note that the sequence is split exactly, as the splitting map i_2 is a Lie homomorphism. The standard direct sum is obtained via the semidirect sum construction if we consider L as a trivial M-module.

Remark 1.12. Central extensions are examples of another construction obtainable from certain other Lie algebras. As they are so "central" to this book we will study their definition, construction, and relations to Lie algebra cohomology in Chapter 6 in great detail.

1.7 Universal enveloping algebras

Given a Lie algebra L, we can construct an associative algebra $U(L)$ with unit which contains in its commutator algebra $(U(L), [.,.])$ the Lie algebra L as subalgebra. Consider L as a vector space and build the tensor algebra $T(L)$ as associative algebra (with unit) with product induced by concatenating the tensors, e.g.,

$$(x \otimes y) \cdot (z \otimes w \otimes u) := x \otimes y \otimes z \otimes w \otimes u. \tag{1.36}$$

Let J be the two-sided ideal generated as

$$J = (\ x_1 \otimes x_2 - x_2 \otimes x_1 - [x_1, x_2] \mid x_1, x_2 \in L\ ). \tag{1.37}$$

The *universal enveloping algebra* is defined as the associative algebra obtained as quotient

$$U(L) = T(L)/J. \tag{1.38}$$

We have the embedding

$$\phi : L \to U(L), \qquad x \mapsto x \bmod J. \tag{1.39}$$

If we equip $U(L)$ with the Lie product given by the commutator, then (by construction) ϕ will be a Lie homomorphism. Hence L can be realized as Lie subalgebra of $U(L)$.

The universal enveloping algebra fulfills the following universality conditions when given an associative algebra A with unit. If there is a Lie homomorphism

$$\psi : L \to (A, [.,.]), \tag{1.40}$$

then there is a unique associative algebra homomorphism

$$\hat{\psi} : U(L) \to A \quad \text{such that} \quad \hat{\psi} \circ \phi = \psi. \tag{1.41}$$

Conversely, every associative algebra homomorphism $\hat{\psi} : U(L) \to A$ will give a Lie homomorphism $\psi = \hat{\psi} \circ \phi$, $\psi : L \to (A, [.,.])$.

If we apply this to Lie representations $\psi : L \to gl(V)$, we see that such an L-representation always induces a $U(L)$ algebra representation on the same space and vice versa. It is obtained by successively applying the operators corresponding to the Lie algebra elements on the vectors $v \in V$.

Later on in the book we will need the *Poincaré–Birkhoff–Witt Theorem,* which states that if L is the Lie algebra, and $B = \{x_\alpha \mid \alpha \in I\}$ a basis which is ordered in a strictly increasing order with ordering set I, then a basis of $U(L)$ is given by sequences of the type

$$x_{\alpha_1}^{i_1} \cdot x_{\alpha_2}^{i_2} \cdots x_{\alpha_r}^{i_r}, \quad r \in \mathbb{N}_0, \quad i_r \in \mathbb{N}, \quad \alpha_1 \le \alpha_2 \le \cdots \le \alpha_r. \tag{1.42}$$

2 The higher genus algebras

In this chapter we will introduce important algebraic structures for certain meromorphic objects defined on an arbitrary compact Riemann surface Σ of genus $g \in \mathbb{N}_0$. Our starting point will be the canonical line bundle $\mathcal{K} = \mathcal{K}_\Sigma$ of the Riemann surface and its (half-)integer powers. The algebras will consist of meromorphic sections of these bundles with certain prescribed regularities. Moreover, we will discuss Lie algebra valued meromorphic functions, i.e., current algebras. In general, the introduced algebras will no longer be graded. In Chapter 3 we will introduce a generalization of the notion of a grading, and almost-grading. Furthermore, we will show that these algebras, at least when they are of Krichever–Novikov type, are almost-graded. In Chapter 6 we will consider their Lie algebraic central extensions.

2.1 Riemann surfaces

Throughout the whole book we will assume the reader has some familiarity with the theory of compact Riemann surfaces. The range of facts needed will vary from topic to topic, and also whether the reader is willing to accept the statements presented about the algebraic structure or whether he wants to see proofs. Hence, I decided not to group the background material together but to recall the necessary facts as needed.

We start by recalling that a *Riemann surface* Σ is a one-dimensional complex manifold. Unless otherwise stated, we always assume that it is compact. From the topological point of view they are classified by their genus $g = g(\Sigma)$. Intuitively, the genus is given by the number of "holes" Σ has. In a more precise manner, the rank of the first simplicial homology group $\mathrm{H}_1(\Sigma, \mathbb{Z})$ is twice the genus g. This says up to homology we have $2g$ independent 1-cycles on Σ. It is possible to introduce a special kind of basis, called *symplectic homology basis*, of 1-cycles α_i, β_i, $i = 1, \ldots, g$, such that with respect to the intersection pairing we have

$$\alpha_i \cdot \alpha_j = 0, \quad \alpha_i \cdot \beta_j = -\beta_j \cdot \alpha_1 = \delta_i^j. \tag{2.1}$$

Figure 2.1 gives an intuitive picture of the homology basis. For more detail see e.g., [203].

By definition, as a complex manifold, Σ admits a covering by coordinate charts $(U_i, z_i)_{i \in J}$. The coordinates z_i are related by holomorphic coordinate change maps $z_j = z_j(z_i)$. Holomorphic functions, differentials, sections, etc., can locally be given by holomorphic functions in the coordinates z_i which transform accordingly under the coordinate change maps. This can easily be extended by allowing the local data to be given by local meromorphic functions in z_i. In this way we also obtain meromorphic differentials, sections, and so on.

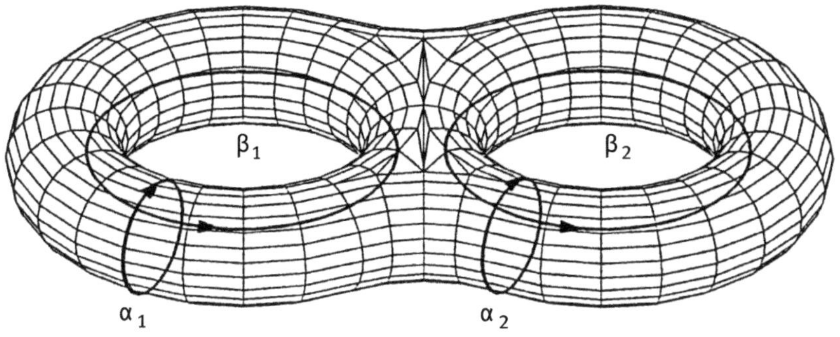

Fig. 2.1. Homology basis in genus two.

Compact Riemann surfaces can equivalently be described as smooth projective curves over the field $\mathbb{C}$. The correspondence is a very strong one in categorical terms, as it also relates vector bundles, sections, differentials, etc. This point of view has certain advantages. It is quite helpful if one studies degenerations of Riemann surfaces. Nevertheless, here we will mainly work in the complex-analytic language.

Recall that up to isomorphy there is only one compact Riemann surface of genus zero. In the analytic picture it is the Riemann sphere S^2, in the curve picture it is the projective line $\mathbb{P}^1(\mathbb{C})$. Compact Riemann surfaces of genus one are complex tori, $\mathbb{C}/L$ with L a lattice, or in algebraic geometric language, elliptic curves, i.e., non-singular cubic curves in the projective plane.

For a gentle introduction to the theory of Riemann surfaces, in particular also their relations to algebraic curves, refer to the book [203].

2.2 Meromorphic forms

Recall that during the whole chapter Σ is a compact Riemann surface of genus $g \geq 0$. Let A be a subset of Σ. In the rest of the book A will typically be a finite set of points of Σ. In this chapter, however, we do not make any restriction on A. This includes the extreme cases $A = \emptyset$ or $A = \Sigma$.

Let $\mathcal{K} = \mathcal{K}_\Sigma$ be the canonical line bundle of Σ. Its local sections are the local holomorphic differentials. If $P \in \Sigma$ is a point and z a local holomorphic coordinate at P, then a local holomorphic differential can be written as $f(z)dz$ with a local holomorphic function f defined in a neighborhood of P. A global holomorphic section can be described locally in coordinates $(U_i, z_i)_{i\in J}$ by a system of local holomorphic functions $(f_i)_{i\in J}$, which are related by the transformation rule induced by the coordinate change map $z_j = z_j(z_i)$ and the condition $f_i dz_i = f_j dz_j$. This yields

$$f_j = f_i \cdot \left(\frac{dz_j}{dz_i}\right)^{-1}. \tag{2.2}$$

A meromorphic section of $\mathcal{K}$, i.e., *meromorphic differential*, is given as a collection of local meromorphic functions $(h_i)_{i\in J}$ with respect to a coordinate covering for which the transformation law (2.2) is true. We will not make any distinction between the canonical bundle and its sheaf of sections, which is a locally free sheaf of rank 1.

In the following, λ is either an integer or a half-integer. If λ is an integer then

(1) $\mathcal{K}^\lambda = \mathcal{K}^{\otimes\lambda}$ for $\lambda > 0$;
(2) $\mathcal{K}^0 = \mathcal{O}$, the trivial line bundle; and
(3) $\mathcal{K}^\lambda = (\mathcal{K}^*)^{\otimes(-\lambda)}$ for $\lambda < 0$.

Here $\mathcal{K}^*$ denotes the dual line bundle to the canonical line bundle. The dual line bundle is the holomorphic tangent line bundle, whose local sections are the holomorphic tangent vector fields $f(z)(d/dz)$.

If λ is a half-integer, then we first have to fix a "square root" of the canonical line bundle, sometimes called a *theta characteristic*. This means we fix a line bundle L, for which $L^{\otimes 2} = \mathcal{K}$. After L has been chosen we set $\mathcal{K}^\lambda := \mathcal{K}^\lambda_L := L^{\otimes 2\lambda}$. In most cases we will not mention L, but we have to keep the choice in mind. Also, the fine-structure of the algebras we are about to define will depend on the choice, but the main properties will remain the same.

Remark 2.1. A Riemann surface of genus g has exactly 2^{2g} non-isomorphic square roots of $\mathcal{K}$. For $g = 0$ we have $\mathcal{K} = \mathcal{O}(-2)$, and $L = \mathcal{O}(-1)$, the tautological bundle, is the unique square root. Already for $g = 1$ we have four non-isomorphic ones. As in this case $\mathcal{K} = \mathcal{O}$, one solution is $L_0 = \mathcal{O}$. But we also have other bundles L_i, $i = 1, 2, 3$. Note that L_0 has a nonvanishing global holomorphic section, whereas this is not the case for L_1, L_2, and L_3. In general, depending on the parity of the dimension of the space of globally holomorphic sections, i.e., $\dim \mathrm{H}^0(\Sigma, L)$, one distinguishes even and odd theta characteristics L. For $g = 1$ the bundle $\mathcal{O}$ is an odd, the others are even theta characteristics. The choice of a theta characteristic is also called a spin structure on Σ [5]. More details on theta characteristics will be given later; see Remark 4.7.

We set

$$\mathcal{F}^\lambda(A) := \{f \text{ is a global meromorphic section of } \mathcal{K}^\lambda \mid f \text{ is holomorphic on } \Sigma\backslash A\}. \tag{2.3}$$

Obviously this is a $\mathbb{C}$-vector space. To avoid cumbersome notation, we will often drop the set A in the notation if A is fixed and/or clear from the context. Recall that in the case of half-integer λ, everything depends on the theta characteristic L, hence, if needed, in this case we sometimes denote this space by $\mathcal{F}^{\lambda,L}(A)$.

Definition 2.2. The elements of the space $\mathcal{F}^\lambda(A)$ are called *meromorphic forms of weight* λ (with respect to the theta characteristic L).

Remark 2.3. In the two extremal cases for the set A we obtain $\mathcal{F}^\lambda(\emptyset)$ the global holomorphic forms, and $\mathcal{F}^\lambda(\Sigma)$ all meromorphic forms. By compactness each $f \in \mathcal{F}^\lambda(\Sigma)$ will have only finitely many poles. In the case that $f \not\equiv 0$ it will also only have a finite number of zeros. Trivially,

$$A \subseteq B \implies \mathcal{F}^\lambda(A) \subseteq \mathcal{F}^\lambda(B). \tag{2.4}$$

Occasionally, we will have to perform local calculations. Let us assume that z_i and z_j are local coordinates for the same point $P \in \Sigma$. For the bundle $\mathcal{K}$ both dz_i and dz_j are frames. If we represent the same form f locally by $f_i dz_i$ and $f_j dz_j$, then we conclude from (2.2) that $f_j = f_i \cdot c_1$, and that the transition function c_1 is given by

$$c_1 = \left(\frac{dz_j}{dz_i}\right)^{-1} = \frac{dz_i}{dz_j}. \tag{2.5}$$

For sections of $\mathcal{K}^\lambda$ with $\lambda \in \mathbb{Z}$, the transition functions are $c_\lambda = (c_1)^\lambda$. The corresponding is true also for half-integer λ. In this case the basic transition function of the chosen theta characteristics L is given as $c_{1/2}$ and all others are integer powers of it. Symbolically, we write $\sqrt{dz_i}$ or $(dz)^{1/2}$ for the local frame, keeping in mind that the sign for the square root is not uniquely defined but depends on the bundle L.

If f is a meromorphic λ-form it can be represented locally by meromorphic functions f_i. If $f \not\equiv 0$ the local representing functions have only a finite number of zeros and poles. Whether a point P is a zero or a pole of f does not depend on the coordinate z_i chosen, as the transition function c_λ will be a nonvanishing function. Moreover, we can define for $P \in \Sigma$ the *order*

$$\operatorname{ord}_P(f) := \operatorname{ord}_P(f_i), \tag{2.6}$$

where $\operatorname{ord}_P(f_i)$ is the lowest nonvanishing order in the Laurent series expansion of f_i around P. It will not depend on the coordinate z_i chosen. The order $\operatorname{ord}_p(f)$ is positive if and only if P is a zero of f. It is negative if and only if P is a pole of f. Moreover, its value gives the orders of zero and pole respectively. By compactness our Riemann surface Σ can be covered by a finite number of coordinate patches. Hence, f can only have a finite number of zeros and poles. We define the *degree* of f as follows:

$$\operatorname{sdeg}(f) := \sum_{P \in \Sigma} \operatorname{ord}_P(f). \tag{2.7}$$

Proposition 2.4. *Let $f \in \mathcal{F}^\lambda$, $f \not\equiv 0$, then*

$$\operatorname{sdeg}(f) = 2\lambda(g-1). \tag{2.8}$$

For this and related results see [203]. [1]

1 In fact, we will discuss the degree again in the frame of the Riemann–Roch theorem in Section 4.1. In particular, the degree will not depend on which form was chosen. The degree could in this way be defined for every line bundle.

Next, we introduce algebraic operations on the vector space obtained by summing over all weights

$$\mathcal{F} := \bigoplus_{\lambda \in \frac{1}{2}\mathbb{Z}} \mathcal{F}^\lambda. \tag{2.9}$$

These operations will allow us to introduce the algebras we are heading for.

2.3 Associative structure

We will drop the subset A in the notation. The natural map of the locally free sheaves of rank one

$$\mathcal{K}^\lambda \times \mathcal{K}^\nu \to \mathcal{K}^\lambda \otimes \mathcal{K}^\nu \cong \mathcal{K}^{\lambda+\nu}, \quad (s,t) \mapsto s \otimes t, \tag{2.10}$$

defines a bilinear map

$$\cdot : \mathcal{F}^\lambda \times \mathcal{F}^\nu \to \mathcal{F}^{\lambda+\nu}. \tag{2.11}$$

With respect to local trivializations, this corresponds to the multiplication of the local representing meromorphic functions

$$(s\,dz^\lambda, t\,dz^\nu) \mapsto s\,dz^\lambda \cdot t\,dz^\nu = s \cdot t\,dz^{\lambda+\nu}. \tag{2.12}$$

If there is no danger of confusion then we will mostly use the same symbol for the section and for the local representing function.

The following is obvious.

Proposition 2.5. *The vector space $\mathcal{F}$ is an associative and commutative graded (over $\frac{1}{2}\mathbb{Z}$) algebra. Moreover, $\mathcal{A} = \mathcal{F}^0$ is a subalgebra and the $\mathcal{F}^\lambda$ are modules over $\mathcal{A}$.*

Of course, $\mathcal{A}$ is the algebra of those meromorphic functions on Σ which are holomorphic outside of A. In case $A = \emptyset$, it is the algebra of global holomorphic functions. By compactness, these are only the constants, hence $\mathcal{A}(\emptyset) = \mathbb{C}$. In case $A = \Sigma$, it is the field of all meromorphic functions $\mathcal{M}(\Sigma)$.

2.4 Lie and Poisson algebra structure

Next, we define a Lie algebraic structure on the space $\mathcal{F}$. Recall that A is fixed. The structure is induced by the map

$$\mathcal{F}^\lambda \times \mathcal{F}^\nu \to \mathcal{F}^{\lambda+\nu+1}, \quad (e,f) \mapsto [e,f], \tag{2.13}$$

which is defined in local representatives of the sections by

$$(e\,dz^\lambda, f\,dz^\nu) \mapsto [e\,dz^\lambda, f\,dz^\nu] := \left((-\lambda)e\frac{df}{dz} + \nu f\frac{de}{dz}\right) dz^{\lambda+\nu+1}, \tag{2.14}$$

and bilinearly extended to $\mathcal{F}$. Of course, we have to show the following proposition.

Proposition 2.6. *The prescription* $[.,.]$ *given by* (2.14) *is well-defined.*

Proof. First, note that the set where poles are allowed will not change. It remains to show that the local function appearing in the result of (2.14) transforms correctly, so that indeed a form of weight $\lambda + \nu + 1$ will result. For this, let z_j and z_i be local coordinates, $e \in \mathcal{F}^\lambda$ and $f \in \mathcal{F}^\nu$ forms of the corresponding weights. For their local representing elements we have $f_j = f_i c_\nu$, and $e_j = e_i c_\lambda$. We calculate

$$\frac{df_j}{dz_j} = \frac{df_i}{dz_i}\frac{dz_i}{dz_j}c_\nu + f\frac{dc_\nu}{dz_i}\frac{dz_i}{dz_j}, \quad \frac{de_j}{dz_j} = \frac{de_i}{dz_i}\frac{dz_i}{dz_j}c_\lambda + e\frac{dc_\lambda}{dz_i}\frac{dz_i}{dz_j}. \tag{2.15}$$

Recall $\frac{dz_i}{dz_j} = c_1$. If we plug in everything in (2.14) expressed in the variable z_j for the local expression we get

$$(-\lambda)e_i\frac{df_i}{dz_i}c_\nu c_\lambda c_1 + (\nu)f_i\frac{de_i}{dz_i}c_\lambda c_\nu c_1 + (-\lambda)e_i f_i c_\lambda\frac{dc_\nu}{dz_i}c_1 + (\nu)f_i e_i c_\nu\frac{dc_\lambda}{dz_i}c_1. \tag{2.16}$$

As $c_\nu c_\lambda c_1 = c_{\nu+\lambda+1}$, the sum of the first two summands gives exactly the desired result. We have to show that the sum of the last two will vanish. As $c_\lambda = (c_1)^\lambda$ and $c_\nu = (c_1)^\nu$ we get

$$\frac{dc_\lambda}{dz_i} = \lambda c_1^{\lambda-1}\frac{dc_1}{dz_i}, \quad \frac{dc_\nu}{dz_i} = \nu c_1^{\nu-1}\frac{dc_1}{dz_i}. \tag{2.17}$$

Hence, the last two terms will cancel. This works also for half-integer weights. In this case we have to use $c_\lambda = (c_{1/2})^{2\lambda}$ and conclude as above. □

Proposition 2.7. *The bilinear map* $[.,.]$ *defines a Lie algebra structure on the vector space* $\mathcal{F}$.

Proof. The antisymmetry is obvious. It remains to check the Jacobi identity. For this it is enough to verify it on triples of "homogeneous" elements $h \in \mathcal{F}^\kappa, e \in \mathcal{F}^\nu$ and $f \in \mathcal{F}^\lambda$. Identifying the forms with local representing functions we get

$$\begin{aligned}[h,[e,f]] = (-\kappa)(\lambda-\nu)h\frac{de}{dz}\frac{df}{dz} + (-\kappa)(-\nu)he\frac{d^2f}{dz^2} + (-\kappa)(\lambda)hf\frac{d^2e}{dz^2} \\ + (\lambda+\nu+1)(-\nu)e\frac{df}{dz}\frac{dh}{dz} + +(\lambda+\nu+1)(\lambda)f\frac{de}{dz}\frac{dh}{dz}.\end{aligned} \tag{2.18}$$

If one cyclically permutes the elements and adds them up, one verifies directly that the cyclic sum will vanish (independently for second order derivatives and for the products of first order derivatives). □

In fact, it is a graded Lie algebra with shifted degree, i.e., the elements of $\mathcal{F}^\lambda$ have to be considered to be of degree $\lambda + 1$, which means that in technical terms one considers $\mathcal{F}[-1]$. As we will not need it in the following we will not introduce this general notation.

Proposition 2.8. *The subspace* $\mathcal{L} = \mathcal{F}^{-1}$ *is a Lie subalgebra, and the* $\mathcal{F}^\lambda$*'s are Lie modules over* $\mathcal{L}$.

Proof. For $\lambda = \nu = -1$ we get as weight of the Lie product $\lambda + \nu + 1 = -1$, hence the subspace is closed under the bracket and a Lie subalgebra. For $e \in \mathcal{L}$ and $h \in \mathcal{F}^\lambda$ the Lie module structure is given by $e \,.\, h := [e, h] \in \mathcal{F}^\lambda$. The Jacobi identity for $e, f \in \mathcal{L}$ and $h \in \mathcal{F}^\lambda$ reads as

$$0 = [[e, f], h] + [[f, h], e] + [[h, e], f] = [e, f] \,.\, h - e \,.\, (f \,.\, h) + f \,.\, (e \,.\, h). \tag{2.19}$$

This is exactly the condition for $\mathcal{F}^\lambda$ being a Lie module. □

Definition 2.9. An algebra $(\mathcal{B}, \cdot, [.,.])$, such that $\cdot$ defines the structure of an associative algebra on $\mathcal{B}$, and $[.,.]$ defines the structure of a Lie algebra on $\mathcal{B}$ is called a *Poisson algebra* if and only if the Leibniz rule is true, e.g.,

$$\forall e, f, g \in \mathcal{B} : [e, f \cdot g] = [e, f] \cdot g + f \cdot [e, g]. \tag{2.20}$$

In other words, via the Lie product $[.,.]$, the elements of the algebra act as derivations on the associative structure. The reader should be warned that $[,.,]$ is not necessarily the commutator of the algebra $(\mathcal{B}, \cdot)$.

Theorem 2.10. *The spaces $\mathcal{F}$ with respect to $\cdot$ and $[.,.]$ are Poisson algebras.*

Proof. Let $e \in \mathcal{F}^\nu$, $g \in \mathcal{F}^\lambda$, and $h \in \mathcal{F}^\kappa$. We have to verify

$$[e, g \cdot h] = [e, g] \cdot h + g \cdot [e, k]. \tag{2.21}$$

Writing the left-hand side in local coordinates (and choosing the same symbol for the representing local functions as for the sections) we get

$$(-\nu) e \frac{d(g \cdot h)}{dz} + (\lambda + \kappa) g \cdot h \frac{de}{dz}. \tag{2.22}$$

This coincides with the expression obtained for the right-hand side. □

Next, we consider important substructures. We already encountered the subalgebras $\mathcal{A}$ and $\mathcal{L}$, but there are more.

2.5 The vector field algebra and the Lie derivative

First, we look again at the Lie subalgebra $\mathcal{L} = \mathcal{F}^{-1}$. Here the Lie action respects the homogeneous subspaces $\mathcal{F}^\lambda$. As forms of weight -1 are vector fields, it could also be defined as the Lie algebra of those meromorphic vector fields on the Riemann surface Σ which are holomorphic outside of A. For vector fields we have the usual Lie bracket and the usual Lie derivative for their actions on forms. For the vector fields we have (again naming the local functions with the same symbol as the section) for $e, f \in \mathcal{L}$

$$[e, f]_| = [e(z)\frac{d}{dz}, f(z)\frac{d}{dz}] = \left(e(z)\frac{df}{dz}(z) - f(z)\frac{de}{dz}(z)\right)\frac{d}{dz}. \tag{2.23}$$

For the Lie derivative we get

$$\nabla_e(f)_{|} = L_e(g)_{|} = e\,.\,g_{|} = \left(e(z)\frac{df}{dz}(z) + \lambda f(z)\frac{de}{dz}(z) \right)\frac{d}{dz}. \tag{2.24}$$

Obviously, these definitions coincide with the definitions already given above. But now we have obtained a geometric interpretation.

2.6 The algebra of differential operators

If we look at $\mathcal{F}$, considered as Lie algebra, more closely, we see that $\mathcal{F}^0$ is an abelian Lie subalgebra and the vector space sum $\mathcal{F}^0 \oplus \mathcal{F}^{-1} = \mathcal{A} \oplus \mathcal{L}$ is also a Lie subalgebra. In the same way, this can also be constructed as semidirect sum of $\mathcal{A}$ considered as abelian Lie algebra and $\mathcal{L}$ operating on $\mathcal{A}$ by taking the derivative.

Definition 2.11. The *Lie algebra of differential operators of degree* ≤ 1 is defined as the semidirect sum of $\mathcal{A}$ with $\mathcal{L}$ and is denoted by $\mathcal{D}^1$.

In terms of elements the Lie product is

$$[(g, e), (h, f)] = (e\,.\,h - f\,.\,g\,,\ [e, f]). \tag{2.25}$$

Using the fact that $\mathcal{A}$ is an abelian subalgebra in $\mathcal{F}$, this is exactly the definition for the Lie product given for this algebra. Hence, $\mathcal{D}^1$ is a Lie algebra. See also Section 1.6

The projection on the second factor $(g, e) \mapsto e$ is a Lie homomorphism and we obtain a short exact sequence of Lie algebras

$$0 \longrightarrow \mathcal{A} \longrightarrow \mathcal{D}^1 \longrightarrow \mathcal{L} \longrightarrow 0. \tag{2.26}$$

Hence $\mathcal{A}$ is an (abelian) Lie ideal of $\mathcal{D}^1$ and $\mathcal{L}$ a quotient Lie algebra. Obviously $\mathcal{L}$ is also a subalgebra of $\mathcal{D}^1$.

Proposition 2.12. *The vector space $\mathcal{F}^\lambda$ becomes a Lie module over $\mathcal{D}^1$ by the operation*

$$(g, e).f := g \cdot f + e.f, \quad (g, e) \in \mathcal{D}^1(A), f \in \mathcal{F}^\lambda(A). \tag{2.27}$$

Proof. As the spaces $\mathcal{F}^\lambda$ are Lie-modules both over $\mathcal{L}$ and $\mathcal{A}$, it is enough to verify the following relation for $e \in \mathcal{L}$, and $h \in \mathcal{A}$, and $f \in \mathcal{F}^\lambda$

$$[e, h]\cdot f = e\,.\,(h\cdot f) - h\cdot(e\,.\,f). \tag{2.28}$$

This can be rewritten (with the help of (2.21)) as

$$(e\,.\,h)\cdot f = (e\,.\,h)\cdot f + h\cdot(e\,.\,f) - h\cdot(e\,.\,f), \tag{2.29}$$

which is obviously true. □

2.7 Differential operators of all degrees

In the following we want to consider differential operators of arbitrary degree acting on $\mathcal{F}^\lambda$. This is obtained via universal constructions. First, we consider the universal enveloping algebra $U(\mathcal{D}^1)$ as described in Section 1.7. We denote its multiplication by $\odot$ and its unit by $\mathbf{1}$.

The universal enveloping algebra contains many elements which act in the same manner on $\mathcal{F}^\lambda$. For example, if h_1 and h_2 are functions different from constants, then $h_1 \cdot h_2$ and $h_1 \odot h_2$ are different elements of $U(\mathcal{D}^1)$. Nevertheless, they act in the same way on $\mathcal{F}^\lambda$.

Hence we will divide out further relations

$$\mathcal{D} = U(\mathcal{D}^1)/J, \quad \text{respectively} \quad \mathcal{D}_\lambda = U(\mathcal{D}^1)/J_\lambda, \tag{2.30}$$

with the two-sided ideals

$$J := (a \odot b - a \cdot b,\ \mathbf{1} - 1 \mid a, b \in \mathcal{A}),$$
$$J_\lambda := (a \odot b - a \cdot b,\ \mathbf{1} - 1,\ a \odot e - a \cdot e + \lambda e\,.\,a \mid a, b \in \mathcal{A}, e \in L).$$

By universality, the $\mathcal{F}^\lambda$ are modules over $U(\mathcal{D}^1)$. The relations in J are fulfilled as $(a \odot b) \cdot f = a \cdot (b \cdot f) = (a \cdot b) \cdot f$. Hence for all λ, the $\mathcal{F}^\lambda$ are modules over $\mathcal{D}$.

If λ is fixed, then the additional relations in J_λ are also true. For this we calculate

$$\begin{aligned}
(a \odot e)\,.\,f &= a \cdot (e\,.\,f) = ae\frac{df}{dz} + \lambda af\frac{de}{dz},\\
(a \cdot e)\,.\,f &= (ae)\frac{df}{dz} + \lambda f\frac{d(ae)}{dz} = (ae)\frac{df}{dz} + \lambda fe\frac{da}{dz} + \lambda fa\frac{de}{dz},\\
\lambda(e\,.\,a) \cdot f &= \lambda\, ef\frac{da}{dz}.
\end{aligned} \tag{2.31}$$

Hence,

$$(a \odot e - a \cdot e + \lambda(e\,.\,a))\,.\,f = 0. \tag{2.32}$$

Consequently, for a fixed λ, the space $\mathcal{F}^\lambda$ is a module over $\mathcal{D}_\lambda$.

Definition 2.13 ([93, IV, 16.8, 16.11] and [15]). A linear map $D : \mathcal{F}^\lambda \to \mathcal{F}^\lambda$ is called an (algebraic) *differential operator* of degree $\le n$ with $n \ge 0$ if and only if

(a) $n = 0$, then $D = b$, the multiplication with a function $b \in \mathcal{A}$;

(b) $n > 0$, then for $a \in \mathcal{A}$ (considered as multiplication operator)

$$[D, a] \quad : \quad \mathcal{F}^\lambda \to \mathcal{F}^\lambda \tag{2.33}$$

is a differential operator of degree $\le (n-1)$.

Let $\mathrm{Diff}^{(n)}(\mathcal{F}^\lambda)$ be the subspace of all differential operators on $\mathcal{F}^\lambda$ of degree $\le n$. By composing the operators

$$\mathrm{Diff}(\mathcal{F}^\lambda) := \bigcup_{n \in \mathbb{N}_0} \mathrm{Diff}^{(n)}(\mathcal{F}^\lambda)$$

becomes an associative algebra, which is a subalgebra of $\mathrm{End}(\mathcal{F}^\lambda)$.

Let $D \in \mathcal{D}$ and assume that D is one of the generators

$$D = a_0 \odot e_1 \odot a_1 \odot e_2 \odot \cdots \odot a_{n-1} \odot e_n \odot a_n, \tag{2.34}$$

with $e_i \in \mathcal{L}$ and $a_i \in \mathcal{A}$ (written as element in $U(\mathcal{D}^1)$).

Proposition 2.14. *Every element $D \in \mathcal{D}$ respectively of $\mathcal{D}_\lambda$ of the form (2.34) operates as (algebraic) differential operator of degree $\le n$ on $\mathcal{F}^\lambda$.*

Proof. Note that the e_i operates as Lie derivative on the forms and the a_i as multiplication. The proof is made via induction over the number n of vector fields e_i in (2.34). If $n = 0$, then $D = a_0$ is just multiplication. Hence, by definition, D is a differential operator of degree 0. Let D be given as (2.34), and let $f \in \mathcal{F}^\lambda$. We set $g = a_n \cdot f$. We calculate with $a \in \mathcal{A}$

$$[a, D](f) = a \cdot a_0 \cdot (e_1 . (.....(e_n . g))) - a_0 \cdot (e_1 . (.....(e_n . (a \cdot g)))). \tag{2.35}$$

For the second term we have

$$e_n . (a_n \cdot a \cdot f) = e_n . (a \cdot g) = (e_n . a) \cdot g + a \cdot (e_n . g). \tag{2.36}$$

$(e_n . a)$ is a function, hence by induction the corresponding part of the second term will be a differential operator of degree $\le n-1$. For the second part we again apply this procedure, and by removing or adding operators of degree $\le n - 1$, we obtain exactly the first term. Hence they cancel. □

In fact, we get an (associative) algebra homomorphism

$$\mathcal{D} \to \mathrm{Diff}(\mathcal{F}^\lambda), \quad \mathcal{D}_\lambda \to \mathrm{Diff}(\mathcal{F}^\lambda). \tag{2.37}$$

In cases where the set A of points where poles are allowed is finite and non-empty, the complement $\Sigma \setminus A$ is affine [100, p. 297]. Hence, as shown in [93], every differential operator can be obtained by successively applying first order operators, i.e., by applying elements from $U(\mathcal{D}^1)$. In other words, the homomorphisms (2.37) are surjective.

2.8 Lie superalgebras of half forms

2.8.1 Lie superalgebras

First, we recall the definition of a Lie superalgebra. Let $\mathcal{S}$ be a vector space which is decomposed into even and odd elements $\mathcal{S} = \mathcal{S}_{\bar{0}} \oplus \mathcal{S}_{\bar{1}}$, i.e., $\mathcal{S}$ is a $\mathbb{Z}/2Z$-graded vector space. Furthermore, let $[.,.]$ be a $\mathbb{Z}/2Z$-graded bilinear map $\mathcal{S} \times \mathcal{S} \to \mathcal{S}$, such that for elements x, y of pure parity

$$[x, y] = -(-1)^{\bar{x}\bar{y}}[y, x]. \tag{2.38}$$

These conditions say that

$$[\mathcal{S}_{\bar{0}}, \mathcal{S}_{\bar{0}}] \subseteq \mathcal{S}_{\bar{0}}, \quad [\mathcal{S}_{\bar{0}}, \mathcal{S}_{\bar{1}}] \subseteq \mathcal{S}_{\bar{1}}, \quad [\mathcal{S}_{\bar{1}}, \mathcal{S}_{\bar{1}}] \subseteq \mathcal{S}_{\bar{0}}, \tag{2.39}$$

and that $[x, y]$ is symmetric for x and y odd, otherwise anti-symmetric. Now $\mathcal{S}$ is a *Lie superalgebra* if in addition the *super-Jacobi identity* (for x, y, z of pure parity)

$$(-1)^{\bar{x}\bar{z}}[x, [y, z]] + (-1)^{\bar{y}\bar{x}}[y, [z, x]] + (-1)^{\bar{z}\bar{y}}[z, [x, y]] = 0 \tag{2.40}$$

is valid. As long as the type of argument is different from *(even, odd, odd)*, all signs can be put to $+1$ and we obtain the form of the usual Jacobi identity. In the remaining case we get

$$[x, [y, z]] + [y, [z, x]] - [z, [x, y]] = 0. \tag{2.41}$$

By the definitions $\mathcal{S}_0$ is a Lie algebra.

With the help of our associative product (2.10), we will obtain examples of Lie superalgebras. We consider

$$\cdot\ \mathcal{F}^{-1/2} \times \mathcal{F}^{-1/2} \to \mathcal{F}^{-1} = \mathcal{L}. \tag{2.42}$$

We introduce the vector space $\mathcal{S}$ with the product

$$\mathcal{S} := \mathcal{L} \oplus \mathcal{F}^{-1/2}, \quad [(e, \varphi), (f, \psi)] := ([e, f] + \varphi \cdot \psi, e \,.\, \varphi - f \,.\, \psi). \tag{2.43}$$

The elements of $\mathcal{L}$ are denoted by $e, f, \ldots$, and the elements of $\mathcal{F}^{-1/2}$ by $\varphi, \psi, \ldots$.

The definition (2.43) can be reformulated as an extension of $[.,.]$ on $\mathcal{L}$ to a super-bracket (denoted by the same symbol) on $\mathcal{S}$ by setting

$$[e, \varphi] := -[\varphi, e] := e \,.\, \varphi = \left(e \frac{d\varphi}{dz} - \frac{1}{2}\varphi \frac{de}{dz}\right)(dz)^{-1/2} \tag{2.44}$$

and

$$[\varphi, \psi] := \varphi \cdot \psi. \tag{2.45}$$

We call the elements of $\mathcal{L}$ elements of even parity, and the elements of $\mathcal{F}^{-1/2}$ elements of odd parity. For such elements x we denote their parity by $\bar{x} \in \{\bar{0}, \bar{1}\}$.

The sum (2.43) can also be described as $\mathcal{S} = \mathcal{S}_{\bar{0}} \oplus \mathcal{S}_{\bar{1}}$, where $\mathcal{S}_{\bar{i}}$ is the subspace of elements of parity $\bar{i}$.

Proposition 2.15. *The space $\mathcal{S}$ with the parity introduced above and product is a Lie superalgebra.*

Proof. By the very definition of (2.43), the equations (2.38) and (2.39) are true. If we consider (2.40) for elements of type *(even, even, even)*, then it reduces to the usual Jacobi identity, which is of course true for the subalgebra of vector fields $\mathcal{L}$. For *(even, even, odd)* it is true, as $\mathcal{F}^{-1/2}$ is a Lie-module over $\mathcal{L}$. For *(even, odd, odd)* we get

$$[e, [\varphi, \psi]] + [\varphi, [\psi, e]] - [\psi, [e, \varphi]] = e \,.\, (\varphi \cdot \psi) - (e \,.\, \psi) \cdot \varphi - (e \,.\, \varphi) \cdot \psi = 0, \tag{2.46}$$

as e acts as derivation on $\mathcal{F}^{-1/2}$. For *(odd, odd, odd)* the super-Jacobi relation writes as

$$[\varphi,[\psi,\chi]] + \quad \text{cyclic permutations} \quad = 0. \tag{2.47}$$

Equivalently,

$$-(\psi \cdot \chi) . \varphi + \quad \text{cyclic permutations} \quad = 0. \tag{2.48}$$

Now (again identifying local representing functions with the element)

$$\begin{aligned}(\psi \cdot \chi) . \varphi &= \left((\psi \cdot \chi) \cdot \varphi' - 1/2((\psi \cdot \chi)'\varphi)\right)(dz)^{-1/2} \\ &= \left(\psi\chi\varphi' - 1/2\,\psi'\chi\varphi - 1/2\,\psi\chi'\varphi\right)(dz)^{-1/2}.\end{aligned} \tag{2.49}$$

Adding up all cyclic permutations yields zero. □

Remark 2.16. The Lie superalgebra introduced above corresponds classically (meaning $g = 0$ and $A = \{0, \infty\}$) to the *Neveu–Schwarz superalgebra*. In string theory, physicists also considered the *Ramond superalgebra* as string algebra (in the two-point case). The elements of the Ramond superalgebra do not correspond to sections of the dual theta characteristics. They are only defined on a 2-sheeted branched covering of Σ, see e.g., [21, 23]. Hence, the elements are only multi-valued sections. As we only consider honest sections of half-integer powers of the canonical bundle we do not deal with the Ramond algebra here.

The choice of the theta characteristics corresponds to choosing a *spin structure* on Σ. For the relation of the Neveu–Schwarz superalgebra to the geometry of graded Riemann surfaces see Bryant [31].

2.8.2 Jordan superalgebras

Leidwanger and Morier-Genoux introduced in [158] a *Jordan superalgebra* in our geometric setting. They put

$$\mathcal{J} := \mathcal{F}^0 \oplus \mathcal{F}^{-1/2} = \mathcal{J}_{\bar{0}} \oplus \mathcal{J}_{\bar{1}}. \tag{2.50}$$

Recall that $\mathcal{A} = \mathcal{F}^0$ is the associative algebra of meromorphic functions. They define the (Jordan) product $\circ$ via the algebra structures for the spaces $\mathcal{F}^\lambda$ by

$$\begin{aligned} f \circ g &:= f \cdot g &&\in \mathcal{F}^0, \\ f \circ \varphi &:= f \cdot \varphi &&\in \mathcal{F}^{-1/2}, \\ \varphi \circ \psi &:= [\varphi, \psi] &&\in \mathcal{F}^0. \end{aligned} \tag{2.51}$$

By rescaling the second definition by the factor 1/2 one obtains a *Lie anti-algebra* as introduced by Ovsienko [191]. See [158] for more details and additional results on representations.

2.9 Higher genus current algebras

We fix an arbitrary finite-dimensional complex Lie algebra $\mathfrak{g}$. Our goal is to generalize the classical current algebra to a higher genus. For this let (Σ, A) be the geometrical data consisting of the Riemann surface Σ, and the subset of points A used to define $\mathcal{A}$ the algebra of meromorphic functions which are holomorphic outside of the set $A \subseteq \Sigma$.

Definition 2.17. The *higher genus current algebra* associated with the Lie algebra $\mathfrak{g}$ and the geometric data (Σ, A) is the Lie algebra $\overline{\mathfrak{g}} = \overline{\mathfrak{g}}(A) = \overline{\mathfrak{g}}(\Sigma, A)$ given as vector space by $\overline{\mathfrak{g}} = \mathfrak{g} \otimes_{\mathbb{C}} \mathcal{A}$ with the Lie product

$$[x \otimes f, y \otimes g] = [x, y] \otimes f \cdot g, \quad x, y \in \mathfrak{g}, f, g \in \mathcal{A}. \tag{2.52}$$

Proposition 2.18. *$\overline{\mathfrak{g}}$ is a Lie algebra.*

Proof. The antisymmetry is clear from the definition. Moreover,

$$[[x \otimes f, y \otimes g], z \otimes h] = [[x, y], z] \otimes ((f \cdot g) \cdot h).$$

As $\mathcal{A}$ is associative and commutative, cyclically summing up the Jacobi identity follows directly from the Jacobi identity for $\mathfrak{g}$. □

As usual we will suppress the mention of (Σ, A) if not needed.

The elements of $\overline{\mathfrak{g}}$ can be interpreted as meromorphic functions $\Sigma \to \mathfrak{g}$ from the Riemann surface Σ to the Lie algebra $\mathfrak{g}$, which are holomorphic outside of A.

In Chapter 9 we will introduce central extensions for these current algebras. They will generalize affine Lie algebras, respectively affine Kac-Moody algebras of untwisted type.

For some applications it is useful to extend the definition by considering differential operators (of degree ≤ 1) associated with $\overline{\mathfrak{g}}$. We define $\mathcal{D}^1_{\mathfrak{g}} := \overline{\mathfrak{g}} \oplus \mathcal{L}$ and take for the summands the Lie product defined there and add the definition

$$[e, x \otimes g] := -[x \otimes g, e] := x \otimes (e.g). \tag{2.53}$$

This operation can be described as the semidirect sum of $\overline{\mathfrak{g}}$ with $\mathcal{L}$; see Section 1.6. Hence, Proposition 1.11 yields the following proposition.

Proposition 2.19. *$\mathcal{D}^1_{\mathfrak{g}}$ is a Lie algebra.*

See Chapter 9 for more details.

2.10 The generalized Krichever–Novikov situation

Until now our set A of possible poles was arbitrary. We have the following obvious result:

Proposition 2.20.

$$A \subseteq B \implies \mathcal{F}^\lambda(A) \subseteq \mathcal{F}^\lambda(B). \tag{2.54}$$

Moreover, these inclusions give inclusions of the introduced algebras.

Even $A = \Sigma$ is allowed. In this case our vector space $\mathcal{F}^\lambda(A)$ is the space of all meromorphic λ-forms and the algebras are the algebras of meromorphic functions, vector fields, meromorphic currents, etc.

2.10.1 The global holomorphic situation

For $A = \emptyset$ the forms are globally holomorphic. The spaces $\mathcal{F}^\lambda(\emptyset)$ are finite-dimensional and the Theorem of Riemann–Roch which will be discussed later (see Section 4.1) gives their dimension. In particular, we obtain.

For $g = 0$:

$$\dim \mathcal{F}^\lambda(\emptyset) = \begin{cases} 1 - 2\lambda, & \lambda \le 0 \\ 0, & \lambda > 0. \end{cases} \tag{2.55}$$

For $g = 1$:

$$\dim \mathcal{F}^\lambda(\emptyset) = \begin{cases} 1, & \lambda \in \mathbb{Z} \\ 1, & \lambda \in \mathbb{Z} + 1/2,\ L = \mathcal{O} \\ 0, & \lambda \in \mathbb{Z} + 1/2,\ L \ne \mathcal{O}. \end{cases} \tag{2.56}$$

Here L is the square root of the canonical bundle and the different cases depend on whether L is equal to the trivial bundle $\mathcal{O}$ or not.

For $g \ge 2$:

$$\dim \mathcal{F}^\lambda(\emptyset) = \begin{cases} 0, & \lambda < 0 \\ 1, & \lambda = 0 \\ l(L), & \lambda = 1/2 \\ (2\lambda - 1)(g - 1), & \lambda \ge 1. \end{cases} \tag{2.57}$$

Here $l(L)$ will take values between 0 and g depending on the choice of the square root L.

The algebra of functions will consist only of the constants, i.e., $\mathcal{A}(\emptyset) = \mathbb{C}$. For $g = 0$ we obtain for the vector field algebra $\mathcal{L}(\emptyset) = \mathfrak{sl}(2, \mathbb{C})$. The superalgebra will have two additional dimensions; see also Section 2.11.

For $g = 1$, $\mathcal{L}(\emptyset) = \mathbb{C}$ is the one-dimensional Lie algebra. For the superalgebra we obtain either an extension by one dimension or nothing new.

For $g \ge 2$ we get $\mathcal{L}(\emptyset) = \{0\}$, and the same for the superalgebra.

2.10.2 The one-point case

In generalizing the holomorphic situation the next choice is that A consists of only one point $A = \{P\}$. As $\Sigma \setminus \{P\}$ will be an open Riemann surface or a smooth affine curve, all our spaces will be infinite dimensional. Our goal later is to introduce an "almost-grading" for these algebras, generalizing the classical graded structure. This we will do in Chapter 3. In order to fix an almost-grading for the one-point situation also, we have to fix a reference point Q and apply the general technique. The algebras for the one-point situation will be described as the two-point subalgebra of objects holomorphic at the reference point. Of course, the algebra as such will not depend on the reference point, only its almost-grading.

2.10.3 The generalized Krichever–Novikov algebras

At least two points are needed to obtain a way to fix an almost-grading directly. This was the original situation studied by Krichever and Novikov [140]. It was extended by the author in [204–207] to the case of A consisting of a finite number of points, $\#A = N \geq 2$.

We will need a decomposition of A into a disjoint union of non-empty subsets $A = I \cup O$, with $\#I = K$ and $\#O = M$. The points in I are called *in-points*, the points in O *out-points* . In case $N > 2$, we have more than one essentially different splitting (i.e., not only an interchange of the role of in- and out-points). The splitting will give our almost-grading.

Definition 2.21. Given Σ a compact Riemann surface and A a finite set of points in Σ, split into two non-empty disjoint subsets $A = I \cup O$ with $\#A \geq 2$, then the algebras $\mathcal{A}$, $\mathcal{L}$, $\mathcal{D}^1$, $\overline{\mathfrak{g}}$, $\mathcal{S}$, $\mathcal{J}$, ... are called *Krichever–Novikov type algebras*.

As we will explain in Chapter 3, the Krichever–Novikov type algebras should be these algebras with a fixed almost-grading, induced by the splitting of $A = I \cup O$.

2.11 The classical situation

Before we deal with arbitrary genus and multi-points in the next chapter we consider the special case when our Riemann surface Σ is of genus zero, i.e., the Riemann sphere $S^2 = \mathbb{P}^1(\mathbb{C})$, and that A consists of two points. We will use the term "classical situation" for this case. The results here are special cases of our general considerations. They can also be obtained directly however, by calculation using certain basic elements. The intention of this section is two-fold. First, we make contact with well-known classical results, and second we want to indicate which kind of results we have to generalize for the arbitrary genus situation.

By some holomorphic automorphism, i.e., a fractional linear transformation on $\mathbb{P}^1(\mathbb{C})$, one point can always be moved to the point $\{0\}$, and the other to the point ∞ with respect to the quasi-global coordinate z. Up to interchanging the roles there is only one choice for the decomposition of A given by $I = \{0\}$ and $O = \{\infty\}$. Let z be the quasi-global variable defined on the affine part $\mathbb{C}$ of $\mathbb{P}^1(\mathbb{C})$. We define the following elements of $\mathcal{F}^\lambda$ with $\lambda \in 1/2\,\mathbb{Z}$

$$f_n^\lambda(z) = z^{n-\lambda}(dz)^\lambda, \quad n \in \mathbb{J}_\lambda, \tag{2.58}$$

where

$$\mathbb{J}_\lambda = \begin{cases} \mathbb{Z}, & \lambda \in \mathbb{Z} \\ \mathbb{Z} + 1/2, & \lambda \in \mathbb{Z} + 1/2. \end{cases} \tag{2.59}$$

For the non-zero orders of f_n^λ we obtain

$$\operatorname{ord}_0(f_n^\lambda) = n - \lambda, \quad \operatorname{ord}_\infty(f_n^\lambda) = -n - \lambda. \tag{2.60}$$

The statement about the order at 0 is obvious. For ∞ we use the local coordinate $w = -1/z$, yielding $dz = (1/w^2)dw$ and hence

$$\tilde{f}_n^\lambda(w) = w^{-n+\lambda}\frac{1}{w^{2\lambda}}dw^\lambda = w^{-n-\lambda}dw^\lambda. \tag{2.61}$$

Furthermore there are no other zeros or poles. In accordance with Proposition 2.4 we have $\operatorname{sdeg}(f_n^\lambda) = -2\lambda$. Note that the behavior at ∞ is uniquely fixed by the form with respect to the variable z. As the orders at 0 are all different, these forms are all linearly independent.

If $\lambda > 0$, then f_n^λ will never be holomorphic. For $\lambda \le 0$ the f_n^λ will be holomorphic exactly for the range

$$f_n^\lambda, \quad \lambda \le n \le -\lambda, n \in \mathbb{J}_\lambda. \tag{2.62}$$

If we compare the length of this range for n with (2.55), we see that for $\lambda \le 0$ the set (2.62) is a basis for $\mathcal{F}^\lambda(\emptyset)$.

Proposition 2.22. *The set of forms* $\{ f_n^\lambda \mid n \in \mathbb{J}_\lambda \}$ *constitutes a basis for* $\mathcal{F}^\lambda = \mathcal{F}^\lambda(\{0, \infty\})$.

Proof. Let $f \in \mathcal{F}^\lambda$, $f \not\equiv 0$. Recall that f can have only a finite number of poles. They might only be located at $z = 0$ and $z = \infty$. By subtracting a finite number of suitable elements f_n^λ from it we obtain an f' without poles at $z = 0$. Next, we subtract elements f_n^λ to reduce the order of the pole at ∞ without introducing new poles at $z = 0$. In this way we obtain f''. We have to distinguish two cases.

Case 1: $\lambda \le 0$. By checking the orders we see that all poles at ∞ can be removed. Hence, f'' is holomorphic and consequently a linear combination of the above given range of f_n^λ. Therefore, f in total is a linear combination.

Case 2: $\lambda > 0$. Poles at ∞ can be removed by subtracting f_n^λ, without introducing poles at 0 if and only if $n - \lambda \ge 0$. Thus the minimal pole order which can still be removed is 2λ. The resulting f'' will be holomorphic at 0 but might have a pole of maximal order $2\lambda - 1$. As a nonvanishing form has to have a degree equal to -2λ, it has to have at least poles in total order of 2λ. Hence f'' has to vanish identically, and consequently f is a linear combination of the f_n^λ too. □

Now we introduce the degree of the element and the homogeneous subspaces of $\mathcal{F}^\lambda$

$$\deg f_n^\lambda := n, \quad \mathcal{F}_n^\lambda = \langle f_n^\lambda \rangle_{\mathbb{C}}. \tag{2.63}$$

In particular, we have

$$\mathcal{F}^\lambda = \bigoplus_{n \in \mathbb{J}_\lambda} \mathcal{F}_n^\lambda. \tag{2.64}$$

Depending on λ the degree will either be half-integer or integer. Also, this degree deg should not be confused with the degree sdeg introduced above. Another important structure we observe is duality. If $f \in \mathcal{F}^\lambda$ and $g \in \mathcal{F}^{1-\lambda}$, then $f \cdot g \in \mathcal{F}^1$ and we define

$$\langle f, g \rangle := \operatorname{res}_0(f \cdot g) = \frac{1}{2\pi \mathrm{i}} \oint_{z=0} fg. \tag{2.65}$$

As $f_n^\lambda(z) \cdot f_m^{1-\lambda}(z) = z^{n+m-1} dz$, we obtain with

$$\langle f_n^\lambda, f_m^{1-\lambda} \rangle = \delta_n^{-m} \tag{2.66}$$

that this is a well-defined total pairing.

It is conventional to use for special weights λ special notations for the elements, e.g.,

$$e_n := f_n^{-1}, \quad A_n := f_n^0, \quad \varphi_n = f_n^{-1/2}, \ldots \tag{2.67}$$

By direct calculations of the operations with the explicit forms of the basic elements we obtain the following structure constants of the algebras and respectively modules:

$$\begin{aligned} [e_n, e_m] &= (m-n) e_{n+m}, & A_n \cdot A_m &= A_{n+m}, \\ [e_n, A_m] &= e_n \,.\, A_m = m A_{n+m}, & A_n \cdot f_m &= f_{n+m}, \\ e_n \,.\, f_m &= (m + \lambda n) f_{n+m}, & [\varphi_n, \varphi_m] &= e_{n+m}. \end{aligned} \tag{2.68}$$

Due to their importance we will discuss the algebras individually after the following remark.

Remark 2.23. Above we used that given two points a and b of $\mathbb{P}^1(\mathbb{C})$, using holomorphic transformations we can always bring a to 0 and b to ∞. It is also easy to give the corresponding basis elements directly without changing the coordinates. The basis elements are

$$f_n^\lambda(z) = (z-a)^{n-\lambda}(z-b)^{-n-\lambda}(a-b)^{n+\lambda} dz^\lambda, \quad n \in \mathbb{J}_\lambda. \tag{2.69}$$

The factor $(a-b)^{n+\lambda}$ is used to normalize the leading coefficient to 1 in the Laurent series expansion with respect to the local coordinate $z - a$. Note that $d(z-a) = dz$.

2.11.1 The vector field algebra – the Witt algebra

The vector field algebra $\mathcal{L}$ has basis and structure equations

$$e_n = z^{n+1}\frac{d}{dz}, \ n \in \mathbb{Z}, \quad [e_n, e_m] = (m-n)e_{n+m}. \tag{2.70}$$

By our definition, it is the algebra of those meromorphic vector fields on the Riemann sphere (i.e., on the projective line) which are holomorphic outside of 0 and ∞. This algebra is the well-known *Witt algebra*, also called *Virasoro algebra without central term*. It (and its central extension to be discussed later) is of fundamental importance in Conformal Field Theory (CFT).

Equally, it can be defined as the complexified algebra of those vector fields on the circle S^1 with a finite number of Fourier modes. The identification is obtained via

$$e_n \iff -\mathrm{i}\ \exp(\mathrm{i}\,\varphi\, n)\frac{d}{d\varphi}, \tag{2.71}$$

with φ the angle coordinate along the circle. The extension of the element on the right-hand side from the circle to the full complex plane yields indeed $z^{n+1}d/dz$. A remark aside: before complexification the algebra can be considered a subalgebra of the algebra of "infinitesimal diffeomorphism" of the sphere S^1.

Remark 2.24. The algebra $\mathcal{L}$ has as 3-dimensional subalgebra

$$\langle e_{-1}, e_0, e_1\rangle_{\mathbb{C}} \tag{2.72}$$

the algebra of globally holomorphic vector fields on $\mathbb{P}^1(\mathbb{C})$. It is isomorphic to $\mathfrak{sl}(2,\mathbb{C})$. The isomorphism is given with the help of the standard generators

$$X = \begin{pmatrix} 0 & 1 \\ 0 & 0 \end{pmatrix}, \quad Y = \begin{pmatrix} 0 & 0 \\ 1 & 0 \end{pmatrix}, \quad H = \begin{pmatrix} 1 & 0 \\ 0 & -1 \end{pmatrix} \tag{2.73}$$

by assigning

$$e_0 \mapsto \frac{1}{2}H, \quad e_1 \mapsto \mathrm{i}\,X, \quad e_{-1} \mapsto \mathrm{i}\,Y. \tag{2.74}$$

We return to the degree introduced above. Obviously

$$\deg([e_n, e_m]) = n + m = \deg(e_n) + \deg(e_m). \tag{2.75}$$

Hence, $\mathcal{L}$ is a graded Lie algebra. Moreover, the modules $\mathcal{F}^\lambda$ are graded modules.

The element e_0 plays a special role. We have

$$[e_0, e_n] = ne_n, \quad e_0 \,.\, f_n^\lambda = nf_n^\lambda. \tag{2.76}$$

This says that the subspace $\mathcal{L}_n$ and $\mathcal{F}_n^\lambda$ of $\mathcal{L}$ and $\mathcal{F}^\lambda$ respectively are eigenspaces of e_0 under the action of $\mathcal{L}$ (via the adjoint action and module action respectively).

Remark 2.25. Due to its importance, the theory of the Witt algebra, its central extension the Virasoro algebra, and their representations is highly developed. A huge number of articles, books, and reviews have appeared. It is impossible to quote a representative number of them. Hence, as a very incomplete list, I refer to the following books which can be used to check for further references [96, 118, 122].

2.11.2 The function algebra

The associative algebra of functions has as a basis $\{A_n = z^n \mid n \in \mathbb{Z}\}$. Obviously,

$$\mathcal{A} = \mathbb{C}[z, z^{-1}] \tag{2.77}$$

the algebra of Laurent polynomials (i.e. polynomials in z and z^{-1}). Clearly, the subalgebra of holomorphic functions consists only of the constants. Both the algebra structure and the module structures are quite simple:

$$A_n \cdot A_m = A_{n+m}, \quad A_n \cdot f_m^\lambda = f_{n+m}^\lambda. \tag{2.78}$$

Obviously, the algebra is a graded algebra and its modules are graded too.

2.11.3 The differential operator algebra

We introduced it as the semidirect sum in Section 2.6. Hence, a basis is given by

$$e_n, \ A_n, \quad n \in \mathbb{Z}, \tag{2.79}$$

with structure equations

$$\begin{aligned} [e_n, e_m] &= (m-n)e_{n+m}, \\ [e_n, A_m] &= mA_{n+m}. \end{aligned} \tag{2.80}$$

The subalgebra of holomorphic differential operators is generated by e_{-1}, e_0, e_1, A_0. In fact, it is the Lie algebra direct sum

$$\langle e_{-1}, e_0, e_1 \rangle_{\mathbb{C}} \oplus A_0 \cdot \mathbb{C}. \tag{2.81}$$

Again, the algebra (and its modules) are graded.

2.11.4 The Lie superalgebra

The superalgebra has as a basis

$$e_n, \ n \in \mathbb{Z}, \quad \varphi_m, \ m \in \mathbb{Z} + \frac{1}{2}, \tag{2.82}$$

with structure equations

$$
\begin{aligned}
[e_n, e_m] &= (m-n)e_{n+m},\\
[\varphi_n, \varphi_m] &= e_{n+m},\\
[e_n, \phi_m] &= e_n \cdot \varphi_m = \left(m - \frac{n}{2}\right)\varphi_{n+m}.
\end{aligned}
\tag{2.83}
$$

The subalgebra of holomorphic elements has as a basis

$$e_{-1},\ e_0\ ,e_1,\ \varphi_{-1/2},\ \varphi_{1/2}, \tag{2.84}$$

and hence is 5-dimensional.

For the Jordan superalgebra (see Section 2.8.2) we obtain as a basis

$$A_n,\ n \in \mathbb{Z}, \quad \varphi_m,\ m \in \mathbb{Z} + \frac{1}{2}, \tag{2.85}$$

with structure

$$
\begin{aligned}
A_n \circ A_m &= A_{n+m},\\
A_n \circ \varphi_m &= \varphi_{n+m},\\
\varphi_n \circ \varphi_m &= \frac{1}{2}(m-n)A_{n+m}.
\end{aligned}
\tag{2.86}
$$

As above, the algebra is graded too.

2.11.5 Current algebras

By general definition the current algebra associated with the finite-dimensional Lie algebra $\mathfrak{g}$ in the classical situation is given by

$$\overline{\mathfrak{g}} = \mathfrak{g} \otimes \mathcal{A} = \mathfrak{g} \otimes \mathbb{C}[z, z^{-1}]. \tag{2.87}$$

We obtain the well-known description of the elements as sums of $x \otimes z^n$ with $x \in \mathfrak{g}$. In cases when $\mathfrak{g}$ is a matrix algebra, the elements can also be described as matrices for which the entries are Laurent polynomials and the defining relations for the matrix algebras are fulfilled. The Lie structure is given as

$$[x \otimes z^n, y \otimes z^m] := [x, y] \otimes z^{n+m}, \quad x, y \in \mathfrak{g},\ n, m \in \mathbb{Z}. \tag{2.88}$$

As graded structure we take the one induced by the grading of $\mathcal{A}$.

3 The almost-grading

Recall that Σ is a compact Riemann surface without any restriction for the genus $g = g(\Sigma)$. Starting with this chapter, let A be a finite subset of Σ. Set $N := \#A$. We will assume that $N \geq 2$, but see Section 3.8 for the one-point situation. We fix the split of A into two non-empty disjoint subsets I and O, i.e., $A = I \cup O$. Set

$$N := \#A, \quad K := \#I, \quad M := \#O, \quad \text{with} \quad N = K + M. \tag{3.1}$$

More precisely, let

$$I = (P_1, \ldots, P_K), \quad \text{and} \quad O = (Q_1, \ldots, Q_M) \tag{3.2}$$

be disjoint ordered tuples of distinct points ("marked points", or "punctures") on the Riemann surface. In particular, we assume $P_i \neq Q_j$ for every pair (i, j). The points in I are called the *in-points* , the points in O the *out-points*. Such a decomposition $A = I \cup O$ we call a *splitting* of A. Sometimes we consider I and O simply as sets. In an interpretation of string theory, Σ corresponds to the world sheet of the string, the set I to points of incoming free strings and the points O to the set of outgoing free strings. We will not need this interpretation for the following.

Recall that by the classical situation we understand

$$\Sigma = \mathbb{P}^1(\mathbb{C}) = S^2, \quad \text{with} \quad I = \{z = 0\}, \quad O = \{z = \infty\}. \tag{3.3}$$

The Figures 3.1, 3.2, and 3.3 give some examples The classical situation was discussed in Section 2.11 and the appearing algebras are the classical algebras of conformal field theory, the Witt, the Virasoro, the affine algebras, etc. As demonstrated in that section, these algebras are graded algebras. In fact, the grading plays an important role in introducing such important concepts as highest weight representations, Verma modules and Fock space representations. Moreover, in case the homogeneous subspaces are finite-dimensional, certain tools of the finite-dimensional situation can be extended to the graded setting. For example, it is possible to introduce a graded

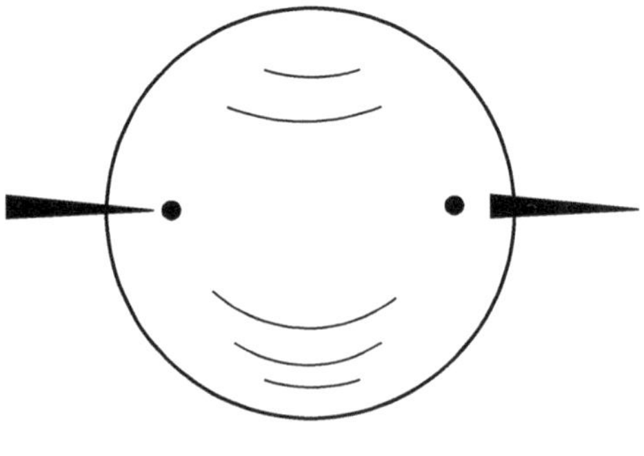

Fig. 3.1. Riemann surface of genus zero with one incoming and one outgoing point.

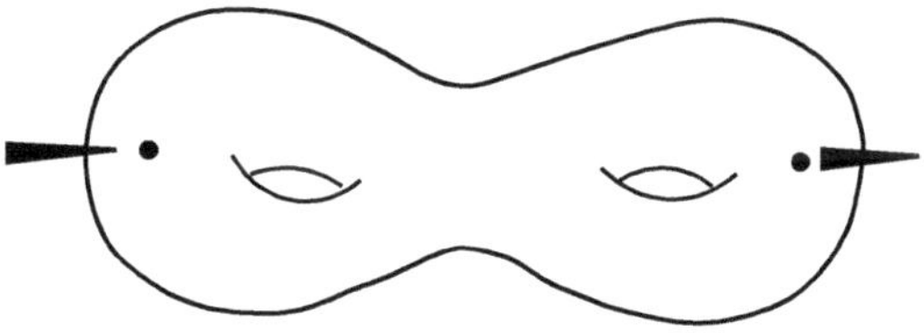

Fig. 3.2. Riemann surface of genus two with one incoming and one outgoing point.

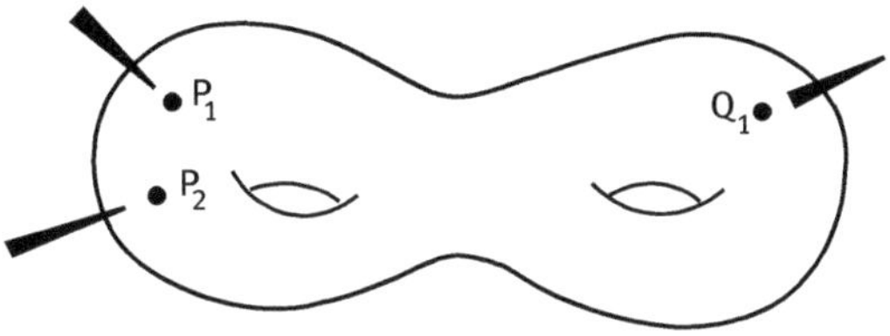

Fig. 3.3. Riemann surface of genus two with two incoming points and one outgoing point.

dimension as formal series with coefficients the dimension of the homogeneous subspaces.

In the higher genus case, and even in the genus zero case with more than two points where poles are allowed, there is no nontrivial grading anymore.

There is a weaker concept, an almost-grading which to a large extent is a valuable replacement for an honest grading. This was realized by Krichever and Novikov [140], and there such a structure was given for the two-point case. Later it was extended to the multi-point situation by the author [206, 207]. In this chapter we will discuss and give the multi-point almost-graded structure. The two-point case will be obtained as a special case. Such a grading will be induced by splitting the set A into $I \cup O$. If A consists only of two points, there is a unique splitting. Some of the technical parts for the almost-grading will be concentrated in Chapter 4, which might be skipped on first reading.

3.1 Definition of an almost-graded structure

Definition 3.1.

(a) Let $\mathcal{L}$ be a Lie or an associative algebra (or even a more general algebraic structure). $\mathcal{L}$ is called an *almost-graded* (Lie-) algebra if
   (i) $\mathcal{L}$ is a vector space direct sum

$$\mathcal{L} = \oplus_{n\in\mathbb{Z}}\mathcal{L}_n. \tag{3.4}$$

   (ii) $\dim \mathcal{L}_n < \infty$.

(iii) Constants $L_1, L_2 \in \mathbb{Z}$ exist such that

$$\mathcal{L}_n \cdot \mathcal{L}_m \subseteq \bigoplus_{h=n+m-L_1}^{n+m+L_2} \mathcal{L}_h, \quad \forall n, m \in \mathbb{Z}. \tag{3.5}$$

The elements in $\mathcal{L}_n$ are called *homogeneous* elements of degree n, and $\mathcal{L}_n$ is called homogeneous subspace of degree n.

(b) Let $\mathcal{L}$ be an almost-graded Lie or associative algebra and $\mathcal{M}$ be a (Lie-) module. $\mathcal{M}$ is called an *almost-graded* (Lie-) module if
   (i) M is a vector space direct sum $\mathcal{M} = \oplus_{n\in\mathbb{Z}}\mathcal{M}_n$;
   (ii) $\dim \mathcal{M}_n < \infty$;
   (iii) constants $M_1, M_2 \in \mathbb{Z}$ exist such that

$$\mathcal{L}_n \cdot \mathcal{M}_m \subseteq \bigoplus_{h=n+m-M_1}^{n+m+M_2} \mathcal{M}_h, \quad \forall n, m \in \mathbb{Z}. \tag{3.6}$$

The elements in M_n are called *homogeneous* elements of degree n, and the M_n are called homogeneous subspaces.

(c) In cases where not all homogeneous subspaces are finite-dimensional, the structure is called *weakly almost-graded* .

(d) In cases where the dimensions are bounded independent of n, the structure is called *strongly almost-graded* .

Krichever and Novikov used the term *quasi-graded*.

Remark 3.2. This definition also makes complete sense for more general index sets $\mathbb{J}$. Of interest to us will be $\mathbb{J} = \mathbb{Z}$, $\mathbb{Z}+1/2$, $(1/2)\mathbb{Z}$. Beside $\mathbb{Z}$, the other sets will only appear if we consider half-integer weights λ. As before, we take $\mathbb{J}_\lambda = \mathbb{Z}$ if λ is an integer, and $\mathbb{J}_\lambda = \mathbb{Z} + 1/2$ if λ is a half-integer.

We will fix an (almost-)grading by exhibiting certain basic elements in the spaces $\mathcal{F}^\lambda$ as homogeneous in Section 3.3. We will show that the Poisson algebra $\mathcal{F}$ is (weakly) almost-graded with respect to both the associative and the Lie structure. Everything else will follow from this. In particular, the vector field algebra, function algebra, differential operator algebra, etc. will be strongly almost-graded. The almost-grading will depend crucially on the splitting of A into $I \cup O$. In fact, a different splitting, not only consisting of swapping the role of I and O, will yield fundamentally different gradings.

3.2 Separating cycle and Krichever–Novikov pairing

A crucial tool will be the Krichever–Novikov pairing, which we will now introduce.

Let C_i be positively oriented (deformed) circles around the points P_i in I, $i = 1, \ldots, K$, and C_j^* positively oriented (deformed) circles around the points Q_j in O, $j = 1, \ldots, M$.

A cycle C_S is called a *separating cycle* if it is smooth, positively oriented, of multiplicity one, and if it separates the in- from the out-points. It might have many components. Next we will integrate meromorphic differentials on Σ without poles in $\Sigma \setminus A$ over closed curves C not meeting A. Hence, we might consider the C and C' as equivalent if $[C] = [C']$ in $\mathrm{H}_1(\Sigma \setminus A, \mathbb{Z})$. In this sense we express the homology class of every separating cycle as

$$[C_S] = \sum_{i=1}^{K} [C_i] = -\sum_{j=1}^{M} [C_j^*]. \tag{3.7}$$

The minus sign in front of the second sum appears due to the opposite orientation. Another way to give such a C_S is via level lines of a "proper time evolution". This will be described in Section 3.9.

Given such a separating cycle C_S (respectively cycle class), we define a linear map

$$\mathcal{F}^1 \to \mathbb{C}, \quad \omega \mapsto \frac{1}{2\pi \mathrm{i}} \int_{C_S} \omega. \tag{3.8}$$

The map will not depend on the separating line C_S chosen, as two of such will be homologous, and the poles of ω are only located in I and O.

Consequently, the integration of ω over C_S can also be described as integration over the special cycles C_i, or equivalently as integration over the cycles C_j^*. For them the integration corresponds to calculating residues

$$\omega \mapsto \frac{1}{2\pi \mathrm{i}} \int_{C_S} \omega = \sum_{i=1}^{K} \operatorname{res}_{P_i}(\omega) = -\sum_{l=1}^{M} \operatorname{res}_{Q_l}(\omega). \tag{3.9}$$

Definition 3.3. The bilinear pairing

$$\mathcal{F}^\lambda \times \mathcal{F}^{1-\lambda} \to \mathbb{C}, \quad (f, g) \mapsto \langle f, g \rangle := \frac{1}{2\pi \mathrm{i}} \int_{C_S} f \cdot g, \tag{3.10}$$

between λ and $1-\lambda$ forms is called *Krichever–Novikov (KN) pairing.*

For the classical situation this coincides exactly with (2.65). Note that the pairing depends not only on A (as the $\mathcal{F}^\lambda$ depends on it), but also critically on the splitting of A into I and O as the integration path will depend on it. But once the splitting is fixed the pairing will be fixed too.

Later, by exhibiting dual basis elements, we will see that the pairing is nondegenerate; see Theorem 3.6

3.3 The homogeneous subspaces

Recall that depending on whether λ is integer or half-integer, we set $\mathbb{J}_\lambda = \mathbb{Z}$ or $\mathbb{J}_\lambda = \mathbb{Z} + 1/2$. For $\mathcal{F}^\lambda$ we will introduce for $m \in \mathbb{J}_\lambda$ subspaces $\mathcal{F}_m^\lambda$ of dimension K, where

$K = \#I$, by exhibiting certain elements $f^\lambda_{m,p} \in \mathcal{F}^\lambda$, $p = 1, \dots, K$ which constitute a basis of $\mathcal{F}^\lambda_m$. Recall that the spaces $\mathcal{F}^\lambda$ for $\lambda \in \mathbb{Z} + 1/2$ depend on the chosen square root L (the theta characteristic) of $\mathcal{K}$.

The elements of $\mathcal{F}^\lambda_m$ are called the elements of degree m. As explained in the following, the degree is related to the zero orders of the elements at the points in I in an essential way.

In the following we will collect the properties of the elements $f^\lambda_{m,p}$. In Chapter 4 we will show the existence of elements with these properties.

Let $I = \{P_1, P_2, \dots, P_K\}$, then we have for the zero-order at the point $P_i \in I$ of the element $f^\lambda_{n,p}$

$$\operatorname{ord}_{P_i}(f^\lambda_{n,p}) = (n + 1 - \lambda) - \delta^p_i, \quad i = 1, \dots, K. \tag{3.11}$$

The prescription at the points in O is made in such a way that the element $f^\lambda_{n,p}$ is essentially uniquely given. Essentially unique means up to multiplication with a constant[1]. After fixing as additional geometric data a system of coordinates z_l centered at P_l for $l = 1, \dots, K$, and requiring that

$$f^\lambda_{n,p}(z_p) = z_p^{n-\lambda}(1 + O(z_p))(dz_p)^\lambda, \tag{3.12}$$

the element $f_{n,p}$ is uniquely fixed.

Proposition 3.4. *The element $f^\lambda_{n,p}$ only depends on the first jet of the coordinate z_p at the point P_p. More precisely, if $z'_p = \alpha_p z_p + O(z_p^2)$ is another coordinate at P_p, and $f^\lambda_{n,s}$ (respectively $f^{\lambda\,\prime}_{n,s}$) are the basis elements as fixed by* (3.12) *with respect to the different coordinates, then*

$$f^{\lambda\,\prime}_{n,p} = (\alpha_p)^{n-\lambda} f^\lambda_{n,p} \quad \textit{and} \quad f^{\lambda\,\prime}_{n,s} = f^\lambda_{n,s}, \; s \neq p. \tag{3.13}$$

Proof. This follows directly from the normalization (3.12). In particular if $z'_p = z_p + O(z_p^2)$ the basis elements will stay the same. □

Example 3.5. The prescriptions at O will be discussed in Chapter 4, or more precisely in Section 4.2.

Here we will consider only the following *model case:*

(1) $O = \{Q\}$ is a one-element set and
(2a) either the genus $g = 0$,
(2b) or $g \geq 2$, $\lambda \neq 0, 1, 1/2$, and the points in A are in generic position.

In this case we require

$$\operatorname{ord}_Q(f^\lambda_{n,p}) = -K \cdot (n + 1 - \lambda) + (2\lambda - 1)(g - 1). \tag{3.14}$$

1 Strictly speaking, there are some special cases where some constants have to be added such that the Krichever–Novikov duality (3.16) is valid: see below.

The construction will also yield

Theorem 3.6. *Set*

$$\mathcal{B}^\lambda := \{ f^\lambda_{n,p} \mid n \in \mathbb{J}_\lambda,\ p = 1, \dots, K \}. \tag{3.15}$$

Then

(a) *$\mathcal{B}^\lambda$ is a basis of the vector space $\mathcal{F}^\lambda$.*

(b) *The introduced basis $\mathcal{B}^\lambda$ of $\mathcal{F}^\lambda$ and $\mathcal{B}^{1-\lambda}$ of $\mathcal{F}^{1-\lambda}$ are dual to each other with respect to the Krichever–Novikov pairing (3.10), i.e.,*

$$\langle f^\lambda_{n,p}, f^{1-\lambda}_{-m,r} \rangle = \delta^r_p\ \delta^m_n, \quad \forall n, m \in \mathbb{J}_\lambda, r, p = 1, \dots, K. \tag{3.16}$$

The space $\mathcal{F}^\lambda_n$ of homogeneous forms of weight λ of degree n is now given by

$$\mathcal{F}^\lambda_n := \langle f^\lambda_{n,p} \mid p = 1, \dots, K \rangle_{\mathbb{C}}, \quad n \in \mathbb{J}_\lambda. \tag{3.17}$$

By Theorem 3.6

$$\mathcal{F}^\lambda = \bigoplus_{n \in \mathbb{J}_\lambda} \mathcal{F}^\lambda_n. \tag{3.18}$$

We define the degree to be

$$\deg(f^\lambda_{n,p}) := n. \tag{3.19}$$

By definition we obtain

$$n = \deg(f^\lambda_{n,p}) = \operatorname{ord}_{P_p}(f^\lambda_{n,p}) - \lambda. \tag{3.20}$$

From part (b) of the theorem it follows that the Krichever–Novikov pairing is non-degenerate. We obtain the following proposition directly from Theorem 3.6 using the duality pairing,

Proposition 3.7. *Let $f \in \mathcal{F}^\lambda$, then v is a finite sum*

$$f = \sum_{n \in \mathbb{J}_\lambda} \sum_{p=1}^{K} \alpha_{n,p} f^\lambda_{n,p}, \quad \alpha_{n,p} \in \mathbb{C}, \tag{3.21}$$

where the coefficients $\alpha_{n,p}$ calculate as

$$\alpha_{n,p} = \langle f, f^{1-\lambda}_{-n,p} \rangle = \frac{1}{2\pi i} \int_{C_S} f \cdot f^{1-\lambda}_{-n,p}. \tag{3.22}$$

We can even say more. Every element $v \in \mathcal{F}^{1-\lambda}$ acts as a linear form on $\mathcal{F}^\lambda$ via

$$\Phi_v : \mathcal{F}^\lambda \mapsto \mathbb{C}, \quad w \mapsto \Phi_v(w) := \langle v, w \rangle. \tag{3.23}$$

Via this pairing $\mathcal{F}^{1-\lambda}$ can be considered as a subspace of $(\mathcal{F}^\lambda)^*$. However, I like to stress the fact that the identification depends on the splitting of A into I and O as the Krichever–Novikov pairing depends on it.

Also, the full dual space $(\mathcal{F}^\lambda)^*$ can be described with the help of the pairing. Consider the series

$$\hat{v} := \sum_{m\in\mathbb{J}_\lambda}\sum_{p=1}^{K} a_{m,p} f_{m,p}^{1-\lambda}, \tag{3.24}$$

a formal series, then $\Phi_{\hat{v}}$ (as a distribution) is a well-defined element of ${\mathcal{F}^\lambda}^*$, as it will only be evaluated for a finite number of basis elements in $\mathcal{F}^\lambda$. Conversely, every element of ${\mathcal{F}^\lambda}^*$ can be given by a suitable $\hat{v}$. In fact, every $\phi \in (\mathcal{F}^\lambda)^*$ is uniquely given by the scalars $\phi(f_{m,r}^\lambda)$. We set

$$\hat{v} := \sum_{m\in\mathbb{J}_\lambda}\sum_{p=1}^{K} \phi(f_{-m,p}^\lambda)\, f_{m,p}^{1-\lambda}. \tag{3.25}$$

Obviously, $\Phi_{\hat{v}} = \phi$. For more information about this "distribution interpretation" see Section 3.10.

The dual elements of the vector field algebra $\mathcal{L}$ will be given by the formal series (3.24) with basis elements from $\mathcal{F}^2$, the quadratic differentials, the dual elements of the function algebra $\mathcal{A}$ correspondingly from $\mathcal{F}^1$, the differentials, and the dual elements of $\mathcal{F}^{-1/2}$ correspondingly from $\mathcal{F}^{3/2}$. The spaces $\mathcal{F}^2$, $\mathcal{F}^1$ and $\mathcal{F}^{3/2}$ themselves can be considered as restricted duals.

In view of this duality, the following convention will be found to be very convenient:

$$\begin{aligned} f_\lambda^{*,(n,p)} &:= f_{(-n,p)}^\lambda, & A_{n,p} &:= f_{(n,p)}^0, & e_{n,p} &:= f_{(n,p)}^{-1}, \\ \omega^{n,p} &:= f_1^{*,(n,p)}, & \Omega^{n,p} &:= f_2^{*,(n,p)}, & \varphi_{n,p} &:= f_{n,p}^{-1/2}. \end{aligned} \tag{3.26}$$

In particular, we have (for all $n, m \in \mathbb{Z}$, $p, r = 1, \dots, K$)

$$\langle e_{n,p}, \Omega^{m,r}\rangle = \langle A_{n,p}, \omega^{m,r}\rangle = \delta_n^m \cdot \delta_r^p. \tag{3.27}$$

In Chapter 4 we will prove the existence of the elements $f_{n,p}^\lambda$ and Theorem 3.6.

3.4 The almost-graded structure for the introduced algebras

The following theorem is the main result of this section.

Theorem 3.8. *There exist constants R_1 and R_2 (depending on the genus g, and on the number and splitting of the points in A), independent of λ and ν, and also of $n, m \in \mathbb{J}$, such that for the basis elements*

$$\begin{aligned} f_{n,p}^\lambda \cdot f_{m,r}^\nu &= f_{n+m,r}^{\lambda+\nu}\delta_p^r + \sum_{h=n+m+1}^{n+m+R_1}\sum_{s=1}^{K} a_{(n,p)(m,r)}^{(h,s)} f_{h,s}^{\lambda+\nu}, \quad a_{(n,p)(m,r)}^{(h,s)} \in \mathbb{C}, \\ [f_{n,p}^\lambda, f_{m,r}^\nu] &= (-\lambda m+\nu n)\, f_{n+m,r}^{\lambda+\nu+1}\delta_p^r + \sum_{h=n+m+1}^{n+m+R_2}\sum_{s=1}^{K} b_{(n,p)(m,r)}^{(h,s)} f_{h,s}^{\lambda+\nu+1}, \quad b_{(n,p)(m,r)}^{(h,s)} \in \mathbb{C}. \end{aligned} \tag{3.28}$$

Before we start with the proof, we introduce

$$M(\lambda) := (2\lambda - 1)(g - 1). \tag{3.29}$$

By direct calculations we obtain

Lemma 3.9.

$$M(\lambda) + M(\mu) = 2(\lambda + \mu - 1)(g - 1) \tag{3.30}$$

$$M(\lambda + \mu) = M(\lambda) + M(\mu) + (g - 1) \tag{3.31}$$

$$M(\lambda) + M(1 - \lambda) = 0 \tag{3.32}$$

$$M(\lambda) + M(\nu) + M(1 - (\lambda + \nu)) = -(g - 1) \tag{3.33}$$

$$M(\lambda) + M(\nu) + M(-(\mu + \lambda)) = -3(g - 1). \tag{3.34}$$

Proof of Theorem 3.8. Here we will provide a proof of the theorem in the model case defined in Example 3.5. In the other situations we first need the description of the basis elements in Chapter 4. From this it will be clear that in this case we also get the corresponding result. The principal idea is as follows: we calculate the objects on the right-hand side locally at the points in I and O. As it will be a well-defined form of weight $\lambda + \nu$, respectively $\lambda + \nu + 1$, it can be expanded with respect to our basis. The coefficients in the expansion are given as described in Proposition 3.7 via Krichever–Novikov duality. The integration over C_s will be done by integration over the in-points in I to obtain a lower bound for the index k, and over the out-points in O to obtain an upper bound.

We start with the associative product. We have to examine

$$\frac{1}{2\pi i} \int_{C_S} f^{\lambda}_{n,p} \cdot f^{\nu}_{m,r} \cdot f^{1-(\lambda+\nu)}_{-k,s}. \tag{3.35}$$

We consider the integrand at the point $P_i \in I$. For its order we calculate

$$\begin{aligned}(n - \lambda + (1 - \delta_i^p) + (m - \nu + (1 - \delta_i^r) + (-k - (1 - (\lambda + \nu)) + (1 - \delta_i^s))\\ = (n + m - k - 1) + 3 - (\delta_i^p + \delta_i^r + \delta_i^s).\end{aligned} \tag{3.36}$$

Hence, the smallest possible k such that there are residues is $k = (n + m)$. But in this case a residue will only be possible if $i = r = p = s$. Hence, maximally one point could contribute to the integration, and this is also only the case if $r = p = s$. The coefficient is then clearly exactly 1. For the upper bound we check when a residue is possible at the point Q in O. We use the order prescription (3.14) and obtain for the order there

$$\begin{aligned}(-K(n + 1 - \lambda) + M(\lambda)) + (-K(m + 1 - \nu) + M(\nu))\\ + (-K(-k + 1 - (1 - (\lambda + \nu)) + M(1 - (\lambda + \nu)))\\ = K(k - (n + m)) - 2K - g + 1.\end{aligned} \tag{3.37}$$

Here we used (3.33). Consequently, there could only be a residue if

$$k \le n + m + \frac{g-2}{K} + 2. \tag{3.38}$$

The upper bound $R_1 = \lfloor \frac{g-2}{K} \rfloor + 2$ depends neither on n and m nor on the weights. For the Lie product we have to consider

$$(-\lambda) f_{n,p}^{\lambda} \frac{df_{m,r}^{\nu}}{dz} + \nu \frac{df_{n,p}^{\lambda}}{dz} \cdot f_{m,r}^{\nu}. \tag{3.39}$$

If we calculate this at the point $P_i \in I$, then as lowest order term of the expansion with respect to the coordinate z_i we obtain

$$(-\lambda(m - \nu + (1 - \delta_i^p)) + \nu(n - \lambda + (1 - \delta_i^r))))z_i^{n+m-(\lambda+\nu)+(1-\delta_i^p)+(1-\delta_i^p)-1} \tag{3.40}$$

If we multiply this with the dual element $f_{-k,s}^{-(\lambda+\nu)}$ we get as order

$$(n + m - k - 1) + (1 - \delta_i^p) + (1 - \delta_i^r) + (1 - \delta_i^s). \tag{3.41}$$

Hence a residue is only possible if $k \ge n+m$. For $k = n+m$ there will only be a residue if $i = r = p = s$. Hence, only at one point and only if $r = p = s$. In this case, the coefficient in (3.40) will be $-\lambda m + \nu n$. Next we consider (3.39) at the point Q. As here we only need an estimate we do not take care of the coefficient. If we use the order description there and calculate the Lie product, by using (3.34) we finally get as order

$$K(k - (n + m)) - 3K - 3g + 3 - 1. \tag{3.42}$$

Hence a residue is only possible if

$$k \le n + m + \frac{3g-3}{K} + 3. \tag{3.43}$$

Therefore the corresponding $R_2 = \lfloor \frac{3g-3}{K} \rfloor + 3$ is independent of n and m and also of the weight. □

Remark 3.10. If we only have 2 points, i.e., $K = 1$, then in the model case we have

$$R_1 = g, \quad R_2 = 3g. \tag{3.44}$$

In the classical situation both constants are zero. We therefore get an honest grading.

We directly obtain the following theorem as a consequence of Theorem 3.8.

Theorem 3.11. *With respect to the grading (3.19), both the multiplicative structure and the Lie structure for the $\mathcal{F}^{\lambda}$ spaces is almost-graded. Moreover, R_1 and R_2 exist, which do not depend on $n, m \in \mathbb{Z}$ and on λ and ν such that*

$$\mathcal{F}_n^{\lambda} \cdot \mathcal{F}_m^{\mu} \subseteq \bigoplus_{h=n+m}^{n+m+R_1} \mathcal{F}_h^{\lambda+\mu}, \tag{3.45}$$

$$[\mathcal{F}_n^{\lambda}, \mathcal{F}_m^{\mu}] \subseteq \bigoplus_{h=n+m}^{n+m+R_2} \mathcal{F}_h^{\lambda+\mu+1}. \tag{3.46}$$

This means in particular that with respect to both the associative and the Lie structures the algebra $\mathcal{F}$ is weakly almost-graded. The reason why it is only weakly almost-graded is that

$$\mathcal{F}^\lambda = \bigoplus_{m \in \mathbb{J}_\lambda} \mathcal{F}^\lambda_m, \quad \text{with} \quad \dim \mathcal{F}^\lambda_m = K. \tag{3.47}$$

If we add up all λ for a fixed m we find that our homogeneous spaces are infinite-dimensional.

In the definition of our Krichever–Novikov type algebra only a finite number of λ are involved, which immediately leads to the following theorem.

Theorem 3.12. *The Krichever–Novikov type vector field algebras $\mathcal{L}$, function algebras $\mathcal{A}$, differential operator algebras $\mathcal{D}^1$, Lie superalgebras $\mathcal{S}$, and Jordan superalgebras $\mathcal{J}$ are all strongly almost-graded.*

We obtain

$$\dim \mathcal{L}_n = \dim \mathcal{A}_n = K, \quad \dim \mathcal{S}_n = \dim \mathcal{J}_n = 2K, \quad \dim \mathcal{D}^1_n = 3K. \tag{3.48}$$

If $\mathcal{U}$ is one of these algebras, with product denoted by [,], then

$$[\mathcal{U}_n, \mathcal{U}_m] \subseteq \bigoplus_{h=n+m}^{n+m+R_i} \mathcal{U}_h, \tag{3.49}$$

with $R_i = R_1$ for $\mathcal{U} = \mathcal{A}$, and $R_i = R_2$ otherwise.

For further reference let us specialize the lowest degree term component in (3.28) for certain special cases.

$$\begin{aligned}
A_{n,p} \cdot A_{m,r} &= A_{n+m,r}\, \delta^p_r + \text{h.d.t.} \\
A_{n,p} \cdot f^\lambda_{m,r} &= f^\lambda_{n+m,r}\, \delta^p_r + \text{h.d.t.} \\
[e_{n,p}, e_{m,r}] &= (m-n) \cdot e_{n+m,r}\, \delta^p_r + \text{h.d.t.} \\
e_{n,p} \cdot f^\lambda_{m,r} &= (m+n\lambda) \cdot f^\lambda_{n+m,r}\, \delta^p_r + \text{h.d.t.} \\
[e_{n,p}, \varphi_{m,r}] &= (m - \frac{n}{2}) \cdot \varphi_{n+m,r}\, \delta^p_r + \text{h.d.t.} \\
[\varphi_{n,p}, \varphi_{m,r}] &= e_{n+m,r}\, \delta^p_r + \text{h.d.t.}
\end{aligned} \tag{3.50}$$

Here h.d.t. denotes linear combinations of basis elements of degree between $n+m+1$ and $n+m+R_i$.

Proposition 3.13. *Let $\overline{\mathfrak{g}} = \mathfrak{g} \otimes \mathcal{A}_n$ be the Krichever–Novikov type current algebra associated with the finite-dimensional Lie algebra $\mathfrak{g}$. Then by setting $\overline{\mathfrak{g}}_n = \mathfrak{g} \otimes \mathcal{A}_n$ we obtain*

$$\overline{\mathfrak{g}} = \bigoplus_{n \in \mathbb{Z}} \overline{\mathfrak{g}}_n, \quad \dim \overline{\mathfrak{g}}_n = K \cdot \dim \mathfrak{g} \tag{3.51}$$

and

$$[\overline{\mathfrak{g}}_n, \overline{\mathfrak{g}}_m] \subseteq \bigoplus_{h=n+m}^{n+m+R_1} \overline{\mathfrak{g}}_h. \tag{3.52}$$

Hence, $\overline{\mathfrak{g}}$ is a strongly almost-graded Lie algebra.

The following simple result will be quite useful.

Proposition 3.14.

$$1 = \sum_{p=1}^{K} A_{0,p}. \tag{3.53}$$

Proof. Following Proposition 3.7, the function $f = 1$ can be written as

$$1 = \sum_{n \in \mathbb{Z}} \sum_{p=1}^{K} \alpha_{n,p} A_{n,p}$$

with

$$\alpha_{n,p} = \frac{1}{2\pi \mathrm{i}} \int_{C_S} f^1_{-n,p} = \frac{1}{2\pi \mathrm{i}} \int_{C_S} \omega^{n,p}.$$

Recall that $\operatorname{ord}_{P_i}(f^1_{-n,p}) = -n - 1 + (1 - \delta_i^P)$. Hence, there is no residue at the points in I if $n < 0$, and if $n = 1$ there is a residue (of value 1) only at the point P_p. If $n > 0$ the description of the orders at the points in O, to be given in the next chapter (see (4.99) and the comments after the equation), will tell that there are no poles at O. Hence, also in this case no residue. Hence the statement. □

3.5 Triangular decomposition and filtrations

Let $\mathcal{U}$ be one of the algebras introduced above (including the current algebra). On the basis of the almost-grading we obtain a triangular decomposition of the algebras[2]

$$\mathcal{U} = \mathcal{U}_{[+]} \oplus \mathcal{U}_{[0]} \oplus \mathcal{U}_{[-]}, \tag{3.54}$$

where

$$\mathcal{U}_{[+]} := \bigoplus_{m>0} \mathcal{U}_m, \quad \mathcal{U}_{[0]} = \bigoplus_{m=-R_i}^{m=0} \mathcal{U}_m, \quad \mathcal{U}_{[-]} := \bigoplus_{m<-R_i} \mathcal{U}_m. \tag{3.55}$$

By the almost-gradedness the $[+]$ and $[-]$ subspaces are (infinite-dimensional) subalgebras. The $[0]$ spaces in general are not. Sometimes we will use the phrase *critical strip* for $\mathcal{U}_{[0]}$.

With respect to the almost-grading of $\mathcal{F}^\lambda$, we introduce a descending filtration

$$\begin{gathered} \mathcal{F}^\lambda_{(n)} := \bigoplus_{m \geq n} \mathcal{F}^\lambda_m, \\ \dots \supseteq \mathcal{F}^\lambda_{(n-1)} \supseteq \mathcal{F}^\lambda_{(n)} \supseteq \mathcal{F}^\lambda_{(n+1)} \dots \end{gathered} \tag{3.56}$$

2 Instead of $[+]$ we will quite often use just $+$, (0) instead of $[0]$, etc., in the following. In this section we have to distinguish this notation from the filtration notation.

In case O has more than one point we will see that there are certain choices, e.g., numbering of the points in O, different rules etc., involved in defining the almost-grading. Hence, if the choices are made differently the subspaces $\mathcal{F}_n^\lambda$ might depend on them, and consequently also the almost-grading. But there is an intrinsic definition of this filtration which is given by the following proposition.

Proposition 3.15.

$$\mathcal{F}_{(n)}^\lambda = \{\, f \in \mathcal{F}^\lambda \mid \operatorname{ord}_{P_i}(f) \geq n - \lambda, \forall i = 1, \dots, K \,\}. \tag{3.57}$$

Proof. By considering the order of $f_{m,p}^\lambda$ for $m \geq n$ at the points P_i, $i = 1, \dots, K$, the inclusion "$\subseteq$" is immediate. Now we consider "$\supseteq$". Let f be an element of the space on the right. Following Proposition 3.7 it can be given as

$$f = \sum_{m \in \mathbb{J}_\lambda} \sum_{p=1}^{K} \alpha_{m,p} f_{m,p}^\lambda \tag{3.58}$$

with

$$\alpha_{m,p} = \frac{1}{2\pi \mathrm{i}} \int_{C_S} f \cdot f_{-m,p}^{1-\lambda}. \tag{3.59}$$

For the orders of the integrand at the points $P \in I$ we calculate

$$\operatorname{ord}_P(f \cdot f_{-m,p}^{1-\lambda}) \geq n - \lambda + (-m - (1 - \lambda)) = n - m - 1. \tag{3.60}$$

A non-vanishing residue can only appear for $m \geq n$. Hence, f will be an element of the space on the left. □

By this proposition the induced filtration is canonically defined via the splitting of A into $I \cup O$.

Moreover, different choices (as long as the splitting is not changed) will give equivalent filtrations. We stress the fact that under a Krichever–Novikov algebra we will always understand one of the introduced algebras (or even some others still to come) together with an almost-grading (respectively equivalence class of almost-grading) introduced by the splitting $A = I \cup O$.

The following statements are automatically true.

Proposition 3.16. *The filtration is compatible with our algebraic structures. This means, for example that:*

$$\mathcal{F}_{(n)}^\lambda \cdot \mathcal{F}_{(m)}^\mu \subseteq \mathcal{F}_{(n+m)}^{\lambda+\mu}, \qquad \mathcal{A}_{(n)} \cdot \mathcal{A}_{(m)} \subseteq \mathcal{A}_{(n+m)}, \tag{3.61}$$

$$[\mathcal{F}_{(n)}^\lambda, \mathcal{F}_{(m)}^\mu] \subseteq \mathcal{F}_{(n+m)}^{\lambda+\mu+1}, \qquad [\mathcal{L}_{(n)}, \mathcal{L}_{(m)}] \subseteq \mathcal{L}_{(n+m)}. \tag{3.62}$$

Proposition 3.17. *With respect to the basis elements exhibited we get*

$$\begin{aligned} f^{\lambda}_{n,r}\cdot f^{\mu}_{m,l} &= \delta^{l}_{r}\, f^{\lambda+\nu}_{n+m,r} \mod \mathcal{F}^{\lambda+\nu}_{n+m+1}, \\ A_{n,r}\cdot A_{m,l} &= \delta^{l}_{r}\, A_{n+m,r} \mod A_{(n+m+1)}, \\ A_{n,r}\cdot f^{\mu}_{m,l} &= \delta^{l}_{r}\, f^{\mu}_{n+m,r} \mod \mathcal{F}^{\mu}_{(n+m+1)}, \\ [f^{\lambda}_{n,r}, f^{\mu}_{m,l}] &= (-\lambda m+\nu n)\cdot\delta^{l}_{r}\, f^{\lambda+\nu+1}_{n+m,r} \mod \mathcal{F}^{\lambda+\nu+1}_{(n+m+1)}, \\ [e_{n,r}, e_{m,l}] &= (m-n)\cdot\delta^{l}_{r}\, e_{n+m,r} \mod \mathcal{L}_{(n+m+1)}, \\ e_{n,r}\cdot f^{\lambda}_{m,l} &= (m+\lambda n)\cdot\delta^{l}_{r}\, f^{\lambda}_{n+m,r} \mod \mathcal{F}^{\lambda}_{(n+m+1)}. \end{aligned} \tag{3.63}$$

From the construction of our grading we obtain directly for the quotients

$$\mathcal{F}^{\lambda}_{(n)}/\mathcal{F}^{\lambda}_{(n+1)} \cong \mathcal{F}^{\lambda}_{n}. \tag{3.64}$$

Hence, the graded object associated with the filtration is

$$\operatorname{gr}\mathcal{F}^{\lambda}_{(.)} = \bigoplus_{n\in\mathbb{J}_\lambda}\mathcal{F}^{\lambda}_{n}. \tag{3.65}$$

3.6 Equivalence of filtrations and almost-gradings

Two filtrations $\{\mathcal{F}^{\lambda}_{(.)}\}$ and $\{\mathcal{F}^{\lambda\,\prime}_{(.)}\}$ are called *equivalent* if and only if for all n there exists n' and n'', and for all m there exists m' and m'' such that

$$\mathcal{F}^{\lambda}_{(n')} \subseteq \mathcal{F}^{\lambda\,\prime}_{(n)} \subseteq \mathcal{F}^{\lambda}_{(n'')} \quad\text{and}\quad \mathcal{F}^{\lambda\,\prime}_{(m')} \subseteq \mathcal{F}^{\lambda}_{(m)} \subseteq \mathcal{F}^{\lambda\,\prime}_{(m'')}. \tag{3.66}$$

We demonstrated above that the filtration $\{\mathcal{F}^{\lambda}_{(.)}\}$ is fixed by the splitting $A = I\cup O$. If $A = I'\cup O'$ is split differently it will yield a filtration $\{\mathcal{F}^{\lambda\,\prime}_{(.)}\}$. These two filtrations will not be equivalent. To see this, take $P\in I'$ with $P\notin I$, and fix n, then forms $f_k\in\mathcal{F}^{\lambda}_{(n)}$ exist, which have every desired high enough pole order k at P. Hence an n' cannot exist such that $\mathcal{F}^{\lambda}_{(n)}\subseteq\mathcal{F}^{\lambda\,\prime}_{(n')}$.

Nevertheless, we have some relations in special cases. Immediately from (3.57) it follows that

$$I'\supseteq I \iff \mathcal{F}^{\lambda\,\prime}_{(n)} \subseteq \mathcal{F}^{\lambda}_{(n)}. \tag{3.67}$$

We now return to almost-gradings. Given the vector space (algebra, ...) $\mathcal{F}^{\lambda}$, we call two almost-gradings $\mathcal{F}^{\lambda}_{n}$ and $\mathcal{F}^{\lambda'}_{n}$ *equivalent* if and only if $\alpha,\beta\in\mathbb{Q}$, $\alpha,\beta>0$, $L_1, L_2, L_3, L_4\in\mathbb{Z}$ exist, all independent of n, such that

$$\mathcal{F}^{\lambda}_{n} \subseteq \bigoplus_{h=\alpha n-L_1}^{\alpha n+L_2}\mathcal{F}^{\lambda'}_{h}, \quad \mathcal{F}^{\lambda'}_{n} \subseteq \bigoplus_{h=\beta n-L_3}^{\beta n+L_4}\mathcal{F}^{\lambda}_{h}. \tag{3.68}$$

If we introduce the associated filtrations to these equivalent almost-gradings then the filtrations are equivalent.

We already mentioned that in the case of $\#O = M > 1$ there are choices before we can uniquely fix the almost-grading. But a different choice will yield an equivalent almost-grading. Note that in both cases the relations given in Section 4.2 have to be true. Hence, via Krichever–Novikov duality the claim follows. The easy details are left to the reader. In fact, in this case $\alpha = \beta = 1$ and $L_1 = L_3 = 0$. Altogether we can say that neither the equivalence class of the almost-grading nor the induced filtration will depend on the choices made.

3.7 Inverted grading

Given a splitting $A = I \cup O$, we introduced an almost-graded structure and a filtration. We also need to consider the almost-graded structure obtained by interchanging the roles of I and O

$$A = I^* \cup O^*, \quad \text{with} \quad I^* = O,\ O^* = I. \tag{3.69}$$

Of course, the algebra will stay the same but the grading will change. We call this grading *inverted grading*. For $N > 2$ it is not only an inversion of the degree. In general, homogeneous elements of the original grading will not stay homogeneous with respect to the inverted grading and vice versa. Nevertheless, there is a certain relation. We denote the objects of the inverted grading by an *, e.g., $f^{\lambda,*}_{m,s}$, etc. By considering the orders at the points P_i and Q_j and using the Krichever–Novikov duality relation (3.16), we obtain

$$\mathcal{F}^\lambda_n \subseteq \bigoplus_{h=-\alpha n-L_1}^{-\alpha n+L_2} F^{\lambda,*}_h, \quad F^{\lambda,*}_n \subseteq \bigoplus_{h=-\beta n-L_3}^{-\beta n+L_4} \mathcal{F}^\lambda_h, \tag{3.70}$$

with $\alpha, \beta > 0$, and L_1, L_2, L_3, L_4 numbers which do not depend on n and m. In particular, the $\mathcal{F}^\lambda_n$ will lie in a subspace given as a finite sum of certain homogeneous subspaces of the inverted grading. Further, if f is an element of degree n in the original grading, it will be a sum of homogeneous elements of a controllable range of the degree.

If we compare (3.70) with (3.68), we obtain up to inverting the degree $n \to -n$ an almost-grading which is equivalent to the original one. In this sense we say that $A = I \cup O = I' \cup O'$ are *essentially different splittings* if and only if neither $I' = I$ nor $I' = O$.

3.8 The one-point situation

If poles are only allowed at one point P of Σ, the algebras and modules are of course still infinite-dimensional. There is no natural way to introduce an almost-grading in this situation. To also use the advantages of an almost-grading we have to choose a reference point $Q \neq P$.

We now obtain almost-graded algebras and modules consisting of meromorphic objects with possible poles at P and Q. The almost-grading is uniquely fixed by these two points and by its induced canonical splitting. The original algebra (or module) we are interested in is given as the subalgebra (submodule) of objects which are holomorphic at the reference point Q. It inherits the almost-grading from the two-point objects. From the orders at P and Q we deduce that with respect to the almost-grading induced by $I = \{P\}$ for large positive degrees the homogeneous components will be trivial, for large negative degrees its components will coincide with the components for the two-point algebra. With respect to the inverse grading just the opposite is true. We stress again the fact that the almost-grading will depend on the reference point chosen.

The situation is quite similar if we consider the situation

$$I := \{P_1, P_2, \dots, P_N\}, \quad O = \emptyset. \tag{3.71}$$

Also in this case we have to choose a reference point P_∞ and enlarge O, before we can start with an almost-graded structure which treats all points in I in the same manner.

Example 3.18. We consider the classical $\mathbb{P}^1(\mathbb{C})$ situation with point $P = \{\infty\}$ and the algebra of vector fields holomorphic outside of $\{\infty\}$. This algebra does not admit a natural (almost)-grading. To endow it with a grading we choose $\{z = 0\}$ as reference point. Our graded algebra is in this case the Witt algebra W. This induces the decomposition into homogeneous subspaces

$$V = \bigoplus_{n=-1}^{\infty} V_n \quad V_n = \left\langle z^{n+1} \frac{d}{dz} \right\rangle_{\mathbb{C}}. \tag{3.72}$$

We could also have chosen $\{z = a\}$ as reference point and obtained the grading

$$V = \bigoplus_{n=-1}^{\infty} V'_n \quad V'_n = \left\langle (z-a)^{n+1} \frac{d}{dz} \right\rangle_{\mathbb{C}}. \tag{3.73}$$

With respect to both degrees, the algebra A is a graded Lie algebra. But the gradings are not the same for $a \neq 0$. The element $z^{n+1} \frac{d}{dz}$ will not be homogeneous in the second grading. The grading is not even equivalent in the sense of the definition (3.68).

3.9 Level lines

The Krichever–Novikov duality is an important concept and tool in our investigations. What we needed was only a smooth "separating cycle" C_S, separating the points in I from O of multiplicity one. In fact, only the cohomology class was of importance and we could take $K = \#I$ deformed circles around the points $P_i \in I$ with positive orientations. For certain applications it will be quite useful (see, e.g., Section 3.10 and the references there) to give such a separating cycle by a more analytic description. This we will do in the following section.

Let $A = I \cup O$. We assign to every point $P_i \in I$ a real number $a_i > 0$, and to every point $Q_j \in O$ a real number $b_j < 0$, such that the condition

$$\sum_{i=1}^{K} a_i + \sum_{j=1}^{M} b_j = 0 \tag{3.74}$$

is fulfilled.

Proposition 3.19. *A unique meromorphic 1-differential $\rho \in \mathcal{F}^1$ exists, holomorphic outside of A, which has poles of order 1 at the points in A and residues*

$$\operatorname{res}_{P_i}(\rho) = \alpha_i, \quad \operatorname{res}_{Q_j}(\rho) = \beta_j, \tag{3.75}$$

for $i = 1, \dots, K$ and $j = 1, \dots, M$, with purely imaginary periods.

Recall that given a differential ρ, the periods of ρ are exactly the values obtained by integrating it over arbitrary closed curves.

Proof. First recall that given two points P and Q on a Riemann surface Σ, there is always a meromorphic differential, holomorphic outside of $\{P, Q\}$ with pole order one at P and Q and residues $+1$, or respectively -1 (see for example [203, Theorem 4.6]). We fix a point $R \notin A$ and consider such differentials ρ^{P_i}, ρ^{Q_j} with respect to the pair of points (P_i, R) and (Q_j, R) respectively. We set

$$\rho := \sum_{i=1}^{K} \alpha_i \rho^{P_i} + \sum_{i=j}^{M} \beta_i \rho^{Q_j}. \tag{3.76}$$

From (3.74) it follows that ρ will not have any residue at R, hence it will be holomorphic there. Furthermore, ρ has exactly the required residue values. For $g \geq 1$ the differentials will not be unique, as globally holomorphic differentials can always be added to the solutions. But by adding a certain linear combination we can normalize ρ such that indeed all periods are purely imaginary; see for example [203, p. 116]. Note that the periods obtained by integrating "circles" around the points in A are purely imaginary as the residues are real.

If ρ and η where two differentials which fulfill the conditions. In this case $\gamma := \rho - \eta$ is a holomorphic differential with imaginary periods. But such a differential has to be zero due to the Riemann bilinear relations [203, p. 56]

$$\operatorname{Im} \sum_{i=1}^{g} \left(\overline{\int_{\alpha_i} \gamma} \cdot \int_{\beta_i} \gamma \right) > 0. \tag{3.77}$$

Here α_i and β_i are a symplectic homology basis; see (2.1). Hence such a differential ρ is uniquely determined. □

Remark 3.20. The differential will depend on the residues chosen. To be definite we might choose as residues

$$a_i = \frac{1}{K}, \quad b_j = -\frac{1}{M}. \tag{3.78}$$

Nevertheless, the additional freedom of choosing arbitrary residues will be quite convenient for applications (which will not be covered here) in string theory. In string theory, the surface Σ corresponds to the *world sheet of string*, the points in I correspond to free incoming strings, the points in O to free outgoing strings. In this case, the a_i can be interpreted as the momentum of the string incoming at the point P_i, and b_j as the momentum of the string outgoing at the point Q_j. The relation $\sum a_i + \sum b_j = 0$ corresponds to momentum conservation. The differential ρ might be considered as *propagation differential.*

Now we fix a distribution for the residue (e.g., the standard one (3.78)) and the corresponding differential. We fix a base point $B \notin A$ and set

$$u(Q) := \operatorname{Re} \int_B^Q \rho. \tag{3.79}$$

As according to the normalization our ρ has only purely imaginary periods, any ambiguity of the path from R to Q will disappear for the function u. Of course, u will depend on the base point B chosen. But the difference will only be a constant.

The function u will be a real-valued harmonic function, as we integrate over a holomorphic differential and take the real part of the (multi-valued) function obtained.

Proposition 3.21. *The function $u(P)$ behaves in the following manner if we approach the points in A*

$$\lim_{R\to P} u(R) = -\infty,\ P \in I \quad \textit{and} \quad \lim_{R\to Q} u(R) = \infty,\ Q \in O. \tag{3.80}$$

Proof. Let z be local coordinate at $P \in A$, then in a neighborhood of P we have

$$\rho = \frac{c}{z} dz + f(z)\, dz, \quad \text{with } c \in \mathbb{R} \tag{3.81}$$

and f a local holomorphic function. Let $S \neq P$ be a point in a disc around P, then

$$\lim_{R\to P} \operatorname{Re} \int_B^R \rho = \lim_{R\to P} \operatorname{Re} \int_S^R \rho + \operatorname{Re} \int_B^S \rho. \tag{3.82}$$

The second term stays finite, hence does not play any role for our considerations. We have to examine the first term.

$$\lim_{R\to P} \operatorname{Re} \int_S^R \frac{c}{z}\, dz = c \cdot \lim_{R\to P} (\log |z(R)| - \log |z(S)|) = (-\operatorname{sign}(c)) \cdot \infty. \tag{3.83}$$

This is the claim. □

With the help of the function u we can define the *level lines*

$$C_\tau := u^{-1}(\tau) = \{\, P \in \Sigma \setminus A \mid u(P) = \tau \,\} \tag{3.84}$$

for $\tau \in \mathbb{R}$. If we vary τ, we obtain a real fibration (also called foliation) of the Riemann surface $\Sigma \setminus A$, i.e.,

$$\Sigma \setminus A = \bigcup_{\tau \in \mathbb{R}} C_\tau \quad \text{and} \quad C_\tau \cap C_{\tau'} = \emptyset \quad \text{for } \tau \neq \tau'. \tag{3.85}$$

A different choice of the base point B changes the value of u, but not the level lines.

Example 3.22. In the classical case with $a_1 = 1$ and $b_1 = -1$, the ρ will be given by $\rho(z) = \frac{1}{z} dz$. Hence,

$$u(z) = \operatorname{Re} \int_1^z \frac{1}{z} dz = \log |z|. \tag{3.86}$$

Obviously (as it has to be)

$$\lim_{z \to 0} u(z) = -\infty, \qquad \lim_{z \to \infty} u(z) = +\infty. \tag{3.87}$$

In this case the level lines will be honest circles around $z = 0$.

See Figure 3.4 for an example of genus 2.

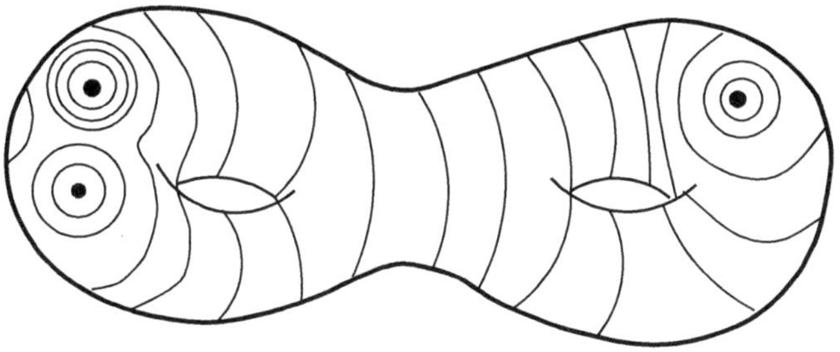

Fig. 3.4. Level lines for two in-points and one out-point in genus 2.

The level lines are not necessarily connected. They decompose into disjoint real curves. Singular points (self intersections, contacts, etc.) can only occur at points where ρ has zeros. As ρ has $K + M = N$ poles of order one, it has to have $2g - 2 + N$ zeroes (counted with multiplicities). For $\tau \to -\infty$ the level line C_τ decomposes into K different components $D_1, \ldots, D_k$. Every D_i is a circle in a suitable coordinate chart at the point $P_i \in I$. This follows from the local form (3.81). For $\tau \to +\infty$ we obtain the similar situation (again circles) around the points $Q_i \in O$.

Remark 3.23. In the interpretations in string theory, mentioned above, τ has the meaning of *proper time* on the "string world sheet". The splitting of level lines and their combination corresponds to interactions of the strings. Given a point $P \in \Sigma \setminus A$, then $\tau = u(P)$ and a local coordinate along C_τ are local world-sheet coordinates. Note that τ is even a global coordinate. As τ is real, we can order the points $R \in \Sigma \setminus A$ with respect to their proper time $\tau = u(R)$. If R and S do not lie on the same level line C_τ, then either R is "earlier" than S or vice versa. This is of importance in field theoretical construction, where one often uses the *time-ordered product*. We have to keep in mind that our definition of the level lines depends on ρ, and that ρ in turn depends on the prescribed residues. Hence, the notion of time ordering too. This is completely in accordance with calling ρ the propagation differential. Only in the case of two points is only one choice possible. Time ordering will play a role in Section 3.10 again.

If we restrict our vector field algebra $\mathcal{L}$ to C_τ we get a homomorphism

$$\mathcal{L} \to \mathcal{L}(C_\tau), \tag{3.88}$$

where the latter denotes the algebra of vector fields on the contour. As for big $|\tau|$, these contours are circles we get by restricting to one component a homomorphism

$$\mathcal{L} \to \mathcal{L}(S^1). \tag{3.89}$$

Krichever and Novikov [140] showed in the higher genus 2-point situation that the image of $\mathcal{L}$ is dense in $\mathcal{L}(S^1)$. Corresponding results are shown also for forms of other weights.

3.10 Delta-distribution

Recall that for $f \in \mathcal{F}^\lambda$ we can express f as a finite sum

$$f = \sum_{n \in \mathbb{J}_\lambda} \sum_{p=1}^{K} \alpha_{n,p} f^\lambda_{n,p}. \tag{3.90}$$

Via the duality (3.16), the coefficients in the sum can be determined by

$$\alpha_{n,p} = \frac{1}{2\pi \mathrm{i}} \int_{C_S} f \cdot f^{*,(n,p)}_{1-\lambda}. \tag{3.91}$$

Here by convention $f^{*,(n,p)}_{1-\lambda} = f^{1-\lambda}_{-n,p}$.

Based on the duality, it will be convenient to make the following definition.

Definition 3.24. The formal infinite sum

$$\Delta_\lambda(Q, Q') := \sum_{n \in \mathbb{J}_\lambda} \sum_{p=1}^{K} f^\lambda_{n,p}(Q) \cdot f^{*,(n,p)}_{1-\lambda}(Q') \tag{3.92}$$

is called *delta-distribution* (of weight λ, respectively associated to the pair of weights $(\lambda, 1-\lambda)$). In the definition Q and Q' are points in the Riemann surface Σ.

According to what we expect from a distribution [3] we get for $f \in \mathcal{F}^\lambda$, (integration over Q')

$$\frac{1}{2\pi i}\int_{C_S} f(Q')\Delta_\lambda(Q,Q') := \sum_{n\in\mathbb{J}_\lambda}\sum_{p=1}^{K} f^\lambda_{n,p}(Q)\frac{1}{2\pi i}\int_{C_S} f(Q')\cdot f^{*,(n,p)}_{1-\lambda}(Q'). \tag{3.93}$$

On the righthand side only a finite number of summands appear. We obtain

$$\frac{1}{2\pi i}\int_{C_S} f(Q')\Delta_\lambda(Q,Q') = \sum_{n\in\mathbb{J}_\lambda}{}' \sum_{p=1}^{K} \alpha_{n,p} f^\lambda_{n,p}(Q) = f(Q). \tag{3.94}$$

In the same way, we obtain for $g \in \mathcal{F}^{1-\lambda}$ (integration over Q)

$$\frac{1}{2\pi i}\int_{C_S} g(Q)\Delta_\lambda(Q,Q') = g(Q'). \tag{3.95}$$

This is a generalization of the "delta distribution" which was introduced by Krichever and Novikov in [141]. For $N = 2$ and $\lambda = 0$ we obtain the form given there

$$\Delta(Q,Q') = \sum_{n\in\mathbb{Z}} A_n(Q)\omega^n(Q'). \tag{3.96}$$

In the classical situation ($g = 0$, $\lambda = 0$, $N = 2$) we get

$$\Delta(z,z') = \sum_{n\in\mathbb{Z}} z^n (z')^{-n-1}dz'. \tag{3.97}$$

Until now our C_S was an arbitrary separating cycle. In case the separating cycles are level lines with respect to the function $u(P)$ (3.79), further analytic properties can be deduced.

Remark 3.25. For example, in the classical case the expression $\Delta(z,z')$ can even be used in a more general context. Let f be a function which is holomorphic in the interior of the annulus $K_{r,R}$ (with radius $0 < r < R$) and can be extended to the boundary, then

$$\frac{1}{2\pi i}\int_{C_\tau} f(z')\Delta(z,z') = f(z)$$

for any circle C_τ with radius τ between r and R, the boundary values included. This follows from the Laurent expansion. Let us denote by

$$i_{|z|<|z'|}\left(\frac{1}{z'-z}\right) = \frac{1}{z'}\sum_{k=0}^{\infty}\left(\frac{z}{z'}\right)^k = \sum_{k=0}^{\infty} z^k(z')^{-k-1} \tag{3.98}$$

3 More precisely, it is the distributional kernel of the distribution defined via the integral.

the convergent power series expansion in the domain $|z| < |z'|$, and by

$$i_{|z'|<|z|}\left(\frac{1}{z'-z}\right) = -\frac{1}{z}\sum_{l=0}^{\infty}\left(\frac{z'}{z}\right)^l = \sum_{k=-\infty}^{-1} z^k (z')^{-k-1} \tag{3.99}$$

the convergent power series expansion in the domain $|z'| < |z|$. Both functions are holomorphic functions for arguments different from 0, ∞ and $|z| \neq |z'|$. The distribution kernel (3.97) can now be written as

$$\Delta(z, z') = i_{|z|<|z'|}\left(\frac{1}{z'-z}\right) - i_{|z'|<|z|}\left(\frac{1}{z'-z}\right) \tag{3.100}$$

with a smooth kernel (smooth outside the above-mentioned points).

Remark 3.26. In case our separating cycle is defined via level lines C_τ, the genus zero two-point situation can be extended to higher genus (still two points). Let C_τ be a smooth but not necessarily connected level line, and f either a smooth form of weight λ (meaning a C^∞ form) on C_τ or a form which is holomorphic on

$$\Sigma_{\tau_1,\tau_2} := \{P \in \Sigma \mid \tau_1 \le u(P) \le \tau_2\}, \quad \text{with } \tau_1 < \tau < \tau_2. \tag{3.101}$$

Then it can be represented as

$$f(Q) = \frac{1}{2\pi \mathrm{i}} \int_{C_\tau} f(Q')\Delta_\lambda(Q, Q') = \sum_{n\in\mathbb{J}_\lambda} f_n^\lambda(Q) \frac{1}{2\pi \mathrm{i}} \int_{C_\tau} f(Q') f_{1-\lambda}^{*,n}(Q'). \tag{3.102}$$

Note that the form f is not necessarily meromorphic on Σ. In the higher genus 2-point situation this generalization was given by Krichever and Novikov [140–142]. The sketches they supplied of their proof were supplemented by a detailed analysis (involving small denominator problems) by Klimek and Lesniewski [130]. Of partial importance is again the splitting of $\Delta_\lambda(Q, Q')$ into two terms involving

$$K^\lambda(Q, Q') = \begin{cases} \sum_{n=0}^{\infty} f_n^\lambda(Q) f_{-n}^{1-\lambda}(Q'), & -\infty < u(Q) < u(Q') < \infty \\ -\sum_{n=-\infty}^{-1} f_n^\lambda(Q) f_{-n}^{1-\lambda}(Q'), & -\infty < u(Q') < u(Q) < \infty. \end{cases} \tag{3.103}$$

In analogy to the genus zero case and with the obvious notation we obtain

$$\Delta_\lambda(Q, Q') = i_{u(Q)<u(Q')}(K^\lambda(Q, Q')) - i_{u(Q')<u(Q)}(K^\lambda(Q, Q')). \tag{3.104}$$

Klimek and Lesniewski showed that $K^\lambda(Q, Q')$ is a holomorphic function for arguments not from A and not lying on the same level line. See Sadov [199] for some multi-point generalizations.

4 Fixing the basis elements

In this chapter I will show the existence of the basis elements as claimed in Chapter 3 such that Theorem 3.6 is true. It was my intention to collect nearly all of what will be needed about the basis and the almost-graded structure there. The current chapter will demonstrate the existence. This will be done via Riemann–Roch type arguments. The presented material is rather technical. The reader might prefer to skip it on first reading.

4.1 The Riemann–Roch theorem

As always, let Σ be a compact Riemann surface of genus g. We assume that the reader knows the basic definitions and results concerning Riemann surfaces. Nevertheless, we will recall some facts about divisors and Riemann–Roch. In particular, as we will have to fix some notation.

4.1.1 The language of divisors

A *divisor* is a formal sum

$$D = \sum_{P\in\Sigma}{}' n_P P, \quad n_P \in \mathbb{Z} \tag{4.1}$$

of points with integer multiplicities such that only a finite number n_P are different from zero. Given two divisors

$$D = \sum_{P\in\Sigma}{}' n_P P, \quad D' = \sum_{P\in\Sigma}{}' n'_P P, \tag{4.2}$$

its sum is defined as

$$D + D' := \sum_{P\in\Sigma}{}' (n_P + n'_P)P. \tag{4.3}$$

With this sum the set of divisors constitute a group. The zero element is $D = 0$, which means that all $n_P = 0$ for all $P \in \Sigma$. The inverse of D is given by $-D = \sum_{P\in\Sigma}{}'(-n_P)P$.

The *degree of the divisor* D is defined by

$$\deg(D) := \sum_{P\in\Sigma}{}' n_P. \tag{4.4}$$

Obviously, $\deg(D + D') = \deg(D) + \deg(D')$ and the divisors of degree zero constitute a subgroup of all divisors.

Given a meromorphic global function f on Σ which does not vanish identically, we assign its divisor (f) in the following way

$$(f) := \sum_{P\in\Sigma} \operatorname{ord}_P(f) \cdot P. \tag{4.5}$$

This says P is a zero of f if and only if $n_P > 0$; P is a pole of f if and only if $n_P < 0$. The absolute value of n_P gives the order of the zero and pole respectively. A zero of negative order is a pole and vice versa. As our Σ is compact, f has only a finite number of zeros and poles. More precisely, counted with multiplicities the number of zeros will coincide with the number of poles. By this the divisor (f) is well-defined. Moreover, it is a divisor of degree zero. A divisor which appears as the divisor of a function is called *principal divisor*. As $(f \cdot g) = (f) + (g)$ and $(1/f) = -(f)$, the principal divisors constitute a subgroup of all divisors (and of the divisors of degree zero).

Two divisors D and D' are called *(linearly) equivalent* if and only if $D' = D + (f)$ with a principal divisor (f). In other words, they are equivalent if they define the same equivalence class of divisor modulo the subgroup of principal divisors. The class of D is denoted by $[D]$.

We set

$$D = \sum_{P\in\Sigma}{}' n_P P, \quad \le \quad D' = \sum_{P\in\Sigma}{}' n'_P P, \tag{4.6}$$

if and only if $n_P \le n'_P, \forall P \in \Sigma$ and $D < D'$ if at least for one P we have $n_p < n'_p$. Obviously,

$$D \le D' \implies \deg(D) \le \deg(D'). \tag{4.7}$$

Given a divisor $D = \sum'_{P\in\Sigma} n_P P$, the *support* of D is defined as the set of points $\{P \mid n_P \neq 0\}$. We can decompose D in a unique way as the sum $D = D^+ + D^-$, where D^+, the positive part, contains the terms with $n_P > 0$, and D^-, the negative part, contains the terms with $n_P < 0$.

4.1.2 Divisors and line bundles

Given a (holomorphic) line bundle L over Σ. Let ψ be a meromorphic section of L, which is not identically zero. Locally around every point P with respect to a holomorphic frame, the section can be given by a local representing meromorphic function f. We define $\operatorname{ord}_P(\psi) = \operatorname{ord}_P(f)$. This definition does not depend on the frame chosen. In this way the divisor (ψ) of the section can be defined via

$$(\psi) := \sum_{P\in\Sigma} \operatorname{ord}_P(\psi) P. \tag{4.8}$$

The degree of ψ is given by the difference of the number of zeros and the number of poles (always counted with multiplicities). The divisor depends on the section chosen, but if we pass over to its linear equivalence class, the class will be uniquely given by the bundle L. The degree of the divisor will not depend on the section chosen. In fact, it will be the degree of the line bundle. We obtain a one-to-one correspondence between the isomorphy classes of line bundles over Σ and the linear equivalence classes of divisors of Σ. If $[D]$ corresponds to L, then the dual line bundle corresponds to the divisor class $[-D]$. The tensor product of line bundles corresponds to taking the sum of the divisor classes.

A special line bundle is the *canonical line bundle* $\mathcal{K}$, whose local sections are the locally holomorphic differentials. It is the bundle which we used to define the spaces $\mathcal{F}^\lambda$. Associated to it via the above correspondence is the *canonical divisor class*. In common abuse of notation we denote it by the same symbol $\mathcal{K}$. If we have a meromorphic differential ω, then it is a meromorphic section of the bundle $\mathcal{K}$, and its divisor (ω) is a *canonical divisor*. In particular $[(\omega)] = \mathcal{K}$. Again it is common to use the symbol $\mathcal{K}$ for an individual canonical divisor.

4.1.3 The theorem

Let $D = \sum_P n_P P$ be a divisor. Let

$$\mathrm{H}^0(\Sigma, D) := \{f \in \mathcal{M}(\Sigma) \mid (f) \geq -D\}. \tag{4.9}$$

In other words, this is the vector space of of those meromorphic functions f on Σ which are holomorphic outside the support of D, and for which $\mathrm{ord}_P(f) \geq -n_P$. At the points appearing in the positive part D^+ of the divisor D, poles are allowed, and at the points in the negative part D^- zeros are forced.

For line bundles L the vector space of global holomorphic sections of the bundle L is denoted by $\mathrm{H}^0(\Sigma, L)$.

If L is represented by a divisor D, then there is an isomorphism $\mathrm{H}^0(\Sigma, L) \cong \mathrm{H}^0(\Sigma, D)$. We will not make a clear distinction between the divisor and the bundle description.

In particular let L be a line bundle corresponding to a divisor D, and L' a line bundle corresponding to D', then the space $\mathrm{H}^0(\Sigma, L \otimes L')$ of holomorphic section of the bundle $L \otimes L'$ can be identified with the space $\mathrm{H}^0(\Sigma, D + D')$ of meromorphic functions f with $(f) \geq -(D + D')$. Equivalently, this space is the space of meromorphic sections f of the trivial bundle with the same conditions on the divisor. Moreover, the space of sections can be identified both with the space of meromorphic sections ψ of L with $(\psi) \geq -D'$ and with the space of meromorphic sections ϕ of L' with $(\phi) \geq -D$.

Theorem 4.1 (Riemann–Roch). *Let Σ be a compact Riemann surface of genus g. Let D be a divisor, L a line bundle, and $\mathcal{K}$ either a canonical divisor or the canonical line bundle, then*

$$\begin{aligned} \dim \mathrm{H}^0(\Sigma, D) - \dim \mathrm{H}^0(\Sigma, \mathcal{K} - D) &= \deg D - g + 1, \\ \dim \mathrm{H}^0(\Sigma, L) - \dim \mathrm{H}^0(\Sigma, \mathcal{K} \otimes L^*) &= \deg L - g + 1. \end{aligned} \tag{4.10}$$

See [203] for more background and for example [61, 79, 92] for proofs.

Let D be a divisor of negative degree. Assuming that $f \in \mathrm{H}^0(\Sigma, D)$, $f \neq 0$ we get as a contradiction

$$0 = \deg(f) \geq \deg(-D) = -\deg(D) > 0. \tag{4.11}$$

Hence,

$$\begin{aligned} \deg(D) < 0 &\implies \dim \mathrm{H}^0(\Sigma, D) = 0, \\ \deg(L) < 0 &\implies \dim \mathrm{H}^0(\Sigma, L) = 0. \end{aligned} \tag{4.12}$$

We will also need the simple facts

$$\deg \mathcal{K} = 2g - 2, \quad \dim \mathrm{H}^0(\Sigma, K) = g, \tag{4.13}$$

which follow from Riemann–Roch.

The following will also be quite useful

$$\dim \mathrm{H}^0(\Sigma, D) \le 1 + \deg D. \tag{4.14}$$

Of course, we are mainly interested in the first term on the left-hand side in (4.10). We call a divisor D *non-special divisor* if the second term $\dim \mathrm{H}^0(\Sigma, K - D)$ vanishes. If this is not the case it is called *special divisor*.

Depending on the degree of the divisor D we obtain the following proposition.

Proposition 4.2.

$$\dim \mathrm{H}^0(\Sigma, D) \quad = \quad \begin{cases} = 0, & \deg(D) < 0 \\ \ge 1 - g + \deg D, & 0 \le \deg(D) < 2g - 1 \\ = 1 - g + \deg D, & \deg(D) \ge 2g - 1. \end{cases} \tag{4.15}$$

Proof. For $\deg(D) < 0$ this follows from (4.12). If $\deg(D) \ge 2g - 1$, then $\deg(K - D) < 0$, hence $\mathrm{H}^0(\Sigma, K - D) = 0$ and the second term in (4.10) will vanish and we get equality. Between this we will have a finite range for the degree where it is possible that from $\dim \mathrm{H}^0(\Sigma, D)$ some number ≥ 0 will be subtracted, leading to the estimate in the second line. □

For genus zero the middle case does not appear, hence the following proposition.

Proposition 4.3. *For the Riemann surface Σ of genus zero we have*

$$\dim \mathrm{H}^0(\Sigma, D) = \max(0, 1 + \deg(D)). \tag{4.16}$$

Proposition 4.4. *If Σ is a Riemann surface of genus one and D a divisor of degree different from zero, then*

$$\dim \mathrm{H}^0(\Sigma, D) = \max(0, \deg(D)). \tag{4.17}$$

For a degree zero divisor D we have $\dim \mathrm{H}^0(\Sigma, D) = 1$ if and only if D is a principal divisor, or equivalently the divisor class is the trivial class, or equivalently the associated bundle is the trivial line bundle. Otherwise $\dim \mathrm{H}^0(\Sigma, D) = 0$. Moreover, the only special non-negative divisor class is the trivial class.

Proof. In the genus one case $2g - 1 = 1$, hence every divisor of degree greater than or equal to one is non-special and $\dim \mathrm{H}^0(\Sigma, D) = \deg(D)$. The statement on the negative degree divisors is clear. It remains to discuss degree zero divisors, or classes. The second line in Proposition 4.2 is valid in this case. In particular we have $\dim \mathrm{H}^0(\Sigma, D) = 0$ or 1. It is 1 if and only if D is a special divisor. If $D = (f)$ is a principal divisor, then $(1/f) \in \mathrm{H}^0(\Sigma, D)$, hence the dimension of the space is 1. Conversely, if there is a nontrivial element $g \in \mathrm{H}^0(\Sigma, D)$, then $(g) \geq -D$. But as divisors of degree zero are on both sides, the divisors have to coincide, i.e. $D = -(g)$, respectively $D = (1/g)$. □

Let L be a line bundle associated with a divisor D'. Let D be another divisor such that $\deg(D' + D) \geq 2g - 1$, then the 3rd line in Proposition 4.2 will give the exact dimension of $\dim \mathrm{H}^0(\Sigma, D' + D)$. In our interpretation the space consists of meromorphic sections of the line bundle L whose orders are better or equal as given by $-D$ (denoted by $\mathrm{H}^0(\Sigma, L + D)$).

If we decrease the degree of the divisor by one, either by reducing the multiplicity of a point P in the support of D, or adding a point P not yet in the support with multiplicity -1 (denote the new divisor by $D'' = D - P$), then additional conditions have to be fulfilled. As long as $\deg(D' + D'') \geq 2g - 1$ we go down in the dimension by exactly one. A problem will appear if we cross the threshold $2g - 1$, as in this case the second line of Proposition 4.2 becomes relevant. This means that the dimension could stay the same (of course it cannot go up). The following proposition states that a case of the dimension not going down is exceptional in the sense that, depending on the line bundle L, there are only a finite number of points P such that it will not go down by one.

Proposition 4.5. *Let D be a divisor and let P be a generic point. If*

$$\dim \mathrm{H}^0(\Sigma, D) = l, \tag{4.18}$$

then we have for all $n \in \mathbb{N}$

$$\dim \mathrm{H}^0(\Sigma, D - nP) = \max(l - n, 0). \tag{4.19}$$

Proof. Let D' be an arbitrary divisor and $\dim \mathrm{H}^0(X, D') = r$, then

$$r - 1 \leq \dim \mathrm{H}^0(X, D' - P) \leq r. \tag{4.20}$$

To see this, let $f, g \in \mathrm{H}^0(X, D')$, but both of them not in $\mathrm{H}^0(X, D' - P)$. This means that $f(P) \neq 0$ and $g(P) \neq 0$. We set $\mu = f(P)/g(P) \in \mathbb{C}$. Now

$$h = (f - \mu \cdot g) \in \mathrm{H}^0(X, D' - P), \tag{4.21}$$

as $h(P) = 0$ by construction. Hence, $\mathrm{H}^0(X, D')/\mathrm{H}^0(X, D' - P)$ is at most one-dimensional. This shows relation (4.20). If we apply this several times we get

$$l - n \leq \dim \mathrm{H}^0(D - nP) \leq l. \tag{4.22}$$

Hence, the claim (4.19) will be true if and only if

$$\dim \mathrm{H}^0(D - lP) = 0, \tag{4.23}$$

as in this case the dimension has to drop exactly by one between $D-rP$ and $D-(r+1)P$. In particular it is enough to show (4.23) for all but a finite number of points. This will be done with the same method used to show that there are only a finite number of Weierstraß-points. In fact, it is a special case of the proposition [61, II.5], [79, 18.4]. Let L be the corresponding line bundle to the divisor D. Let $\mathcal{U} = (U_i)_i$ be a finite covering of Σ by trivializing coordinate charts (for the bundle L). Let z_i be a local coordinate on U_i. We can take a covering such that a shrinking $\mathcal{V} = (V_i)_i$ with V_i open and $\overline{V_i} \subset U_i$ still covers Σ. Let $s_1, s_2, \ldots, s_l$ be a basis for the space $\mathrm{H}^0(X, L)$. With respect to one U_i we represent the sections as local holomorphic functions $f_1, f_2, \ldots, f_l$ in the variable z_i. As the sections are linearly independent this is true for the functions too. Let

$$s \in \mathrm{H}^0(X, L), \quad s \neq 0, \ s = \sum_{j=1}^{l} c_j s_j, \tag{4.24}$$

then the section corresponds to the holomorphic function $f = \sum_{j=1}^{l} c_j f_j$. The point $P \in U_i$ is a zero of s of at least order k if

$$f^{(m)}(z_i(P)) = 0, \quad m = 0, 1, \ldots, k-1. \tag{4.25}$$

The *Wronski determinant* is defined as

$$W = W(f_1, f_2, \ldots, f_l) = \det \begin{pmatrix} f_1 & f_2 & \cdots & f_l \\ f_1' & f_2' & \cdots & f_l' \\ \vdots & \vdots & \vdots & \vdots \\ f_1^{(l-1)} & f_2^{(l-1)} & \cdots & f_l^{(l-1)} \end{pmatrix}. \tag{4.26}$$

The function W is a local holomorphic function. As the f_i are linearly independent it does not vanish identically [79, 18.4]. Hence, $W(z_i(Q)) = 0$ if and only if nontrivial $c_j \in \mathbb{C}, \ j = 1, \ldots, l$ exists with

$$0 = \sum_{j=1}^{l} c_j f_j^{(m)}(z_i(Q)), \quad m = 0, 1, \ldots, l-1. \tag{4.27}$$

If $W(z_i(Q)) = 0$, then with the same coefficients $s = \sum_{j=1}^{l} c_j s_j$ will be a nonvanishing global section which has a zero of at least order l at Q. Conversely, the existence of such a section with a zero of order l at $Q \in U_i$ implies that W has a zero of this order. The set of zeros is a discrete subset of U_i and hence cut down to V_i a finite subset of V_i. As a finite number of V_i already cover Σ, only a finite number of points $Q_1, Q_2, \ldots, Q_r$ exist (depending on the bundle L), such that there is a nontrivial section of L which has at one of these points a zero of at least order l. For the choice of P these have to be avoided. □

From the proof we be more precise about the meaning of "generic'. It is true for all points with the exception of a finite number. Hence, it is true for a Zariski open and dense subset.

Here we are particularly interested in the canonical line bundle and its tensor powers.

Proposition 4.6. *Let $\mathcal{K}$ be the canonical divisor, respectively canonical bundle, $\lambda \in \mathbb{Z}$ (or $\lambda \in (1/2)\mathbb{Z}$ after the choice of a square root L of the canonical bundle), then*

$$\dim \mathrm{H}^0(\Sigma, \lambda K + D) - \dim \mathrm{H}^0(\Sigma, (1-\lambda)K - D) = (2\lambda - 1)(g-1) + \deg D. \tag{4.28}$$

Proof. This follows directly from Theorem 4.1 as $\deg(\lambda K) = 2\lambda(g-1)$. □

The following are useful results depending on the genus.

(1) In the case of genus zero, the canonical line bundle has degree −2 and there are no global holomorphic differentials. For $\lambda \in \mathbb{Z}$ and $\lambda \in \mathbb{Z} + 1/2$ we obtain

$$\begin{aligned} \dim \mathrm{H}^0(\Sigma, \lambda\mathcal{K}) &= 1 - 2\lambda, \quad \text{for } \lambda \le 0, \\ \dim \mathrm{H}^0(\Sigma, \lambda\mathcal{K}) &= 0, \quad \text{for } \lambda > 0. \end{aligned} \tag{4.29}$$

Note that in genus zero, the square root L of the canonical bundle is uniquely fixed as the bundle of degree −1.

(2) In the case of genus one (i.e., the torus), $\mathcal{K}$ will be the trivial line bundle $\mathcal{O}$, meaning the line bundle corresponding to the holomorphic function. This is therefore also the case for all integer powers, e.g.,

$$\dim \mathrm{H}^0(\Sigma, \lambda\mathcal{K}) = 1, \quad \text{for all } \lambda \in \mathbb{Z}. \tag{4.30}$$

We have to take the choice of L into account for the half-integer λ. For $g = 1$ we have 4 different choices. As $\deg L = 1/2 \, \deg \mathcal{K} = 0$, we get either $L = \mathcal{O}$ and

$$\dim \mathrm{H}^0(\Sigma, L) = \dim \mathrm{H}^0(\Sigma, \mathcal{O}) = 1, \tag{4.31}$$

or for the 3 other choices

$$\dim \mathrm{H}^0(\Sigma, L) = 0. \tag{4.32}$$

Thus for $L = \mathcal{O}$, Equation (4.30) is also true for λ, a half-integer. For $L \neq \mathcal{O}$ we have for λ a half-integer $\dim \mathrm{H}^0(\Sigma, \lambda K) = 0$.

(3) For $g \ge 2$ the degree of $\lambda\mathcal{K}$ is negative for negative λ, hence there are no nontrivial sections. For $\lambda > 1$ the degree is above the threshold, hence the bundle is nonspecial, only $\mathcal{K}$ itself and $1/2\,\mathcal{K}$ are special. We obtain

$$\begin{aligned} \dim \mathrm{H}^0(\Sigma, \lambda\mathcal{K}) &= 0 \quad \text{for } \lambda < 0, \\ \dim \mathrm{H}^0(\Sigma, 0 \cdot \mathcal{K}) &= 1 \\ \dim \mathrm{H}^0(\Sigma, \mathcal{K}) &= g \\ \dim \mathrm{H}^0(\Sigma, \lambda\mathcal{K}) &= (2\lambda - 1)(g-1) \quad \text{for } \lambda > 1. \end{aligned} \tag{4.33}$$

For $\lambda = 1/2$ the result will depend on the choice of L, i.e.,

$$0 \le \dim \mathrm{H}^0(\Sigma, L) = l(L) \le g. \tag{4.34}$$

The bounds are obtained from the fact that $\deg 1/2\,\mathcal{K} = (g-1)$, using (4.14) and (4.15).

Remark 4.7. Recall that a theta characteristic was called even or odd depending on the parity of $\dim \mathrm{H}^0(\Sigma, L)$. A Riemann surface of genus g has in total 2^{2g} theta characteristics. From them $2^{g-1}(2^g - 1)$, will be odd and $2^{g-1}(2^g + 1)$ will be even. A theta characteristic will be called *non-singular* if in the odd case $\dim \mathrm{H}^0(\Sigma, L) = 1$, and in the even case $\dim \mathrm{H}^0(\Sigma, L) = 0$[1]. For genus $g \le 2$ all theta characteristics are non-singular. This is immediate for odd theta characteristics as $\dim \mathrm{H}^0(\Sigma, L) \le g \le 2$. For $g \le 1$ the same follows for the even theta characteristics. For even ones in cases where $g = 2$ more involved techniques are needed, see e.g., [173]. For a generic Riemann surface (generic in the sense of the moduli space of Riemann surfaces) of genus $g \ge 3$, all theta characteristics are non-singular. The reader should be warned that important classes of Riemann surfaces, e.g., the hyperelliptic Riemann surfaces of genus $g \ge 3$, have also singular theta characteristics. These Riemann surfaces are called *Riemann surfaces with vanishing theta nulls*. See also [173, 176, 177].

For later reference, in Section 4.3 we state and prove the following result.

Proposition 4.8. *Let $P_1, P_2, \ldots, P_N$ be generic points on Σ, $n_1, \ldots, n_N \in \mathbb{Z}$.*
(a) *If at least one $n_i < 0$, then*

$$\dim \mathrm{H}^0\left(\Sigma, K - \sum_{i=1}^{N} n_i P_i\right) = \max\left(g - \sum_{i=1}^{N} n_i - 1\,,\, 0\right). \tag{4.35}$$

(b) *If all $n_i \ge 0$, then*

$$\dim \mathrm{H}^0\left(\Sigma, K - \sum_{i=1}^{N} n_i P_i\right) = \max\left(g - \sum_{i=1}^{N} n_i\,,\, 0\right). \tag{4.36}$$

(c) *If at least one $n_i > 0$, then*

$$\dim \mathrm{H}^0\left(\Sigma, - \sum_{i=1}^{N} n_i P_i\right) = \max\left(-g - \sum_{i=1}^{N} n_i + 1\,,\, 0\right). \tag{4.37}$$

(d) *If all $n_i \le 0$, then*

$$\dim \mathrm{H}^0\left(\Sigma, - \sum_{i=1}^{N} n_i P_i\right) = \max\left(-g - \sum_{i=1}^{N} n_i + 1\,,\, 1\right). \tag{4.38}$$

Proof. (a): Without restriction let $n_1 < 0$. Then due to the fact that

$$\deg(K - n_1 P_1) = (2g - 2) + |n_1| \ge 2g - 1, \tag{4.39}$$

$(K - n_1 P_1)$ is a non-special divisor, i.e., we have

$$\dim \mathrm{H}^0(\Sigma, K - n_1 P_1) = g - 1 - n_1. \tag{4.40}$$

1 The reason for this terminology will become clearer with Remark 5.7.

Now for the points $P_2, \ldots, P_N$ we can use Proposition 4.5 and obtain the result (a).

(b): Here we start with K, which is a special divisor, and we have $\dim \mathrm{H}^0(\Sigma, K) = g$. Now Proposition 4.5 gives the result (b).

(c): With Riemann–Roch we calculate

$$\dim \mathrm{H}^0\left(\Sigma, -\sum_{i=1}^{N} n_i P_i\right) - \dim \mathrm{H}^0\left(\Sigma, K + \sum_{i=1}^{N} n_i P_i\right) = -\sum_{i=1}^{N} n_i - g + 1. \tag{4.41}$$

If one $n_i > 0$, then according to part (a) we have

$$\dim \mathrm{H}^0\left(\Sigma, -\sum_{i=1}^{N} n_i P_i\right) = -\sum_{i=1}^{N} n_i - g + 1 + \max\left(g + \sum_{i=1}^{N} n_i - 1\,,\,0\right), \tag{4.42}$$

hence (c).

(d): If all $n_i \leq 0$, then from part (b) it follows

$$\dim \mathrm{H}^0\left(\Sigma, -\sum_{i=1}^{N} n_i P_i\right) = -\sum_{i=1}^{N} n_i - g + 1 + \max\left(g + \sum_{i=1}^{N} n_i\,,\,0\right), \tag{4.43}$$

which is (d). □

4.2 Choice of a basis for the generic case

4.2.1 Axiomatic characterisation

Let A be the finite set of points where poles are allowed and $A = I \cup O$ its splitting. As always

$$I = \{P_1, P_2, \ldots, P_K\} \quad O = \{Q_1, Q_2, \ldots, Q_M\}. \tag{4.44}$$

We will consider $\mathcal{K}^\lambda$ with λ fixed and $n \in \mathbb{J}_\lambda$ and the divisor

$$D_{n,p} = D_{n,p,I} + D_{n,p,O}, \tag{4.45}$$

with

$$\begin{aligned} D_{n,p,I} &= -\sum_{i=1}^{K} (n - \lambda + (1 - \delta_i^p))\, P_i \\ D_{n,p,O} &= \sum_{j=1}^{M} (a_j(n - \lambda) + b_{j,n}^\lambda)\, Q_j. \end{aligned} \tag{4.46}$$

Here the a_j and $b_{j,n}^\lambda$ are rational numbers fulfilling the conditions that

$$a_j > 0, \quad a_j(n - \lambda) + b_{j,n}^\lambda \in \mathbb{Z}, \tag{4.47}$$

and that the latter expression is increasing with n (not necessarily strictly). Furthermore,

$$\sum_{j=1}^{K} a_j = K, \quad \sum_{j=1}^{M} b_{j,n}^\lambda = K - M(\lambda). \tag{4.48}$$

Recall $M(\lambda) = (2\lambda - 1)(g - 1)$. We require also that the constants B^λ, C_j and E_j exist such that $|b^\lambda_{j,n}| \le B^\lambda$ for all j and n, and that for all n, m, k, λ, ν we have

$$|b^\lambda_{j,n} + b^\mu_{j,m} + b^{1-(\mu+\nu)}_{j,-k}| \le C_j, \quad |b^\lambda_{j,n} + b^\mu_{j,m} + b^{-(\mu+\nu)}_{j,-k}| \le E_j. \tag{4.49}$$

Finally, we require

$$(b^\lambda_{j,n} + b^{1-\lambda}_{j,m}) \le 2a_j, \quad \text{for } n + m < 0. \tag{4.50}$$

We will see in the following section why we have these technical requirements. But we will also show in Section 4.2.2 that in all possible cases for the splitting we can find such a_j and $b^\lambda_{j,n}$.

In this section we will consider the generic situation that either $g = 0$ without any further restriction or that $g \ge 2$, $\lambda \ne 0, 1, 1/2$, and the points in A are in generic position. In a later section we will deal with the remaining cases. We already point out now that the divisor part $D_{n,p,I}$ will always remain the same, only $D_{n,p,O}$ needs to be modified.

Recall from Lemma 3.9 the relations

$$M(\lambda) + M(\nu) + M(1 - (\lambda + \nu)) = -(g - 1), \tag{4.51}$$

$$M(\lambda) + M(\nu) + M(-(\lambda + \nu)) = -3(g - 1), \tag{4.52}$$

$$M(\lambda) + M(1 - \lambda) = 0. \tag{4.53}$$

Remark 4.9. To illustrate the above requirements let us consider the model situation, i.e. when $O = \{Q_M\}$. In this case we have

$$a_M = K, \quad b^\lambda_M = K - M(\lambda). \tag{4.54}$$

We have boundedness $|b^\lambda_M| \le K + |M(\lambda)|$, and using (4.51) and (4.52) we calculate

$$\begin{aligned} b^\lambda_M + b^\nu_M + b^{1-(\lambda+\nu)}_M &= 3K + (g - 1), \\ b^\lambda_M + b^\nu_M + b^{-(\lambda+\nu)}_M &= 3K + 3(g - 1). \end{aligned} \tag{4.55}$$

From (4.53) we conclude

$$(b^\lambda_M + b^{-\lambda}_M) = 2K = 2a_M. \tag{4.56}$$

Theorem 4.10. *Under the general assumption above for the genus g and the weight $\lambda \in \mathbb{Z} + 1/2$, for every λ, $n \in \mathbb{J}_\lambda$, $p = 1, \dots, K$ a unique element $f^\lambda_{n,p} \in \mathcal{F}^\lambda$ exists, up to multiplication with a non-zero scalar, such that*

$$\begin{aligned} \operatorname{ord}_{P_r}(f^\lambda_{n,p}) &= (n - \lambda + (1 - \delta^p_r)), \quad r = 1, \dots, K \\ \operatorname{ord}_{Q_j}(f^\lambda_{n,p}) &= -a_j(n - \lambda) - b^\lambda_{j,n}, \quad j = 1, \dots, M. \end{aligned} \tag{4.57}$$

Let z_p be a fixed coordinate at P_p, then by imposing additionally

$$f^\lambda_{n,p}(z_p)_| = z_p^{n-\lambda}(1 + O(z_p))(dz_p)^\lambda, \tag{4.58}$$

the element will be uniquely fixed.

Proof. For a given $n \in \mathbb{J}_\lambda$ and $p = 1, \ldots, K$ we consider the divisor $D_{n,p}$ given by (4.45). Its degree calculates to

$$\deg(D_{n,p}) = -(K-1) + \sum_{j=1}^{M} b_{n,j}^{\lambda} = -M(\lambda) + 1 = -(2\lambda - 1)(g-1) + 1. \tag{4.59}$$

Our aim is to calculate the dimension of the space $\mathrm{H}^0(\Sigma, \lambda\mathcal{K} + D)$ for $D = D_{n,p}$. The elements of this space correspond to sections ψ of the bundle $\mathcal{K}^\lambda$ which have orders

$$\begin{aligned} &\mathrm{ord}_{P_r}(\psi) \geq (n + 1 - \lambda),\ r \neq p, \\ &\mathrm{ord}_{P_p}(\psi) \geq (n - \lambda), \\ &\mathrm{ord}_{Q_j}(\psi) \geq -(a_j(n-\lambda) + b_{j,n}^{\lambda}),\ j = 1, \ldots, M. \end{aligned} \tag{4.60}$$

Using Proposition 4.6 to calculate the dimension $\dim \mathrm{H}^0(\Sigma, \lambda\mathcal{K} + D)$, we obtain

$$\dim \mathrm{H}^0(\Sigma, \lambda K + D) - \dim \mathrm{H}^0(\Sigma, (1-\lambda)K - D) = 1. \tag{4.61}$$

First we study $g \geq 2$ and $\lambda > 1$. In this case (see (4.33)), the $\lambda\mathcal{K}$ is non-special and we obtain

$$\dim \mathrm{H}^0(\Sigma, \lambda\mathcal{K}) = (2\lambda - 1)(g-1). \tag{4.62}$$

Assuming that the points in A are generic, we obtain from Proposition 4.5

$$\begin{aligned} &\dim \mathrm{H}^0(\Sigma, \lambda\mathcal{K} + D) = 1, \\ &\dim \mathrm{H}^0(\Sigma, \lambda\mathcal{K} + D - P) = 0, \quad \text{for every } P \in A. \end{aligned} \tag{4.63}$$

To see this we decompose D into its positive and negative parts, $D = D^+ + D^-$. First we take the positive part of the divisor D (respectively $D - P$) into consideration. As we stay in the non-special region we obtain

$$\dim \mathrm{H}^0(\Sigma, \lambda\mathcal{K} + D^+) = (2\lambda - 1)(g-1) + \deg(D^+). \tag{4.64}$$

Next we use Proposition 4.5 to reduce by the degree of D^- to obtain

$$\dim \mathrm{H}^0(\Sigma, \lambda\mathcal{K} + D) = (2\lambda - 1)(g-1) + \deg(D^+) + \deg(D^-) = 1. \tag{4.65}$$

Similar for the divisor $D - P$, with $P \in A$, we get

$$\dim \mathrm{H}^0(\Sigma, \lambda\mathcal{K} + D - P) = 0. \tag{4.66}$$

This says that there is up to multiplication with a non-zero scalar a nontrivial element $f_{n,p}^{\lambda} \in \mathrm{H}^0(\Sigma, \lambda\mathcal{K} + D)$. In particular this $f_{n,p}^{\lambda}$ has orders better or equal as required by (4.60). But for every point P in A, the spaces in (4.66) are zero. Hence $f_{n,p}^{k}$ cannot have a higher order at any of these points, otherwise it would be an element of at least one of the spaces (4.66). We get the exact order required. The local condition (4.58) can be used to normalize the element and it will be uniquely given by the (first order jet of the) local coordinate z_p.

If $\lambda < 0$, then $(1-\lambda) > 1$ and $(1-\lambda)\mathcal{K}$ will not be special. With the same kind of genericity arguments as above we now obtain

$$\begin{aligned} \dim \mathrm{H}^0(\Sigma, (1-\lambda)\mathcal{K} - D) &= \max(0, -1) = 0, \\ \dim \mathrm{H}^0(\Sigma, (1-\lambda)\mathcal{K} - (D-P)) &= 0. \end{aligned} \tag{4.67}$$

Hence on the left-hand side of (4.28), only the first term will remain and we get the same results as above.

In the genus zero case we always have (see Proposition 4.3)

$$\dim \mathrm{H}^0(\Sigma, \lambda\mathcal{K} + D) = \max(0, 1 - 2\lambda + \deg D) \tag{4.68}$$

and we obtain directly

$$\dim \mathrm{H}^0(\Sigma, \lambda\mathcal{K} + D) = 1, \quad \dim \mathrm{H}^0(\Sigma, \lambda\mathcal{K} + D - P) = 0. \tag{4.69}$$

We continue as above without any restriction on the weight λ and the positions of the points. □

Remark 4.11. In the proof of the proposition above we used the fact that either $\lambda\mathcal{K}$ or $(1-\lambda)\mathcal{K}$ is non-special. For genus $g \geq 1$ the classes $\mathcal{K}$, $0 \cdot \mathcal{K} = \mathcal{O}$, and possibly $(1/2)\mathcal{K}$ are special divisor classes. Hence the arguments above need a modification. Furthermore, in the case of genus one $\mathcal{K} \cong \mathcal{O}$, hence all $\lambda\mathcal{K} \cong \mathcal{O}$ (at least for $\lambda \in \mathbb{Z}$) and we have to make modifications for all weights λ. The proof indicates in which directions a modification is needed. If we consider for example $g \geq 2$ and $\lambda = 1$, then as long as our divisor D has a nonvanishing positive part D^+ (corresponding to poles of the form), we can calculate as above without any change. Only if the divisor has only negative parts do we have to modify the description. This can only be the case for a finite number n. Details will be given in Section 4.3.

Proposition 4.12. *Outside A the elements $f^\lambda_{n,p}$ have exactly g zeros counted with multiplicities.*

Proof. The element $f^\lambda_{n,p}$ as nonvanishing section of $\mathcal{K}^\lambda$ has to have the section degree $2\lambda(g-1)$, i.e., the sum of the orders at all points in Σ. If we add up only the points in A we obtain the degree $(2\lambda-1)(g-1)-1$. Hence, the element will have

$$2\lambda(g-1) - (2\lambda-1)(g-1) + 1 = g \tag{4.70}$$

additional zeros outside of A. Note that there are no poles outside A. □

Next we will verify that the elements defined above are dual with respect to the Krichever–Novikov pairing (3.10).

Proposition 4.13. *Given $f^{\lambda}_{n,p} \in \mathcal{F}^{\lambda}$ and $f^{1-\lambda}_{m,r} \in \mathcal{F}^{1-\lambda}$, then*

$$\langle f^{\lambda}_{n,p}, f^{1-\lambda}_{m,r} \rangle = \frac{1}{2\pi \mathrm{i}} \int_{C_S} f^{\lambda}_{n,p} \cdot f^{1-\lambda}_{m,r} = \delta^{-m}_{n} \delta^{r}_{p},$$
$$n, m \in \mathbb{J}_{\lambda}, \; p, r = 1, \ldots, K. \tag{4.71}$$

Proof. We calculate the orders at the points in I and O of the differential obtained as the product. We obtain

$$\begin{aligned} \operatorname{ord}_{P_s}(f^{\lambda}_{n,p} \cdot f^{1-\lambda}_{m,r}) &= (n+m-1) + (1-\delta^{p}_{s}) + (1-\delta^{r}_{s}), \\ \operatorname{ord}_{Q_j}(f^{\lambda}_{n,p} \cdot f^{1-\lambda}_{m,r}) &= -a_j(n-\lambda) - a_j(m-(1-\lambda)) - b^{\lambda}_{j,n} - b^{1-\lambda}_{j,m} \\ &= -a_j\left((n+m-1) + \frac{1}{a_j}(b^{\lambda}_{j,n} + b^{1-\lambda}_{j,m})\right). \end{aligned} \tag{4.72}$$

Integration over a separating cycle can equivalently be done either by calculating the residues at the points in I or in O. First we consider the points in I. If $(n+m) > 0$ there is no residue at the points in I. If $n+m=0$, i.e., $m=-n$, then there is only a residue if $r=p$ and it will be at the point P_s. The residue will be 1. Hence, the statement of the proposition will be true for $n+m \geq 0$. Next we consider $n+m<0$ (i.e., $n+m \leq -1$) and the points in O. We use the fact that we have

$$(b^{\lambda}_{j,n} + b^{1-\lambda}_{j,m}) \leq 2a_j, \quad \text{for } n+m<0. \tag{4.73}$$

Hence,

$$(n+m-1) + \frac{1}{a_j}(b^{\lambda}_{j,n} + b^{1-\lambda}_{j,m}) \leq (n+m+1) \leq 0, \tag{4.74}$$

in the range $n+m<0$. This implies that the order at the point Q_j is ≥ 0. Consequently, for pairs (n,m) with $n+m<0$ there will be no residue. And in total we obtain the relation (4.71). □

Proposition 4.14. *The set*

$$\{f^{\lambda}_{n,p} \mid n \in \mathbb{J}_{\lambda}, \; p = 1, \ldots, K\} \tag{4.75}$$

is a basis of $\mathcal{F}^{\lambda}$.

Proof. By the duality (4.71) the elements are linearly independent. It remains to show that they generate $\mathcal{F}^{\lambda}$. For the following proof it is more convenient to use an index shift. We denote by $g^{\lambda}_{m,r}$ the element $f^{\lambda}_{m+\lambda,r}$. Note that $m+\lambda$ will always be an integer. The original set will be generating if and only if the new set of elements is generating.

We take for $n \in \mathbb{N}$ the divisor

$$D(n) := \sum_{r=1}^{K} nP_r + \sum_{j=1}^{M} (a_j n + b^{\lambda}_{j,n+\lambda}) Q_j, \tag{4.76}$$

and consider the space

$$V(n) := \mathrm{H}^0(\Sigma, \lambda\mathcal{K} + D(n)). \tag{4.77}$$

For the degree we calculate

$$\deg(D(n)) = K \cdot (2n+1) - (2\lambda - 1)(g-1). \tag{4.78}$$

Hence,

$$\deg(\lambda\mathcal{K} + D(n)) = K \cdot (2n+1) + g - 1. \tag{4.79}$$

If n is big enough, then the divisor $\lambda\mathcal{K} + D(n)$ has degree $\geq 2g-1$ and hence is non-special. Via Riemann–Roch we calculate

$$\dim V(n) = K \cdot (2n+1). \tag{4.80}$$

The divisor $D(n)$ is the maximal polar divisor allowed for the elements in $V(n)$. By checking their orders at the points in I and O we see that the elements

$$g^\lambda_{m,p}, \quad -n \leq m \leq n, \ p = 1, \dots, K \tag{4.81}$$

lie in $V(n)$. Here we use the fact that $a_j n + b^\lambda_{j,n+\lambda}$ is increasing. As these are $K \cdot (2n+1)$ elements, they constitute the basis of $V(n)$. An arbitrary $v \in \mathcal{F}^\lambda$ has finite pole orders at the finite number of points in A, hence there is an n such that $v \in V(n)$. Hence the set (4.75) is also generating and consequently a basis. □

Under the assumption that for every splitting of A we find such a_j and $b^\lambda_{j,n}$ and that we are in the generic situation we have proved nearly everything formulated in Chapter 3. There is only one small point to complete. In the proof of Theorem 3.8 we used duality. But in the proof we only considered the model situation (meaning $M = 1$). This we still have to extend for arbitrary M.

Proof of the missing point for $M > 1$ in Theorem 3.8. We have to consider the points Q_j and make calculations as in (3.37) for the orders at every such point:

$$-a_j(n+m-k-1) - (b^\lambda_{j,n} + b^\nu_{j,m} + b^{1-(\lambda+\nu)}_{j,-k}). \tag{4.82}$$

By the conditions in (4.49) we have that the second term is bounded independently of n, m, λ, and ν by a constant C_j. In particular there can only be a residue at the point Q_j if $k \leq (n+m) + L_j$ with the bound $L_j = 1 + (1/a_j)(C_j - 1)$. Hence, in total there could only be a contribution at any of the points if

$$k \leq (n+m) + R_1, \quad \text{with} \quad R_1 = \max_{j=1,\dots,M} \{L_j\}. \tag{4.83}$$

Exactly the same kind of argument can be used for the Lie product. Now we use (3.42) and the constants E_j from (4.49) yield L'_j and finally a bound R_2. □

4.2.2 Realizing all splittings

Until now we have been using axiomatic characterisation of a_j and $b_{j,n}^{\lambda}$. Here we will show that we can find such a system associated with every possible splitting of $A = I \cup O$. I will present results obtained in [207] and [206]. See also the work of Sadov [199] for some related partial results. The original settings (up to an index shift) of Krichever and Novikov [140, 141] are obtained as special cases for $K = M = 1$.

I would like to point out that with the exception of the model situation the prescription will not be unique. Different schemes are possible, and hence different homogeneous subspaces are possible. Examples of such schemes are given by changing the numeration of the points. Even more involved changes are possible. But in any case the induced filtrations will be the same, as they are defined in a way independent of the points in O. Moreover, the different almost-gradings obtained in this way will be equivalent; see Section 3.6 for the definition.

In the following section we will drop the n and the λ in the notation $b_{j,n}^{\lambda}$ if the element does not depend on them.

K = M

In this case $a_j = a = 1$, $j = 1, \ldots, M$, and

$$b_j = 1,\ j = 1, \ldots, M-1, \quad b_M = 1 - M(\lambda). \tag{4.84}$$

This prescription fulfills all criteria and is thus a possible choice. To see this we have to use (4.51), (4.52), and (4.53).

K > M

A solution is given by

$$\begin{aligned} a_j = 1,\ j = 1, \ldots, M-1, \quad a_M &= (K-M) + 1 =: a \geq 1, \\ b_j = 1,\ j = 1, \ldots, M-1, \quad b_M &= 1 - M(\lambda). \end{aligned} \tag{4.85}$$

The summation conditions are clearly true. Moreover, we have

$$(b_j^{\lambda} + b_j^{1-\lambda}) = 2 \leq 2a. \tag{4.86}$$

Also the sum of 3 elements fulfills the boundedness criteria, as we have

$$b_j^{\lambda} + b_j^{\nu} + b_j^{1-(\lambda+\nu)} = \begin{cases} 3, & j \neq M \\ g+2, & j = M. \end{cases} \tag{4.87}$$

$$b_j^{\lambda} + b_j^{\nu} + b_j^{-(\lambda+\nu)} = \begin{cases} 3, & j \neq M \\ 3g, & j = M. \end{cases} \tag{4.88}$$

K > M

Here the fixing is a little bit more involved. Let $a = (M - K) + 1 > 1$. We introduce the residue class representatives $\epsilon_n, \overline{\epsilon}_n$ by

$$\begin{aligned} \epsilon_n^\lambda &\equiv (n - \lambda + 1) \mod a, \quad \epsilon_n \in \{0, 1, \ldots, a-1\} \\ \overline{\epsilon}_n &\equiv (n - \lambda + 1) \mod a, \quad \epsilon_n \in \{-a+1, \ldots, -1, 0\}. \end{aligned} \tag{4.89}$$

We have to keep in mind that these values also depend on λ, not only on n. For $j = 1, \ldots, K-1$ we set

$$a_j = 1, \quad b_j = 1. \tag{4.90}$$

For $j = K, \ldots, M$ we set

$$\begin{aligned} a_j &= \frac{1}{a} \\ b_{j,n}^\lambda &= \begin{cases} \frac{1}{a}(1 - \epsilon_n), & j = K, \ldots, K + |\overline{\epsilon}_n| - 1, \\ \frac{1}{a}(1 - \overline{\epsilon}_n), & j = K + |\overline{\epsilon}_n|, \ldots, M-1, \\ \frac{1}{a}(1 - \overline{\epsilon}_n) - M(\lambda), & j = M. \end{cases} \end{aligned} \tag{4.91}$$

With this description we have also for $j = K, \ldots, M$

$$a_j(n - \lambda) + b_{j,n}^\lambda = \frac{1}{a}(n - \lambda + 1 - \tilde{\epsilon}_n) \in \mathbb{Z}. \tag{4.92}$$

Here $\tilde{\epsilon}_n$ should denote either ϵ_n or $\overline{\epsilon}_n$ depending on the case. Writing down the expressions we see that the orders are increasing as required.

Also,

$$\sum_{j=1}^{M} a_j = (K-1) + \frac{1}{a}(M - K + 1) = K, \tag{4.93}$$

$$\sum_{j=1}^{M} b_{j,n}^\lambda = K - M(\lambda) + \frac{1}{a}(|\overline{\epsilon}_n| \cdot (-\epsilon_n) + (-\overline{\epsilon}_n)(a - |\overline{\epsilon}_n|)). \tag{4.94}$$

We claim that the last term will vanish. This is clear for $(n - \lambda + 1) \equiv 0 \mod a$, as then $\tilde{\epsilon}_n = 0$. If $(n - \lambda + 1) \not\equiv 0 \mod a$, then $\epsilon_n = \overline{\epsilon}_n + a$ and $|\overline{\epsilon}_n| = -\overline{\epsilon}_n$, hence it will also vanish in this case.

Obviously $|b_{j,n}^\lambda| \leq B_\lambda|$ with a bound B_λ. For the first of the 3-term expressions we get

$$|b_{j,n}^\lambda + b_{j,n}^\nu + b_{j,n}^{1-(\lambda+\nu)}| \leq \begin{cases} 3, & j = 1, \ldots, K-1 \\ \frac{1}{a}(3 + 3a), & j = K, \ldots, M-1 \\ \frac{1}{a}(3 + 3a) + (g-1), & j = M. \end{cases} \tag{4.95}$$

For the first alternative this is immediate. For the second alternative we know that the absolute value of the residue class representative is bounded by a. In the last case the sum (4.51) appears. For the second 3-term expression things stay the same, we only have to use (4.52). This yields as bound $E_M = \frac{1}{a}(3 + 3a) + 3(g - 1)$.

It remains to verify

$$(b_{j,n}^{\lambda} + b_{j,m}^{1-\lambda}) \le 2a_j \tag{4.96}$$

for $n + m < 0$. In the first range for $j = 1, \ldots, K - 1$ this is clear. In the second range $j = K, \ldots, M - 1$ we have

$$(b_{j,n}^{\lambda} + b_{j,m}^{1-\lambda}) = 2 - \tilde{\epsilon}_{n,\lambda} - \tilde{\epsilon}_{m,1-\lambda}. \tag{4.97}$$

In any case, $|\tilde{\epsilon}| \le a - 1$. Hence indeed (4.96). For $j = M$ we have the additional term $M(\lambda) + M(1 - \lambda)$, which by (4.53) is zero.

4.3 The remaining cases

Before we turn to the remaining cases I would like to point out what we did from the point of view of the filtrations introduced in Section 3.5. The defined filtrations $\mathcal{F}_{(n)}^{\lambda}$ do not make any reference to a choice of a basis. The quotient space

$$\mathcal{F}_{(n)}^{\lambda} / \mathcal{F}_{(n+1)}^{\lambda} \tag{4.98}$$

is K-dimensional. Our constructed $f_{n,p}^{\lambda}$, $p = 1, \ldots, K$ define $\mathrm{mod}\mathcal{F}_{(n+1)}^{\lambda}$ a basis of this quotient.

Conversely, we could have started differently. Starting from a basis of the quotient, we lift them to elements of $\mathcal{F}_{(n)}^{\lambda}$ and they will be a basis of the homogeneous subspace $\mathcal{F}_{n}^{\lambda}$. But the lifts will be highly non-unique. To make them unique (always up to multiplication with a scalar) we try to formulate conditions to force them to behave as nicely as possible at the points in O. Another property to realize is the Krichever–Novikov duality. This is exactly what we did with our prescription in the generic situation. In the special situation, either corresponding to the special values for λ or non-genericity of the points in A, we have to modify the generic prescription minimally and remain as optimal as possible.

In the generic situation we used the fact that either $\lambda\mathcal{K}$ or $(1 - \lambda)\mathcal{K}$ is non-special. For genus $g \ge 1$ the classes $\mathcal{K}$, $0 \cdot \mathcal{K} = \mathcal{O}$, and possibly $(1/2)\mathcal{K}$ are special divisor classes. Hence the arguments presented above need to be modified for these λ values. Furthermore, in the case of genus one $\mathcal{K} \cong \mathcal{O}$, hence all $\lambda\mathcal{K} \cong \mathcal{O}$ (at least for $\lambda \in \mathbb{Z}$) and we have to make modifications for all weights λ. The proof in the generic situation indicates in which directions a modification is needed. If we consider, for example, $g \ge 2$ and $\lambda = 1$, then as long as our divisors $D_{n,p}$ and $D_{n,p} - P$ $(P \in A)$ have nonvanishing positive parts (corresponding to poles of the form), we calculate as above without any change. Only when one of these divisors has only negative parts do we have to modify the description; see Proposition 4.8. This can only be the case for a finite range of n.

In this range the necessary modification will be bounded, meaning that the change of orders will only be finite. Hence, the changes are controllable.

Moreover, we can make the modification in such a way that duality remains established. Note that $\lambda = 0$ is dual to $\lambda = 1$, and $\lambda = 1/2$ is dual to itself. This guarantees that all statements about the almost-gradedness remain true. Only the boundaries R_1 and R_2 for the almost-graded structure need some readjustment, taking into account the finite number of modified elements.

For $\lambda = 1/2$ we have to take into account that now things will depend on the chosen theta characteristics L, $L^2 = \mathcal{K}$. More precisely, it will depend on $\dim \mathrm{H}^0(\Sigma, L)$. But the principal system remains the same.

4.3.1 Genus greater or equal to two

First, we will discuss the case $\lambda = 0$ and $\lambda = 1$. Recall the divisor $D_{n,p}$ (4.45), (4.46) in our generic description. The A-part of the divisor $(f^\lambda_{n,p})$ is given by $-D_{n,p}$. In particular, if we add the orders of $f^\lambda_{n,p}$ over the points in A we obtain $(2\lambda - 1)(g - 1) - 1$.

Let us consider first $\lambda = 1$. In this case, this sum is $g - 2$. Note that $M(1) = g - 1$. As an illustration take the example $K = M$. In this case we have the following generic order prescription for $f^1_{n,p}$:

$$\begin{aligned} \operatorname{ord}_{P_p}(f^1_{n,p}) &= n - 1, \qquad & \operatorname{ord}_{P_r}(f^1_{n,p}) &= n, \ r \neq p, \\ \operatorname{ord}_{Q_j}(f^1_{n,p}) &= -n, \qquad & \operatorname{ord}_{Q_M}(f^1_{n,p}) &= -n + g - 1. \end{aligned} \tag{4.99}$$

Modification ($\lambda = 1$ and for all cases of K and M). The modifications are guided by Proposition 4.8 (a) and (b), which we apply with respect to the divisor $D_{n,p}$ and $D_{n,p} - P$.

(1) Assume there is just one order -1 and it is at a point in I. Then necessarily $n = 0$. According to Riemann–Roch a section which is better or equal to the prescription exists. Indeed (4.35) gives us dimension one. If we consider now the divisor minus this point we are in the situation where (4.36) is valid. Again we obtain a dimension of one. Hence having a pole at the point is not possible. Indeed, this is also clear from the the residue theorem as a differential with just one pole of order 1 does not exist.
We will make the following modification.
(1a) If $M > 1$, then we put -1 as order at Q_1 and increase the order at Q_M by 1. The degree of the divisor remains unchanged. Now we have a unique element which has exactly the required orders as from the passage of this new divisor to subtracting a point the dimension is reduced by one.
(1b) If $M = 1$ we set the order at Q_M also to -1. Such a differential exists, but it is not unique. To make it unique we require that all real periods vanish (see Section 3.9).

(2) If the -1 order is at a point Q in O then the existing section which is better or equal to the prescription has no pole. Hence we set the order at the point Q equal to 0 and take this section.
(3) If all orders are ≥ 0, then according to Riemann–Roch (see (4.36)) the dimension of the space of sections greater or equal to the prescription is 2 (see (4.36) and note that $\sum_{i=1}^{N} n_i = g-2$). By increasing the order at Q_M by 1 we make the element unique.

For example, for $K = M = 1$ we obtain elements with the following orders

$$\begin{aligned} &\mathrm{ord}_P(f_0^1) = -1, \quad &&\mathrm{ord}_Q(f_0^1) = -1, \\ &\mathrm{ord}_P(f_n^1) = n-1, \quad &&\mathrm{ord}_Q(f_n^1) = g-n, \quad 1 \leq n \leq g. \end{aligned} \tag{4.100}$$

These elements are uniquely given modulo rescaling.

Modification ($\lambda = 0$ and for all cases of K and M). We have to take Proposition 4.8 (c) and (d) into account. Note that $M(0) = -g+1$ and we use the notation $A_{n,p} = f_{n,p}^0$. The generic scheme for $K = M$ is

$$\begin{aligned} &\mathrm{ord}_{P_p}(A_{n,p}) = n, \qquad &&\mathrm{ord}_{P_r}(A_{n,p}) = n+1,\ r \neq p, \\ &\mathrm{ord}_{Q_j}(A_{n,p}) = -(n+1), \qquad &&\mathrm{ord}_{Q_M}(A_{n,p}) = -n-g. \end{aligned} \tag{4.101}$$

(1) If all orders are 0 and the order at Q_M equals $-g$ (this is only possible for $K = L = 1$ and $n = 0$), then the only solutions will be the constants. Hence we set the order at Q_M to be 0 too. In particular $A_0 = 1$.
(2) If all orders are ≤ 0 and we are not in case 1, then necessarily $n \leq -1$. In this case we decrease the order at Q_M by 1, i.e., we increase the order of the pole there. By Equation (4.38), the dimension of the solution space will be two. Without the modification only the constants will fulfill the conditions. But now the elements obtained by this modification will not be uniquely fixed. The addition of a constant will always be possible.

For $K = M = 1$ we obtain elements with the following orders

$$\begin{aligned} &\mathrm{ord}_P(A_0) = 0, \quad &&\mathrm{ord}_Q(A_0) = 0, \\ &\mathrm{ord}_P(A_n) = n, \quad &&\mathrm{ord}_Q(A_n) = -g-n-1, \quad -g \leq n \leq -1. \end{aligned} \tag{4.102}$$

To fix the elements coming from the modification (2) we use the duality requirements. Let $A'_{m,r}$ be one of these special elements determined by the modification rules. We calculate

$$\gamma_{r,p}^m := \frac{1}{2\pi \mathrm{i}} \int_{C_S} A'_{m,r} \cdot f_{0,p}^1, \tag{4.103}$$

and set

$$A_{m,r} := A'_{m,r} - \sum_{p=1}^{K} \gamma_{r,p}^m \cdot A_{0,p}. \tag{4.104}$$

We still have

$$\langle A_{0,r}, f^1_{0,s}\rangle = \delta^s_r, \tag{4.105}$$

and hence by construction

$$\langle A_{m,r}, f^1_{0,s}\rangle = 0. \tag{4.106}$$

By considering all cases we see that all other duality relations remain undisturbed. For the detailed calculations see [207].

Now we consider $\lambda = 1/2$, i.e., the chosen theta characteristic L, which says the square root of the canonical bundle itself. If we specialize Riemann–Roch (4.28), then with $\mathcal{K} \otimes L^{-1} = L$

$$\dim \mathrm{H}^0(\Sigma, L + D) - \dim \mathrm{H}^0(\Sigma, L - D) = \deg D. \tag{4.107}$$

In this case $M(1/2) = 0$. If L is an even non-singular theta characteristic, then $\mathrm{H}^0(\Sigma, L) = 0$ and we are in the non-special situation for Riemann–Roch. This means that we can make exactly the same description which we did for arbitrary λ. In the two-point situation we have for example

$$\mathrm{ord}_P(f^{1/2}_n) = n - 1/2, \quad \mathrm{ord}_Q(f^{1/2}_n) = -n - 1/2, \quad n \in \mathbb{Z} + 1/2. \tag{4.108}$$

This statement is also true in the genus 1 situation for all even theta characteristics.

In the odd non-singular situation we have

$$\dim \mathrm{H}^0(\Sigma, L) = 1. \tag{4.109}$$

Hence, we are in the regime of special divisors and we have to correct the prescriptions for those finite number n for which all orders are negative (similar to the $\lambda = 1$ situation). For example, in the two-point situation corrections of (4.108) are necessary for $n = 1/2$ and $n = -1/2$

$$\begin{aligned} \mathrm{ord}_P(f^{1/2}_{1/2}) &= 0, & \mathrm{ord}_Q(f^{1/2}_{1/2}) &= 0, \\ \mathrm{ord}_P(f^{1/2}_{-1/2}) &= -1, & \mathrm{ord}_Q(f^{1/2}_{-1/2}) &= -1. \end{aligned} \tag{4.110}$$

For the other n we take (4.108). In particular, $f^{1/2}_{1/2}$ will be the (up to rescaling) unique nonvanishing holomorphic section of L. The element $f^{1/2}_{-1/2}$ will not be fixed alone by the order condition, but only by imposing the Krichever–Novikov duality

$$\langle f^{1/2}_{-1/2}, f^{1/2}_{1/2}\rangle = 0. \tag{4.111}$$

The $\lambda = 1/2$ case with singular theta characteristics can be treated similar to the $\lambda = 0$ and $\lambda = 1$ case. We will not give the details here.

Remark 4.15. Similar techniques for the modification must also be used for the case when the points A are in non-generic position. We have only to observe that from an ideal generic distribution of zeros (respectively poles) for the $f^\lambda_{n,p}$ we have to make modifications involving a bounded number of orders to reach the existence of a unique element. There is a bound which does not depend on n. Hence, in all our considerations only the bounds R_1 and R_2 have to be adapted. The almost-graded structure will be achieved again.

4.3.2 Genus one

The genus one needs a slightly different treatment. As $\mathcal{K} \cong \mathcal{O}$, a basis of $\mathcal{F}^0$ will give a basis of $\mathcal{F}^\lambda$ at least for all $\lambda \in \mathbb{Z}$. As half-integer λ are concerned, the situation will depend on the chosen theta characteristic. Recall that we have the choice $L = \mathcal{O}$ or $L = L_1, L_2$, or L_3 with $\dim \mathrm{H}^0(\Sigma, L_i) = 0$. In the latter case we can use the description given above,

Here we concentrate on $\mathcal{K}^\lambda \cong \mathcal{O}$. We have to take into account the λ-dependent shift of the degree. Let

$$\{A_{n,p} \mid n \in \mathbb{Z},\ p = 1, \dots .K\} \tag{4.112}$$

be a basis of $\mathcal{F}^0 = \mathcal{A}$, then a basis of $\mathcal{F}^\lambda$ is given by

$$f_{n,p}^\lambda := A_{n-\lambda,p} dz^\lambda, \quad n \in \mathbb{J}_\lambda,\ p = 1, \dots, K. \tag{4.113}$$

In a first step we make the necessary modifications as described for arbitrary genus without establishing Krichever–Novikov duality. Note that

$$f_{0,p}^1 := A_{-1,p} dz \tag{4.114}$$

is not completely fixed yet. Let $A'_{m,p}$ be the modified basis elements from the first step. We set

$$\gamma_{r,p}^{-1} := \frac{1}{2} \frac{1}{2\pi \mathrm{i}} \int_{C_S} A'_{-1,r} \cdot A'_{-1,p} dz = \gamma_{p,r}^{-1}, \tag{4.115}$$

and

$$A_{-1,r} := A'_{-1,r} - \sum_{s=1}^{K} \gamma_{r,s} A_{0,s}. \tag{4.116}$$

Now we can take (4.114) as definition. We still have

$$\langle f_{0,p}^1, A_{0,r} \rangle = \delta_p^r. \tag{4.117}$$

By (4.115) we obtain

$$\langle f_{0,p}^1, A_{1,r} \rangle = 0. \tag{4.118}$$

If there are more special values to be modified we can use the same expressions (4.103) and (4.104) as for the $g \geq 2$ case.

5 Explicit expressions for a system of generators

For the general structure of the Krichever–Novikov type algebras the knowledge of the content of the previous chapters is enough. We used Riemann–Roch type arguments to show the existence of certain basis elements which give us the almost-graded structure of the algebras and the modules. Recall that e.g., for the vector field algebra $\mathcal{L}$, the structure equations with respect to the basis given by the determined Krichever–Novikov basis reads as

$$[e_{n,p}, e_{m,r}] = (m-n)e_{n,p}\delta_p^r + \sum_{h=n+m+1}^{n+m+S} C_{(n,p)(m,r)}^{(h,t)} e_{h,t}, \tag{5.1}$$

with a constant S independent of n and m. Hence, for the lowest degree coefficients we always obtain

$$C_{(n,p)(m,r)}^{(n+m,t)} = (m-n)\,\delta_p^t\,\delta_r^t. \tag{5.2}$$

The geometric situation, meaning the genus of the Riemann surface, the complex structure, and the positions of the points are encoded in the other coefficients $C_{(n,p)(m,r)}^{(h,t)}$. To study an individual algebra given by this geometric data one needs to determine them. For this goal we need to find explicit realizations of the basis elements of the space $\mathcal{F}^\lambda$ of forms of weight λ. We will show in this chapter that the basis elements, whose existence was established in Chapter 4 via Riemann–Roch type arguments, can be given with the help of rational function in the genus zero ($g = 0$) case and by theta functions and prime forms for $g \geq 1$. In addition, we give expressions with the help of the Weierstraß σ function for $g = 1$. More precisely, for $g \geq 1$ we will only deal with the case that the points in A are in generic position.

By observing the dependence on the complex moduli of the Riemann surface and the position of the points, at least in the generic situation, we will obtain smooth analytic families of basis elements over an open dense subset of the moduli space of Riemann surfaces with marked points. This point of view will be of importance in Chapter 11 on Wess–Zumino–Novikov–Witten models.

However, apart from the analytic dependence the explicit form will not be used. The reader in a hurry might skip the details of the construction and consider it a collection of results to be referred to when needed. Of course, individual examples are of importance, e.g., in the context of integrable systems. Of particular interest are geometric configurations carrying a lot of symmetry. In Chapter 12 we will consider a special application related to the deformation of the Witt and Virasoro algebra. Due to the symmetry there, it will be more convenient to use the Weierstraß $\wp$ function.

Before we start with the genus dependent description we introduce and recall some notation. Let $\lambda \in \frac{1}{2}\mathbb{Z}$ be the (integer or half-integer) weight of the forms we consider. We recall the genus- and weight-dependent integer

$$M(\lambda) = (2\lambda - 1)(g - 1). \tag{5.3}$$

Let $A = \{P_1, P_2, \ldots, P_N\}$ be the set of points where poles are allowed. Without further mentioning for $g \geq 1$ we will assume that the points in A are in generic position. In this chapter we do not make any reference to the splitting $A = I \cup O$. As usual, $\mathcal{F}^\lambda = \mathcal{F}^\lambda(A)$ will denote the vector space of meromorphic forms of weight λ holomorphic outside of A.

Let $\underline{n} \in \mathbb{Z}^N$, $\underline{n} = (n_1, n_2, \ldots, n_N)$ be given, then by

$$f^l(\underline{n}) = f^\lambda(n_1, n_2, \ldots, n_N) \tag{5.4}$$

we denote a meromorphic form which is holomorphic outside A and satisfies for the points in A the condition

$$\operatorname{ord}_{P_i}(f^\lambda(\underline{n})) = n_i, \quad i = 1, \ldots, N. \tag{5.5}$$

In particular, $f^\lambda(\underline{n}) \in \mathcal{F}^\lambda$. If a form $f^\lambda(\underline{n})$ exist, then for all $\alpha \in \mathbb{C}$, $\alpha \neq 0$ the form $\alpha(f^\lambda(\underline{n}))$ is another such form. If we speak about "the" form in the following sections we always mean up to this rescaling ambiguity. We set

$$B_{g,\lambda} := \{\underline{n} \in \mathbb{Z}^N \mid \sum_{i=1}^{N} n_i = M(\lambda) - 1\}. \tag{5.6}$$

Proposition 5.1. *Let $g = 0$ or $g \geq 2$ and $\lambda \neq 0, 1, 1/2$, then*
(a) *For $\underline{n} \in B_{g,\lambda}$ a unique $f^\lambda(\underline{n})$ exists (up to rescaling).*
(b) *If one of the n_i is increased by 1 to obtain $\underline{n}'$, then there will be no nonvanishing $f^\lambda(\underline{n}')$.*
(c) *The set*

$$\{f^\lambda(\underline{n}) \mid \underline{n} \in B_{g,\lambda}\} \tag{5.7}$$

is a generating set of $\mathcal{F}^\lambda$.

Proof. The proof is based on our consideration in Chapter 4. For the cases of the genus and λ considered, using Riemann–Roch type arguments, we showed that up to rescaling just one nontrivial element exists which has order $\geq n_i$ at the point P_i. If we increase one of the orders there will be no nontrivial element; see Section 4.2. This shows that the existing element has exactly the order n_i. From these arguments (a) and (b) follow. Moreover, with respect to an arbitrary splitting of A, we exhibited a basis of $\mathcal{F}^\lambda$ by imposing order conditions fulfilling (5.6). Hence, in particular the set of all such elements will be a generating set. □

For $N > 2$ the set (5.7) will not be a basis. It will, however, contain as subsets all basis sets with respect to any possible splitting of A into $I \cup O$.

Remark 5.2. The proposition and its proof show that if we are able to determine all elements $f^\lambda(\underline{n})$ for all possible $\underline{n} \in B_{g,\lambda}$, then we will also have an almost-graded basis with respect to any intended splitting. We only have to go through the prescription which we made in Chapter 4. In addition we have to rescale the elements accordingly.

Remark 5.3. In Proposition 5.1 we excluded $\lambda = 0, 1, 1/2$ respectively all λ for $g = 1$. The only crucial point was the statement using the Riemann–Roch theorem in Chapter 4. We saw there that the distribution of the n_i, including $\sum n_i$ has to be adjusted for a finite number of combinations $\underline{n}$. After we do this we will obtain again a set of generating elements of $\mathcal{F}^\lambda$. For the passage from the generating set to a basis certain additional modifications for the finite amount of special cases have to be made to fulfill the Krichever–Novikov duality, as explained in Section 4.3.

A warning is in order. Let $f^\lambda(\underline{n})$, then we do not claim that this element will not have any other zero outside A. In fact, as $f^\lambda(\underline{n})$ is a meromorphic section of $\mathcal{K}^\lambda$, we have the section degree

$$\operatorname{sdeg}(f^\lambda(\underline{n})) = \lambda(2g-2). \tag{5.8}$$

The part of the degree supported at A equals $(2\lambda - 1)(g-1) - 1$. Hence, there are g additional zeros (counted with multiplicities) outside A; see also Proposition 4.12.

In the following section we will consider all integer weights λ. For half-integers we will restrict ourselves to a special choice (to be seen later) of the theta-characteristics, i.e., of a square root L of the canonical line bundle $\mathcal{K}$.

5.1 The construction via rational functions in the $g = 0$ case

In the genus zero case

$$M(\lambda) - 1 = (2\lambda - 1)(g-1) - 1 = -2\lambda. \tag{5.9}$$

Let z be the (quasi-)global coordinate on $\mathbb{P}^1(\mathbb{C})$. Assume first that none of the points lie at ∞. Then the points $P_1, P_2, .., P_N$ are given by

$$z = a_1, \quad z = a_2, \quad \cdots z = a_N, \tag{5.10}$$

with $\alpha_i \in \mathbb{C}$, $i = 1, \ldots, N$. For $\underline{n} = (n_1, n_2, \ldots, n_N)$ with $\sum n_i = M(\lambda) - 1 = -2\lambda$ we obtain

$$f^\lambda(\underline{n})(z) = \sum_{i=1}^{N} (z - a_i)^{n_i} (dz)^\lambda. \tag{5.11}$$

At the point $z = \infty$, with respect to the local coordinate $w = 1/z$, we have

$$(dz)^\lambda = (-1)^{2\lambda} w^{-2\lambda} (dw)^\lambda. \tag{5.12}$$

Hence, if we express (5.11) in the variable $z(w)$ we obtain

$$\begin{aligned} f^\lambda(\underline{n})(z(w)) &= (-1)^{2\lambda} w^{-\sum n_i - 2\lambda} \sum_{i=1}^{N} (1 - a_i w)^{n_i} (dw)^\lambda \\ &= (-1)^{2\lambda} \sum_{i=1}^{N} (1 - a_i w)^{n_i} (dw)^\lambda. \end{aligned} \tag{5.13}$$

In particular, at $z = \infty$ there is no pole (and also no zero).

If one of the points $P_l = \infty$, then in (5.11) the term $(z - a_l)^{n_l}$ is left out. The transformation $w = 1/z$ will exactly produce the required order n_l at the point ∞, as $\sum_{i \neq l} = -2\lambda - n_l$.

The above expressions also work for half-integer weights. Of course, in this case there is a unique square root of the canonical bundle which is the tautological bundle $\mathcal{O}(-1)$.

Remark 5.4. For the basis elements of Chapter 4 we have to normalize them with respect to a chosen local coordinate. In this way the ambiguity is removed. This normalization is be made with respect to a splitting $A = I \cup O$. As we do not make any reference to a splitting in this chapter, we will not make any such normalization. This remark also concerns the following sections.

5.2 The construction via theta functions and prime forms in the case $g \geq 1$ (general case)

Here the presentation essentially follows [204, 205, 207], which were inspired by the two-point treatment in [22]; see also [142]. For a similar treatment see also [55]. We will consider here only the case of integer weights λ in detail. For half-integer see Section 5.4.

We start by collecting the relevant facts on the building blocks from [61, 62, 177, 203].

Let Σ be a Riemann surface of genus $g \geq 1$ and

$$\alpha_1, \alpha_2, \ldots, \alpha_g, \quad \beta_1, \beta_2, \ldots, \beta_g \tag{5.14}$$

a symplectic homology basis (see (2.1)). With respect to this homology basis there exists a unique set of basis elements $\omega_1, \omega_2, \ldots, \omega_g$ of the space of global holomorphic (1-)differentials with (see [203, Theorem 4.6.])

$$\int_{\alpha_i} \omega_j = \delta_{i,j}, \qquad i, j = 1, \ldots, g. \tag{5.15}$$

If we integrate now the differentials over the cycles β_i, we obtain

$$\int_{\beta_i} \omega_j = \pi_{i,j}, \quad \Pi = (\pi_{i,j}) \quad i, j = 1, \ldots, g, \tag{5.16}$$

with a $n \times n$ complex matrix Π. This matrix is called *period matrix* of Σ. The *Jacobian* $\mathrm{Jac}(\Sigma)$ of Σ is given as

$$\mathbb{C}^g / L, \quad \text{with} \quad L = \mathbb{Z}^g \oplus \Pi \cdot \mathbb{Z}^g. \tag{5.17}$$

Here L is called the *period lattice* . By its very definition, $\mathrm{Jac}(\Sigma)$ is an abelian group. It is in $1:1$ correspondence with the set of isomorphy classes of line bundles of degree zero over Σ. Moreover, this identification gives a group isomorphism with the group structure on the line bundles given by their tensor product. The element $0 \bmod L$ corresponds to the trivial line bundle $\mathcal{O}$.

The first building block is the **theta function** which is defined as

$$\vartheta(z,\Pi) := \sum_{n\in\mathbb{Z}^g} \exp(\pi \mathrm{i}^t n\cdot\Pi\cdot n + 2\pi\mathrm{i}^t n\cdot z) \tag{5.18}$$

for $z \in \mathbb{C}^g$. The function ϑ is holomorphic on $\mathbb{C}^g$.

We will also need the *theta functions with characteristics* in the following, hence I will define them already. A characteristic is given by $a, b \in \mathbb{R}^g$.

$$\vartheta\begin{bmatrix} a \\ b \end{bmatrix}(z,\Pi) := \exp\left(\pi\mathrm{i}^t a\cdot\Pi\cdot a + 2\pi\mathrm{i}^t a\cdot(z+b)\right)\cdot\vartheta\left(z+\Pi\cdot a+b,\Pi\right). \tag{5.19}$$

The basic theta function is obtained via

$$\vartheta(z,\Pi) = \vartheta\begin{bmatrix} 0 \\ 0 \end{bmatrix}(z,\Pi).$$

These functions have the following quasi-periodic behavior under translation of z with vectors from the lattice L ($m \in \mathbb{Z}^g$)

$$\begin{aligned} \vartheta\begin{bmatrix} a \\ b \end{bmatrix}(z+m,\Pi) &= \exp(2\pi\mathrm{i}^t a\cdot m)\cdot\vartheta\begin{bmatrix} a \\ b \end{bmatrix}(z,\Pi)\,, \\ \vartheta\begin{bmatrix} a \\ b \end{bmatrix}(z+\Pi\cdot m,\Pi) &= \exp(-2\pi\mathrm{i}^t b\cdot m) \\ &\quad\times\exp(-\pi\mathrm{i}^t m\cdot\Pi\cdot m - 2\pi\mathrm{i}^t m\cdot z)\vartheta\begin{bmatrix} a \\ b \end{bmatrix}(z,\Pi). \end{aligned} \tag{5.20}$$

For the basic theta function the exponential term in the first expression will always be 1. With respect to the characteristics we have

$$\vartheta\begin{bmatrix} a \\ b \end{bmatrix} = \vartheta\begin{bmatrix} a+p \\ b+q \end{bmatrix}\cdot\exp(2\pi\mathrm{i}^t q\cdot a), \quad p,q\in\mathbb{Z}^g.$$

In particular, we obtain $\vartheta\begin{bmatrix} a \\ b \end{bmatrix} = \vartheta\begin{bmatrix} 0 \\ 0 \end{bmatrix}$ for $a, b \in \mathbb{Z}^g$.

We fix Σ and our homology basis. Hence we will drop the period matrix Π in the notation. But we should keep in mind that our final expressions will depend via Π on our complex structure of the Riemann surface.

The Riemann surface can be embedded via the Jacobi map J into its Jacobian. For this we choose a base point $Q \in \Sigma \setminus A$ and set

$$\Sigma \to \mathrm{Jac}(\Sigma)\,, \quad P \mapsto J(P) := \left(\int_Q^P \omega_1, \int_Q^P \omega_2, \dots, \int_Q^P \omega_g\right) \mod L. \tag{5.21}$$

Here $\int_Q^P$ is an arbitrary path from Q to P. As the value of the integral will depend on this path, the result will only be well-defined modulo the period lattice. I will use $J(P)$ for both the image of P in $\mathrm{Jac}(\Sigma)$ and the multi-valued image in $\mathbb{C}^g$. The pullback $P \mapsto \vartheta(J(P))$ of $J(P)$ composed with ϑ is a multi-valued function on Σ.

Proposition 5.5 (Riemann's Theorem [177, p. 149]). *A vector $\Delta \in \mathbb{C}^g$ (the Riemann vector) exists such that for every $w \in C^g$ either $\vartheta(J(P) + w)$ vanishes identically on Σ or it has exactly g zeros $Q_1, Q_2, \ldots, Q_g$ (not necessarily distinct) with*

$$\sum_{i=1}^{g} J(Q_i) = -w + \Delta \mod L. \tag{5.22}$$

In the following, we will use as vector w certain values which will depend on our points $P_1, P_2, \ldots, P_N$. If we choose them generically, the first case will never occur [61, Theorem VI.3.3].

Proposition 5.6. *The (multi-)valued function*

$$\vartheta(J(P) - gJ(P_N) + \Delta) \tag{5.23}$$

has a zero of order g at P_N and vanishes nowhere else.

Proof. With $w = -gJ(P_N) + \Delta$ we consider

$$\vartheta(-gJ(P_N) + J(P)). \tag{5.24}$$

Its zeros $\mod L$ (using Proposition 5.5) are given by $-w + \Delta = gJ(P_N)$, which was the claim. □

Remark 5.7. Before we can introduce the next building block we have to relate the theta characteristics introduced as square roots of the canonical bundle with the theta functions with half-integer characteristics. As mentioned above, $\mathrm{Jac}(\Sigma)$ can be interpreted as the group of line bundles of degree zero. The two-torsion points correspond to line bundles H with $H^{\otimes 2} \cong \mathcal{O}$. If we fix one theta characteristic L, then the set of all theta characteristics can be given as

$$L' = L \otimes M, \quad \text{with} \quad M^{\otimes 2} \cong \mathcal{O}. \tag{5.25}$$

This also shows that the number of theta characteristics equals 2^{2g}, as we have that many two-torsion points in $\mathrm{Jac}(\Sigma)$.

Now there is a deep relation with the theta functions with characteristics (see Mumford [177] for details and further references). There is a $1:1$ correspondence between theta functions with half-integer characteristics $a, b \in (\frac{1}{2}\mathbb{Z})^g$ and theta characteristics. Given a characteristic $c = \begin{bmatrix} a \\ b \end{bmatrix}$ for the theta function and denote the associated bundle by L_c, then

$$\dim \mathrm{H}^0(\Sigma, L_c) = \mathrm{ord}_{z=0}(\vartheta[c](z)). \tag{5.26}$$

Moreover, L_c is even (odd) if $\vartheta[c]$ is an even (odd) function. This might also be expressed via $c = \begin{bmatrix} a \\ b \end{bmatrix}$ as

$$\dim \mathrm{H}^0(\Sigma, L_c) \equiv \frac{1 - (-1)^{4\,{}^t a b}}{2} \mod 2. \tag{5.27}$$

A theta characteristic L_c will be non-singular if in the even case $\vartheta[c]$ does not have a zero at $z = 0$, and in the odd case the zero is of multiplicity 1. It can also be shown that at least one odd non-singular theta characteristic always exists.

The next building block is the **prime form** $E(P, R)$. We recall its definition and properties from [177, II, p. 3.210].

To define it, one fixes a theta function with characteristics $c = \begin{bmatrix} a \\ b \end{bmatrix}$ given by numbers 0 or 1/2, such that

$$\vartheta\begin{bmatrix} a \\ b \end{bmatrix}(0) = 0 \quad \text{and} \quad d_z\vartheta\begin{bmatrix} a \\ b \end{bmatrix}(0) \neq 0.$$

By the above remark such a characteristic determines an odd non-singular theta characteristic, i.e., a line bundle L_c, such that

$$L^{\otimes 2} = K, \quad \text{and} \quad \dim \mathrm{H}^0(X, L) = 1.$$

Let h_c be a nontrivial section of L_c, hence a $\left(\frac{1}{2}\right)$-form on Σ. The product h_c^2 is a differential. Moreover,

$$h_c^2(P) = \sum_{i=1}^{g} \frac{\partial\vartheta[c]}{\partial z_i}(0)\, \omega_i(P)$$

after suitable normalization. The *prime form* is defined as

$$E(P, R) := \frac{\vartheta[c]\,(J(P) - J(R))}{h_c(P) \cdot h_c(R)}. \tag{5.28}$$

As the denominator depends on the integration path from Q to P, the form $E(P, R)$ will be a multi-valued form on $\Sigma \times \Sigma$ of weight $-(1/2)$ in each argument. [1]

If we move P along a homology cycle back to P again, this can be written as

$$P \mapsto P' = P + \sum_{i=1}^{g} n_i a_i + \sum_{i=1}^{g} m_i b_i \quad (\text{“}= P\text{”}). \tag{5.29}$$

The prime form has the following properties.

1 Of course, it can also be considered a single-valued form on some covering $\tilde{\Sigma} \times \tilde{\Sigma}$, or as some section of a suitable line bundle on $\Sigma \times \Sigma$.

Proposition 5.8 ([177, II, p. 3.210]).
(a) $E(P,R) = -E(R,P)$;
(b) $E(P,R) = 0 \Longleftrightarrow P = R$;
(c) *the zero along the diagonal is of order one.*
(d) *If P is moved around a homology cycle* (5.29) *we get*

$$E(P',R) = \epsilon \cdot \exp\left(-\pi \mathrm{i}^t m \cdot \Pi \cdot m + 2\pi \mathrm{i}^t m \cdot (J(R) - J(P))\right) \cdot E(P,R). \qquad (5.30)$$

Here ϵ is a sign factor which depends onthe cycle and the characteristics $c = \begin{bmatrix} a \\ b \end{bmatrix}$ of the theta function used to define the prime form. More precisely, we have

$$\epsilon = (-1)^{2({}^t a \cdot n - {}^t b \cdot m)}.$$

(e) *If R is moved around a homology cycle similar to* (5.29) *we get*

$$E(P,R') = \epsilon \cdot \exp\left(-\pi \mathrm{i}^t m \cdot \Pi \cdot m - 2\pi \mathrm{i}^t m \cdot (J(R) - J(P))\right) \cdot E(P,R), \qquad (5.31)$$

with

$$\epsilon = (-1)^{2({}^t b \cdot m - {}^t a \cdot n)}.$$

(f) *E does not depend on the chosen odd non-singular theta characteristics.*

The third building block is the σ-**differential**. We define it by using the other two ingredients

$$\sigma(P) = \vartheta(J(P) - gJ(P_N) + \Delta) \cdot E(P,P_N)^{-g} \qquad (5.32)$$

Proposition 5.9. *The form σ is a multi-valued form of weight $g/2$ without zeros. Under* (5.29) *it transforms as*

$$\sigma(P') = \epsilon^g \cdot \exp\left(\mathrm{i}\pi(g-1)^t m \cdot \Pi \cdot m - \mathrm{i}2\pi^t m \cdot (\Delta - (g-1)J(P))\right) \cdot \sigma(P). \qquad (5.33)$$

Proof. The weight comes from the product definition. The pole of the prime form at $P = P_N$ is annihilated by the zero of the theta term (see Proposition 5.6). Also, from the proposition it follows that there will be no further zero. The calculation of the behavior (5.33) under (5.29) is straightforward and shows that it is multi-valued. □

Remark 5.10. If σ' is another $g/2$ form with the same transformation (5.33), then σ'/σ will be a global holomorphic function on Σ, hence a constant. Consequently, up to multiplication with a constant, our σ agrees with the σ in [62, p. 31] and [22].

Now we are ready to give the explicit expressions of our generators.

Proposition 5.11. *Let $g \geq 2$, and $\lambda \neq 0, 1$, and*

$$n_1, n_2, \ldots, n_N \in \mathbb{Z}, \quad \sum_{i=1}^{N} n_i = M(\lambda) - 1 = (2\lambda - 1)(g-1) - 1, \qquad (5.34)$$

then

$$f^\lambda(n_1, n_2, \ldots, n_N)(P) = \prod_{i=1}^{N} E(P, P_i)^{n_i} \cdot \sigma(P)^{(2\lambda-1)} \\ \times \vartheta\Big(J(P) + \sum_{i=1}^{N} n_i J(P_i) - (2\lambda - 1)\Delta\Big), \tag{5.35}$$

or equivalently

$$f^\lambda(n_1, n_2, \ldots, n_N)(P) = \prod_{i=1}^{N-1} E(P, P_i)^{n_i} \cdot E(P, P_N)^{n_N-(2\lambda-1)g} \\ \times \vartheta\Big(J(P) - gJ(P_N) + \Delta\Big)^{(2\lambda-1)} \cdot \vartheta\Big(J(P) + \sum_{i=1}^{N} n_i J(P_i) - (2\lambda - 1)\Delta\Big). \tag{5.36}$$

Proof. (1) Both expressions are the same and the difference is only that in one form the σ-differential is replaced by its definition.

(2) Obviously, the expressions cannot be identically zero.

(3) Every factor $E(P, P_i)$ has weight $(-\frac{1}{2})$ in the variable P, the σ has weight $(\frac{g}{2})$, and ϑ has weight 0. Hence, altogether we obtain the weight

$$\sum_{i=1}^{N} n_i \cdot \left(-\frac{1}{2}\right) + (2\lambda - 1) \cdot \frac{g}{2} = \lambda.$$

(4) Next we have to show that it is a well-defined object on Σ, i.e., that it is single-valued. This says that the combinations of the automorphic factors appearing in (5.20), (5.30), (5.33) under (5.29) multiply to 1. The multiplication yields

$$\prod_{i=1}^{N}\Big(\epsilon^{n_i} \cdot \exp\Big(\big(-\pi i^t m \cdot \Pi \cdot m + 2\pi i^t m \cdot (J(P_i) - J(P))\big) \cdot n_i\Big)\Big) \\ \times \epsilon^{g\cdot(2\lambda-1)} \cdot \exp\Big(\big(\pi i(g-1)^t m \cdot \Pi \cdot m - 2\pi i^t m \cdot (\mathfrak{D}^1 - (g-1)J(P))\big) \cdot (2\lambda - 1)\Big) \\ \times \exp\Big(-\pi i^t m \cdot \Pi \cdot m - 2\pi i^t m \cdot (J(P) + \sum_{i=1}^{N} n_i J(P_i) - (2\lambda - 1)\mathfrak{D}^1)\Big) = 1.$$

(5) By the $E(P, P_i)^{n_i}$ factors the right-hand side has at least the required orders n_i at the points $P_i \in A$. By the ϑ factor at these points we do not have additional zeros. If the order were greater than n_i there, then by Proposition 5.1 (2) the form would be identically zero.[2] □

The constant K depends on the points P_i, the multiplicities n_i, and the weight λ. It can be calculated by calculating the lowest coefficient of the Laurent series expansion

2 The nonvanishing of the last term can also be proven using Riemann's Theorem (Proposition 5.5).

of the right-hand side of (17), with respect to the coordinate z_k around P_k which was chosen to fix $f^\lambda(n_1, n_2, \ldots, n_N)$ (see [22] in the case $k = 2$).

Using the definition of σ we can simply rewrite (5.36), which reduces in the case $k = 2$ to the form given in [22].

Remark 5.12.
(1) The additional g zeros of $f^\lambda(\underline{n})$ will be exactly the zeros of the theta function factor.
(2) The expressions obtained for $N = 2$ are the same as given by [22].

Remark 5.13. The expressions (5.35), (5.36) are well-defined with respect to the point P. This says they will be independent of the lift $J(P) \in \mathbb{C}^g$ chosen. As far as the points P_k are concerned it will depend on their lift. As long as we work with fixed points or only locally varying them it will not disturb us. But if we want to have the possibility of varying them globally this will disturb us. If we move a point P_i around a homology cycle in general we will obtain of course again such an $f^\lambda(\underline{n})$, but now with a different scaling. By multiplying the above forms with a suitable compensation constant $C(J(P_1), \ldots, J(P_N))$, depending on the lift chosen for the points P_i the result will be independent of the lifts. To fix such a constant we choose a generic point Q which neither coincides with the points P_i, the g additional zeros of (5.35), or the zeros of the section h_c used in the construction of the prime form. We set

$$C = \frac{h_c^{2\lambda}(Q)}{f^\lambda(n_1, n_2, \ldots, n_N)(Q)}. \tag{5.37}$$

Here $f^\lambda(\ldots)$ is the expression on the left-hand side of (5.35). The C is a well-defined constant which depends on the lifts $J(P_i)$.

Proposition 5.14. *The product of* (5.35) *and the constant* (5.37) *are independent of the lifts for* $J(P_i)$, $i = 1, \ldots, N$.

Proof. We calculate the behavior of (5.35) under moving the point P_j with a homology cycle (5.29). We will only write down the automorphy factor. Let first $j \neq N$, then (attention, n_j is not the cycle coefficient but the zero order)

$$\begin{aligned}\epsilon^{n_j} \cdot \exp\Big((-\pi \mathrm{i}^t m \cdot \Pi \cdot m - 2\pi \mathrm{i}^t m \cdot (J(P_j) - J(P)) n_j\Big) \\ \times \exp\Big(-\pi n_j^2 \mathrm{i}^t m \cdot \Pi \cdot m - 2\pi n_j \mathrm{i}^t m \cdot (J(P) + \sum_i n_i J(P_i) - (2\lambda - 1)\mathfrak{D}^1)\Big).\end{aligned}$$

For $j = N$ we have additionally

$$\begin{aligned}\epsilon^{-g} \cdot \exp\Big((-\pi \mathrm{i}^t m \cdot \Pi \cdot m - 2\pi \mathrm{i}^t m \cdot (J(P_N) - J(P))(-g)\Big) \\ \times \exp\Big(-\pi g^2 \mathrm{i}^t m \cdot \Pi \cdot m + 2\pi g \mathrm{i}^t m \cdot (J(P) - g J(P_N) + \mathfrak{D}^1)\Big).\end{aligned}$$

In both cases the dependence of $J(P)$ vanishes. Hence, the factor (5.32) transforms by definition just in the inverse manner and the product is invariant. □

5.3 The construction via theta functions and prime forms in the case $g \geq 1$ (exceptional cases)

We consider first $g \geq 2$ and $\lambda = 1$, hence $M(1) - 1 = g - 2$. As described in Section 4.3, we have to consider different types of generators.

The **first** type is the generic one

$$f^1(n_1, n_2, \dots, n_N), \tag{5.38}$$

at least one $n_i \leq -2$, all $n_i \neq -1$, $\sum_{i=1}^N n_i = g - 2$.

Here again (5.35) and (5.36) will work

$$f^1(n_1, n_2, \dots, n_N)(P) := \prod_{i=1}^N E(P, P_i)^{n_i} \cdot \sigma(P) \times \vartheta\Big(J(P) + \sum_{i=1}^N n_i J(P_i) - \Delta\Big), \tag{5.39}$$

or equivalently

$$\begin{aligned} f^1(n_1, n_2, \dots, n_N)(P) := &\prod_{i=1}^{N-1} E(P, P_i)^{n_i} \cdot E(P, P_N)^{n_N - g} \\ &\times \vartheta(J(P) - gJ(P_N) + \Delta) \cdot \vartheta\Big(J(P) + \sum_{i=1}^N n_i J(P_i) - \Delta\Big). \end{aligned} \tag{5.40}$$

For the **second** type we consider

$$f^1(n_1, n_2, \dots, n_N), \quad \text{all } n_i \geq 0, \ \sum_{i=1}^N n_i = (g - 1). \tag{5.41}$$

These elements are globally holomorphic. The generic prescription $\sum_{i=1}^N n_i = (g - 2)$ would not single out a unique element, hence we have to increase the order requirement. To construct such a form we have to take an additional point $R \in X$, different from all P_i, and set

$$\begin{aligned} f^1(n_1, n_2, \dots, n_N)(P) := &\prod_{i=1}^N E(P, P_i)^{n_i} \cdot \sigma(P) \cdot E(P, R)^{-1} \\ &\times \vartheta\Big(J(P) + \sum_{i=1}^N n_i J(P_i) - J(R) - \Delta\Big). \end{aligned} \tag{5.42}$$

This can be rewritten as

$$\begin{aligned} f^1(n_1, n_2, \dots, n_N)(P) = &\prod_{i=1}^{N-1} E(P, P_i)^{n_i} \cdot E(P, P_N)^{n_N - g} \cdot E(P, R)^{-1} \\ &\times \vartheta(J(P) - gJ(P_N) + \Delta) \cdot \vartheta\Big(J(P) + \sum_{i=1}^N n_i J(P_i) - J(R) - \Delta\Big). \end{aligned} \tag{5.43}$$

As in the general case, this is a differential on X. The term $E(P,R)^{-1}$ takes care that the total weight remains one. The pole of $E(P,R)^{-1}$ at $P = R$ is annihilated by a zero of the last theta function term. Otherwise, the right-hand side would be a differential with exactly one pole of order one, but this is impossible due to the residue theorem. The quotient of two such forms with different choices of R will be a global holomorphic function, hence a constant. Consequently, up to a rescaling, the form does not depend on the point R chosen. A possible choice could be the base point Q of the Jacobi map.

For the **third** type of generators the order prescription is given that for exactly a pair (i, j) with $i \neq j$ we have

$$n_i = n_j = -1, \quad n_k = 0, \ k \neq i, j. \tag{5.44}$$

We set

$$\omega_{i,j} := d\left(\log \frac{E(P, P_i)}{E(P, P_j)}\right). \tag{5.45}$$

It is a meromorphic (1-)differential on Σ, which is holomorphic on $\Sigma \backslash \{P_i, P_j\}$. It has pole order 1 at the points P_i and P_j with residues $+1$ and -1. If integrated along the α_i cycles it has zero periods [177, 3.212], [203, p. 116]. By adding a suitable linear combination of holomorphic differentials ω_l, we can achieve that the resulting differential has pure imaginary periods [207, p. 116]. The result is

$$f^1(0,\ldots,-1,0,\ldots,-1,\ldots,0) = \omega_{i,j} + \mathrm{i}\sum_r \left(\sum_s ((\operatorname{Im}\Pi)^{-1})_{rs} \cdot \left(\operatorname{Re}\int_{\beta_s} \omega_{i,j}\right)\right)\omega_r. \tag{5.46}$$

Here i and j denote the indices l with $n_l = -1$.

This kind of differential also appears for example in the definition of the level lines. In fact, we will only need this kind of generators for $N = 2$.

Now we consider $g \geq 2$, $\lambda = 0$, or $g = 1$ and all $\lambda \in \mathbb{Z}$. We have $M(0) - 1 = -g$. Again, there are 3 types of generators as described in Section 4.3.

The **first** type is the constant

$$f^0(0,\ldots,0) \equiv 1. \tag{5.47}$$

The **second** type is the generic one

$$f^0(n_1, n_2, \ldots, n_N), \qquad \text{at least one } n_i > 0, \ \sum_{i=1}^N n_i = -g. \tag{5.48}$$

We obtain

$$f^0(n_1, n_2, \ldots, n_N)(P) := \prod_{i=1}^N E(P, P_i)^{n_i} \cdot \sigma(P)^{-1} \times \vartheta\left(J(P) + \sum_{i=1}^N n_i J(P_i) + \Delta\right). \tag{5.49}$$

or

$$\begin{aligned} f^0(n_1, n_2, \ldots, n_N)(P) := {} & \prod_{i=1}^{N-1} E(P, P_i)^{n_i} \cdot E(P, P_N)^{n_N + g} \\ & \cdot \vartheta(J(P) - gJ(P_N) + \Delta)^{-1} \cdot \vartheta\left(J(P) + \sum_{i=1}^N n_i J(P_i) + \Delta\right) \end{aligned} \tag{5.50}$$

The **third** case remains

$$f^0(n_1, n_2, \dots, n_N), \qquad \text{all } n_i \le 0, \quad \sum_{i=1}^{N} n_i = -g - 1. \tag{5.51}$$

In contrast to the generic prescription, we have to allow one pole order more otherwise only the constant would be greater than or equal to the prescribed order. Recall that our points are in generic position, hence Weierstraß points are excluded. Again we choose a point R and set

$$\begin{aligned} f^0(n_1, n_2, \dots, n_N)(P) := \prod_{i=1}^{N} E(P, P_i)^{n_i} \cdot \sigma(P)^{-1} \cdot E(P, R) \\ \times \vartheta\Big(J(P) + \sum_{i=1}^{N} n_i J(P_i) + J(R) + \Delta\Big), \end{aligned} \tag{5.52}$$

or

$$\begin{aligned} f^0(n_1, n_2, \dots, n_N)(P) := \prod_{i=1}^{N-1} E(P, P_i)^{n_i} \cdot E(P, P_N)^{n_N + g} \cdot E(P, R) \\ \times \vartheta(J(P) - gJ(P_N) + \Delta)^{-1} \cdot \vartheta\Big(J(P) + \sum_{i=1}^{N} n_i J(P_i) + J(R) + \Delta\Big). \end{aligned} \tag{5.53}$$

The weight of the addition $E(P, P_i)^{-1}$ term will be compensated by the weight of the $E(P, R)$ term. The function has a zero at the point R. By varying R we get different functions. This reflects the fact that in this case we were only able to fix the generator up to addition of a constant function and multiplication with a constant. The function can be fixed by requiring the duality relations (3.10) to be valid.

Note that Remark 5.13 and Proposition 5.14 apply also in the exceptional cases.

Remark 5.15. Recall that in the case of genus one, the canonical bundle is the trivial line bundle $\mathcal{K} = \mathcal{O}$. Given a generating set for the forms of weight λ we obtain by

$$f^0(n_1, n_2, \dots, n_N) \mapsto f^\lambda(n_1, n_2, \dots, n_N) = f^0(n_1, n_2, \dots, n_N)(dz)^\lambda \tag{5.54}$$

a generating set of $\mathcal{F}^\lambda$. This is also true for half-integer weights if our chosen theta characteristic (i.e. square root of the canonical bundle) is the trivial bundle.

Furthermore, in the genus zero case we have $\Sigma = \mathrm{Jac}(\Sigma)$. With respect to the homology basis given by the lattice basis vectors $1, \tau \in \mathbb{C}$, $\operatorname{Im} \tau > 0$, the normalized holomorphic differential is $\omega_1 = dz$. If the base point Q corresponds to $z = 0$, then the Jacobi map J will be given by $\bar{z} \mapsto z$.

5.4 Half-integer weights

For half-integer weights we have to modify our description slightly. We follow closely [22], of course now extended to the multi-point case. See also [55] for a similar treatment. Let L_c be the theta characteristic chosen as a basis for our half-integer powers, and $\vartheta[c]$ the corresponding theta function with characteristics $c = \begin{bmatrix} a \\ b \end{bmatrix}$; see Remark 5.7. The L_c will not necessary be the same as the one to be used for the definition of the prime form.

The expression (5.35) will also work for generic cases if we replace the theta function there with $\vartheta[c]$.

$$\begin{aligned} f^\lambda(n_1, n_2, \dots, n_N)(P) &= \prod_{i=1}^{N} E(P, P_i)^{n_i} \cdot \sigma(P)^{(2\lambda-1)} \\ &\quad \times \vartheta[c]\Big(J(P) + \sum_{i=1}^{N} n_i J(P_i) - (2\lambda - 1)\Delta\Big), \end{aligned} \tag{5.55}$$

with $\sum_i n_i = (2\lambda - 1)(g - 1) - 1$. For $\lambda \neq 1/2$ and $g > 1$ this expression will always work. For $\lambda = 1/2$, L_c a non-singular even characteristic and arbitrary genus by our analysis in Section 4.3 there are no exceptions from the generic situation, and again (5.55) specialises (with $\sum_i n_i = -1$) to

$$f^{1/2}(n_1, n_2, \dots, n_N)(P) = \prod_{i=1}^{N} E(P, P_i)^{n_i} \times \vartheta[c]\Big(J(P) + \sum_{i=1}^{N} n_i J(P_i)\Big). \tag{5.56}$$

Remark 5.16. Both expressions (5.35) and (5.55) will be the same if we replace in (5.35) the theta function ϑ by a theta function with a λ-depending characteristic, i.e., $\vartheta\begin{bmatrix} 2\lambda a \\ 2\lambda b \end{bmatrix}$, where $\begin{bmatrix} a \\ b \end{bmatrix}$ is the characteristic chosen as basis for the half-integer powers. By the periodicity we have

$$\vartheta\begin{bmatrix} 2\lambda a \\ 2\lambda b \end{bmatrix} = \begin{cases} \vartheta\begin{bmatrix} 0 \\ 0 \end{bmatrix} = \vartheta, & \lambda \in \mathbb{Z}, \\ \pm\vartheta\begin{bmatrix} a \\ b \end{bmatrix}, & \lambda \in \mathbb{Z} + \frac{1}{2}. \end{cases} \tag{5.57}$$

See also [55].

In other cases there will be a finite number of modifications to be made corresponding to the modification also made in Section 5.3. For example, in the case of an odd non-singular theta characteristic L_c, in a situation similar to the one discussed in (5.51), if we have $\sum_i n_i = -1$, with $n_i \le 0$, then we take instead generators which have one more point as pole which is then compensated by a moving additional point R, e.g.

$$f^{1/2}(-1, -1, 0, \dots, 0) = \frac{E(P, R)}{E(P, P_1)E(P, P_2)} \vartheta[c](J(P) - J(P_1) - J(P_2) + J(R)).$$

Additionally, we might need the generator $f^{1/2}(0, 0, \dots, 0)$ given by the global nonvanishing holomorphic section of L_c.

For $g = 1$ in the odd characteristics case we simply take

$$f^\lambda(\underline{n}) = f^0(\underline{n})(dz)^\lambda. \tag{5.58}$$

5.5 The construction via the Weierstraß σ-function in the $g = 1$ case

We will describe in the genus 1 case a different way to obtain explicit expressions of generators. Again, the presentation is based on [205, 207]. For $N = 2$ see also [140, 171].

Let T be a one-dimensional complex torus (i.e., a genus 1 Riemann surface) given as $\mathbb{C}/L$ with L a lattice in $\mathbb{C}$. We can restrict ourselves to the case

$$L = \mathbb{Z} \oplus \tau \cdot \mathbb{Z}, \qquad \tau \in \mathbb{C},\ \operatorname{Im} \tau > 0. \tag{5.59}$$

An arbitrary torus is always complex analytic isomorphic to such a torus. The Weierstraß σ-function is defined as

$$\sigma(z) := z \cdot \prod_{w \in L \setminus \{0\}} \left(\left(1 - \frac{z}{w}\right) \exp\left(\frac{z}{w} + \frac{1}{2}\left(\frac{z}{w}\right)^2\right)\right). \tag{5.60}$$

The reader should not confuse the Weierstraß σ-function with the σ-differential introduced in the previous sections. The function σ is holomorphic. It is odd and has only zeros at the lattice points. The order of each zero is equal to one [111]. Of course, σ is not a doubly periodic function, hence not a function on the torus. But if we take

$$f(z) = \prod_{i=1}^{m} (\sigma(z - a_i))^{n_i} \tag{5.61}$$

with

$$\text{(a)} \quad \sum_{i=1}^{m} n_i = 0 \qquad \text{(b)} \quad \sum_{i=1}^{m} n_i a_i = 0\,, \tag{5.62}$$

then f is a doubly periodic meromorphic function (hence a function on the torus), which has order n_i at the points $a_i + L$ and is holomorphic elsewhere. Let $P_1, P_2, \ldots, P_N$ be the points on the torus chosen to define the Krichever–Novikov algebra. We choose $a_i \in \mathbb{C}$ with

$$a_i \bmod L = P_i, \quad i = 1, \ldots, N \tag{5.63}$$

and fix them. As the canonical bundle $\mathcal{K}$ on T is trivial we have

$$f^\lambda(n_1, n_2, \ldots, n_N) = f^0(n_1, n_2, \ldots, n_N)\,(dz)^\lambda. \tag{5.64}$$

Hence, it is enough to determine $f^0(n_1, n_2, \ldots, n_N)$.

Again we have the 3 different types of generators. The **first** type

$$f^0(0, \ldots, 0) = 1 \tag{5.65}$$

remains the same.

For the **second** type, which is the generic situation (5.48) with $\sum_{i=1}^N n_i = -1$ and at least one $n_i > 0$, we take

$$b = -\left(\sum_{i=1}^{N} n_i a_i\right) \tag{5.66}$$

and set

$$f^0(n_1, n_2, \dots, n_N) := \prod_{i=1}^{N} (\sigma(z - a_i))^{n_i} \cdot \sigma(z - b). \tag{5.67}$$

The conditions (5.62) are fulfilled. Hence the right-hand side of (5.67) is a doubly periodic function. If the P_i are chosen to be in general position the point $b \bmod L$ will never coincide with any $a_i \bmod L$. Otherwise $\sigma(z - b)$ would increase the order of the zero at the point P_i. As in the generic case, we know there is no such nontrivial function. Hence (5.67) is in fact the generator we are looking for. Note that the additional zero is given by the point $b \bmod L$.

If we want to normalize it, e.g., at the point P_N with respect to the local coordinate $z_N = z - a_N$, around the point P_N we calculate take the constant

$$C = \prod_{i=1}^{N-1} (\sigma(a_N - a_i))^{n_i} \cdot \sigma(a_N - b)\,, \tag{5.68}$$

and divide (5.67) by it.

The **third** and remaining case (5.51) consists of two possibilities:

$$f^0(0, \dots, 0, -2, 0, \dots, 0) \quad \text{and} \quad f^0(0, \dots, -1, 0, \dots, -1, 0, \dots, 0). \tag{5.69}$$

Let the points of non-zero order be a_i, respectively a_i and a_j, with $i < j$. We determine w_1 and w_2 such that

$$w_1 + w_2 = -2a_i, \quad \text{respectively} \quad w_1 + w_2 = -(a_i + a_j), \tag{5.70}$$

and that

$$w_1, w_2 \neq a_l \bmod L \quad \text{for} \quad l = 1, \dots, k. \tag{5.71}$$

Generators of the types in (5.69) are now given by

$$\begin{aligned} f^0(0, \dots, 0, -2, 0, \dots, 0) &:= C\sigma(z - a_i)^{-2} \cdot \sigma(z - w_1) \cdot \sigma(z - w_2) + D \\ f^0(0, \dots, -1, 0, \dots, -1, 0, \dots, 0) &:= C\sigma(z - a_i)^{-1} \cdot \sigma(z - a_j)^{-1} \\ &\quad \times \sigma(z - w_1) \cdot \sigma(z - w_2) + D. \end{aligned} \tag{5.72}$$

Here C and D are constants which can be suitably fixed in case we are looking for a basis (see Section 4.3.2), fulfilling the Krichever–Novikov duality.

In case the chosen theta characteristic is the unique odd theta characteristic $L = \mathcal{O}$, the above formula works also for half-integer weights λ.

Remark 5.17. In Chapter 12 we will give for a certain symmetric choice of points a representation with the help of Weierstraß $\wp$ function and its derivative $\wp'$. The σ-function and $\wp$ are related as

$$\wp = -(\log \sigma)''. \tag{5.73}$$

Remark 5.18. Krichever and Novikov calculated the structure equations for the vector field algebra for two points in generic position by representing the basis elements as above in [140]. In the expressions the *Weierstraß ζ function* shows up. It is defined as logarithmic derivative of sigma

$$\zeta = (\log\sigma)' = \frac{\sigma'}{\sigma}, \quad \wp = -\zeta'. \tag{5.74}$$

6 Central extensions of Krichever–Novikov type algebras

In this chapter we will study central extensions of the introduced Lie algebras. Central extensions appear naturally in the context of quantizations of classical field theories. They are quite often related to “regularization procedures”. To give an example: for our Lie algebras certain natural representations exist, e.g., the representation of the Lie algebra $\mathcal{L}$ given by the Lie module structure of $\mathcal{F}^\lambda$. In applications in quantum field theory, one searches for representations which have a vacuum state starting from which the representation can be generated. Unfortunately, this is not the case for our natural representations. Certain standard techniques exist for obtaining representations with the required properties from these natural representations. However, the action of $\mathcal{L}$ has to be “regularized”, with the consequence that the obtained “representation” fails to be an honest Lie algebra representation. It will only be a projective representation. This can be corrected by extending the Lie algebra $\mathcal{L}$ to a Lie algebra $\widehat{\mathcal{L}}$, which has $\mathcal{L}$ as subalgebra, such that the projective representation of $\mathcal{L}$ becomes an honest representation of $\widehat{\mathcal{L}}$. An example of such a procedure will be discussed in Chapter 7.

In this chapter we will explain what a central extension is, how it is related to Lie algebra cohomology, and how to construct central extensions of the geometrically induced Lie algebras. We will review uniqueness and classification results for these central extensions.

6.1 Lie algebra cohomology

Let L be an arbitrary Lie algebra L over the field $\mathbb{C}$ and M a Lie-module over L. We will consider the Lie algebra cohomology of L with values in the module M. The definition which we give will work for Lie algebras over arbitrary fields $\mathbb{K}$ of characteristics different to two.

For $q \in \mathbb{Z}$ let $C^q(L, M)$ be the vector space of alternating q-multi-linear maps from L to M. The elements in $C^q(L, M)$ are called q-cochain. By convention we set

$$C^0(L, M) := M, \quad C^q(L, M) := \{0\},\ q < 0. \tag{6.1}$$

The coboundary operator

$$d = d_q \ :\ C^q(L, M) \to C^{q+1}(L, M) \tag{6.2}$$

is defined as follows. Let $\gamma \in C^q(L, M)$ then we set

$$\begin{aligned}(d_q\gamma)(g_1, g_2, \ldots, g_{q+1}) := &\sum_{1\le s<t\le q+1} (-1)^{s+t-1}\gamma([g_s, g_t], g_1, \ldots, \check{g}_s, \ldots, \check{g}_t, \ldots, g_{q+1}) \\ &+ \sum_{1\le s\le q+1} (-1)^s g_s \cdot \gamma(g_1, \ldots, \check{g}_s, \ldots, g_{q+1}).\end{aligned} \tag{6.3}$$

Here $g_1, g_2, \ldots, g_{q+1} \in L$, and $\check{g}$ denotes as usual that g was left out of the list of arguments. We set $d_q := 0$ for $q < 0$.

Proposition 6.1.

$$d_{q+1} \circ d_q = 0. \tag{6.4}$$

Proof. This is a straightforward but rather awkward calculation. The reader might do certain low degree cases as exercise. □

The elements of the subspaces of $C^q(L, M)$

$$Z^q(L, M) := \ker d_q, \quad \text{and} \quad B^q(L, M) := \operatorname{im} d_{q-1} \tag{6.5}$$

are called q-cocycles and q-coboundaries respectively. By Proposition 6.1 it is clear that $B^q(L, M)$ is a subspace of $Z^q(L, M)$. The quotient space

$$H^q(L, M) := Z^q(L, M)/B^q(L, M) \tag{6.6}$$

is called the space of q-cohomology classes of L with values in M.

Remark 6.2. We will deal exclusively with *algebraic cohomology*. This means that beside the fact of multi-linearity we do not put any other requirements on the maps. This remark is important, as we are considering also infinite-dimensional Lie algebra like the Lie algebras of vector fields. In this context it is sometimes quite useful to put additional conditions on the maps, like continuity with respect to a certain topology on the cochains.

The Lie algebra cohomology of a Lie algebra L contains very important information about the Lie algebra itself, its representations, its deformations, and further properties.[1]

In this chapter we are only dealing with central extensions of L. As we will show in Section 6.2, they are related to the second Lie algebra cohomology with values in the trivial module $\mathbb{C}$ of L. In fact, this cohomology space will classify them up to "equivalence".

Given a Lie algebra over the field $\mathbb{C}$, the field $\mathbb{C}$ becomes a one-dimension module by setting

$$L \times \mathbb{C} \to \mathbb{C}, \quad (v, \alpha) \mapsto v\,.\,\alpha := 0. \tag{6.7}$$

This module is called the *trivial module*. If we consider cohomology with values in the trivial module, then only the first sum in (6.3) will survive.

For $\boxed{q = 0}$ we see that $d_0 = 0$. Hence $\mathrm{H}^0(L, \mathbb{C}) = C^0(L, \mathbb{C}) = \mathbb{C}$.

1 The Lie algebra cohomology with values in the adjoint module is related to the deformation of Lie algebras and will appear in Chapter 12.

For $\boxed{q = 1}$ we obtain for a linear form $\gamma : L \to \mathbb{C}$

$$(d_1\gamma)(g_1, g_2) = \gamma([g_1, g_2]). \tag{6.8}$$

Hence γ will be a 1-cocycle if and only if it vanishes on the derived subalgebra $[L, L]$ of L. Recall that $[L, L]$ is the subspace generated by all Lie products of pairs of elements of L. As $\operatorname{im} d_0 = \{0\}$, the cohomology space $\mathrm{H}^1(L, \mathbb{C})$ is the space of all linear forms vanishing on $[L, L]$. Equivalently,

$$\mathrm{H}^1(L, \mathbb{C}) \cong (L/[L, L])^*. \tag{6.9}$$

A Lie algebra is called *perfect* if $[L, L] = L$. Hence, if L is perfect then $\mathrm{H}^1(L, \mathbb{C}) = \{0\}$. (Semi-)simple Lie algebras are always perfect.

For $\boxed{q = 2}$ we have to consider alternating bilinear forms $\gamma \in C^2(L, \mathbb{C})$ fulfilling the cocycle condition. The coboundary operator now reads as

$$d_2\gamma(f, g, h) = \gamma([f, g], h) - \gamma([f, h], g) + \gamma([g, h], f). \tag{6.10}$$

From $d_2\gamma(f, g, h) = 0$ with the antisymmetry we obtain the cocycle condition

$$\gamma([f, g], h) + \gamma([g, h], f) + \gamma(h, f], g) = 0. \tag{6.11}$$

A 2-cocycle γ is a coboundary if it can be written with the help of a linear form $\kappa : L \to \mathbb{C}$ as

$$\gamma(f, g) = \kappa([f, g]). \tag{6.12}$$

In the next section we will show that $\mathrm{H}^2(L, \mathbb{C})$ classifies central extension up to equivalence.

As a result aside let me present the following theorem.

Theorem 6.3 (Whitehead). *Let L be a finite-dimensional (semi-)simple Lie algebra over $\mathbb{C}$, and M a finite-dimensional L-module, then*

$$\mathrm{H}^1(L, M) = \mathrm{H}^2(L, M) = \{0\}. \tag{6.13}$$

For a text book reference, see [104, Prop. 6.1, Prop. 6.3]. Hence (in view of the next section), finite-dimensional (semi-)simple Lie algebras do not have nontrivial central extensions. The situation will be completely different in our infinite dimensional context.

6.2 Central extensions and 2-cocycles

In this section we will recall the relation between central extensions of a Lie algebra L and the 2nd cohomology space $\mathrm{H}^2(L, \mathbb{C})$ of L with values in the trivial module $\mathbb{C}$.

Let L be any Lie algebra (over $\mathbb{C}$). A (one-dimensional) central extension $\widehat{L}$ of L is the middle term of a short exact sequence of Lie algebras

$$0 \longrightarrow \mathbb{C} \xrightarrow{\ i\ } \widehat{L} \xrightarrow{\ p\ } L \longrightarrow 0, \qquad (6.14)$$

such that $i(\mathbb{C})$ is central in $\widehat{\mathfrak{L}}$. Two central extensions $\widehat{L}_1$ and $\widehat{L}_2$ of L are called equivalent if there is a Lie isomorphism $\varphi : \widehat{L}_1 \to \widehat{L}_2$ such that the diagram

$$\begin{array}{ccccccccc} 0 & \longrightarrow & \mathbb{C} & \xrightarrow{i_1} & \widehat{L}_1 & \xrightarrow{p_1} & L & \longrightarrow & 0 \\ \| & & \| & & \varphi\downarrow & & \| & & \| \\ 0 & \longrightarrow & \mathbb{C} & \xrightarrow{i_2} & \widehat{L}_2 & \xrightarrow{p_2} & L & \longrightarrow & 0 \end{array} \qquad (6.15)$$

commutes.

If we consider the sequence (6.14), we can always find $s : L \to \widehat{L}$ a linear splitting map (i.e. fulfilling $p \circ s = id_L$). We take such a splitting map s. Set $t = i(1) \in \widehat{L}$. We denote for the moment the Lie product in L by $[.,.]$ and the Lie product in $\widehat{L}$ by $[.,.]^\wedge$.

Proposition 6.4. *A Lie algebra 2-cocycle γ_s exists with values in $\mathbb{C}$ such that*

$$[s(x), s(y)]^\wedge = s([x, y]) + \gamma_s(x, y) \cdot t. \qquad (6.16)$$

Proof. For a given x and y set $z = [s(x), s(y)]^\wedge - s([x, y])$. As p is a Lie homomorphism with $p \circ s = id_L$, we calculate

$$p(z) = [p \circ s(x), p \circ s(y)] - p \circ s([x, y]) = 0.$$

Hence $z \in \ker p = \operatorname{im} i$ and we get a relation as claimed in (6.16) with a constant $\gamma_s(x, y)$. Next we consider its dependence on x and y. Obviously $\gamma_s(x, y)$ is an alternating bilinear form. In $\widehat{L}$ the Jacobi identity has to be true. We calculate

$$\begin{aligned} [[s(x), s(y)]^\wedge, s(z)]^\wedge &= [s([x, y]) + \gamma_s(x, y)\, t, s(z)]^\wedge \\ &= [s([x, y]), s(z)]^\wedge = s([[x, y], z]) + \gamma_s([x, y], z)\, t. \end{aligned} \qquad (6.17)$$

The sum over all cyclic permutations will be equal to zero.

$$\begin{aligned} 0 = s\,&([[x, y], z] + [[y, z], x] + [[z, x], y]) \\ &+ \gamma_s([x, y], z) + \gamma_s([y, z], x) + \gamma_s([z, x], y). \end{aligned} \qquad (6.18)$$

The first term is zero due to the Jacobi identity in L. Hence, the remaining sum has to be zero. Consequently, $(\widehat{L}, [.,.]^\wedge)$ is a Lie algebra if and only if γ_s fulfills the 2-cocycle condition (6.11). □

Proposition 6.5. *Let s_1 and s_2 be two linear splitting maps of* (6.14), *then*

$$\gamma_{s_1} - \gamma_{s_2} \qquad (6.19)$$

will be a 2-coboundary.

Proof. As $p \circ s_1 = p \circ s_2$ we get

$$s_1(x) - s_2(x) \quad \in \ker p = \operatorname{im} i, \quad \forall x \in L. \tag{6.20}$$

This says that

$$s_1(x) - s_2(x) = \alpha(x)t, \quad \alpha(x) \in \mathbb{C}. \tag{6.21}$$

Varying x, we obtain a linear form $\alpha : L \to \mathbb{C}$. If we plug this into the defining formulas for the cocycles γ_{s_1} and γ_{s_2} we obtain

$$\begin{aligned}\gamma_{s_2}(x, y) \cdot t &= [s_2(x), s_2(y)]^\wedge - s_2([x, y]) \\ &= [s_1(x), s_1(y)]^\wedge - s_1([x, y]) + \alpha([x, y])t = (\gamma_{s_1}(x, y) + \alpha([x, y]))t \\ &= (\gamma_{s_1}(x, y) + d_1\alpha([x, y]))t.\end{aligned} \tag{6.22}$$

Hence γ_{s_1} and γ_{s_2} are cohomologous. □

If we have two equivalent central extensions (6.15) with linear splitting maps

$$s_1 : L \to \widehat{L}_1, \quad s_2 : L \to \widehat{L}_2, \tag{6.23}$$

then $\varphi \circ s_1$ will be another splitting map for $L \to \widehat{L}_2$. Hence with the above proposition, the two cocycles γ_{s_1} and γ_{s_2} defining the first and second central extension in (6.15) will be cohomologous.

Now we will give an explicit description of the central extension. For this we fix a linear splitting map s. Hence the sequence (6.14) will split as a vector space direct sum

$$\widehat{L} \cong \mathbb{C} \oplus L. \tag{6.24}$$

With respect to this realization the splitting map φ is given by

$$s(x) = \hat{x} = (0, 1), \quad t = (1, 0). \tag{6.25}$$

The Lie structure on $\widehat{L}$ is given by

$$[\hat{x}, \hat{y}] = \widehat{[x, y]} + \gamma(x, y) \cdot t, \quad [t, \widehat{L}] = 0, \quad x, y \in L. \tag{6.26}$$

The formula (6.26) can be used as starting point to define an explicitly given central extension. The only condition which we will need is that $\gamma(x, y)$ has to be a Lie algebra 2-cocycle. In this way we see that every element of $Z^2(L, \mathbb{C})$ yields a central extension. A cohomologous cocycle corresponds to choosing a different splitting map s_1. With respect to s it will be given as

$$s_1(s) = (\alpha(x), x) = s(x) + \alpha(x)t, \tag{6.27}$$

with $\alpha : L \to \mathbb{C}$ a linear form. As explained above, it yields an equivalent central extension. In such a way the space $\mathrm{H}^2(L, \mathbb{C})$ classifies equivalence classes of central extensions.

If the cocycle γ is zero, the central extension $\widehat{L}$ defined by it will be the Lie-direct sum of the trivial algebra $\mathbb{C}$ and L. Moreover, in this case the splitting map $x \mapsto (0, x)$ will be a Lie homomorphism. If $\gamma = d_1\alpha$ is a coboundary, then $\widehat{L}$ defined by it will be equivalent to the Lie direct sum. To see this, denote by $[.,.]^{(0)}$ the Lie product of the Lie-direct sum and by $[.,.]^{(1)}$ the product defined via $\psi = d_1\alpha$. The two extensions are modeled on the same vector space $\mathbb{C} \oplus L$. We identify x with $(0, x)$ and define $\phi : \widehat{L}_0 \to \widehat{L}_1$ by

$$\phi(t) = t, \quad \phi(x) = x + \alpha(x)t. \tag{6.28}$$

Then $[x, y]^{(1)} = [x, y]^{(0)} + \alpha([x, y])t$, and

$$\begin{aligned}[\phi(x), \phi(y)]^{(1)} &= [x + \alpha(x)t, y + \alpha(y)t]^{(1)} \\ &= [x, y]^{(1)} = [x, y]^{(0)} + \alpha([x, y])t = \phi([x, y]^{(0)}).\end{aligned} \tag{6.29}$$

Products of elements where at least one factor is a multiple of t will vanish in any case. This shows ϕ is a Lie isomorphism. Furthermore, $\phi_{|\mathbb{C}} = id$ and $p_2 \circ \phi_{|L} \to L$ is also the identity. Hence, ϕ gives an equivalence of central extensions.

Clearly, by the definition, two central extensions of L which are equivalent will be isomorphic. But non-equivalent central extensions might also be isomorphic. The problem of determining central extensions up to isomorphy is not that simple in the general context. But there are always special isomorphisms which could be taken care of. To see them, assume that we have a central extension defined via the cocycle γ. Let $\beta \in \mathbb{C}^*$ be a non-zero scalar. The cocycle $\beta \cdot \gamma$ will also define a central extension. If γ is not a coboundary, the two-cocycle will not be cohomologous if $\beta \neq 1$. Hence the central extensions are not equivalent. But they are isomorphic, as they correspond only to a rescaling of the central element.

Quite often one is only interested in a classification of central extensions up to equivalence and rescaling the central element. If we take out the trivial (meaning splitting) central extensions corresponding to the zero class $[0] \in \mathrm{H}^2(L, \mathbb{C})$, then the set of nontrivial central extensions, up to equivalence and rescaling, is parameterized by the projective space $\mathbb{P}(\mathrm{H}^2(L, \mathbb{C}))$. In particular, if $\dim \mathrm{H}^2(L, C) = 1$, then L has only one such class of nontrivial central extensions.

Remark 6.6. After fixing a vector space basis $\{e_\rho \mid \rho \in \mathcal{R}\}$ of L, a vector space basis of $\widehat{L}$ will be given by $\{\hat{e}_\rho := (0, e_\rho) \mid \rho \in \mathcal{R}\} \cup \{t := (1, 0)\}$. An equivalent and rescaled central extension can be described as a change of basis obtained by adding multiples of the central element, with a rescaling of the central element. In detail:

$$\hat{e}_\rho' = \hat{e}_\rho + \varphi(e_\rho)\,t, \quad t' = \alpha \cdot t, \tag{6.30}$$

with $\alpha \in \mathbb{C}^*$ and $\varphi : L \to \mathbb{C}$ a linear form.

In Section 6.10.1 we will consider universal central extensions.

For further reference we close this section with the following simple proposition.

Proposition 6.7.
(a) *Let L be an abelian Lie algebra, then*

$$\mathrm{H}^2(L, \mathbb{C}) = Z^2(L, \mathbb{C}), \tag{6.31}$$

coincides with the vector space of all alternating bilinear forms.
(b) *If additionally L is finite-dimensional of dimension n, then*

$$\dim \mathrm{H}^2(L, \mathbb{C}) = \frac{n(n-1)}{2}. \tag{6.32}$$

6.3 Projective actions and central extensions

If a Lie algebra L operates on a vector space V,

$$L \times V \to V, \quad (x, v) \mapsto x \,.\, v \tag{6.33}$$

in a linear manner, then this can be described via a linear map

$$\begin{aligned} \Phi &: L \to gl(V), \\ \Phi(x) &:= \{ V \to V, \quad v \mapsto \Phi(x)(v) := x \,.\, v. \end{aligned} \tag{6.34}$$

If the map Φ is a Lie homomorphism, i.e.,

$$\Phi([x, y]) = [\Phi(x), \Phi(y)], \tag{6.35}$$

where the bracket on the right is given by the commutator of linear maps, then the operation of L on V is called a *Lie action*. See also Section 1.4. Equivalently, one says V carries a representation of the Lie algebra L, respectively V is an L module. In terms of elements this is written as

$$[x, y] \,.\, v = x \,.\, (y \,.\, v) - y \,.\, (x \,.\, v). \tag{6.36}$$

Definition 6.8. A linear action of L on V is called a *projective Lie action* if (6.35), respectively (6.36) is true up to multiples of the identity. This says

$$\forall x, y \in L, \ \exists \psi(x, y) \in \mathbb{C} \quad \text{such that} \quad \Phi([x, y]) = [\Phi(x), \Phi(y)] - \psi(x, y) \cdot id. \tag{6.37}$$

In terms of elements this is written as

$$[x, y] \,.\, v = x \,.\, (y \,.\, v) - y \,.\, (x \,.\, v) - \psi(x, y) \cdot v. \tag{6.38}$$

The space V is also called *projective representation space* of L or a *projective L-module*.

Proposition 6.9. *Given a projective Lie action* (6.37), *then the map*

$$L \times L \to \mathbb{C}, \quad (x, y) \mapsto \psi(x, y) \tag{6.39}$$

is a Lie algebra 2-cocycle for L with values in the trivial module.

Proof. That ψ is bilinear and antisymmetric is obvious. We calculate from (6.37) by using the fact that multiples of the identity will vanish under the commutator

$$\begin{aligned}\Phi([[x, y], z]) &= [\Phi([x, y], \Phi(z)] - \psi([x, y], z) \cdot id \\ &= [[\Phi(x), \Phi(y)], \Phi(z)] - \psi([x, y], z) \cdot id.\end{aligned} \tag{6.40}$$

Cyclically permuting the arguments x, y, z and adding up all 3 expressions obtained, yields

$$\Phi\left(\sum_{cycl.} [[x, y], z] \right) = \sum_{cycl.} [[\Phi(x), \Phi(y)], \Phi(z)] - \sum_{cycl.} \psi([x, y], z). \tag{6.41}$$

The left-hand side vanishes due to the Jacobi identity in L, the first term on the right vanishes due to the Jacobi identity in $gl(V)$. Hence, the 3rd term has to vanish too. But this is the cocycle condition. □

Given such a projective representation (6.37), then on the basis of the previous proposition we construct the central extension $\widehat{L}_\psi$ given by the cocycle ψ. As usual, we denote by $\hat{x} = (0, x)$ and $t = (1, 0)$ the corresponding elements in $\widehat{L}_\psi$. We have

$$[\hat{x}, \hat{y}]^\wedge = \widehat{[x, y]} + \psi(x, y) \cdot t. \tag{6.42}$$

We set

$$\widehat{\Phi}(\hat{x}) := \Phi(x), \quad \widehat{\Phi}(t) = id, \tag{6.43}$$

and verify directly

$$\begin{aligned}\widehat{\Phi}([\hat{x}, \hat{y}]^\wedge) &= \widehat{\Phi}(\widehat{[x, y]} + \psi(x, y) \cdot t) = \Phi([x, y]) + \psi(x, y) \cdot id \\ &= [\widehat{\Phi}(\hat{x}), \widehat{\Phi}(\hat{y})],\end{aligned} \tag{6.44}$$

as on the other hand

$$[\widehat{\Phi}(\hat{x}), \widehat{\Phi}(\hat{y})] = [\Phi(x), \Phi(y)] = \Phi([x, y]) + \psi(x, y) \cdot id. \tag{6.45}$$

Proposition 6.10. *Every projective representation of L can be made to an honest Lie representation of a certain central extension $\widehat{L}$.*

Conversely, every Lie representation of a central extension $\widehat{L}$ of L yields a projective representation of L.

Proof. The first part was shown above. For the opposite, given a central extension $\widehat{L}$ of L and a representation of $\widehat{L}$. With respect to the standard splitting (6.26), we set $\Phi(x) := \widehat{\Phi}(\hat{x})$. Obviously this defines a projective representation. In case $\widehat{\Phi}(t) = 0$ or the defining cocycle ψ for $\widehat{L}$ will vanish it will be an honest Lie action. □

6.4 Projective and affine connections

Recall that the Witt algebra is the vector field algebra $\mathcal{L}$ for the classical situation $g = 0$ and $A = \{0, \infty\}$. It is generated by the vector space basis

$$e_n = z^{n+1}\frac{d}{dz}, \qquad n \in \mathbb{Z}, \tag{6.46}$$

with the Lie bracket

$$[e_n, e_m] = (m - n)e_{n+m}. \tag{6.47}$$

As explained in Section 6.2, its central extensions are generated as vector space by elements $\hat{e}_n$, $n \in \mathbb{Z}$ and t, and Lie structure given with the help of 2-cocycles $\psi \in Z^2(W, \mathbb{C})$

$$[\hat{e}_n, \hat{e}_m] = (m - n)\hat{e}_{n+m} + \psi(e_n, e_m) \cdot t. \tag{6.48}$$

A classical result is that $\mathrm{H}^2(W, \mathbb{C})$ is one-dimensional and that a representing element of a generator is given by

$$\psi(e_n, e_m) = \frac{1}{12}(n^3 - n)\,\delta_n^{-m}. \tag{6.49}$$

Hence, up to equivalence and rescaling there is only one nontrivial central extension.

This result will be obtained also as a special case of our general classification of central extensions of $\mathcal{L}$. The explicit form (in particular the factor $1/12$) is partly by convention. In (6.49) the term $-n$ can be made to zero by cohomological changes. Nevertheless, the form $(n^3 - n)$ has the advantage that if the cocycle is evaluated for the subalgebra

$$W_{hol} =< e_{-1}, e_0, e_1 >_{\mathbb{C}} \cong \mathfrak{sl}(2, \mathbb{C}) \tag{6.50}$$

of holomorphic vector fields it will vanish identically not only up to coboundary. This is in accordance with the fact that $\mathrm{H}^2(\mathfrak{sl}(2, \mathbb{C}), \mathbb{C}) = \{0\}$; see Theorem 6.3.

We want to generalize the cocycle to Riemann surfaces of arbitrary genus. To do this we have to find a geometric description of (6.49). Sometimes, in the physics literature for the above cocycle, one also finds the expression

$$\psi(e, f) = \frac{1}{24\pi \mathrm{i}} \int_C \frac{1}{2}(e''' f - e f''')\,dz. \tag{6.51}$$

Here C is a circle around 0 and the same symbol for the vector fields and their representing local meromorphic functions was used. Indeed, this definition will reproduce (6.49). Nevertheless, despite its look, the expression is not really geometric as the integrand is not a 1-differential. A counter-term has to be added which is related to a projective connection. Later on for one of our algebras we will also need affine connections. Hence we will introduce both now.

6.4.1 The definitions

Definition 6.11. Let $(U_\alpha, z_\alpha)_{\alpha\in J}$ be a covering of the Riemann surface Σ by holomorphic coordinates with transition functions $z_\beta = f_{\beta\alpha}(z_\alpha)$.

(a) A system of local (holomorphic, meromorphic) functions $R = (R_\alpha(z_\alpha))$ is called a (holomorphic, meromorphic) *projective connection* if it transforms under coordinate transformations as

$$R_\beta(z_\beta)\cdot(f'_{\beta,\alpha})^2 = R_\alpha(z_\alpha) + S(f_{\beta,\alpha}), \tag{6.52}$$

with

$$S(h) = \frac{h'''}{h'} - \frac{3}{2}\left(\frac{h''}{h'}\right)^2, \tag{6.53}$$

the Schwartzian derivative. Here $'$ denotes differentiation with respect to the coordinate z_α.

(b) A system of local (holomorphic, meromorphic) functions $T = (T_\alpha(z_\alpha))$ is called a (holomorphic, meromorphic) *affine connection* if it transforms as

$$T_\beta(z_\beta)\cdot(f'_{\beta,\alpha}) = T_\alpha(z_\alpha) + \frac{f''_{\beta,\alpha}}{f'_{\beta,\alpha}}. \tag{6.54}$$

From the definitions it follows that the difference of two projective connections will be a quadratic differential, and that the difference of two affine connections will be a differential. Hence, after fixing one affine (respectively projective) connection, all others are obtained by adding differentials (respectively quadratic differentials).

Proposition 6.12. *Every Riemann surface admits a holomorphic projective connection.*

Proof. This is a classical result, see [97, 101]. In Section 6.4.2 we will give a sketch of a different proof. □

In the case of the sphere ($g = 0$) and the torus ($g = 1$), with respect to the standard coordinates z and $w = 1/z$, respectively $z - a$, the projective connection $R = 0$ can be chosen. This is immediate for $g = 1$. For $g = 0$ one calculates directly that $S(h)$ will vanish for the coordinate transformation $w = 1/z$. In fact, $S(h)$ will vanish for all projective linear transformations of $\mathbb{P}^1(\mathbb{C})$.

Proposition 6.13. *Given a point Q, a meromorphic affine connection always exists, holomorphic outside Q, and having maximally a pole of order one there.*

We will supply a proof of this fact in Section 6.4.2. For the case of the torus with standard coordinates $T_0 = 0$ is possible. Hence, a holomorphic affine connection exists. For the sphere (with standard coordinates $(z, w = 1/z)$)

$$T = (t_\alpha, t_\beta) = \left(0, \frac{2}{w}\right) \tag{6.55}$$

will do. In particular, T has a pole of order one at ∞.

Remark 6.14. In fact the integrand in (6.51) would be independent of the coordinate chosen if we allow as coordinate transformations $z_\beta = f_{\beta\alpha}(z_\alpha)$ only projective transformations. They are given by

$$z_\beta = f_{\beta\alpha}(z_\alpha) = \frac{az_\alpha + b}{cz_\alpha + d}, \quad a, b, c, d \in \mathbb{C}, \; ad - bc \neq 0. \tag{6.56}$$

Alternatively, if we consider only atlases such that all coordinate transformations are of this type. In fact, for such $f_{\beta\alpha}$ the Schwartzian derivative (6.53) will vanish. Hence, with respect to such a covering, $R_\alpha(z_\alpha) \equiv 0$ will be a solution of (6.52). The choice of such an atlas is called a *projective structure* on Σ. The existence of a holomorphic projective connection on Σ is equivalent to the existence of a projective structure on Σ. Hence, every Riemann surface admits a projective structure. We prefer to work with the projective connection instead of the projective structure. Note also that only for genus one the corresponding object, an *affine structure*, will exist, as only in this case do we have a holomorphic affine connection.

6.4.2 Proof of existence of an affine connection

Here we will use some techniques from algebraic geometry. The reader might skip this part and just accept the statement. As in the definition of an affine connection, let (U_α, z_α), $\alpha \in J$ be a system of coordinate charts, with $z_\beta = h(z_\alpha)$ the change of the coordinates defined if $U_\beta \cap U_\alpha \neq \emptyset$. The following transition matrix is defined

$$C_{\beta\alpha} = \begin{pmatrix} (h')^{-1} & (h'')(h')^{-2} \\ 0 & 1 \end{pmatrix}. \tag{6.57}$$

Here $'$ denotes differentiation with respect to z_α. Obviously,

$$C_{\beta\alpha} \in GL(2, \mathcal{O}(U_\beta \cap U_\alpha)), \tag{6.58}$$

where the latter set denotes the set of invertible 2×2 matrices with holomorphic functions on $U_\beta \cap U_\alpha$ as entries.

Proposition 6.15. *$C_{\beta\alpha}$ defines a cocycle of rank 2 on Σ, and hence an associated rank 2 vector bundle E.*

Proof. Clearly, $C_{\alpha\alpha}$ is the identity matrix. The cocycle condition has to be shown on $U_\alpha \cap U_\beta \cap U_\gamma \neq \emptyset$

$$C_{\gamma\alpha} = C_{\gamma\beta} \cdot C_{\beta\alpha}. \tag{6.59}$$

Let $z_\gamma = g(z_\beta)$, hence $z_\gamma = k(z_\alpha) = (g \circ h)(z_\alpha)$. Let * denote differentiation with respect to z_β. We get

$$C_{\gamma\beta} \cdot C_{\beta\alpha} = \begin{pmatrix} (g^*)^{-1} \cdot (h')^{-1} & (g^*)^{-1}(h'')(h')^{-1} + (g^{**})(g^*)^{-2} \\ 0 & 1 \end{pmatrix}. \tag{6.60}$$

Obviously, $k' = g^* \cdot h'$. We have to check the element at the position $(1, 2)$. We get

$$k'' = (g^* \cdot h')' = g^{**}(h')^2 + g^* h''. \tag{6.61}$$

Hence, also the elements at position $(1, 2)$ coincide. □

Let v be a global holomorphic section of the bundle E, then it can be locally represented by a pair of holomorphic functions, $v_\alpha = {}^t(s_{\alpha,1}, s_{\alpha,2})$ with $s_{\alpha,1}, s_{\alpha,2} \in \mathcal{O}(U_\alpha)$. The v_α transform as

$$\begin{aligned} v_\beta &= C_{\beta\alpha} \cdot v_\alpha, \\ (s_{\beta,1}, s_{\beta,2}) &= \left(s_{\alpha,1} \cdot (h')^{-1} + \frac{h''}{(h')^2} \cdot s_{\alpha,2} \, , \, s_{\alpha,2} \right). \end{aligned} \tag{6.62}$$

If the second component does not vanish identically then the quotient

$$t_\alpha = \frac{s_{\alpha,1}}{s_{\alpha,2}} \tag{6.63}$$

will define a meromorphic affine connection, as it transforms exactly as required. The zeros of $s_{\alpha,2}$ will become poles of t_α.

Based on the form of the cocycle (6.60), one obtains a short exact sequence of vector bundles

$$0 \to \mathcal{K} \to E \to \mathcal{O} \to 0 \tag{6.64}$$

($\mathcal{K}$ is the canonical bundle, $\mathcal{O}$ the trivial bundle). Sections in $\mathcal{K}$ are automatically sections of $\mathcal{E}$. They correspond to sections with vanishing second component. Let Q be a fixed point and L_Q the associated point bundle, i.e., the bundle which has exactly one linearly independent section s_Q. This section has a zero at Q (see [203, p. 106]). If W is a vector bundle and w a holomorphic section of $W \otimes L_Q$, then w/s_Q defines a section of W which is holomorphic over $\Sigma \setminus \{Q\}$. For this section the local component functions might have poles of at most order 1 at Q.

We tensorise (6.64) with L_Q and pass over to the long cohomology sequence. After applying Serre duality [203] we obtain

$$0 \to \mathrm{H}^0(\Sigma, \mathcal{K} \otimes L_Q) \to \mathrm{H}^0(\Sigma, E \otimes L_Q) \to \mathrm{H}^0(\Sigma, L_Q) \to \mathrm{H}^0(\Sigma, L_Q^*) \to \cdots . \tag{6.65}$$

We have $\mathrm{H}^0(\Sigma, L_Q^*) = 0$ and $\dim \mathrm{H}^0(\Sigma, L_Q) = 1$, hence

$$\dim \mathrm{H}^0(\Sigma, E \otimes L_Q) = 1 + \dim \mathrm{H}^0(\Sigma, \mathcal{K} \otimes L_Q) = 1 + g = 1 + \dim \mathrm{H}^0(\Sigma, \mathcal{K}). \tag{6.66}$$

Recall that by Riemann-Roch we have $\dim \mathrm{H}^0(\Sigma, \mathcal{K} \otimes L_Q) = g$. In particular, from the dimension formula (6.66) we conclude that there is always a meromorphic section v of E for which the second component (in the local representing elements) will not identically vanish, as otherwise we would have $g + 1$ linearly independent holomorphic sections of $\mathcal{K}$, which is impossible.

Let $(s_{\alpha,2})_\alpha$ be the collection of the second components of this v. Equation (6.60) says that the $s_{\beta,2}$ and $s_{\alpha,2}$ on $U_\alpha \cap U_\beta$ coincide. Hence, they define a global meromorphic function s by $s_{|U_\alpha} = s_{\alpha,2}$. The function s has at most a pole of order 1 at Q and is otherwise holomorphic. We now take the (t_α) as defined by (6.63). Possible poles are the point Q and the zeros of s. In the case where $g \geq 1$, the function s has to be constant as there is no meromorphic function with total pole order 1. Hence in this case we are done.

For $g = 0$ the s might be nontrivial. If it is non-constant then it has exactly one point P a zero (as it might have maximally one pole of order 1 at Q). This zero is of order 1. Hence, the affine connection which we constructed has at most poles of order 1 at Q and P. If T is an affine connection and ω a meromorphic differential, then $T+\omega$ is also a meromorphic affine connection. By subtracting meromorphic differentials with poles of order 1 at Q and P, the pole at P can be removed without increasing the pole order 1 at Q. This shows Proposition 6.13.

With the same methods it is possible to show the existence of a holomorphic projective connection for $g \geq 2$. I would like to clarify this in a few words. The notation is as above.

$$D_{\beta\alpha} = \begin{pmatrix} (h')^{-2} & \dfrac{h'''}{(h')^3} - \dfrac{3}{2}\left(\dfrac{h''}{(h')^2}\right)^2 \\ 0 & 1 \end{pmatrix} \tag{6.67}$$

fulfills the cocycle condition as verified by direct calculation. Hence, the existence of an associated rank 2 vector bundle F follows. Based on the form (6.67) we obtain the short exact sequence

$$0 \to \mathcal{K}^2 \to F \to \mathcal{O} \to 0. \tag{6.68}$$

The long cohomology sequence terminates after level zero:

$$0 \to \mathrm{H}^0(\Sigma, \mathcal{K}^2) \to \mathrm{H}^0(\Sigma, F) \to \mathrm{H}^0(\Sigma, \mathcal{O}) \to \mathrm{H}^0(\Sigma, \mathcal{K}^*) = 0. \tag{6.69}$$

Hence,

$$\dim \mathrm{H}^0(\Sigma, F) = 1 + \dim \mathrm{H}^0(\Sigma, \mathcal{K}^2), \tag{6.70}$$

and we can conclude as above that there is a holomorphic projective connection $R_\alpha = s_{\alpha,1}/s_{\alpha,2}$.

6.5 Geometric cocycles

Proposition 6.16. *Let R be a meromorphic projective connection and T an affine meromorphic connection both without poles outside A. The following bilinear maps are then* *well-defined:*

$$\widehat{\gamma}^{\mathcal{A}} : \mathcal{A} \times \mathcal{A} \to \mathcal{F}^1, \quad \widehat{\gamma}^{\mathcal{A}}(g,h) = g(dh), \tag{6.71}$$

$$\widehat{\gamma}_R^{\mathcal{L}} : \mathcal{L} \times \mathcal{L} \to \mathcal{F}^1, \quad \widehat{\gamma}_R^{\mathcal{L}}(e,f) = \left(\frac{1}{2}(e'''f - ef''') - R(e'f - ef')\right) dz \tag{6.72}$$

$$\widehat{\gamma}_T^{(m)} : \mathcal{A} \times \mathcal{L} \to \mathcal{F}^1, \quad \widehat{\gamma}_T^{(m)}(g,e) = (eg'' + T \cdot eg')dz. \tag{6.73}$$

Proof. As our R and T will be holomorphic outside A, the expressions will not have poles outside A. The only property to show is that they are indeed differentials. This is true for $\widehat{\gamma}^{\mathcal{A}}$ by definition. For the others we have to make direct calculations. Here adding the terms with the connections will be of importance.

Let (U_α, z_α) and (U_β, z_β) be coordinate covering with non-empty intersection. We have to show that (6.72) (and correspondingly (6.73)) expressed in z_α and z_β are the same. Let $z_\beta = h(z_\alpha)$ be the coordinate transformation. First we deal with (6.72). Denote by f_α and f_β, respectively e_α and e_β the local representing functions of the vector fields f and e. We have

$$f_\beta(z_\beta) = \frac{dh}{dz_\alpha}(z_\alpha) \cdot f_\alpha(z_a), \quad z_\beta = h(z_\alpha)$$

and correspondingly for e_β. Furthermore,

$$dz_\beta = \frac{dh}{dz_\alpha}(z_\alpha)dz_\alpha$$

$$R_\beta(z_\beta) = \left(\frac{dh}{dz_\alpha}(z_\alpha)\right)^{-2} (R_\alpha(z_\alpha) + S(h)).$$

Now we write (6.72) with respect to the variable z_β and insert the expressions above. We have to apply several times the chain rule, as now $'$ denotes the derivative with respect to z_β. After some long but easy calculation the claim follows. Just to note one important intermediate step. The expression $e'''f - ef'''$ with respect to z_β, gives after change to the variable z_α

$$\frac{1}{h'}(e'''f - e'''f) + \left(\frac{2h'''}{(h')^2} - \frac{3(h'')^2}{(h')^3}\right)(e'f - f'e). \tag{6.74}$$

Here $'$ denotes the derivative with respect to z_α. The second term will be compensated exactly by the Schwartzian derivative.

In a similar way we consider (6.73). As above, let e_α and e_β be local representing elements for the vector field e, and g_α and g_β the expressions for a fixed function g in the local coordinates. We have (where $'$ denotes the derivative with respect to z_α)

$$e_\beta(z_\beta) = h'(z_\alpha) \cdot e_\alpha(z_\alpha), \quad dz_\beta = h'(z_\alpha)\, dz_\alpha, \quad \frac{dg_\beta}{dz_\beta}(z_\beta) = \frac{dg_\alpha}{dz_\alpha}(z_\alpha) \cdot (h')^{-1}(z_\alpha),$$

$$\frac{d^2 g_\beta}{dz_\alpha^2}(z_\beta) = \frac{d^2 g_\alpha}{dz_\alpha^2}(z_\alpha) \cdot (h')^{-2}(z_\alpha) - \frac{dg_\alpha}{dz_\alpha}(z_\alpha) \cdot (h')^{-3}(z_\alpha) \cdot h''(z_\alpha), \tag{6.75}$$

$$T_b(z_\beta) = T_\alpha(z_\alpha)(h')^{-1} + \frac{h''}{(h')^2}.$$

Writing the integrand in the variable z_β and making the above replacement we obtain

$$(h')^2 e_\alpha dz_\alpha \left(g_\alpha''(h')^{-2} - g_\alpha' h''(h')^{-3} + T_\alpha (h')^{-2} g_\alpha' - g_\alpha' h''(h')^{-3} \right) \\ = \left(e_\alpha g_\alpha'' + T_\alpha \cdot e_\alpha g_\alpha' \right) dz_\alpha, \quad (6.76)$$

indeed a well-defined differential. □

By Proposition 6.16 we can integrate the given forms over closed curves on $\Sigma \setminus A$. In this way we will construct in the following section 2-cocycles for the algebras under consideration.

First we will need some simple technical facts. We collect them in the form of a lemma.

Lemma 6.17. *Let $h \in \mathcal{A}$, $e \in \mathcal{L}$, $\omega \in \mathcal{F}^1$, $f \in \mathcal{F}^\lambda$, $g \in \mathcal{F}^{1-\lambda}$, and C any closed curve in $\Sigma \setminus A$. Then*

(a)

$$\int_C d(h) = 0, \quad (6.77)$$

and $d(h)$ does not have any residue on Σ.

(b)

$$e \,.\, \omega = d(\omega(e)). \quad (6.78)$$

(c)

$$e \,.\, dh = d(e \,.\, h). \quad (6.79)$$

(d)

$$\int_C (e \,.\, dh) = 0, \quad \int_C (e \,.\, \omega) = 0, \quad (6.80)$$

and both integrands do not have any residue on Σ.

(e)

$$\int_C (e \,.\, f) \cdot g = - \int_C f \cdot (e \,.\, g). \quad (6.81)$$

Proof. In the notation there might appear on first sight to be a certain ambiguity, as we do not distinguish between the deRham differential and the (holomorphic) complex differential. But as all our objects are holomorphic or meromorphic, both coincide. The statement (a) is true by Stokes' theorem as we integrate an exact differential over a closed curve. Also, that dh does not have any residue can be concluded equivalently from the local Laurent series expansion, respectively by using the residue formula $0 = \oint dh = 2\pi \mathrm{i} \cdot \mathrm{res}_P(dh)$.

(b) In the local coordinate z we write locally $\omega = \hat{\omega} dz$ and $e = \hat{e} \frac{d}{dz}$. Hence, for the left-hand side (locally)

$$e \,.\, \omega = \left(\hat{e} \frac{d\hat{\omega}}{dz} + 1 \cdot \frac{d\hat{e}}{dz} \hat{\omega} \right) dz.$$

As $d(\omega(e)) = d(\hat{\omega} dz(\hat{e}\frac{d}{dz})) = d(\hat{\omega} \cdot \hat{e})$ according to the Leibniz rule, this is equal to the right-hand side.

(c) We apply (b) to $\omega = dh$ and note that $dh(e) = e\,.\,h$.

(d) The elements on the right-hand side of (b) and (c) are differentials. Hence, the claim follows from part (a).

(e) According to the Leibniz rule for the e-action we have $e\,.\,(f \cdot g) = (e\,.\,f) \cdot g + f \cdot (e\,.\,g)$. From (d) it follows that the integral over the left-hand side vanishes. Hence, the claim. □

6.5.1 Geometric cocycles for function algebra

The function algebra considered as Lie algebra is abelian. Hence any alternating bilinear form will define a 2-cocycle. For any $f, g \in \mathcal{A}$ and any linear form $\phi : \mathcal{A} \to \mathbb{C}$ we obtain $\phi([f, g]) = 0$. Consequently, there will be no coboundary, i.e., $\mathrm{H}^2(\mathcal{A}, \mathbb{C})$ is equal to the space of all alternating bilinear forms.

In the following section we will consider cocycles which are of geometric origin. For C, a differentiable curve in $\Sigma \setminus A$, we set

$$\gamma_C^{\mathcal{A}} : \mathcal{A} \times \mathcal{A} \to \mathbb{C}, \quad \gamma_C^{\mathcal{A}}(g, h) := \frac{1}{2\pi \mathrm{i}} \int_C \widehat{\gamma}^{\mathcal{A}} = \frac{1}{2\pi \mathrm{i}} \int_C g dh. \tag{6.82}$$

As

$$0 = \int_C d(gh) = \int_C g dh + \int_C h dg, \tag{6.83}$$

(see (6.77)), the expression is alternating. Hence, (6.82) is a cocycle.

If we replace C by any other cycle C' which is homologous in $\mathrm{H}_1(\Sigma \setminus A, \mathbb{Z})$ to C, i.e., we have $C - C' = \delta D$ with D a 2-cycle of $\Sigma \setminus A$, then

$$\int_C g dh - \int_{C'} g dh = \iint_D d(g dh) = 0,$$

as already $d(gdh) = 0$. This says that the cocycle value will not depend on the representing cycle chosen, i.e.,

$$\gamma_{[C]}^{\mathcal{A}} := \gamma_C^{\mathcal{A}} = \gamma_{C'}^{\mathcal{A}}, \tag{6.84}$$

where $[C]$ denotes the cycle class. Moreover, we can even define $\gamma_{[C]}^{\mathcal{A}}$ for cycle classes with multiplicities. Also, for the cocycle defined for the other algebras we will without further mention change between the curve and the cycle notation.

Any cocycle obtained by choosing a cycle C in (6.82) is called a *geometric cocycle*. As we have a lot of freedom in choosing an alternating bilinear form we will have cocycles which will not be geometric cocycles. To distinguish the geometric ones we will look for some natural additional properties which geometric cocycles have, but others possibly do not.

Definition 6.18.
(a) A cocycle γ for $\mathcal{A}$ is called *$\mathcal{L}$-invariant* if

$$\gamma(e.g, h) = \gamma(e.h, g), \quad \forall e \in \mathcal{L}, \ \forall g, h \in \mathcal{A}. \tag{6.85}$$

(b) A cocycle γ for $\mathcal{A}$ is called *multiplicative* if

$$\gamma(f \cdot g, h) + \gamma(g \cdot h, f) + \gamma(h \cdot f, g) = 0, \quad \forall f, g, h \in \mathcal{A}. \tag{6.86}$$

Both properties are of importance. If the cocycle is multiplicative, this means that it fulfills the "cocycle condition" for the associative algebra. Moreover, in this case we can introduce a cocycle for the current algebra by the technique used in Section 6.5.5. Below we will show that a cocycle of the function algebra which is obtained via restriction from the differential operator algebra will be $\mathcal{L}$-invariant. Further down we will show that for bounded cocycles (see Section 6.6) both conditions are equivalent.

Remark 6.19. In the context of Connes' cyclic cohomology [39], the fact that γ is a multiplicative cocycle for a commutative algebra can also be formulated as that it is a 1-cocycle in cyclic cohomology $\mathrm{HC}^1(\mathcal{A}, \mathbb{C})$.

Proposition 6.20. *The cocycle $\gamma_C^{\mathcal{A}}$ (6.82) for the abelian Lie algebra $\mathcal{A}$ is multiplicative and $\mathcal{L}$-invariant.*

Proof. That $\gamma_C^{\mathcal{A}}$ is multiplicative follows from $\int_C d(fgh) = 0$ and the Leibniz rule. For the $\mathcal{L}$-invariance we consider

$$\int_C (e\,.\,g)dh = \int_C e\,.\,(gdh) - \int_C g \cdot (e\,.\,dh) = -\int_C g \cdot (d(e\,.\,h)) = \int_C (e\,.\,h)dg.$$

In the first step we used $e.(a \otimes b) = (e\,.\,a) \otimes b + a \otimes (e\,.\,b)$ for $a \in \mathcal{F}^\lambda$ and $b \in \mathcal{F}^\mu$, in the second step the fact that the first integral vanishes as the integrand is an exact differential (using $e\,.\,\omega = d(\omega(e))$, see (6.78)), and in the last step the antisymmetry of the cocycle. □

6.5.2 Geometric cocycles for vector field algebra

As already described above, in the classical situation there is up to equivalence and rescaling only one nontrivial central extension of the Witt algebra: the Virasoro algebra. In terms of generators e_n, the standard form of the cocycle is

$$\gamma(e_n, e_m) = \frac{1}{12}(n^3 - n)\delta^n_{-m}. \tag{6.87}$$

For the higher genus multi-point situation we consider for each cycle C (or cycle class) with respect to a chosen meromorphic projective connection R which is holomorphic outside A

$$\gamma_{C,R}^{\mathcal{L}}(e,f) := \frac{1}{24\pi \mathrm{i}} \int_C \widehat{\gamma}_R^{\mathcal{L}}(e,f) = \frac{1}{24\pi \mathrm{i}} \int_C \left(\frac{1}{2}(e'''f - ef''') - R\cdot(e'f - ef')\right) dz. \tag{6.88}$$

Recall that we use the same letter for the vector field and its local representing function. As mentioned above, to add the term with the connection R is important, as otherwise the integrand will not be a well-defined differential.

This cocycle was introduced for the $N = 2$ case and with respect to a separating cycle as integration path (see below) by Krichever and Novikov [140, 141]. In [206, 207] it was extended to the multi-point situation.

As $\widehat{\gamma}_R^{\mathcal{L}}$ is a holomorphic form on $\Sigma \setminus A$, the integration value will not depend on the chosen C within the cycle class $[C]$.

Proposition 6.21. *The bilinear form $\gamma_{C,R}^{\mathcal{L}}$ defines a 2-cocycle for the vector field Lie algebra $\mathcal{L}$.*

Proof. It has already been shown that the integrand is a well-defined 1-differential (Proposition 6.16). It remains to show the cocycle condition

$$\chi := \gamma_{C,R}^{\mathcal{L}}([e,f],h) + \gamma_{C,R}^{\mathcal{L}}([f,h],e) + \gamma_{C,R}^{\mathcal{L}}([h,e],f) = 0. \tag{6.89}$$

Let $\tilde{\chi}$ be the integrand of the integral χ. We will show that $\tilde{\chi}$ is a differential of a meromorphic function, which is holomorphic outside A. Hence, the integral over the closed curve C will vanish; see Lemma 6.17 (a).

For simplicity we will identify the vector field with its local representing functions. Let us denote the individual integrands as

$$\begin{aligned}\tilde{\chi}([e,f],h) = &\frac{1}{2}\left[(ef' - e'f)^{(3)}h - h^{(3)}(ef' - e'f)\right] \\ &- R\cdot\left[(ef'' - e''f)h + (e'f - f'e)h'\right].\end{aligned} \tag{6.90}$$

The other two summands are given by cyclic permutations. The factor coming with R will vanish after adding all cyclic permutations. Only the first term F remains. As $\frac{1}{2}$ is unimportant we will drop it. By direct calculation we obtain

$$F = \left((ef' - e'f)h\right)^{(3)} - 3\left((ef' - e'f)'h'\right)' - 2\left(h^{(3)}(ef' - e'f)\right). \tag{6.91}$$

Note that our Fs after the cyclic summations adds up to a well-defined differential. The first two terms are already derivatives of functions and hence can be ignored individually. The last term rewrites as (ignoring again scalar factors)

$$h^{(3)}(ef' - e'f) = \left(h^{(2)}(ef' - fe')\right)' - h^{(2)}(ef' - fe')'. \tag{6.92}$$

Also here, the first term is a derivative of a function and can be ignored. After cyclic summation it remains

$$h^{(2)}(ef' - fe')' + e^{(2)}(fh' - hf')' + f^{(2)}(hg' - gh')' = 0. \tag{6.93}$$

Hence, in total

$$\tilde{\chi} = \frac{dg}{dz}dz = dg, \tag{6.94}$$

with a global meromorphic function g which is holomorphic outside A. This yields $\int_C \tilde{\chi} = 0$. □

We will call a cocycle a *geometric cocycle* if it can be represented as (6.88) with a suitable cycle C and projective connection R.

In the genus zero situation, with respect to the standard coordinates $(z, w = 1/z)$, if we choose $R = 0$ as projective connection and C a circle around $z = 0$, the expression (6.88) will reproduce exactly (6.87).

In the definition we have fixed a projective connection R. If we choose another projective connection R', which only has poles at A, then $R' - R = \Omega$ with a meromorphic quadratic differential Ω with poles only at A, i.e., $\Omega \in \mathcal{F}^2$. We calculate

$$\gamma_{C,R}^{\mathcal{L}}(e,f) - \gamma_{C,R'}^{\mathcal{L}}(e,f) = \frac{1}{24\pi i}\int_C \Omega(ef' - fe')dz = \frac{1}{24\pi i}\int_C \Omega \otimes [e,f]. \tag{6.95}$$

This implies that the two cocycles are cohomologous, leading to the following proposition.

Proposition 6.22. *The cohomology class $[\gamma_{C,R}^{\mathcal{L}}]$ does not depend on the chosen projective connection R.*

Remark 6.23 (On coboundaries). Recall that a 2-coboundary is given via a linear form $\phi : \mathcal{L} \to \mathbb{C}$ on $\mathcal{L}$ by $d_1\phi(e,f) = \phi([e,f])$. In Section 3.2 we discussed a way to describe the dual space $\mathcal{L}^*$ of $\mathcal{L}$. For this we have to choose a splitting of $A = I \cup O$ and the corresponding homogeneous basis elements $e_{n,p}$. The dual basis elements with respect to the Krichever–Novikov pairing (3.10) given by this splitting are the quadratic differentials $\Omega^{n,p}$

$$\langle \Omega^{n,p}, e_{m,r}\rangle = \delta_n^m \cdot \delta_p^r. \tag{6.96}$$

Every linear form ϕ can be described as

$$\phi(e) = \langle W, e\rangle, \quad \text{with} \quad W = \sum_{n\in\mathbb{Z}}\sum_{r=1}^{K} \beta_{n,r}\Omega^{n,r}, \quad \beta_{n,r} \in \mathbb{C}. \tag{6.97}$$

Here the outer sum can reach indeed from $-\infty$ to $+\infty$. Nevertheless, evaluated for a fixed e only a finite number of terms in (6.97) will be non-zero. In this way we obtain every coboundary by choosing such an infinite sum W. Let us denote this coboundary by

$$D_W(e,f) = \langle W, [e,f]\rangle. \tag{6.98}$$

The dual space $\mathcal{L}^*$ of $\mathcal{L}$ depends only on $\mathcal{L}$, hence on the set A and not on the almost-grading. But the identification with the formal series of elements in $\mathcal{F}^2$ is based on the Krichever–Novikov duality, and consequently on the splitting $A = I \cup 0$. A different splitting will yield again an identification which in general will be different. We will always get a representation as in (6.97), but it is possible that with respect to one splitting the sum will be finite and for another splitting infinite.

6.5.3 Geometric cocycles for the differential operator algebra

Recall that for the differential operator algebra $\mathcal{D}^1$ we have the exact sequence of Lie algebras

$$0 \longrightarrow \mathcal{A} \xrightarrow{i_1} \mathcal{D}^1 \xrightarrow{p_2} \mathcal{L} \longrightarrow 0. \tag{6.99}$$

Both $\mathcal{L}$ and A are Lie subalgebras of $\mathcal{D}^1$.

Let L be an arbitrary Lie algebra and M a subalgebra. Given a 2-cocycle γ of L, then by restricting γ to M, i.e., $\gamma_{|M} := \gamma_{|M\times M}$, one obtains a 2-cocycle for M. If γ was a coboundary of L, then its restriction will be a coboundary of M too. A warning is in order, it is possible that $\gamma_{|M}$ is a coboundary of M but γ is not a coboundary of L. Also, given a cocycle ψ for M, the question of whether it is possible to extend it to L is not easy to answer.

In our differential operator algebra case we can examine the situation in detail. First, every cocycle $\gamma^{\mathcal{L}}$ of $\mathcal{L}$ will define via pullback a cocycle $p_2^*(\gamma^{\mathcal{L}})$ on $\mathcal{D}^1$. Restricting the pullback to the subalgebra $\mathcal{L}$ in $\mathcal{D}^1$ will give exactly the cocycle $\gamma^{\mathcal{L}}$ and it will vanish if one of the arguments is from $\mathcal{A}$. Hence, the pullback can also be described as extension of $\gamma^{\mathcal{L}}$ by zero outside the subalgebra $\mathcal{L}$ in $\mathcal{D}^1$. We will denote the extended, respectively pulled back cocycle again by $\gamma^{\mathcal{L}}$.

The situation is slightly more complicated for the subalgebra $\mathcal{A}$ of functions in $\mathcal{D}^1$. As explained above, every cocycle of $\mathcal{D}^1$ defines by restriction a cocycle for $\mathcal{A}$. The problem is, given a cocycle on $\mathcal{A}$, can it be extended to $\mathcal{D}^1$?

Proposition 6.24.
(a) *A cocycle $\tilde{\gamma}$ of $\mathcal{D}^1$ restricted to the subalgebra $\mathcal{A}$ defines an $\mathcal{L}$-invariant cocycle for $\mathcal{A}$.*
(b) *A cocycle γ of $\mathcal{A}$ can be extended to a cocycle of $\mathcal{D}^1$ if and only if it is $\mathcal{L}$-invariant, i.e.,*

$$\gamma(e.g, h) = \gamma(e.h, g), \quad \forall e \in \mathcal{L},\ \forall g, h \in \mathcal{A}. \tag{6.100}$$

Proof. (a) Let $\tilde{\gamma}$ be a cocycle for $\mathcal{D}^1$ and γ its restriction to $\mathcal{A}$. If we write out the cocycle condition for the elements $e \in \mathcal{L}$ and $g, h \in \mathcal{A}$ we obtain

$$0 = \tilde{\gamma}([e, g], h) + \tilde{\gamma}([g, h], e) + \tilde{\gamma}([h, e], g). \tag{6.101}$$

By the definition of the Lie product in $\mathcal{D}^1$, and using $[g, h] = 0$, we get

$$\gamma(e.g, h) - \gamma(e.h, g) = 0, \tag{6.102}$$

which is (6.100). Hence, $\tilde{\gamma}_{|\mathcal{A}} = \gamma$ is necessarily $\mathcal{L}$-invariant.

(b) The statement (a) states that $\mathcal{L}$-invariance is a necessary condition for be extendability. Now given a cocycle γ for $\mathcal{A}$ we define the extended bilinear map

$$\tilde{\gamma} : \mathcal{D}^1 \times \mathcal{D}^1 \to \mathbb{C}, \quad \tilde{\gamma}((g,e),(h,f)) := \gamma(g,h). \tag{6.103}$$

Clearly it is alternating. We have to check the cocycle condition. By linearity it is enough to do this for "pure" elements (e,f,g). If at least 2 of them are vector fields or all of them are functions, then each of the terms in the cocycle relation vanishes separately. This leaves $e \in \mathcal{L}$ and $g,h \in \mathcal{A}$. Because $[g,h] = 0$, the cocycle condition is equivalent to (6.100). Hence, γ can be extended if it is $\mathcal{L}$-invariant. □

By Proposition 6.20 the geometric cocycles fulfill (6.100), which leads to the next proposition.

Proposition 6.25. *The geometric cocycles* $\gamma_C^{\mathcal{A}}(g,h) = \frac{1}{2\pi i}\int_C g dh$ *can be extended to* $\mathcal{D}^1$ *by zero outside* $\mathcal{A} \times \mathcal{A}$.

Next we examine whether there are other cocycles which are not of the types obtained by extensions of a vector field cocycle or of a function algebra cocycle as discussed above. Let γ be an arbitrary cocycle of $\mathcal{D}^1$, and let $\gamma^{\mathcal{A}}$ be its restriction to $\mathcal{A}$ and $\gamma^{\mathcal{L}}$ its restriction to $\mathcal{L}$, and both of them extended by zero to $\mathcal{D}^1$ again. Then $\gamma^{(m)} = \gamma - \gamma^{\mathcal{A}} - \gamma^{\mathcal{L}}$ will again be a cocycle. It will only have non-zero values for arguments which are of mixed type. I.e., for $e \in \mathcal{L}$ and $f \in \mathcal{A}$, and it will fulfill $\gamma^{(m)}(e,f) = -\gamma^{(m)}(f,e)$. We call $\gamma^{(m)}$ a *mixing cocycle*. This decomposition of $\gamma = \gamma^{\mathcal{A}} + \gamma^{\mathcal{L}} + \gamma^{(m)}$ is unique.

Coboundaries for $\mathcal{D}^1$ are given again by choosing linear forms ϕ on $\mathcal{D}^1$; see Remark 6.23. The dual spaces to the functions (vector fields) are given by the differentials (quadratic differentials) with the Krichever–Novikov pairing as duality. Hence, let

$$V = \sum_{n\in\mathbb{Z}} \sum_{r=1}^{K} \alpha_{n,r}\omega^{n,r}, \quad W = \sum_{n\in\mathbb{Z}} \sum_{r=1}^{K} \beta_{n,r}\Omega^{n,r} \tag{6.104}$$

be possibly both-sided infinite sums, then

$$\phi((f,e)) = \langle V, f\rangle + \langle W, e\rangle. \tag{6.105}$$

The corresponding coboundary is given as

$$\begin{aligned}\phi([(g,e),(h,f)]) &= \phi((e\,.\,h - f\,.\,g, [e,f]) \\ &= E_V([e,h] - [f,g]) + D_W([e,f]) \\ &= \langle V, e\,.\,h - f\,.\,g\rangle + \langle W, [e,f]\rangle.\end{aligned} \tag{6.106}$$

This implies that splitting into the three types remains if we pass to cohomology. The coboundary for $\gamma^{\mathcal{L}}$ will be given by W, the coboundary for $\gamma^{(m)}$ will be given by V, and there is of course no coboundary for $\gamma^{\mathcal{A}}$.

Next we study mixing cocycles in more detail.

Proposition 6.26. *Every bilinear form $\gamma : \mathcal{L} \times \mathcal{A} \to \mathbb{C}$ fulfilling*

$$\gamma([e,f],g) - \gamma(e,f\,.\,g) + \gamma(f,e\,.\,g) = 0\,, \quad \forall e,f \in \mathcal{L}, \forall g \in \mathcal{A} \tag{6.107}$$

defines by alternating extension and by setting it to zero on $\mathcal{A}\times\mathcal{A}$ and on $\mathcal{L}\times\mathcal{L}$ a mixing cocycle for $\mathcal{D}^1$.

Proof. Let γ be a bilinear form extended as described. Per construction it is alternating. The only cocycle condition evaluated for elements of pure type which does not trivially vanish is the one involving the two vector fields e and f and one function g. But this cocycle condition is exactly (6.107). □

Proposition 6.27. *Let C be any cycle on $\Sigma \setminus A$ and T a meromorphic affine connection which has at most poles at A. Then the integral of $\widehat{\gamma}_T^{(m)}$ over C given by*

$$\begin{aligned} \gamma_{C,T}^{(m)}(e,g) = -\gamma_{C,T}^{(m)}(g,e) &= \frac{1}{2\pi \mathrm{i}} \int_C \widehat{\gamma}_T^{(m)} \\ &= \frac{1}{2\pi \mathrm{i}} \int_C \left(e \cdot g'' + T \cdot (e \cdot g')\right) dz \end{aligned} \tag{6.108}$$

defines a mixing cocycle.

Proof. We have to verify (6.107). Again, we know that the integrand is indeed a well-defined differential. We identify the vector fields with their local representing functions. With these representing functions we have

$$[e,f] = e \cdot f' - f \cdot e', \quad [e,g] = e \cdot g', \quad [f,g] = f \cdot g'.$$

First we discuss all 3 terms together coming with the connection T. A direct calculation shows

$$(ef' - ge')g' - e(fg')' + f(eg')' = 0.$$

For the sum over the first term we get

$$(ef' - fe')g'' - e(fg')'' + f(eg')'' = (-ef'g' + e'fg')'.$$

This is a differential of a function, and consequently the integral over C vanishes. □

Cocycles obtained via (6.108) are called geometric cocycles.

Again, homologous cycles will give the same cocycle.

Proposition 6.28. *The cohomology class $[\gamma_{C,T}^{(m)}]$ does not depend on the chosen affine connection T (with possible poles only at A).*

Proof. The difference of two affine connections T, T' is a one-differential $\omega = T - T'$. Hence,

$$\gamma_{C,T}^{(m)}(e,g) - \gamma_{C,T'}^{(m)}(e,g) = \frac{1}{2\pi \mathrm{i}} \int_C \omega(e\,.\,g), \tag{6.109}$$

which is a coboundary. □

Similar to the vector field case, the coboundaries can be given via $E_V(e,g) := \langle V, e.g\rangle$, with

$$V = \sum_{n\in\mathbb{Z}}\sum_{r=1}^{K} \alpha_{n,r}\omega^{n,r}, \quad \alpha_{n,r} \in \mathbb{C}. \tag{6.110}$$

Remark 6.29. In Section 2.7 we constructed the Lie algebra of differential operators $\mathcal{D}$ and $\mathcal{D}_\lambda$ of all degrees as the universal enveloping algebra $\mathcal{U}(\mathcal{D}^1)$ with the commutator as Lie product. The algebra $\mathcal{D}^1$ is a subalgebra. It is not at all clear whether it is possible to extend any of the above cocycles of $\mathcal{D}^1$ to $\mathcal{D}_\lambda$. Via the representation theoretical methods presented in Chapter 7 we will show that a certain linear combination can be extended. The obtained algebra $\widehat{\mathcal{D}}_\lambda$ is the higher genus multi-point generalization of the $W_{1+\infty}$ algebra; see Remark 7.26.

6.5.4 Special integration curves

In the previous sections, a different choice of a projective or affine connection did not change the cohomology class. If we change the integration cycle by staying in the same cycle class we do not even change the cocycle. But replacing C by a non-homologous C' will give cocycles which are essentially different. Hence, there is no chance of unicity for cohomology if $\dim \mathrm{H}_1(\Sigma \setminus A, \mathbb{R}) > 1$. With the exception of the classical situation, genus zero and two points, this dimension will always be larger than one. See also Section 6.10. We will introduce the notion of local cocycles (see Definition 6.31) in Section 6.6. We will show that every local cocycle (see Definition 6.31), and more generally every cocycle which is bounded from above, will be a sum of geometric cocycles obtained by integrating over circles around the points in I.

Hence these integration curves will be of special importance.

Let $C_1, C_2, \ldots, C_K$ be circles around the points P_i, $i = 1, \ldots, K$, and $C_S = \sum_i C_i$ the separating cycle as discussed in Section 3.2. In abuse of notation we will not distinguish between the curves and their cycle class. The corresponding cocycles will sometimes be denoted by

$$\gamma_i^{\mathcal{A}},\ \gamma_{i,T}^{(m)},\ \gamma_{i,R}^{\mathcal{L}}, \quad i = 1,\ldots,K, \qquad \gamma_S^{\mathcal{A}},\ \gamma_{S,T}^{(m)},\ \gamma_{S,R}^{\mathcal{L}}, \tag{6.111}$$

e.g., $\gamma_S^{\mathcal{A}} := \gamma_{C_S}^{\mathcal{A}}$, $\gamma_i^{\mathcal{A}} := \gamma_{C_i}^{\mathcal{A}}$, etc. Recall that they do not depend on the chosen cycle inside the homology class. The S stands for the separating cycle C_S. If the connection is fixed it will even sometimes be dropped in the notation. A cocycle obtained via integration over a separating cycle (class) will be called a *separating cocycle*. An important observation is that for these integration paths, the cocycle values can be calculated with the help of residues, e.g.,

$$\begin{aligned} \gamma_i^{\mathcal{A}}(f,g) &= \operatorname{res}_{P_i}(fdg) \\ \gamma_S^{\mathcal{A}}(f,g) &= \sum_{P\in I}\operatorname{res}_P(fdg) = -\sum_{Q\in O}\operatorname{res}_Q(fdg). \end{aligned} \tag{6.112}$$

In the last equation we used the residue theorem and the fact that the separating cocycle is cohomologous to the sum of circles around the points in I and to the sum of circles around the points in O with negative orientation.

For vector field algebra cocycles and mixing cocycles we get exactly the same kind of residue description:

$$\begin{aligned} \gamma_{i,R}^{\mathcal{L}} &= \frac{1}{12}\operatorname{res}_{P_i}\left(\frac{1}{2}(e'''f - ef''') - R(e'f - ef')\right)dz \\ \gamma_{S,R}^{\mathcal{L}} &= \frac{1}{12}\sum_{P\in I}\operatorname{res}_{P}\left(\frac{1}{2}(e'''f - ef''') - R(e'f - ef')\right)dz \\ &= -\frac{1}{12}\sum_{Q\in O}\operatorname{res}_{Q}\left(\frac{1}{2}(e'''f - ef''') - R(e'f - ef')\right)dz, \end{aligned} \tag{6.113}$$

and

$$\begin{aligned} \gamma_{i,T}^{(m)} &= \operatorname{res}_{P_i}(eg'' + T\cdot eg')dz \\ \gamma_{S,T}^{(m)} &= \sum_{P\in I}\operatorname{res}_P(eg'' + T\cdot eg')dz = -\sum_{Q\in O}\operatorname{res}_Q(eg'' + T\cdot eg')dz. \end{aligned} \tag{6.114}$$

6.5.5 Geometric cocycles for the current algebra $\overline{\mathfrak{g}}$

To give cocycles for the current algebra $\overline{\mathfrak{g}} = \mathfrak{g}\otimes\mathcal{A}$, we first have to fix a symmetric, invariant, bilinear form β on $\mathfrak{g}$ (not necessarily non-degenerate). Invariance means

$$\beta([x,y],z) = \beta(x,[y,z]), \quad \forall x,y,z\in\mathfrak{g}. \tag{6.115}$$

Proposition 6.30. *The expression*

$$\gamma_{C,\beta}^{\overline{\mathfrak{g}}}(x\otimes g, y\otimes h) := \beta(x,y)\cdot\gamma_C^{\mathcal{A}}(g,h) = \beta(x,y)\cdot\frac{1}{2\pi\mathrm{i}}\int_C g\,dh, \tag{6.116}$$

for all $x, y\in\mathfrak{g}$, $g,h\in\mathcal{A}$, defines a 2-cocycle for the current algebra $\overline{\mathfrak{g}}$.

Proof. As $\gamma_C^{\mathcal{A}}$ is alternating and β symmetric the expression will be alternating. We evaluate

$$\gamma_{C,\beta}^{\overline{\mathfrak{g}}}([x\otimes g, y\otimes h], z\otimes f) = \beta([x,y],z)\cdot\gamma_C^{\mathcal{A}}(gh,f).$$

Adding up all cyclic permutation, using the invariance and the fact that β is symmetric we have for the cocycle expression

$$\beta([x,y],z)\cdot\left(\gamma_C^{\mathcal{A}}(gh,f) + \gamma_C^{\mathcal{A}}(hf,g) + \gamma_C^{\mathcal{A}}(fg,h)\right). \tag{6.117}$$

As the cocycle $\gamma_C^{\mathcal{A}}$ is multiplicative (see Proposition 6.20), this calculates to zero, hence the cocycle condition. □

From the proof we conclude that every multiplicative cocycle for the function algebra will do as partner for β; see Section 9.2.

The central extensions obtained in this way will be called affine algebras of Krichever–Novikov type. We will discuss them in Chapter 9 in more detail.

6.6 Uniqueness and classification of central extensions

In Section 6.5 we introduced cocycles for our algebras in a geometric manner. Our cocycles depend on the choice of the connections R and T. But different choices will not change the equivalence class. Hence, this ambiguity does not disturb us. What really matters is that they depend on the integration curve C chosen.

In contrast to the classical situation, for the higher genus and/or multi-point situation there are many essentially different closed curves, and hence many non-equivalent central extensions defined by the integration.

However. we should take into account that we want to extend the almost-grading from our algebras to the centrally extended ones. This means we take $\deg \hat{x} := \deg x$ and assign a degree $deg(t)$ to the central element t, and obtain almost-gradedness for the central extension.

This is possible if and only if our defining cocycle γ is “local” in the sense of the following definition.

Definition 6.31. Let $W = \bigoplus_{m\in\mathbb{Z}} W_n$ be an almost-graded Lie algebra, and γ a Lie algebra cocycle for W.

(a) The cocycle γ is called *local* if and only if $M_1, M_2 \in \mathbb{Z}$ exists such that

$$\forall n, m: \quad \gamma(W_n, W_m) \neq 0 \implies M_1 \leq n+m \leq M_2. \tag{6.118}$$

(b) The cocycle is called *bounded (from above)* if and only if $M \in \mathbb{Z}$ exists such that

$$\forall n, m: \quad \gamma(W_n, W_m) \neq 0 \implies n+m \leq M. \tag{6.119}$$

(c) The cocycle is called *bounded from below* if and only if $M \in \mathbb{Z}$ exists such that

$$\forall n, m: \quad \gamma(W_n, W_m) \neq 0 \implies n+m \geq M. \tag{6.120}$$

Clearly, locality is equivalent to being bounded from above and below. Of course, the corresponding definition works if we consider an almost-grading over the half-integers, as in the superalgebra case.

The notion of locality was introduced in the two-point case by Krichever and Novikov in [140]. We have to stress the fact that “local” and “bounded” are defined in terms of the almost-grading, and the grading itself depends on splitting $A = I \cup O$. Hence what is “local” or bounded depends on splitting too.

Definition 6.32. A cocycle class is called bounded (respectively local) if and only if it contains a representing cocycle which is bounded (respectively local).

Note that not all cocycles in a bounded class have to be bounded.

Proposition 6.33. *Let R be a meromorphic projective connection and T a meromorphic affine connection both holomorphic outside O. Let C_i be circles around the points in I, and C_j^* circles around the points in O, then the geometric cocycles*

$$\gamma_C^{\mathcal{A}}, \quad \gamma_C^{\mathcal{L}}, \quad \gamma_C^{(m)}, \quad \gamma_C^{\overline{\mathfrak{g}}} \tag{6.121}$$

are
(a) *bounded from above if* $C = C_i$, $i = 1, \dots, K$, *and* 0 *will be an upper bound;*
(b) *bounded from below if* $C = C_j^*$, $j = 1, \dots, L = N - K$;
(c) *local if* $C = C_S$ *is a separating cycle, and* 0 *will be an upper bound.*

Proof. Note that for these cases the calculation of the integrals can be made in terms of residues; see (6.112), (6.113) and (6.114). We have to consider the order of the homogeneous elements at the points $P_i \in I$ and $Q_j \in O$. The claim (c) follows from (a) and (b), as the integration over a separating cycle can be done equivalently by summation over the integrals obtained via integration over the C_i, or by the negative of the sum obtained by integration over the C_j^*. Hence, the cocycle will be bounded from above and from below. This gives locality. Let us consider the case (a) for the vector field algebra $\mathcal{L}$ in more detail. For $e \in \mathcal{L}_n$, $f \in \mathcal{L}_m$, we have

$$\operatorname{ord}_{P_i}(e) \geq n + 1, \quad \operatorname{ord}_{P_i}(f) \geq m + 1, \qquad P_i \in I \tag{6.122}$$

and

$$\operatorname{ord}_{Q_j}(e) \geq -a \cdot n + b, \quad \operatorname{ord}_{Q_j}(f) \geq -a \cdot m + b, \qquad Q_j \in O, \tag{6.123}$$

with coefficients a, b independent of n and m and $a > 0$. We consider the differential (6.72) locally at P_i. We obtain

$$\begin{aligned} \operatorname{ord}_{P_i}(e'''f),\ \operatorname{ord}_{P_i}(ef''') &\geq n + m - 1, \\ \operatorname{ord}_{P_i}(R \cdot (e'f),\ \operatorname{ord}_{P_i}(R \cdot (ef')) &\geq 0 + (n + m + 1). \end{aligned} \tag{6.124}$$

Hence, a residue at any of the points P_i can only occur if $n + m \leq 0$. This shows that $\gamma_{C_l}^{\mathcal{L}}$, $l = 1, \dots, K$ is bounded from above by 0. Now we consider the situation at Q_j. Let r be the maximal pole order of R at the points in O. We get

$$\begin{aligned} \operatorname{ord}_{Q_j}(e'''f - ef''') &\geq -a(n + m) + 2b - 3, \\ \operatorname{ord}_{Q_j}(R \cdot (e'f - ef')) &\geq -r - a(n + m) + 2b - 1. \end{aligned} \tag{6.125}$$

Hence, a residue at any of the points Q_j is only possible if at least one of the expressions on the right in (6.125) is less than or equal to -1. Hence,

$$n + m \geq \frac{1}{a}(2b - \max(2, r)). \tag{6.126}$$

This gives a lower bound for the possible nonvanishing of the cocycles $\gamma_{C_l^*}^{\mathcal{L}}$, $l = 1, \dots, L$. The proofs for the other cases are completely analogous. □

Remark 6.34. On one hand we can strengthen the estimate on the lower bound by allowing only holomorphic projective connections, respectively affine connections which have only a pole of order 1 at one of the points in O. From Section 6.4 we know that such types of connections exist. On the other hand, we can relax these conditions and consider also meromorphic connections with poles in I. The Laurent expansion of R and T at the points in I will also appear in the estimates (6.124) and (6.125) for the orders. The value 0 in the second equation in (6.124) might be replaced by some negative value and the upper bound might increase, but the statements about boundedness and locality will remain valid. Only the bounds might change. If we allow poles of maximal order two for R, respectively order one for T at the points P_i, the upper bound still will be zero. Also, if needed, given the geometric situation, the bounds can be explicitly calculated.

Remark 6.35. We would like to return to Remark 6.23, that the coboundaries are given exactly in terms of the elements D_V and E_W defined there. They are given in terms of infinite series of differentials, respectively of quadratic differentials which are added to the connections. In case we have chosen a splitting of $A = I \cup O$, it is convenient to use the same splitting for the Krichever–Novikov duality. In this case, we know that the cocycle will remain local if the sum will be a finite sum. Boundedness will be the case if the sum is one-sided finite. In the vector field case or mixing cocycle case, by taking infinite sums we can easily produce for a given local (or bounded) cocycle a cohomologous one which is not local or bounded any more.

With exactly the residue calculus presented in the proof of Proposition 6.33 we can show the following proposition.

Proposition 6.36. *Under the same assumption as in Proposition 6.33, on the connections we have for the value at the upper bound*

$$\begin{aligned}
\gamma_i^{\mathcal{A}}(A_{m,p}, A_{-m,r}) &= (-m)\,\delta_i^p\delta_i^r, \\
\gamma_{i,R}^{\mathcal{L}}(e_{m,p}, e_{-m,r}) &= \frac{1}{12}(m^3 - m)\,\delta_i^p\delta_i^r, \\
\gamma_{i,T}^{(m)}(e_{m,p}, A_{-m,r}) &= m(m+1)\,\delta_i^p\delta_i^r, \\
\gamma_S^{\mathcal{A}}(A_{m,p}, A_{-m,r}) &= (-m)\,\delta_p^r, \\
\gamma_{S,R}^{\mathcal{L}}(e_{m,p}, e_{-m,r}) &= \frac{1}{12}(m^3 - m)\,\delta_p^r, \\
\gamma_{S,T}^{(m)}(e_{m,p}, A_{-m,r}) &= m(m+1)\,\delta_p^r.
\end{aligned} \tag{6.127}$$

Proof. We use the local form of our elements at the points P_i (with coordinate z_i). As an example we give the vector field algebra $\mathcal{L}$ case. We have locally

$$e_{m,p|} = \left(z_i^{m+1}\,\delta_i^p + O(z_i^{m+2})\right)\frac{d}{dz_i}. \tag{6.128}$$

Hence, if we differentiate this expression three times and multiply it with $e_{-m,r}$ we obtain

$$\left((m-1)m(m+1)z_i^{-1}\,\delta_i^p\delta_i^r + O(1)\right).\frac{d}{dz_i}. \tag{6.129}$$

This term will only contribute to the residue if $i = r = p$. The term with the connection R will not contribute to the residue. For the term $-e_{m,p}(e_{-m,r})'''$ we get the same expression as (6.129). Hence exactly the claimed form. As $\gamma_S^{\mathcal{L}} = \sum_{i=1}^K \gamma_i^{\mathcal{L}}$, we calculate

$$\gamma_S^{\mathcal{L}}(e_{m,p}, e_{-m,r}) = \sum_{i=1}^{K} \gamma_i^{\mathcal{L}}(e_{m,p}, e_{-m,r}) = \frac{1}{12}(m^3 - m)\sum_{i=1}^{K} \delta_i^p \delta_i^r = \frac{1}{12}(m^3 - m)\delta_p^r. \quad (6.130)$$

This had to be shown. The other formulas are calculated in exactly the same manner. □

If our projective connection R has maximally a pole of order two at the points P_i, i.e.,

$$R_| = \alpha_i z_i^{-2} + O(z_i^{-1}), \quad (6.131)$$

then the calculation above shows

$$\gamma_S^{\mathcal{L}}(e_{m,p}, e_{-m,r}) = \frac{1}{12}(m^3 - m + 2\alpha_r m)\ \delta_p^r. \quad (6.132)$$

For the rest of this section we will collect classification results about bounded and local cocycles for the algebras $\mathcal{A}$, $\mathcal{L}$, and $\mathcal{D}^1$. These results were obtained in [214, 215, 222]. The proofs will be mainly postponed to Section 6.8. The proofs presented there are simplified versions of the original proofs. Still, they are are partly technically quite involved and might be skipped in the first reading. In Section 6.9 we deal with the Lie superalgebra $\mathcal{S}$. The results for the current algebras $\overline{\mathfrak{g}}$ will be given and discussed in Chapter 9.

Proposition 6.37. *In the following, let γ be either the function algebra cocycle* (6.82), *the vector field algebra cocycle* (6.88), *or the mixing cocycle* (6.108).

(a) *The cocycles $\gamma_i = \gamma_{C_i}$ for $i = 1, \dots, K$ are linearly independent cohomology classes. In particular, they will not be cohomologous to zero.*

(b) *The separating cocycle γ_S is not cohomologous to zero.*

Proof. The claim (b) follows from (a) because $\gamma_S = \sum_i \gamma_i$. For (a) see Section 6.8. □

We will fix as reference connection a global holomorphic projective connection R and a global meromorphic affine connection T which is either holomorphic or has a pole of maximally order 1 at the point $Q_L \in O$.

Let us denote the subspace of local (respectively bounded) cohomology classes by H^2_{loc} (respectively H^2_b), and in the function algebra case the subspace of local (respectively bounded) and $\mathcal{L}$-invariant by $H^2_{\mathcal{L},loc}$ (respectively $H^2_{\mathcal{L},b}$). Recall that the condition is only required for at least one representative in the cohomology class.

Theorem 6.38. *Let $\mathcal{A}$ be the function algebra of Krichever–Novikov type considered as abelian Lie algebra with chosen almost-grading induced by the splitting of $A = I \cup O$, $K = \#I$.*

(a) *The space of multiplicative (or equivalently) $\mathcal{L}$-invariant cocycles for the function algebra which are bounded from above is K-dimensional. A basis is given by the cocycles*

$$\gamma_{C_i}^{\mathcal{A}}(f, g) = \frac{1}{2\pi i}\int_{C_i} f dg, \quad i = 1, \dots, K. \quad (6.133)$$

(b) *Every bounded multiplicative cocycle is $\mathfrak{L}$-invariant and vice versa.*

(c) *A cocycle γ for the function algebra $\mathcal{A}$ which is either multiplicative or $\mathfrak{L}$-invariant is local if and only if it is a multiple of the separating cocycle, i.e.,$\alpha \in \mathbb{C}$ exists such that*

$$\gamma^{\mathcal{A}}(f,g) = \alpha\gamma^{\mathcal{A}}_{C_s}(f,g) = \frac{\alpha}{2\pi i}\int_{C_S} f dg. \tag{6.134}$$

(d) $\dim \mathrm{H}^2_{\mathfrak{L},loc}(\mathcal{A},\mathbb{C}) = 1$ *and* $\dim \mathrm{H}^2_{\mathfrak{L},b}(\mathcal{A},\mathbb{C}) = K$.

(e) *An $\mathfrak{L}$-invariant local cocycle will be bounded by zero, The values at the upper bound are given by*

$$\gamma(A_{n,r}, A_{-n,s}) = -\alpha\cdot n\cdot\delta_r^s, \quad \textit{with} \quad \alpha = -\gamma(A_{1,r}, A_{-1,r}) \tag{6.135}$$

for any $r = 1,\ldots,K$ (it will not depend on r).

(f) *Up to rescaling there is only one nontrivial one-dimensional central extension $\widehat{\mathcal{A}}$ of the function algebra $\mathcal{A}$ given by an $\mathfrak{L}$-invariant cocycle which allows an extension of the almost-grading.*

Proof. The statements (a), (b), (c), and (e) are shown in Section 6.8.1. Both statements (d) and (f) follow from (a) and (b). Recall that in the abelian case there are no nontrivial coboundaries. □

Corollary 6.39.

(a) *If K (the number of points in I) is equal to 1 and γ is a bounded (from above) cocycle for $\mathcal{A}$, then γ is local.*

(b) *If L (the number of points in O) is equal to 1 and γ is a bounded from below cocycle for $\mathcal{A}$, then γ is local.*

(c) *If N (the number of points in A) is equal to 2 then every bounded cocycle (either from above or from below) γ for $\mathcal{A}$ is local.*

Proof. If $K = 1$ then there is only one C_1, which will automatically be a separating cycle. Hence, for a γ bounded from above, Theorem 6.38, part (a), says that γ will be a multiple of the separating cocycle and consequently local. Statement (b) follows accordingly by taking the inverted grading obtained by switching the roles of I and O. As $N = 2 = 1+1$ is the only possible split, the statement (c) is immediate from (a). □

Remark 6.40. The central extension $\widehat{\mathcal{A}}$ obtained via the separating cocycle (6.134) is the *generalized infinite-dimensional Heisenberg algebra* of Krichever–Novikov type, also called *oscillator algebra*. If we set

$$\gamma_{(n,r)(m,s)} := \frac{1}{2\pi i}\int_{C_S} A_{n,r} dA_{m,s}, \tag{6.136}$$

then in $\widehat{\mathcal{A}}$

$$[A_{n,r}, A_{m,s}] = \gamma_{(n,r)(m,s)}\cdot t, \quad [A_{n,r}, t] = 0. \tag{6.137}$$

Let us specialize this to the classical situation ($g = 0$, $N = 2$). We obtain

$$[A_n, A_m] = m \cdot \delta_n^{-m} t = -n \cdot \delta_{-n}^m t, \quad [A_n, t] = 0. \tag{6.138}$$

This is the classical infinite-dimensional Heisenberg algebra (oscillator algebra); see e.g. [122]. Note that in the classical case A_0 is also central in $\widehat{\mathcal{A}}$. This is not necessarily the case in general.

Theorem 6.41. *Let $\mathcal{L}$ be the Krichever–Novikov vector field algebra with given almost-graded structure induced by splitting $A = I \cup O$, $\#I = K$.*

(a) *The space of bounded cohomology classes is K-dimensional. A basis is given by the classes of the cocycles obtained by taking as integration path in (6.88) C_i, $i = 1, \ldots, K$, the little (deformed) circles around the points $P_i \in I$, i.e.,*

$$\gamma_{C_i,R}^{\mathcal{L}}(e, f) = \frac{1}{24\pi \mathrm{i}} \int_{C_i} \left(\frac{1}{2}(e''' f - e f''') - R \cdot (e' f - e f') \right) dz, \quad i = 1, \ldots, K. \tag{6.139}$$

(b) *The space of local cohomology classes is one-dimensional. A generator is given taking the class obtained by integrating (6.88) over a separating cocycle C_S, i.e.,*

$$\gamma_{C_S,R}^{\mathcal{L}}(e, f) = \frac{1}{24\pi \mathrm{i}} \int_{C_S} \left(\frac{1}{2}(e''' f - e f''') - R \cdot (e' f - e f') \right) dz. \tag{6.140}$$

(c) $\dim \mathrm{H}^2_{loc}(\mathcal{L}, \mathbb{C}) = 1$ *and* $\dim \mathrm{H}^2_b(\mathcal{L}, \mathbb{C}) = K$.

(d) *If γ is a local cocycle which is bounded from above by zero, then at level zero the cocycle is given by*

$$\begin{aligned} \gamma(e_{n,r}, e_{-n,s}) &= \left(\frac{(n+1)n(n-1)}{12} \cdot \alpha + n b_r \right) \delta_s^r, \\ \text{with} \quad \alpha &:= 2\gamma(e_{2,r}, e_{-2,r}) - 4\gamma(e_{1,r}, e_{-1,r}), \\ \text{and} \quad b_r &:= \gamma(e_{1,r}, e_{-1,r}). \end{aligned} \tag{6.141}$$

Here α can be calculated with respect to any r and is independent of r. In case $\gamma = \gamma_{C_S,R}^{\mathcal{L}}$ with R holomorphic also in I, then $b_r = 0$.

(e) *Up to equivalence and rescaling there is only one nontrivial one-dimensional central extension $\widehat{\mathcal{L}}$ of the vector field algebra $\mathcal{L}$ which allows an extension of the almost-grading.*

As far as local cocycles are concerned, part (c) says that the result looks quite similar to the case of the Witt algebra. Here I like to repeat the fact that depending on the set A and its possible splits into two disjoint subsets, there are different almost-gradings for $\mathcal{L}$. Hence, the "unique" central extension finally obtained will also depend on the splitting. Only in the two- point case is there only one splitting possible.

Proof. The statements (a), (b), and (d) are shown in Section 6.8.2. The statements (c) and (d) are consequences. □

Corollary 6.42.
(a) *If K (the number of points in I) is equal to 1 and $[\gamma]$ is a bounded (from above) cohomology class for $\mathcal{L}$, then the class $[\gamma]$ is local.*
(b) *If L (the number of points in O) is equal to 1 and $[\gamma]$ is a bounded (from below) cohomology class for $\mathcal{L}$, then the class $[\gamma]$ is local.*
(c) *If N (the number of points in A) is equal to 2, then every bounded class $[\gamma]$ (either from above or from below) for $\mathcal{L}$ is a local class.*

Proof. The proof works as for Corollary 6.39 with the only modification that in this case we have to take into consideration that our results are only valid for the cohomology classes. In particular, a bounded cocycle γ, even in the $N = 2$ case, does not have to be local. We can only conclude that there will be a local cocycle cohomologous to γ. □

Proposition 6.43. *If γ is a local cocycle for the vector field algebra which is not cohomologous to zero, then a meromorphic projective connection R' holomorphic outside A and $\alpha \in \mathbb{C}^*$ exists, such that*

$$\gamma(e, f) = \alpha\gamma^{\mathcal{L}}_{S,R'}(e, f). \tag{6.142}$$

Proof. From Theorem 6.41 we conclude that γ is cohomologous to an $\alpha\gamma^{\mathcal{L}}_{S,R}$ with our holomorphic reference connection R. Hence, we can write

$$\gamma = \alpha\gamma^{\mathcal{L}}_{S,R'} + D_W, \quad W = \sum_{n\in\mathbb{Z}}\sum_{r=1}^{K} \beta_{n,r}\Omega^{n,r}. \tag{6.143}$$

As both γ and $\gamma^{\mathcal{L}}_{S,R'}$ are local, the coboundary D_W will be local too. By Proposition 6.74 the outer sum will be a finite sum. Hence, $R' = R + 1/\alpha W$ is another projective connection (which might have poles at the points in A). This shows $\gamma = \alpha\gamma^{\mathcal{L}}_{S,R'}$ □

Theorem 6.44. *Let $\mathcal{D}^1$ be the Krichever–Novikov vector differential operator algebra with given almost-graded structure induced by the splitting $A = I \cup O$, $\#I = K$.*
(a) *The cohomology space of mixing cocycles bounded from above is K-dimensional and generated by the classes $[\gamma^{(m)}_{r,T}]$, $r = 1, \ldots, K$, i.e.,*

$$\gamma^{(m)}_{C_i,T}(e, g) = \frac{1}{2\pi \mathrm{i}}\int_{C_i} \left(e \cdot g'' + T \cdot (e \cdot g')\right) dz. \tag{6.144}$$

(b) *The subspace of local cohomology classes which are given by mixing cocycles is one-dimensional and generated by the class*

$$\gamma^{(m)}_{C_S,T}(e, g) = \frac{1}{2\pi \mathrm{i}}\int_{C_S} \left(e \cdot g'' + T \cdot (e \cdot g')\right) dz. \tag{6.145}$$

(c) *If γ is a local mixing cocycle which is bounded from above by zero, then at level zero the cocycle is given by*

$$\begin{aligned} \gamma(e_{n,r}, A_{-n,s}) &= (n(n+1)\alpha + n \cdot b_r) \cdot \delta_s^r. \\ \alpha &:= 1/2\left(\gamma(e_{1,r}, A_{-1,r}) + \gamma(e_{-1,r}, A_{1,r})\right), \qquad \text{and} \\ b_r &:= \gamma(e_{-1,r}, A_{1,r}). \end{aligned} \tag{6.146}$$

Here α can be calculated with respect to any r and is independent from r. If $\gamma = \gamma^{(m)}_{C_S,T}$ with the reference affine connection T, then $b_r = 0$.

Corollary 6.45.
(a) *If K (the number of points in I) is equal to 1 and $[\gamma]$ is a bounded (from above) mixing cohomology class, then the class $[\gamma]$ is local.*
(b) *If L (the number of points in O) is equal to 1 and $[\gamma]$ is a bounded (from below) mixing cohomology class, then the class $[\gamma]$ is local.*
(c) *If N (the number of points in A) is equal to 2 then every bounded (either from above or from below) mixing cohomology class $[\gamma]$ is a local class.*

The proof stays word for word nearly the same as for Corollary 6.42. This is also true for the following proposition (this time we add 1-differentials).

Proposition 6.46. *If γ is a mixing cocycle for the differential operator algebra which is not cohomologous to zero, then a meromorphic projective connection T' holomorphic outside A and $\alpha \in \mathbb{C}^*$ exists such that*

$$\gamma(e, f) = \alpha\gamma^{(m)}_{S,T'}(e, f). \tag{6.147}$$

We have shown in Section 6.5.3 that every cocycle for the differential operator algebra can be decomposed as a sum of a vector field cocycle, a function cocycle, and a mixing cocycle, and that this is also true for the coboundaries. Hence, the following theorem can be concluded from the above theorems.

Theorem 6.47.
(a) *A cocycle γ for the differential operator algebra is local if and only if it is a linear combination of the cocycles obtained by extension of the separating cocycle for the function algebra, the cocycle obtained by pulling back the separating vector field cocycle, and the separating mixing cocycle*

$$\gamma = a_1\gamma_S^{\mathcal{A}} + a_2\gamma^{(m)}_{S,T} + a_3\gamma^{\mathcal{L}}_{S,R} + E_V + D_W, \tag{6.148}$$

and coboundary terms $E_V + D_W$. If $a_2, a_3 \neq 0$, then the coboundary terms can be included in the connections R and T.
(b) *The subspace of $\mathrm{H}^2_{loc}(\mathcal{D}^1, \mathbb{C})$ of cocycles cohomologous to local cocycles is three-dimensional and is generated by the cohomology classes of the separating cocycles of function, mixing, and vector field type.*

(c) *The corresponding statement is also true for bounded cocycles, in particular we have* $\dim \mathrm{H}^2_b(\mathcal{D}^1, \mathbb{C}) = 3K$.

We collect the results about the dimension of the subspace of local cocycle classes for future reference.

Corollary 6.48.
(a) $\dim \mathrm{H}^2_{\mathcal{L},loc}(\mathcal{A}, \mathbb{C}) = 1$,
(b) $\dim \mathrm{H}^2_{loc}(\mathcal{L}, \mathbb{C}) = 1$,
(c) $\dim \mathrm{H}^2_{loc}(\mathcal{D}^1, \mathbb{C}) = 3$,
(d) $\dim \mathrm{H}^2_{loc}(\overline{\mathfrak{g}}, \mathbb{C}) = 1$ *for* $\mathfrak{g}$ *a simple finite-dimensional Lie algebra,*
(e) $\dim \mathrm{H}^2_{loc}(\mathcal{S}, \mathbb{C}) = 1$.
A basis of the cohomology spaces is given by taking the cohomology classes of the cocycles (6.134), (6.140), (6.145), (6.116), (6.276) *obtained by integration over a separating cycle* C_S.

Proof. The statements (a), (b), and (c) follow directly from the results which are given above. The statement (d) will be treated in Chapter 9, the statement (e) in Section 6.9. □

Remark 6.49.
(1) We obtain also for these algebras the corresponding result about uniqueness, respectively classification, of almost-graded central extensions.
(2) For the bounded cocycle classes we have to multiply the dimensions above by K.
(3) For the supercase with odd central elements the bounded cohomology vanishes.
(4) Let $\overline{\mathfrak{g}}$ be the current algebra of a reductive Lie algebra $\mathfrak{g}$. A complete classification of local cocycle classes which, restricted to the abelian part, are $\mathcal{L}$-invariant can be found in Chapter 9.
(5) In most applications the cocycles coming from representations, regularizations, etc. are local. Hence, the uniqueness or classification results presented here can be employed.

6.7 The classical situation

In this section we consider the classical situation ($g = 0$ and $N = 2$). In this case the corresponding results about the cocycle classes are true without boundedness or locality assumption for the cocycles. Let us start with the cocycles for the function algebra.

Proposition 6.50. *Let* γ *be either a multiplicative or an* $\mathcal{L}$*-invariant cocycle, then*

$$\gamma(f, g) = \alpha \frac{1}{2\pi \mathrm{i}} \int_{C_S} f dg, \quad \text{with} \quad \alpha \in \mathbb{C}, \tag{6.149}$$

with C_S a circle around 0. Moreover, every $\mathcal{L}$-invariant cocycle is multiplicative and vice versa.

Proof. Assume that γ is $\mathcal{L}$-invariant. Recall this means that

$$\gamma(e_n \,.\, A_m, A_p) + \gamma(A_m, e_n \,.\, A_p) = 0. \tag{6.150}$$

We are in the graded case and we get

$$e_0 \,.\, A_m = m \cdot A_m. \tag{6.151}$$

Hence, (for $n = 0$)

$$m \cdot \gamma(A_m, A_p) + p\gamma(A_m, A_p) = 0. \tag{6.152}$$

This shows $(m + p)\gamma(A_m, A_p) = 0$, and consequently

$$\gamma(A_m, A_p) \neq 0 \implies (m + p) = 0. \tag{6.153}$$

Which means the cocycle is automatically local and we apply Theorem 6.38.

For the multiplicative case we will give the proof as a supplement to the higher genus case proof in Section 6.8.1; see Remark 6.64. □

Proposition 6.51. *Let γ be cocycle for the vector field algebra $\mathcal{L}$ in the classical situation, i.e., for the Witt algebra, then γ is cohomologous to a local cocycle and is cohomologous to a multiple of the the cocycle $\gamma_S^{\mathcal{L}}$.*

Proof. Of course this is a classical result, but it also follows from our classification. We have only to show that an arbitrary γ can be made local by a cohomological change. We define a linear map $\phi : \mathcal{L} \to \mathbb{C}$ by

$$\phi(e_n) := \frac{\gamma(e_0, e_n)}{n}, \quad n \neq 0, \quad \phi(e_0) := 0. \tag{6.154}$$

We will show that the cohomologous cocycle $\tilde{\gamma} = \gamma - d_1\phi$ will be local. Using $[e_0, e_n] = ne_n$, we calculate

$$\tilde{\gamma}(e_0, e_n) = \gamma(e_0, e_n) - \phi([e_0, e_n]) = 0. \tag{6.155}$$

Next we consider the cocycle condition for the triple (e_n, e_m, e_0):

$$\tilde{\gamma}([e_n, e_m], e_0) + \tilde{\gamma}([e_m, e_0], e_n) + \tilde{\gamma}([e_0, e_n], e_m) = 0. \tag{6.156}$$

The first summand will vanish and after using the structure equation we obtain

$$(m + n)\tilde{\gamma}(e_n, e_m) = 0. \tag{6.157}$$

Hence, for $m + n \neq 0$ we obtain $\tilde{\gamma}(e_n, e_m) = 0$, which says that the cocycle $\tilde{\gamma}$ is local; and we can apply Theorem 6.41. □

Proposition 6.52. *The space of cohomology classes for the differential operator algebra $\mathcal{D}^1$ is 3-dimensional. A basis is given by the cohomology classes represented by $\gamma_S^{\mathcal{A}}$, $\gamma_S^{\mathcal{L}}$ and $\gamma_S^{(m)}$.*

Proof. This statement for the classical situation was shown in [4]. Again it can also be obtained via our classification results. The cocycles coming from $\mathcal{A}$ and $\mathcal{L}$ were treated above. For the mixing part we make a cohomological change to make it local. Recall that

$$[e_0, A_m] = e_0 \,.\, A_m = mA_m. \tag{6.158}$$

As in the previous proof, we define for the summand $\mathcal{A}$ in $\mathcal{D}^1$ a linear form

$$\phi_1(A_n) = \frac{1}{n}\gamma(e_0, A_n), \quad n \neq 0, \quad \phi_1(A_0) = 0. \tag{6.159}$$

For the cohomologous cocycle $\tilde{\gamma} = \gamma - d_1\phi_1$ we obtain that $\tilde{\gamma}(e_0, A_n) = 0$. Writing down the Jacobi relation for the triple (e_0, A_m, e_n), we obtain as above $(m+n)\tilde{\gamma}(e_n, A_m) = 0$. Hence again locality, and the claim follows from Theorem 6.44. □

Remark 6.53. The arguments used in the proofs of Proposition 6.51 and Proposition 6.52 are very similar. In fact, there is a common picture behind them. In the graded Lie algebra case one can decompose the cohomology (with values in a graded Lie module) into different degree terms. A cocycle will be of degree $d \in \mathbb{Z}$ if it maps pairs of homogeneous elements of degree n and m to elements of degree $n+m+d$ in the module.

A Lie algebra L (and the module M) is internally graded if an element $e_0 \in L$ exists such that the eigenspace decomposition of the action of e_0 on L and M gives their decomposition into homogeneous subspaces. Here the action on e_0 on L will be the adjoint action. In this case, one can show that the cohomology of degree $d \neq 0$ will vanish. See [85] for a general proof. Hence, only degree zero cohomology will be nontrivial.

In our case, the module M is the trivial module for which all elements have degree zero. The grading element for both algebras $\mathcal{L}$ and $\mathcal{D}^1$ is the element e_0, as we have

$$[e_0, e_n] = ne_n, \quad [e_0, A_n] = nA_n. \tag{6.160}$$

We have $\mathcal{L}_n = \langle e_n \rangle$ and $\mathcal{D}^1_n = \langle e_n, A_n \rangle$ and obtain that the degree decomposition coincides with the eigenspace decomposition. Up to coboundary terms, only the degree zero cocycles will be of relevance. Hence, by cohomologous changes we can replace any cocycle (and this is what we did in our proofs above) by a degree zero one. The subspace of degree zero cocycles will be exactly given by the cocycles which have nonvanishing values only for level zero.

As Krichever–Novikov type algebras are in general only almost-graded we cannot use such arguments for them.

6.8 Proofs for the classification results

In this section we will prove the classification results presented in Section 6.6. The reader only interested in the results might skip it on first reading or might be satisfied with an indication of the strategy as described in the following remark.

Remark 6.54. We start with a remark on the general technique. For preciseness let us formulate it for the Lie algebra $\mathcal{A}$ of functions with the homogeneous basis elements $\{A_{n,p}\}$. Let γ be an $\mathcal{L}$-invariant cocycle which is bounded from above. For a pair $(A_{n,p}, A_{m,r})$ we call the sum $l = n+m$ of the first indices the *level* of the pair. The cocycle values of level l are given as $\gamma(A_{n,p}, A_{-n+l,r})$, $n \in \mathbb{Z}$, $p = 1, \ldots, K$.

We will make descending recursion on the level l. First, we will show that starting at a level $l > 0$ for which the values of the cocycle will be zero for l and all higher levels, the values will also be zero for all levels between $l-1$ and 1. Then we will show that for all levels less than zero the cocycle values are determined by its values at level 0. Finally, we analyse the level zero. In particular it will turn out that all possible values for level zero can be realized by suitable linear combinations of the geometric cocycles $\gamma_r^{\mathcal{A}}$, $r = 1, \ldots, K$ related to the points in I. We conclude that γ itself is this linear combination.

If the cocycle is local, meaning also bounded from below, we continue by considering the inverted grading obtained by switching the roles of the points in I and O. In this way we obtain another representation of the same cocycle, now with respect to the geometric cocycles related to points in O. Some refined analysis shows that this implies that in the linear combination all coefficients have to be the same.

In the vector field or differential operator case, cohomological changes have to be taken into account to fix the linear combination.

Remark 6.55. In the following we will use the phrases "can be expressed by elements of higher level", "determined by higher level", or simply "= h.l.", to denote that it is a universal linear combination of cocycle values for pairs of homogeneous elements of level higher than the level under consideration. The coefficients appearing in this linear combination only depend on the geometric situation, i.e., on the structure constants of the algebra and not on the cocycle under consideration. This should be understood by the term "universal linear combination". In particular, if the values of two cocycles a and b on the level l under consideration coincide up to cocycle values of higher level and also coincide there, then they will coincide also on level l. For simplicity we quite often use also $a \equiv b$ (for cocycle expressions a and b), to indicate that both sides coincide up to universal combinations of cocycle values of pairs of higher level than the level under consideration.

Furthermore, if a and b are elements from our algebras the notion $a \equiv b$ will mean that they coincide up to elements of the algebra of higher degree than the degree under consideration.

6.8.1 The function algebra

We introduced in Section 6.5.1 two subclasses of cocycles of the function algebra, the $\mathcal{L}$-invariant and the multiplicative cocycles. Finally, for bounded cocycles both classes will be the same, as the classification results below, obtained for each class separately, will show. As we do not know this a priori we have to do the proof for both classes separately.

6.8.1.1 $\mathcal{L}$-invariant cocycles

In this subsection we consider cocycles of the function algebra which are $\mathcal{L}$-invariant, i.e.,

$$\gamma(e.g, h) + \gamma(g, e.h) = 0, \quad e \in \mathcal{L}, g, h \in \mathcal{A}. \tag{6.161}$$

The cocycles which are obtained via restriction of cocycles of the algebra $\mathcal{D}^1$ of differential operators are of this type (Proposition 6.24). Also, the geometric cocycles have this property (Proposition 6.20).

By the almost-graded action of $\mathcal{L}$ on $\mathcal{A}$ we have

$$e_{n,r}.A_{m,s} = \delta_r^s \cdot m \cdot A_{n+m,r} + \sum_{h=n+m+1}^{n+m+L_2} \sum_t b_{(n,r),(m,s)}^{(h,t)} A_{h,t}, \tag{6.162}$$

with $b_{(n,r),(m,s)}^{(h,t)} \in \mathbb{C}$, and L_2 the upper bound for the almost-graded structure. In shorter notation

$$e_{n,r}.A_{m,s} \equiv \delta_r^s \cdot m \cdot A_{n+m,r}. \tag{6.163}$$

Using (6.161) we get

$$\gamma(e_{n,r}.A_{m,s}, A_{p,t}) + \gamma(A_{m,s}, e_{n,r}.A_{p,t}) = 0,$$

and with the almost-graded structure we obtain

$$\delta_s^r \cdot m \cdot \gamma(A_{m+n,r}, A_{p,t}) + \delta_r^t \cdot p \cdot \gamma(A_{m,s}, A_{p+n,r}) \equiv 0. \tag{6.164}$$

For $r = t \neq s$ we obtain $p \cdot \gamma(A_{m,s}, A_{p+n,r}) \equiv 0$ for every p. This implies the following lemma.

Lemma 6.56. *The elements $\gamma(A_{n,r}, A_{-n+l,s})$ of level l for $r \neq s$ are universal linear combinations of elements of level $\geq (l+1)$.*

It remains to consider $s = r = t$. For notational simplicity we will drop the second index in (6.164) and obtain

$$m \cdot \gamma(A_{m+n}, A_p) + p \cdot \gamma(A_m, A_{p+n}) \equiv 0. \tag{6.165}$$

If we set $n = 0$, we obtain

$$(m + p) \cdot \gamma(A_m, A_p) \equiv 0. \tag{6.166}$$

Note that $m + p$ is the level. Hence, for level $l \neq 0$ everything is determined by higher levels. This is demonstrated by the following lemma.

Lemma 6.57. *The level l for $l \neq 0$ is completely determined by higher levels.*

For level 0 we set $p = -(n + 1)$ and $m = 1$ in (6.165) and obtain

$$\gamma(A_{n+1}, A_{-(n+1)}) \equiv (n + 1) \cdot \gamma(A_1, A_{-1}), \tag{6.167}$$

which corresponds to (6.187). At this point we remark that we did not use the fact that the bilinear form γ was anti-symmetric, see Proposition 6.60.

Proposition 6.58. *Let γ be an $\mathcal{L}$-invariant cocycle which is bounded from above then:*
(a) *0 is also an upper bound, i.e., $\gamma(\mathcal{A}_{n,r}, \mathcal{A}_{m,s}) = 0$ for $n + m > 0$.*
(b) *It is determined by its value at level 0.*
(c) *The level zero is given as*

$$\gamma(A_{n,r}, A_{-n,s}) = n \cdot \delta_s^r \cdot \alpha_r, \quad \textit{with } \alpha_r := \gamma(A_{1,r}, A_{-1,r}) \tag{6.168}$$

for $\quad n \in \mathbb{Z},\ r, s = 1, \dots, K.$

Proof. Assume γ to be bounded. If $M > 0$ is an upper bound, then by Lemma 6.57 and Lemma 6.56 its values at the level M are linear combinations of levels $> M$. Hence they are also vanishing on level M, and finally 0 is also an upper bound. This proves (a). Part (b) follows again from the above lemmas. For $r \neq s$ the equation (6.168) follows from Lemma 6.56. For $r = s$ this is (6.167), which has to be applied for each r separately. □

Theorem 6.59. *The space of $\mathcal{L}$-invariant cocycles for the function algebra which are bounded from above is K-dimensional. A basis is given by the cocycles*

$$\gamma_r^{\mathcal{A}}(f, g) = \frac{1}{2\pi \mathrm{i}} \int_{C_r} f dg, \quad r = 1, \dots, K. \tag{6.169}$$

Proof. From the Proposition 6.58 it follows that the space is at most K-dimensional. By Proposition 6.33 and Proposition 6.20, the K geometric cocycles γ_i are bounded from above and $\mathcal{L}$-invariant. Hence they are elements of this space. We have to show that they are linearly independent then they will constitute a basis. Assume $0 = \sum_{r=1}^{K} \alpha_r \gamma_r^{\mathcal{A}}$. We evaluate this relation for the pairs $(A_{-1,s}, A_{1,s})$, $s = 1, \dots, K$. As

$$\gamma_r^{\mathcal{A}}(A_{-1,s}, A_{1,s}) = \frac{1}{2\pi} \int_{C_r} A_{-1,s} dA_{1,s} = \delta_r^s, \tag{6.170}$$

we obtain $\alpha_r = 0$ for all r, hence they are linearly independent. □

Proof of Theorem 6.38 ($\mathcal{L}$-invariant case). Part (a) is Theorem 6.59. Next we consider (c). If γ is a multiple of the separating cocycle then it is local (see Proposition 6.33).

Now assume that γ is $\mathcal{L}$-invariant and local. Hence it is bounded from above and can be written as $\gamma = \sum_i \alpha_i \gamma_i$. (We dropped the mentioning of $\mathcal{A}$ in the notation.) We have to show that $\alpha_1 = \alpha_2 = \cdots = \alpha_K$. For $K = 1$ the claim is immediate. If we interchange the role of I and O we obtain the inverted grading (see Section 3.7) which we denote by *. A cocycle which is bounded from below with respect to the original grading is bounded from above with respect to the inverted grading. Denote by C_i^* circles around the points Q_i for $i = 1, \ldots, N-K$, and by $\gamma_i^*(f,g) = \frac{1}{2\pi \mathrm{i}} \int_{C_i^*} f dg$ the corresponding geometric cocycle. Using Theorem 6.59 for the inverted grading we obtain $\gamma = \sum_{i=1}^{N-K} \alpha_i^* \gamma_i^*$ with certain $\alpha_i^* \in \mathbb{C}$. Hence, we get

$$0 = \sum_{i=1}^{K} \alpha_i \gamma_i - \sum_{i=1}^{N-K} \alpha_i^* \gamma_i^* . \tag{6.171}$$

Furthermore, by the residue theorem we have the relation

$$0 = \sum_{i=1}^{K} \gamma_i + \sum_{i=1}^{N-K} \gamma_i^* . \tag{6.172}$$

If we take the first relation and subtract α_1 times the second relation we obtain

$$0 = \sum_{k=2}^{K} (\alpha_k - \alpha_1) \gamma_k - \sum_{k=1}^{N-K} (\alpha_k^* - \alpha_1) \gamma_k^* . \tag{6.173}$$

For each $k = 2, \ldots, K$ separately we take a pair of functions f_n and g_n for n big enough with

$$\operatorname{ord}_{P_k}(f_n) = -n, \quad \operatorname{ord}_{P_k}(g_n) = n, \tag{6.174}$$

possibly with poles at P_1 and holomorphic elsewhere. Such elements exist by Riemann–Roch. Furthermore, we normalize them so that with respect to the chosen local coordinate z_k at P_k, the leading coefficient is 1. If we evaluate (6.173) for the pair (f_n, g_n) nearly all terms will vanish. It remains

$$0 = (\alpha_k - \alpha_1) \gamma_k (f_n, g_n) = (\alpha_k - \alpha_1)\, n . \tag{6.175}$$

This implies that $\alpha_k = \alpha_1$. As is true for every k, we get the claim (c).

Next we consider (e). By Proposition 6.58 zero is an upper bound for the cocycle. As γ is a local cocycle by (c) it can be written as

$$\gamma = \sum_{s=1}^{K} \alpha \gamma_r^{\mathcal{A}} = \alpha \gamma_S^{\mathcal{A}} . \tag{6.176}$$

We evaluate (see (6.127))

$$\alpha_r = \gamma(A_{1,r}, A_{-1,r}) = \sum_{s=1}^{K} \alpha \gamma_r^{\mathcal{A}}(A_{1,r}, A_{-1,r}) = \sum_{s=1}^{K} \alpha(-1)\delta_r^s = -\alpha . \tag{6.177}$$

Hence, α_r does not depend on r and furthermore we get the explicit expression (6.135) from (6.168). □

Also, Proposition 6.37 is a corollary of the above proof.

As pointed out after Lemma 6.57, the fact that γ is anti-symmetric was not necessary in the assumptions. It follows via the above proofs from boundedness, locality and $\mathcal{L}$-invariance. Hence,

Proposition 6.60. *Let γ be a bilinear form for $\mathcal{A}$ which is $\mathcal{L}$-invariant and bounded (respectively local), then γ will be anti-symmetric and consequently a bounded, respectively local cocycle. In particular, it will be of the form given by Theorem 6.38, (6.134).*

This proposition will be of use later.

6.8.1.2 Multiplicative cocycles

For the multiplicative cocycles we also use recursion arguments. In this case they are more involved. First we prove that Lemma 6.56 is also true in the multiplicative case.

Proof of Lemma 6.56 in the multiplicative case. By the multiplicativity we have

$$\gamma(A_{0,r} \cdot A_{n,r}, A_{-n+l,s}) + \gamma(A_{n,r} \cdot A_{-n+l,s}, A_{0,r}) + \gamma(A_{-n+l,s} \cdot A_{0,r}, A_{n,r}) = 0.$$

We replace the products with the help of the almost-grading for $\mathcal{A}$, i.e., by

$$A_{n,r} \cdot A_{m,s} = \delta_r^s \cdot A_{n+m,r} + \sum_{h=n+m+1}^{n+m+L} \sum_t a_{(n,r),(m,s)}^{(h,t)} A_{h,t}, \tag{6.178}$$

where $a_{(n,r),(m,s)}^{(h,t)} \in \mathbb{C}$, and L is the upper bound for the almost-grading. As usual, any summation range over the second index is $\{1, \ldots, K\}$. Hence for $r \neq s$

$$\gamma(A_{n,r} + h.d.t., A_{-n+l,s}) + \gamma(h.d.t., A_{0,r}) + \gamma(h.d.t., A_{n,r}) = 0.$$

Recall that $h.d.t.$ denotes linear combinations of elements of degree which do not contribute to the levels under consideration. This implies $\gamma(A_{n,r}, A_{-n+l,s})$ can be expressed as linear combinations of values of the cocycle of higher level than l. □

Lemma 6.61. *The value $\gamma(A_{0,r}, A_{l,r})$ can be expressed by elements of level $\geq l+1$.*

Proof. From the multiplicativity

$$\gamma(A_{0,r} \cdot A_{0,r}, A_{l,r}) + \gamma(A_{0,r} \cdot A_{l,r}, A_{0,r}) + \gamma(A_{l,r} \cdot A_{0,r}, A_{0,r}) = 0\,,$$

with the almost-grading we obtain

$$\gamma(A_{0,r}, A_{l,r}) + 2 \cdot \gamma(A_{l,r}, A_{0,r}) \equiv 0.$$

By the antisymmetry of the cocycle the claim follows. □

Lemma 6.62. $\gamma(1, f) = 0, \quad \forall f \in \mathcal{A}.$

Proof. From $\gamma(1\cdot 1, f)+\gamma(1\cdot f, 1)+\gamma(f\cdot 1, 1)=0$ we conclude $0=\gamma(1,f)+2\gamma(f,1)=\gamma(f,1)$. □

By Lemma 6.56 only the case $r=s$ is of importance at the level l. Again, to simplify notation we will suppress in the following the second index. Starting from

$$\gamma(A_k\cdot A_n, A_m)+\gamma(A_n\cdot A_m, A_k)+\gamma(A_m\cdot A_k, A_n)=0$$

we obtain

$$\gamma(A_{k+n}, A_m)+\gamma(A_{n+m}, A_k)+\gamma(A_{m+k}, A_n)\equiv 0. \tag{6.179}$$

We specialize this for $m=-1$ and $m=1$:

$$\gamma(A_{k+n}, A_{-1})+\gamma(A_{n-1}, A_k)+\gamma(A_{k-1}, A_n\equiv 0, \tag{6.180}$$

$$\gamma(A_{k+n}, A_1)+\gamma(A_{n+1}, A_k)+\gamma(A_{k+1}, A_n)\equiv 0, \tag{6.181}$$

and set in (6.180) $k=l-n+1$, and in (6.181) $k=l-n-1$ (l denotes the level) to obtain

$$\gamma(A_{l+1}, A_{-1})+\gamma(A_{n-1}, A_{l-(n-1)})+\gamma(A_{l-n}, A_n)\equiv 0, \tag{6.182}$$

$$\gamma(A_{l-1}, A_1)+\gamma(A_{n+1}, A_{l-(n+1)})+\gamma(A_{l-n}, A_n)\equiv 0. \tag{6.183}$$

Subtracting (6.182) from (6.183) we obtain the recursion formula

$$\gamma(A_{n+1}, A_{l-(n+1)})\equiv\gamma(A_{n-1}, A_{l-(n-1)})-\gamma(A_{-1}, A_{l+1})+\gamma(A_1, A_{l-1}). \tag{6.184}$$

If we set $n=-m$ and $k=l$ in (6.179) we obtain

$$\gamma(A_{l-m}, A_m)+\gamma(A_0, A_l)+\gamma(A_{l+m}, A_{-m})\equiv 0.$$

From Lemma 6.61 it follows that $\gamma(A_0, A_l)$ is of higher level, hence

$$\gamma(A_m, A_{l-m})\equiv-\gamma(A_{-m}, A_{l+m}). \tag{6.185}$$

For $m=1$ we obtain $\gamma(A_1, A_{l-1})\equiv-\gamma(A_{-1}, A_{l+1})$, which we can insert into (6.184) to obtain

$$\gamma(A_{n+1}, A_{l-(n+1)})\equiv\gamma(A_{n-1}, A_{l-(n-1)})+2\gamma(A_1, A_{l-1}). \tag{6.186}$$

Hence, the knowledge of $\gamma(A_0, A_l)$ and $\gamma(A_1, A_{l-1})$ will fix the complete cocycle at level l by the knowledge of the higher levels. But $\gamma(A_0, A_l)$ itself is fixed by higher level (Lemma 6.61), hence $\gamma(A_1, A_{l-1})$, or equivalently $\gamma(A_{-1}, A_{l+1})$ will fix everything.

First we consider the level $l=0$ and obtain the recursion

$$\gamma(A_{n+1}, A_{-(n+1)})\equiv\gamma(A_{n-1}, A_{-(n-1)})+2\gamma(A_1, A_{-1}).$$

This implies

$$\gamma(A_n, A_{-n})\equiv n\cdot\gamma(A_1, A_{-1}). \tag{6.187}$$

Lemma 6.63. *The values at the level l for $l \neq 0$ are completely determined by higher levels.*

Proof. First consider $l > 0$. We have to show that $\gamma(A_1, A_{l-1})$ is determined by higher levels. For $l = 1$ we obtain $\gamma(A_1, A_{l-1}) = \gamma(A_1, A_0)$, which is determined by higher level (see Lemma 6.61). For $l = 2$ we obtain $\gamma(A_1, A_{l-1}) = \gamma(A_1, A_1) = 0$ by the antisymmetry. Hence we can assume $l > 2$. We insert (6.179) $k = l - r - 1, n = 1, m = r$ and obtain

$$\gamma(A_1, A_{l-1}) + \gamma(A_r, A_{l-r}) - \gamma(A_{r+1}, A_{l-r-1}) \equiv 0. \tag{6.188}$$

Set $m := \frac{l-2}{2}$ for l even, or $m := \frac{l-1}{2}$ for l odd. We let r run through $1, 2, \dots, m$ and obtain from (6.188) m equations. The first equation will always be

$$2 \cdot \gamma(A_1, A_{l-1}) - \gamma(A_2, A_{l-2}) \equiv 0.$$

The last equation will depend on the parity of l. For l even and $r = m$ the last term on the left of (6.188) will be $\gamma(A_{\frac{l}{2}}, A_{\frac{l}{2}})$, which vanishes. For l odd, the last term of the last equation will coincide with the second term. Hence

$$\gamma(A_1, A_{l-1}) + 2 \cdot \gamma(A_{\frac{l-1}{2}}, A_{\frac{l+1}{2}}) \equiv 0$$

will be the last equation. In this case we divide it by 2. All these equations are added up. As result we obtain

$$(m + \epsilon) \cdot \gamma(A_1, A_{l-1}) \equiv 0,$$

where ϵ is 1 for l even $1/2$ for l odd. This shows the claim for $l > 0$.

For $l < 0$ note that we can equally determine $\gamma(A_{-1}, A_{l+1})$ to fix the cocycle. Now the arguments work in exactly the same way as above. The claim for $l = -1, -2$ follows immediately. We insert $k = l - r + 1, n = -1, m = r$ into (6.179) and obtain

$$\gamma(A_{-1}, A_{l+1}) + \gamma(A_r, A_{l-r}) - \gamma(A_{r-1}, A_{l-r+1}) \equiv 0. \tag{6.189}$$

We set $m := \frac{-l-2}{2}$ for l even, and $m := \frac{-l-1}{2}$ for l odd and consider the equation (6.189) for $r = -1, -2, \dots, -m$. They have similar structure to $l > 0$ and we can add them up again to obtain the statement about $\gamma(A_{-1}, A_{l+1})$. □

Remark 6.64. In the classical situation we have $A_n \cdot A_m = A_{n+m}$ and above in all equations $\equiv$ can be replaced by $=$. Hence, both Lemma 6.57 and Lemma 6.63 can be sharpened saying that every $\mathcal{L}$-invariant or multiplicative cocycle (not necessarily bounded) will vanish for level $l \neq 0$. This completes the proof of Proposition 6.50.

The proofs of Proposition 6.58, Theorem 6.59 and Theorem 6.38 rely only on these lemmas and the relations (6.187) and (6.167). Hence we find that they are also valid if we use “multiplicative” instead of “$\mathcal{L}$-invariant” and we obtain the corresponding statements.

In particular, we obtain that in both cases we get exactly the same cocycles, which leads to the next proposition.

Proposition 6.65.

(a) *A bounded cocycle for function algebra is multiplicative if and only if it is $\mathcal{L}$-invariant.*

(b) *In the classical situation this is true without the boundedness condition.*

This shows the remaining claim (b) in Theorem 6.38.

6.8.2 Vector field algebra

For the classification of bounded cocycles for vector field algebra we have to take coboundaries into account. As a first step for bounded cocycles for vector field algebra we will change a given bounded cocycle by a coboundary to obtain a "normalized" cocycle. For this normalized cocycle we will show in a recursive manner that it is fixed by its values at level zero. More precisely, it is fixed by the knowledge of $K(=\#I)$ individual special values. Then we will show that these values can be realized by our $\gamma_1^{\mathcal{L}}, \gamma_2^{\mathcal{L}}, \ldots \gamma_K^{\mathcal{L}}$. With some additional consideration this yields that the classes $[\gamma_1^{\mathcal{L}}], [\gamma_2^{\mathcal{L}}], \ldots [\gamma_K^{\mathcal{L}}]$ constitute a basis of $\mathrm{H}^2_b(\mathcal{L}, \mathbb{C})$, the space of bounded cohomology classes.

For local cocycles we use the same principle technique as for function algebra. By switching the set of points I and O we will show that a linear combination of the above basis cocycle classes can only be local if all linear coefficients are the same. This means that a local cocycle class has to be a multiple of $[\gamma_S^{\mathcal{L}}]$, and moreover that the space $\mathrm{H}^2_{loc}(\mathcal{L}, \mathbb{C})$ is one-dimensional and generated by it.

6.8.2.1 Cohomological change

First we recall that a coboundary γ is given with the help of a linear map $\phi : \mathcal{L} \to \mathbb{C}$ by

$$\gamma(e, f) = d_1\phi(e, f) = \phi([e, f]). \tag{6.190}$$

The linear map will be fixed if we prescribe values

$$\phi(e_{n,p}) := \alpha_{n,p} \in \mathbb{C}, \tag{6.191}$$

for the basis elements $e_{n,p}$.

With respect to the Krichever–Novikov pairing $\langle\,,\,\rangle$ induced by splitting $A = I \cup O$ under consideration, we set

$$V := \sum_{n\in\mathbb{Z}} \sum_{p=1}^{K} \alpha_{n,p}\Omega^{n,p} \tag{6.192}$$

as (formal) infinite sum and the linear map ϕ is given

$$\phi(e) = \langle V, e\rangle = \sum_{n\in\mathbb{Z}} \sum_{p=1}^{K} \frac{1}{2\pi \mathrm{i}} \int_{C_S} \Omega^{n,p} e. \tag{6.193}$$

In particular, we obtain for the coboundary γ the expression

$$\gamma(e, f) = \phi([e, f]) = \sum_{n\in\mathbb{Z}} \sum_{p=1}^{K} \frac{1}{2\pi i} \int_{C_S} \Omega^{n,p}(ef' - fe')\frac{d}{dz}. \tag{6.194}$$

By almost-grading for the algebra $\mathcal{L}$ we have

$$\begin{aligned} [e_{0,p}, e_{n,p}] &= n \cdot e_{n,p} + Y(n, p), \\ [e_{-1,p}, e_{1,p}] &= 2 \cdot e_{0,p} + Z(p), \end{aligned} \tag{6.195}$$

with $Y(n, p)$ and $Z(p)$ finite linear combinations of elements of degrees higher than n and 0.

Let γ be an arbitrary cocycle bounded from above with bound M (which we assume for the moment to be > 0). In particular, we have $\gamma(e_{n,p}, e_{m,r}) = 0$ for $n + m > M$. We set

$$\phi(e_{n,p}) := 0, \quad \forall n > M, \; p = 1, \ldots, K. \tag{6.196}$$

Next we define recursively in descending order for $n = M, M - 1, \ldots, 1$

$$\phi(e_{n,p}) := \frac{1}{n}\left(\gamma(e_{0,p}, e_{n,p}) - \phi(Y(n, p))\right), \quad p = 1, \ldots, K. \tag{6.197}$$

For $n = 0$ we set

$$\phi(e_{0,p}) := \frac{1}{2}\left(\gamma(e_{-1,p}, e_{1,p}) - \phi(Z(p))\right), \quad p = 1, \ldots, K. \tag{6.198}$$

For $n < 0$ we use, again in descending order, (6.197). If $M = 0$ the first step is obsolete and we start with (6.198). If $M < 0$ we start with (6.197) in the negative range. Note that $\phi(Y(n, p))$ (respectively $\phi(Z(p))$) are at every step well-defined by the previous prescriptions.

We consider the cohomologous cocycle $\tilde{\gamma} = \gamma - d_1\phi$. For $n \neq 0$ we calculate

$$\begin{aligned} \tilde{\gamma}(e_{0,p}, e_{n,p}) &= \gamma(e_{0,p}, e_{n,p}) - \phi([e_{0,p}, e_{n,p}]) \\ &= \gamma(e_{0,p}, e_{n,p}) - \phi(n\, e_{n,p} + Y(n, p)) = 0. \end{aligned} \tag{6.199}$$

Similarly,

$$\tilde{\gamma}(e_{-1,p}, e_{1,p}) = 0. \tag{6.200}$$

We call a cocycle γ *normalized* if

$$\gamma(e_{0,p}, e_{n,p}) = 0, \quad \forall n \in \mathbb{Z}, \quad \gamma(e_{-1,p}, e_{1,p}) = 0, \quad p = 1, \ldots, K. \tag{6.201}$$

Note that $\gamma(e_{0,p}, e_{0,p}) = 0$ is automatic. Obviously, the cocycle $\tilde{\gamma}$ obtained from γ as above is a normalized and bounded cocycle, which brings us to the following proposition.

Proposition 6.66. *Every bounded cocycle for $\mathcal{L}$ is cohomologous to a bounded normalized one.*

Remark 6.67. In the classical case ($g = 0$ and $N = 2$) there are no additional terms in the relations (6.195). We can therefore make the cohomological changes given by (6.197) and (6.198) without assuming that γ is a bounded cocycle. In particular we obtain the statement of Proposition 6.66 for every cocycle for $\mathcal{L}$.

6.8.2.2 Induction

We use the almost-graded structure and perform induction on the level of the pairs of elements. Recall

$$[e_{k,r}, e_{n,s}] = \delta_r^s \cdot (n-k) \cdot e_{k+n,r} + \sum_{h=n+m+1}^{n+m+L_3} \sum_{t=1}^{K} c_{(k,r),(n,s)}^{(h,t)} e_{h,t}, \tag{6.202}$$

with $c_{(k,r),(n,s)}^{(h,t)} \in \mathbb{C}$, and L_3 the upper bound for the almost-graded structure of $\mathcal{L}$. In abbreviated form we denote this by

$$[e_{k,r}, e_{n,s}] \equiv \delta_r^s \cdot (n-k) \cdot e_{k+n,r}. \tag{6.203}$$

If we insert the almost-grading (6.202) into the cocycle condition

$$\gamma([e_{n,r}, e_{m,s}], e_{p,t}) + \gamma([e_{m,s}, e_{p,t}], e_{n,r}) + \gamma([e_{p,t}, e_{n,r}], e_{m,s}) = 0 \tag{6.204}$$

for triples of basis elements we obtain

$$\delta_r^s \cdot (m-n)\gamma(e_{n+m,r}, e_{p,t}) + \delta_t^s \cdot (p-m)\gamma(e_{m+p,s}, e_{n,r}) \\ + \delta_r^t \cdot (n-p)\gamma(e_{n+p,r}, e_{m,s}) \equiv 0. \tag{6.205}$$

Recall that $\equiv$ means up to universal combinations of cocycle values of higher level. If s, t, r are mutually distinct this does not produce any relation on the level $n+m+p$.

For $s = r \neq t$ we obtain

$$(m-n)\gamma(e_{n+m,r}, e_{p,t}) \equiv 0. \tag{6.206}$$

For $k \in \mathbb{Z}$ we set $m := \frac{k+1}{2}$, $n := \frac{k-1}{2}$ if k is odd, and $m := \frac{k+2}{2}$, $n := \frac{k-2}{2}$ if k is even. In both cases we obtain that $\gamma(e_{k,r}, e_{p,t}) \equiv 0$, which means it is of higher level.

Lemma 6.68. *For $r \neq t$ the value of the cocycle $\gamma(e_{k,r}, e_{p,t})$ is given as a universal linear combination of values of the cocycle at higher level.*

Hence, only $r = s = t$ remains to be examined. To simplify the notation we will drop the second index. We obtain

$$(m-n) \cdot \gamma(e_{n+m}, e_p) + (p-m) \cdot \gamma(e_{m+p}, e_n) + (n-p) \cdot \gamma(e_{n+p}, e_m) \equiv 0. \tag{6.207}$$

If we set $n = 0$ and use antisymmetry we obtain

$$(m+p)\gamma(e_m, e_p) + (p-m)\gamma(e_{m+p}, e_0) \equiv 0. \tag{6.208}$$

Hence,

Lemma 6.69. *If the level $l = m + p \neq 0$, then*

$$\gamma(e_m, e_p) \equiv \frac{m-p}{m+p} \cdot \gamma(e_{m+p}, e_0). \tag{6.209}$$

We point out that (6.209) is true for arbitrary cocycles (not necessarily bounded).

Proposition 6.70. *If γ is a normalized cocycle, then*
(a) *for all levels $l \neq 0$ the cocycle at level l is fixed by the cocycle values at higher level.*
(b) *Moreover, if γ is bounded, then it is bounded by zero.*

Proof. By definition, normalized implies that $\gamma(e_l, e_0) = 0$, hence claim (a). If we apply this in the bounded case for positive levels, then the cocycle will vanish there. Hence, it will be bounded by zero. □

Next we consider the level zero case. Clearly, $\gamma(e_0, e_0) = 0$ due to antisymmetry. Setting $p = -(n+1)$ and $m = 1$ in (6.207) we get

$$(n-1)\gamma(e_{n+1}, e_{-(n+1)}) \equiv (n+2)\gamma(e_n, e_{-n}) - (2n+1)\gamma(e_1, e_{-1}). \tag{6.210}$$

This recursion fixes the level zero starting from higher level and the values of $\gamma(e_1, e_{-1})$ and $\gamma(e_2, e_{-2})$.

Proposition 6.71.
(a) *For a cocycle γ, not necessarily normalized, for the vector field algebra the level zero is given by the data α_r and β_r for $r = 1, \dots, K$, fixed by*

$$\alpha_r = 1/6\left(\gamma(e_{2,r}, e_{-2,r}) - 2\gamma(e_{1,r}, e_{-1,r})\right), \quad \beta_r = \gamma(e_{1,r}, e_{-1,r}) \tag{6.211}$$

and higher level values via

$$\gamma(e_{k,r}, e_{-k,s}) \equiv \left((k+1)k(k-1)\alpha_r + k\beta_r\right)\delta_r^s. \tag{6.212}$$

(b) *If the cocycle γ is bounded from above it is uniquely given by the collection of values*

$$\begin{aligned} &\gamma(e_{1,r}, e_{-1,r}),\ \gamma(e_{2,r}, e_{-2,r}),\ r = 1, \dots, K, \\ &\gamma(e_{n,r}, e_{0,r}),\ n \in \mathbb{Z} \setminus \{0\},\ r = 1, \dots, K. \end{aligned} \tag{6.213}$$

Proof. (a) For $r \neq s$ Lemma 6.68 gives the claim. For $r = s$ Equation (6.210) gives the recursive relation. It remains to show the explicit formula. By antisymmetry it is enough to consider $k > 0$ and there it follows from induction starting with $k = 1$ and $k = 2$.

For part (b) let γ_1 and γ_2 be two cocycles bounded from above, which have the same set of values (6.213). Let M be a common upper bound. Recall that at level $l \neq 0$ the elements of this level are fixed as certain universal linear combinations of elements of level higher than l and the element $\gamma(e_0, e_l)$. Hence, the two cocycles coincide for every level $l > 0$. For level 0 we use part (a), hence they coincide on level 0 and later on every level. □

We specialize the above proposition to the bounded normalized case.

Proposition 6.72. *Let γ be a bounded normalized cocycle. Then γ is zero for all levels $l > 0$. At level zero it is fixed by the values of $\alpha_r = \frac{1}{6}\gamma(e_{2,r}, e_{-2,r})$, $r = 1, \ldots K$ and evaluates to*

$$\gamma(e_{k,r}, e_{-k,s}) = (k+1)k(k-1)\alpha_r \delta_r^s. \qquad (6.214)$$

Negative levels are fixed by higher level.

Proof. For normalized cocycles we have $\gamma(e_{-1,r}, e_{1,r}) = \gamma(e_{n,r}, e_{0,r}) = 0$, hence the claim. □

Proposition 6.73. *Two cocycles γ_1 and γ_2 which coincide on all levels $l \geq 0$ are cohomologous.*

Proof. The cocycle $\gamma = \gamma_1 - \gamma_2$ will be a bounded cocycle with bound $M < 0$. Hence, it is trivially normalized on levels ≥ 0. By the cohomological changes on negative levels made at the beginning of this section the cocycle γ will be cohomologous to a normalized $\tilde{\gamma}$. In Proposition 6.72 all α_r will be zero and the cocycle $\tilde{\gamma}$ will be zero. In other words, γ will be a coboundary. □

Proposition 6.74. *Let γ be a coboundary.*
(a) *If γ is bounded by M, then $\gamma = d_1 D_W$ with*

$$W = \sum_{n=-\infty}^{M} \sum_{r=1}^{K} \beta_{m,r} \Omega^{m,r}, \quad \beta_{m,r} \in \mathbb{C}. \qquad (6.215)$$

(b) *If γ is local, then $\gamma = d_1 D_W$ with W a finite sum.*

Proof. As γ is a coboundary, $\gamma = d_1\phi$, with $\phi : L \to \mathbb{C}$ a linear form. As shown earlier, ϕ can be given as a D_W, where $D_W(e) = \langle W, e\rangle$ with W a sum of the type (6.215). Now assume that γ is bounded by M. By the recursion given at the beginning of this section, e.g., by (6.196), we see that the outer sum is indeed of the form claimed in (6.215). This was claim (a). In the case of locality, we consider also the inverted grading obtained by interchanging the roles of I and O. Bounded from below corresponds now to bounded from above (see Section 3.7). Hence, the outer summation can be restricted to a finite strip and claim (b) follows. □

Remark 6.75. By the recursion argument we did not directly prove that for a bounded cocycle given as a both-sided infinite outer sum, the large index coefficients will vanish. Nevertheless, this will be true. Assume that the same coboundary is given by two (possibly different) sums. Consequently, the difference of these sums will be the zero coboundary. Hence, it will vanish on the derived subalgebra $[\mathcal{L}, \mathcal{L}]$. As $\mathcal{L}$ is perfect (see Proposition 6.99), it has to vanish everywhere. As the Krichever–Novikov pairing is non-degenerate (see Theorem 3.6), the difference must be identically zero. Hence, the representation as sum is unique.

Proposition 6.76. *Let γ be a normalized bounded coboundary, then $\gamma = 0$.*

Proof. As γ is normalized and bounded it will be bounded by zero. According to Proposition 6.74 it can be given by $\gamma = d_1\phi$ with $\phi(e) = 0$ for all $e \in \mathcal{L}_+$. Let the notation be as in Proposition 6.71. By the normalization condition we have $\beta_r = 0$. We also show that $\alpha_r = 0$. We calculate

$$\gamma(e_{2,r}, e_{-2,s}) = \phi([e_{2,r}, e_{-2,s}]) = (-4)\phi(e_{0,r})\delta_r^s.$$

But

$$0 = \gamma(e_{1,r}, e_{-1,s}) = \phi([e_{1,r}, e_{-1,s}] = (-2)\phi(e_{0,r})\delta_r^s.$$

Hence, $\phi(e_{0,r}) = 0$, and consequently $\alpha_r = \gamma(e_{2,r}, e_{-2,r}) = 0$. □

Theorem 6.77. *The cohomology space of cocycles bounded from above is K-dimensional and generated by the classes of the geometric cocycles $[\gamma_{r,R}^{\mathcal{L}}]$, $r = 1, \ldots, K$.*

Proof. By Proposition 6.66 every bounded cocycle is cohomologous to a normalized one. Hence, with Proposition 6.72 it follows that the space of bounded cohomology classes is at most K-dimensional.

We might assume that the projective connection R, used in the definition of the geometric cocycles $\gamma_{r,R}^{\mathcal{L}}$, is holomorphic. Recall that the choice of another connection (possibly with poles at the points in A) will only change the cohomology class. For R holomorphic the $\gamma_{r,R}^{\mathcal{L}}$ will be bounded from above by zero. Moreover,

$$\gamma_{r,R}^{\mathcal{L}}(e_{-1,p}, e_{1,p}) = 0, \quad \forall p = 1, \ldots, K. \tag{6.216}$$

Let γ be an arbitrary bounded cocycle. By cohomological changes we can assume that it is normalised (in fact the normalization requirement for all levels $l \geq 0$ will suffice). Hence (as normalized cocycle), it is bounded by zero. For the geometric cocycles we calculate (see (6.127))

$$\gamma_{r,R}^{\mathcal{L}}((e_{2,r}, e_{-2,p}) = \frac{1}{2}\delta_r^p. \tag{6.217}$$

We consider the cocycle

$$\tilde{\gamma} = \gamma - \sum_{l=1}^{K} 2\gamma(e_{2,l}, e_{-2,l}) \cdot \gamma_{l,R}^{\mathcal{L}}. \tag{6.218}$$

It is also bounded by zero and by construction and normalization

$$\tilde{\gamma}(e_{2,p}, e_{-2,p}) = \tilde{\gamma}(e_{1,p}, e_{-1,p}) = 0, \quad p = 1, \ldots, K. \tag{6.219}$$

This means that γ and the sum coincide at all levels $l \geq 0$ (see Proposition 6.71), and by Proposition 6.73 they are cohomologous. Consequently, $[\gamma]$ is a linear combination of the geometric cocycle classes $[\gamma_{i,R}^{\mathcal{L}}]$.

It remains to show that the geometric cocycles are linearly independent as cocycle classes. Assume that we have a linear combination

$$\sum_{i=1}^{K} \alpha_i [\gamma_{i,R}^{\mathcal{L}}] = [0], \tag{6.220}$$

which says

$$\sum_{i=1}^{K} \alpha_i \gamma_{i,R}^{\mathcal{L}} = d_1\phi, \tag{6.221}$$

with ϕ a linear form on $\mathcal{L}$. We have to show that all $\alpha_i = 0$. First note that all $\gamma_{i,R}^{\mathcal{L}}$ are bounded by zero. Hence, this has to be the case for the coboundary on the right-hand side too. Recall that a coboundary which is bounded by zero can be described as

$$d_1\phi(e,f) = \sum_{k\le 0}\sum_{s=1}^{K} \alpha_{k,s} \frac{1}{2\pi \mathrm{i}} \int_{C_S} \Omega^{k,s}([e,f]). \tag{6.222}$$

Here we used Proposition 6.74. We evaluate (6.221) for pairs $(e_{n,p}, e_{-n,p})$ with $n > 0$ and p arbitrary but fixed. This yields

$$\begin{aligned}
\sum_{i=1}^{K} \alpha_i \frac{n^3-n}{12}\,\delta_i^p &= \sum_{k\le 0}\sum_{s=1}^{K} \alpha_{k,s} \frac{1}{2\pi \mathrm{i}} \int_{C_S} \Omega^{k,s}([e_{n,p}, e_{-n,p}]) \\
&= \sum_{k\le 0}\sum_{s=1}^{K} \alpha_{k,s} \frac{1}{2\pi \mathrm{i}} \int_{C_S} \Omega^{k,s}(-2ne_{0,p} + h.t.) \\
&= \sum_{k\le 0}\sum_{s=1}^{K} \alpha_{k,s}(-2n)\delta_k^0\delta_s^p = \alpha_{0,p}(-2n).
\end{aligned} \tag{6.223}$$

Here we used the Krichever–Novikov duality $\langle \Omega^{k,s}, h.t.\rangle = 0$ for $k \le 0$. The relation (6.223) is valid for all n. This is only possible if all α_i (and all $\alpha_{0,p}$) are zero. Hence, the cocycle classes are linearly independent. □

This also supplies a proof of Proposition 6.37 for the vector field case.

Proof of Theorem 6.41. Part (a) is Theorem 6.77. Next we show (b). If γ is a cocycle which is a multiple of the separating cocycle, then it will be local; see Proposition 6.33. Here we have to show the opposite. Given a local cocycle γ', then by Theorem 6.77 is can be represented up to coboundary as a linear combination

$$\gamma = \sum_{i=1}^{K} \alpha_i \gamma_i, \quad \alpha_i \in \mathbb{C}. \tag{6.224}$$

We have to show that if γ is a local cocycle, then $\alpha_1 = \alpha_2 = \cdots = \alpha_K$. As reference connection R we fix a holomorphic projection connection. The strategy of the proof is similar to the function case, i.e., we consider the inverted grading obtained by switching I and O. The steps are slightly more complicated as we have to deal with coboundaries too. Again, we denote the cocycles obtained by integrating around the point Q_j in O by γ_j^* (always with respect to the fixed projective connection R). If $K = 1$, then nothing has to be shown. Hence, we assume that $K > 1$. The cocycle γ is also bounded from below. We invert the grading by switching I and O and with respect to this new grading γ is bounded from above. Hence, we can again use our result about bounded cocycles and write (with $L = N - K$)

$$\gamma = \sum_{j=1}^{L} \alpha_j^* \gamma_j^* + D_{W^*} \quad \alpha_j^* \in \mathbb{C}. \tag{6.225}$$

Here D_{W^*} is the coboundary needed in the presentation by the standard geometric cocycles (always with respect to inverted grading). The pole orders in W^* at the points $Q_j \in O$ are bounded. The bounds only depend on γ. We have the identity

$$0 = \sum_{i=1}^{K} \alpha_i \gamma_i - \sum_{j=1}^{L} \alpha_j^* \gamma_j^* - D_{W^*}. \tag{6.226}$$

Recall that by the residue theorem

$$0 = \sum_{i=1}^{K} \gamma_i - \sum_{j=1}^{L} \gamma_j^*. \tag{6.227}$$

Now we make reference to the first index $i = 1$. Subtracting α_1 times (6.227) from (6.226) yields

$$0 = \sum_{i=2}^{K} (\alpha_i - \alpha_1)\gamma_i - \sum_{j=1}^{L} (\alpha_j^* - \alpha_1)\gamma_j^* - D_{W^*}. \tag{6.228}$$

Take k from $\{2, 3, \ldots, K\}$. According to Riemann–Roch a vector field e exists such that $\operatorname{ord}_{P_k}(e) = 3$. e might have poles at the point P_1, and is holomorphic elsewhere. In addition, we require that the e has zeros of high enough orders at the points in O to compensate for possible poles coming from W^*. Similarly, let f be an element with $\operatorname{ord}_{P_k}(f) = -1$ fulfilling the same conditions at the other points as e. We normalize e and f so that they have leading coefficients 1 at the point P_k. Note that freedom at the point P_1 is essential for existence. Also, via W^* the elements might depend on the γ. We insert the pair (e, f) into (6.228)

$$0 = \sum_{i=2}^{K} (\alpha_i - \alpha_1)\gamma_i(e, f) - \sum_{j=1}^{L} (\alpha_j^* - \alpha_1)\gamma_j^*(e, f) - D_{W^*}(e, f). \tag{6.229}$$

As both e and f are holomorphic at the points in O, the second sum in (6.229) will vanish. Also the zero orders are high enough that the third term will not contribute. From the first sum only

$$0 = (\alpha_k - \alpha_1)\gamma_k(e, f) \tag{6.230}$$

survives. A direct calculation shows $\gamma_k(e, f) = 1/2$. Hence, $\alpha_k = \alpha_1$ and this is true for all k. This shows the claim that γ is cohomologous to a multiple of the separating cocycle. This was (b).

(d) The explicit form (6.141) for the cocycle values at level zero for a cocycle which is local and bounded by zero follows from (6.211) and (6.212). Note that if γ is essentially equal to a multiple of the separating cocycle, all α_r have the same value; see also (6.132). Recall that the b_r come from possible poles of the connection of order 2 at the points P_r. □

6.8.3 Mixing cocycle for the differential operator algebra

For the differential operator algebra we demonstrated in Section 6.5.3 that the cocycles can be written as the sum of three different types: $\mathcal{L}$-invariant cocycles for function algebra, vector field algebra cocycles, and mixing cocycles. The classification results of the first two cocycles have been shown in the previous sections. The mixing cocycles remain, i.e., those cocycles defined for the differential operator algebra $\mathcal{D}^1$ which vanish on the subalgebra $\mathcal{A}$ of functions and the subalgebra $\mathcal{L}$ of vector fields.

Our strategy is similar to the case of the vector field algebra.

6.8.3.1 Cohomological change

First we again discuss a cohomological change. By the almost-grading for the algebra $\mathcal{D}^1$ we have

$$\begin{aligned} [e_{0,p}, A_{n,p}] &= n \cdot A_{n,p} + Y(n,p), \\ [e_{-1,p}, A_{1,p}] &= A_{0,p} + Z(p), \end{aligned} \tag{6.231}$$

with $Y(n,p)$ and $Z(p)$ finite linear combinations of elements of degrees greater than n and 0.

Let γ be a mixing cocycle bounded from above with bound M (which we assume for the moment to be > 0). In particular we have $\gamma(e_{n,p}, A_{m,r}) = 0$ for $n+m > M$. We set

$$\phi(A_{n,p}) := 0, \quad \forall n > M, \; p = 1, \ldots, K. \tag{6.232}$$

Next we define recursively in descending order for $n = M, M-1, \ldots, 1$

$$\phi(A_{n,p}) := \frac{1}{n}\left(\gamma(e_{0,p}, A_{n,p}) - \phi(Y(n,p))\right), \quad p = 1, \ldots, K. \tag{6.233}$$

For $n = 0$ we set

$$\phi(A_{0,p}) := \left(\gamma(e_{-1,p}, A_{1,p}) - \phi(Z(p))\right), \quad p = 1, \ldots, K. \tag{6.234}$$

For $n < 0$ we use (6.233), again in descending order. If $M = 0$ the first step is obsolete and we start with (6.234). If $M < 0$ we start with (6.233) in the negative range. Every step is well-defined by the previous prescriptions.

We consider the cohomologous cocycle $\tilde{\gamma} = \gamma - d_1\phi$. For $n \neq 0$ we calculate as above

$$\tilde{\gamma}(e_{0,p}, A_{n,p}) = 0, \qquad \tilde{\gamma}(e_{-1,p}, A_{1,p}) = 0. \tag{6.235}$$

Again, we call a cocycle γ normalized if

$$\begin{aligned} \gamma(e_{0,p}, A_{n,p}) &= 0, \quad \forall n \in \mathbb{Z}\, n \neq 0, \\ \gamma(e_{-1,p}, A_{1,p}) &= 0, \quad p = 1, \ldots, K. \end{aligned} \tag{6.236}$$

The cocycle $\tilde{\gamma}$ obtained from γ as above is a normalized and bounded cocycle, which brings us to the next proposition.

Proposition 6.78. *Every bounded mixing cocycle is cohomologous to a bounded normalized one.*

Remark 6.79. In the classical case ($g = 0$ and $N = 2$) in the relations (6.231) there are no additional terms. Hence, we can make the cohomological changes given by (6.233) and (6.234) without assuming that γ is a bounded cocycle. Consequently, in the classical case we can drop the boundedness condition in the previous proposition.

6.8.3.2 Induction

For the moment let γ be an arbitrary cocycle. We start with the cocycle relation for $e, f \in \mathcal{L}$ and $g \in \mathcal{A}$

$$\gamma([e,f],g) - \gamma(e, f.g) + \gamma(f, e.g) = 0, \tag{6.237}$$

which we evaluate for the basis elements

$$\gamma([e_{k,r}, e_{n,s}], A_{m,t}) - \gamma(e_{k,r}, e_{n,s}.A_{m,t}) + \gamma(e_{n,s}, e_{k,r}.A_{m,t}) = 0. \tag{6.238}$$

We use the almost-graded structure

$$[e_{k,r}, A_{n,s}] \equiv \delta_r^s \, n \, A_{k+n,r}. \tag{6.239}$$

If we insert the almost-grading (6.239) into the cocycle condition we get

$$\delta_r^s \cdot (n-k) \cdot \gamma(e_{k+n,r}, A_{m,t}) - \delta_s^t \cdot m \cdot \gamma(e_{k,r}, A_{m+n,s}) + \delta_r^t \cdot m \cdot \gamma(e_{n,s}, A_{m+k,t}) \equiv 0. \tag{6.240}$$

If all r, s, t are mutually distinct, this does not produce any relation on this level. For $s = t \neq r$, $m = -1$, and $n = p + 1$ we obtain $\gamma(e_{k,r}, A_{p,s}) \equiv 0$. Hence,

Lemma 6.80. *The cocycle values $\gamma(e_{k,r}, A_{p,s})$ for $r \neq s$ can be expressed as universal linear combinations of cocycle values of higher level.*

Again, we use the phrase "universal linear combination" to denote the situation explained in Remark 6.55.

This shows that it is enough to consider elements with the same second index. We will drop it in the notation. The equation (6.240) will now be written as follows:

$$(n-k) \cdot \gamma(e_{k+n}, A_m) - m \cdot \gamma(e_k, A_{m+n}) + m \cdot \gamma(e_n, A_{m+k}) \equiv 0. \tag{6.241}$$

Setting $k = 0$ in (6.241) yields

$$(n+m)\gamma(e_n, A_m) \equiv m\gamma(e_0, A_{m+n}). \tag{6.242}$$

Lemma 6.81.

(a) *If the level $l = (n+m) \neq 0$, then*

$$\gamma(e_n, A_m) \equiv \frac{m}{n+m} \cdot \gamma(e_0, A_{n+m}). \tag{6.243}$$

(b) $\gamma(e_n, A_0) \equiv 0$ *for all* $n \in \mathbb{Z}$.

Proof. Part (a) is obtained by dividing (6.242) by $n + m \neq 0$. For $n \neq 0$ we obtain (b) by setting $m = 0$. We set $m = 1$ and $n = -1$ in (6.242) and get (b), also for $n = 0$. □

In case γ is a normalized cocycle, then at level $l \neq 0$ everything is fixed by higher level. Recall that if the cocycle is bounded, respectively if we are in the classical situation every cocycle is cohomologous to a normalized one.

Dealing with the level 0 case remains. We set $k = -n - m$ in (6.241) and obtain

$$(2n+m) \cdot \gamma(e_{-m}, A_m) - m \cdot \gamma(e_{-(n+m)}, A_{n+m}) + m \cdot \gamma(e_n, A_{-n}) \equiv 0. \tag{6.244}$$

We specialize further to $m = 1$ and $m = -1$

$$(2n+1)\gamma(e_{-1}, A_1) - \gamma(e_{-(n+1)}, A_{n+1}) + \gamma(e_n, A_{-n}) \equiv 0 \tag{6.245}$$
$$(2n-1)\gamma(e_1, A_{-1}) + \gamma(e_{-(n-1)}, A_{n-1}) - \gamma(e_n, A_{-n}) \equiv 0. \tag{6.246}$$

Adding these equations yields

$$\gamma(e_{-(n+1)}, A_{n+1}) \equiv \gamma(e_{-(n-1)}, A_{n-1}) + (2n-1) \cdot \gamma(e_1, A_{-1}) + (2n+1) \cdot \gamma(e_{-1}, A_1). \tag{6.247}$$

Recall that $\gamma(e_0, A_0) \equiv 0$, hence the values on level zero are uniquely fixed by $\gamma(e_1, A_{-1})$ and $\gamma(e_{-1}, A_1)$. We use

$$\alpha := 1/2\left(\gamma(e_1, A_{-1}) + \gamma(e_{-1}, A_1)\right), \quad \beta := -\gamma(e_{-1}, A_1), \tag{6.248}$$

and obtain

$$\gamma(e_{-(n+1)}, A_{n+1}) = \gamma(e_{-(n-1)}, A_{n-1}) + 2(2n-1)\alpha - 2\beta. \tag{6.249}$$

The starting elements of the recursion are

$$\gamma(e_0, A_0) \equiv 0, \quad \gamma(e_{-1}, A_1) = -\beta. \tag{6.250}$$

By induction this implies

$$\gamma(e_{-n}, A_n) \equiv n(n-1)\alpha - n\beta. \tag{6.251}$$

Proposition 6.82.

(a) *For a mixing cocycle γ (not necessarily normalized), the level 0 is fixed by the data*

$$\begin{aligned} \alpha_r &:= 1/2\left(\gamma(e_{1,r}, A_{-1,r}) + \gamma(e_{-1,r}, A_{1,r})\right), \\ \beta_{0,r} &= -\gamma(e_{-1,r}, A_{1,r}), \end{aligned} \tag{6.252}$$

for $r = 1, \dots, K$ and higher level values via

$$\gamma(e_{-n,r}, A_{n,s}) \equiv (n(n-1)\alpha_r - n\beta_{0,r}) \cdot \delta_s^r. \tag{6.253}$$

(b) *If γ is mixing cocycle which is bounded from above it is uniquely given by the collection of values*

$$\begin{aligned} &\gamma(e_{1,r}, A_{-1,r}),\ \gamma(e_{-1,r}, A_{1,r}),\ r = 1, \dots, K, \\ &\gamma(e_{0,r}, A_{n,r}),\ n \in \mathbb{Z} \setminus \{0\}. \end{aligned} \tag{6.254}$$

Proof. Our analysis above works for every r separately. Lemma 6.80 gives the statement for $r \neq s$. This shows (a).

Let γ_1 and γ_2 be two cocycles bounded from above which have the same set of values (6.254). Let M be a common upper bound. Recall that at level $l \neq 0$ the elements of this level are fixed as certain universal linear combinations of elements of levels higher than l and the elements $\gamma(e_0, A_l)$. Hence, the two cocycles coincide for every level $l > 0$. For level 0 we use part (a), hence they coincide on level 0 and later on every level. □

We specialize the above proposition to the bounded normalized case.

Proposition 6.83. *Let γ be a bounded normalized cocycle. Then γ is zero for all levels $l > 0$.*

At level zero it is fixed by the values of $\alpha_r = \frac{1}{2}\gamma(e_{1,r}, A_{-1,r})$, $r = 1, \ldots, K$ and evaluates to

$$\gamma(e_{k,r}, A_{-k,s}) = k(k+1)\alpha_r \delta_r^s. \tag{6.255}$$

Negative levels are fixed by higher level.

Proof. For normalized cocycles we have $\gamma(e_{-1,r}, A_{1,r}) = 0$, hence the claim □

The proof of Proposition 6.73 works also word by word for the following proposition.

Proposition 6.84. *Two cocycles γ_1 and γ_2 which coincide on all levels $l \geq 0$ are cohomologous.*

Proposition 6.85. *Let γ be coboundary.*
(a) *If γ is bounded by M, then $\gamma = d_1 E_V$ with*

$$V = \sum_{n=-\infty}^{M} \sum_{r=1}^{K} \beta_{m,r} \omega^{m,r}, \quad \beta_{m,r} \in \mathbb{C}. \tag{6.256}$$

(b) *If γ is local then $\gamma = d_1 E_V$ with V a finite sum.*

The proof of the above proposition works completely analogous to the proof of Proposition 6.74.

Proposition 6.86. *Let γ be a normalized bounded coboundary, then $\gamma = 0$.*

Proof. Let $\gamma = d_1\phi$. As γ is normalized, we may assume with Proposition 6.85 that $\gamma = d_1\phi$ with $\phi = 0$ on $\mathcal{A}_+$. Furthermore, $\gamma_r(e_{-1,r}, A_{1,r}) = 0$. We calculate

$$\gamma(e_{1,r}, A_{-1,r}) = \phi(e_{1,r} \,.\, A_{-1,r}) = -\phi(A_{0,r}).$$

But

$$0 = \gamma(e_{-1,r}, A_{1,s}) = \phi(e_{-1,r} \,.\, A_{1,s}) = \phi(A_{0,r}).$$

Hence, $\phi(A_{0,r}) = 0$ and consequently $\gamma(e_{-1,r}, A_{1,r}) = 0$. Which results in the cocycle vanishing identically. □

Theorem 6.87. *The cohomology space of mixing cocycles bounded from above is K-dimensional and generated by the classes $[\gamma_{r,T}^{(m)}]$, $r = 1, \ldots, K$.*

Recall that $\gamma_{r,T}^{(m)}$ denotes the cocycle obtained by (6.108), where we integrate over C_r using our fixed reference affine connection T.

Proof. From Proposition 6.83 it follows that the space of bounded cohomology classes is at most K-dimensional. We might assume that the affine connection T used in the definition of the geometric cocycles $\gamma_{r,T}^{(m)}$ has maximally a pole of order 1 at the point $Q_L \in O$ and is holomorphic elsewhere. Recall that the choice of another connection (possibly with poles at the points in A) will only change the cohomology class. For such a T the $\gamma_{r,T}^{(m)}$ will be bounded from above by zero. Moreover,

$$\gamma_{r,T}^{(m)}(e_{-1,p}, A_{1,p}) = 0, \quad \forall p = 1, \ldots, K. \tag{6.257}$$

Let γ be an arbitrary bounded cocycle. By cohomological changes we can assume that it is normalized (in fact the normalization requirement for all levels $l \geq 0$ will suffice). For the geometric cocycles we calculate (see (6.127))

$$\gamma_{r,T}^{(m)}(e_{1,r}, A_{-1,p}) = 2\delta_r^p. \tag{6.258}$$

We consider the cocycle

$$\tilde{\gamma} = \gamma - \sum_{l=1}^{K} \frac{1}{2}\gamma(e_{1,l}, A_{-1,l}) \cdot \gamma_{l,T}^{(m)}. \tag{6.259}$$

It is also bounded by zero and by construction

$$\tilde{\gamma}(e_{1,p}, A_{-1,p}) = \tilde{\gamma}(e_{-1,p}, A_{1,p}) = 0, \quad p = 1, \ldots, K. \tag{6.260}$$

This means that γ and the sum coincide at all levels $l \geq 0$, and by Proposition 6.84 they are cohomologous. Consequently, $[\gamma]$ is a linear combination of the geometric cocycle classes $\gamma_{i,R}^{\mathcal{L}}$.

It remains to show that the geometric cocycles are linearly independent as cocycle classes. Assume that we have a linear combination

$$\sum_{i=1}^{K} \alpha_i [\gamma_{i,T}^{(m)}] = [0], \tag{6.261}$$

which says

$$\sum_{i=1}^{K} \alpha_i \gamma_{i,T}^{(m)} = d_1\phi. \tag{6.262}$$

We have to show that all $\alpha_i = 0$. First note that all $\gamma_{i,R}^{\mathcal{L}}$ are bounded by zero. Hence, this has to be the case for the coboundary on the right-hand side too. Recall that such a coboundary can be described as (see Proposition 6.85)

$$d_1\phi(e, A) = \sum_{k \leq 0} \sum_{s=1}^{K} \alpha_{k,s} \frac{1}{2\pi \mathrm{i}} \int_{C_S} \omega^{k,s}(e \,.\, A), \tag{6.263}$$

using the same kind of arguments as in the vector field case.

We evaluate (6.262) for the pairs $(e_{n,p}, A_{-n,p})$ for $n > 0$. This yields

$$\sum_i \alpha_i n(n+1)\,\delta_i^p = \sum_{k\le 0}\sum_{s=1}^{K} \alpha_{k,s}\frac{1}{2\pi \mathrm{i}}\int_{C_S} \omega^{k,s}([e_{n,p}, A_{-n,p}])$$

$$= \sum_{k\le 0}\sum_{s=1}^{K} \alpha_{k,s}\frac{1}{2\pi \mathrm{i}}\int_{C_S} \omega^{k,s}(-nA_{0,p} + h.t.) = \alpha_{0,p}(-n). \tag{6.264}$$

The relation (6.264) is valid for all n. This is only possible if all α_i (and all $\alpha_{0,p}$) are zero. Hence, the cocycle classes are linearly independent. □

Proof of Theorem 6.44. Theorem 6.87 shows the statement (a). The proof of statement (b) is completely analogous in its structure to the vector field case. Now we fix T an affine connection which might have a pole of order 1 at one point in O. Instead of the linear form D_{W^*} on $\mathcal{L}$, we have a linear form E_{V^*} on $\mathcal{A}$ to describe the coboundary. To show that all α_k are equal to α_1, we have to consider a pair (e, A). Here e is a vector field e such that $\operatorname{ord}_{P_k}(e) = 2$, e might have poles at the point P_1, and is holomorphic elsewhere. In addition, we require that the e has zeros of high enough orders at the points in O to compensate for possible poles coming from V^* and from the affine reference connection. Similarly, let A be a function with $\operatorname{ord}_{P_k}(A) = -1$, fulfilling the same conditions at the other points as e. We normalize e and A so that they have leading coefficients 1 at the point P_k. Again, such elements exist by Riemann–Roch and the fact that we allow suitable high pole orders at the point P_1. Exactly the same kind of arguments as in the vector field case with $\gamma_k(e, A) = 2$, showing that $\alpha_k = \alpha_1$.

Also, the arguments for the explicit form on level zero formulated in statement (c) work as in the vector field case. In particular, (6.146) follows from (6.253). □

6.9 Central extensions – the supercase

In this section we consider central extensions of our Lie superalgebra $\mathcal{S}$. Such a central extension is given by a bilinear map

$$\gamma : \mathcal{S} \times \mathcal{S} \to \mathbb{C} \tag{6.265}$$

via an expression completely analogous to (6.26). Additional conditions for γ follow from the fact that the resulting extension should again be a superalgebra. This implies that for the homogeneous elements $x, y, z \in \mathcal{S}$ ($\mathcal{S}$ might be an arbitrary Lie superalgebra), we have

$$\gamma(y, x) = -(-1)^{\bar{x}\bar{y}}\gamma(x, y). \tag{6.266}$$

Here $\bar{x}$ denotes as usual the parity of x, etc.

The bilinear map γ will be symmetric if x and y are odd, otherwise it will be alternating. The super-cocycle condition reads in complete analogy with the super-Jacobi relation as

$$(-1)^{\bar{x}\bar{z}}\gamma(x,[y,z]) + (-1)^{\bar{y}\bar{x}}\gamma(y,[z,x]) + (-1)^{\bar{z}\bar{y}}\gamma(z,[x,y]) = 0. \tag{6.267}$$

As we will need it anyway I will write it out for the different types of arguments. For *(even,even,even)*, *(even,even, odd)*, and *(odd,odd,odd)* it will be of the "usual form" of the cocycle condition

$$\gamma(x,[y,z]) + \gamma(y,[z,x]) + \gamma(z,[x,y]) = 0. \tag{6.268}$$

For *(even,odd,odd)* we obtain

$$\gamma(x,[y,z]) + \gamma(y,[z,x]) - \gamma(z,[x,y]) = 0. \tag{6.269}$$

Now we have to decide which parity our central element should have. In our context the natural choice is that the central element should be even, as we want to extend the central extension of the vector field algebra to the superalgebra. This implies that our bilinear form γ has to be an even form. Consequently,

$$\gamma(x,y) = \gamma(y,x) = 0, \quad \text{for } \bar{x} = 0,\ \bar{y} = 1. \tag{6.270}$$

In this case, from the super-cocycle conditions only (6.269) for the *(even,odd,odd)*, and (6.268) for the *(even,even,even)* case will give relations which are not nontrivially zero. In Section 6.9.2 we will consider the case when the central element is of odd parity.

Given a linear form $\kappa : \mathcal{S} \to \mathbb{C}$ we assign to it

$$d_1 k(x,y) = k([x,y]). \tag{6.271}$$

As in the classical case, $d_1\kappa$ will be a super-cocycle. A super-cocycle will be a coboundary if and only if a linear form $\kappa : \mathcal{S} \to \mathbb{C}$ exists such that $\gamma = d_1\kappa$. As κ is a linear form, it can be written as $\kappa = \kappa_{\bar{0}} \oplus \kappa_{\bar{1}}$, where $\kappa_{\bar{0}} : \mathcal{S}_{\bar{0}} \to \mathbb{C}$ and $\kappa_{\bar{1}} : \mathcal{S}_{\bar{1}} \to \mathbb{C}$. Again we have the two cases of the parity of the central element. Let γ be a coboundary $d_1\kappa$. If the central element is even then γ will also be a coboundary of a κ with $\kappa_{\bar{1}} = 0$. In other words κ is even. In the odd case we have $\kappa_{\bar{0}} = 0$ and κ is odd.

After fixing a parity of the central element we consider the quotient spaces

$$\mathrm{H}^2_{\bar{0}}(\mathcal{S},\mathbb{C}) := \{\text{even cocycles}\}/\{\text{even coboundaries}\}, \tag{6.272}$$

$$\mathrm{H}^2_{\bar{1}}(\mathcal{S},\mathbb{C}) := \{\text{odd cocycles}\}/\{\text{odd coboundaries}\}. \tag{6.273}$$

These cohomology spaces classify central extensions of $\mathcal{S}$ with even (or respectively odd) central elements up to equivalence. Equivalence is defined as in the non-super setting.

For the rest of this section our algebra $\mathcal{S}$ will be the Lie superalgebra introduced in Section 2.8. Moreover, for the moment we concentrate on the case of an even central element t. Recall our convention to denote vector fields by $e, f, g, \ldots$, and (-1/2)-forms by $\varphi, \psi, \chi, \ldots$. From the discussion above we know

$$\gamma(e,\varphi) = 0, \quad e \in \mathcal{L},\ \varphi \in \mathcal{F}^{-1/2}. \tag{6.274}$$

The super-cocycle condition for the even elements is just the cocycle condition for the Lie subalgebra $\mathcal{L}$. The only other nonvanishing super-cocycle condition is for the *(even,odd,odd)* elements, and reads as

$$\gamma(e,[\varphi,\psi]) - \gamma(\varphi, e\,.\,\psi) - \gamma(\psi, e\,.\,\varphi) = 0. \tag{6.275}$$

Here the definition of the product $[e,\psi] := e\,.\,\psi$ was used to rewrite (6.269).

In particular, if we have a cocycle γ for the algebra $\mathcal{S}$ we obtain by restriction a cocycle for the algebra $\mathcal{L}$. For the mixing term we know that $\gamma(e,\psi) = 0$. A naive attempt to put just anything for $\gamma(\varphi,\psi)$ will not work, as (6.275) relates the restriction of the cocycle on $\mathcal{L}$ with its values on $\mathcal{F}^{-1/2}$.

Proposition 6.88. *Let C be any closed (differentiable) curve on Σ not meeting the points in A, and let R be any (holomorphic) projective connection, then the bilinear extension of*

$$\begin{aligned}
\Phi_{C,R}(e,f) &:= \frac{1}{24\pi\mathrm{i}}\int_C \left(\frac{1}{2}(e'''f - ef''') - R\cdot(e'f - ef')\right)dz \\
\Phi_{C,R}(\varphi,\psi) &:= -\frac{1}{24\pi\mathrm{i}}\int_C \left(\varphi''\cdot\psi + \varphi\cdot\psi'' - R\cdot\varphi\cdot\psi\right)dz \\
\Phi_{C,R}(e,\varphi) &:= 0
\end{aligned} \tag{6.276}$$

gives a Lie superalgebra cocycle for $\mathcal{S}$, hence defines a central extension of $\mathcal{S}$.

Proof. First, one has to show that the integrands are well-defined differentials. It is exactly this point for which R had to be introduced. This was done for the vector field case in Proposition 6.16. The case for the second part of (6.276) is completely analogous, and follows from straightforward calculations. We will not write them down here.

Next the super-Jacobi identities have to be verified. The first one (6.268) was shown in Proposition 6.21. For the other super-Jacobi identity (6.275), we write out the three integrands and add them up (before we integrate over C). By direct calculation we find that the term coming with the projective connection will identically vanish. Hence the rest will be a well-defined meromorphic differential, It will not necessarily vanish identically. We only claim that the sum will vanish after integration over an arbitrary closed curve. In fact, the meromorphic differential will be a total differential of a meromorphic function with poles only at A. Hence, by (6.77) the integral will vanish.

In a first step we calculate

$$\begin{aligned}
e'''f &= \frac{1}{2}(e'''f - ef''') + 1/2((ef)'' - 3(e'f'))' \\
-2(\varphi'\psi') &= (\varphi''\psi + \varphi\psi'') - \left((\varphi\psi)'\right)'.
\end{aligned}$$

Hence, we can replace the corresponding integrands in the cocycle expressions by integrands given by the left-hand side. The total integrand of (6.275) can now be written as Bdz with

$$B := e'''(\varphi\psi) - 2\varphi'(e\psi' - 1/2e'\psi)' - 2\psi'(e\varphi' - 1/2e'\varphi)'. \tag{6.277}$$

From this we calculate

$$B\,dz = (e''(\varphi\psi) - 2(e\varphi'\psi'))'\,dz. \tag{6.278}$$

Consequently, Bdz integrated over any closed curve C will vanish. This shows that the cocycle condition (6.275) is true. □

A formula similar to (6.276) was given by Bryant in [31]. By adding the projective connection in the second part of (6.276) he corrected some formula appearing in the work of Bonora and collaborators [21]. In both articles only the two-point case and the integration over a separating cycle were considered. See also Kreusch [137] for the multipoint case, where still only the integration over a separating cycle is considered.

How will the central extension depend on C and R? Obviously, two cycles lying in the same homology class class of $\Sigma \setminus A$ will define the same cocycle.

Proposition 6.89. *If R and R' are two projective connections, then $\Phi_{C,R}$ and $\Phi_{C,R'}$ are cohomologous. Hence the cohomology class will not depend on the choice of R.*

Proof. The difference between two projective connections is a quadratic differential, $\Omega = R' - R$. We calculate (as always, using the same symbol for the section and its representing function)

$$\begin{aligned}\Phi_{C,R'}(e,f) - \Phi_{C,R}(e,f) &= \frac{1}{24\pi\mathrm{i}}\int_C \Omega\cdot(ef' - e'f)dz = \frac{1}{24\pi\mathrm{i}}\int_C \Omega\cdot[e,f],\\ \Phi_{C,R'}(\varphi,\psi) - \Phi_{C,R}(\varphi,\psi) &= \frac{1}{24\pi\mathrm{i}}\int_C \Omega\cdot(\varphi\cdot\psi)dz = \frac{1}{24\pi\mathrm{i}}\int_C \Omega\cdot[\varphi,\psi].\end{aligned} \tag{6.279}$$

If we fix the quadratic differential Ω, then the map

$$\kappa_C : \mathcal{L} \to \mathbb{C}, \quad e \mapsto \frac{1}{24\pi\mathrm{i}}\int_C \Omega\cdot e, \tag{6.280}$$

is a linear map. We extend this map by zero on $\mathcal{F}^{-1/2}$ and using (6.279) find that

$$\Phi_{C,R'} - \Phi_{C,R} = d_1\kappa_C. \tag{6.281}$$

Hence both cocycles are cohomologous. □

In the following section we will approach the classification problem. We recall here the results published in [222]. As in the pure vector field case for the non-classical situation, there will be many inequivalent central extensions given by different cycle classes as integration paths. Recall that classical means $g = 0$ and $N = 2$.

We have to restrict our attention to bounded, respectively local cocycles with respect to an almost-grading. We will need the special integration paths C_i, (C_j^*), the circles around the points $P_i \in I$ ($Q_j \in O$), and C_S a separating cycle. Recall from (3.7) the identity

$$[C_S] = \sum_{i=1}^{K} [C_i] = -\sum_{j=1}^{M} [C_j^*], \tag{6.282}$$

as homology classes.

Proposition 6.90.
(a) *The cocycles $\Phi_{C_i,R}$ are bounded (from above) by zero.*
(b) *The cocycle $\Phi_{C_S,R}$ obtained by integrating over a separating cycle is a local cocycle. For level zero the cocycle values are*

$$\begin{aligned}
\Phi_{C_i,R}(\phi_{n,r}, \phi_{-n,s}) &= -\frac{1}{6}(n^2 - \frac{1}{4})\delta_r^s \delta_r^i, \quad n \in \mathbb{Z} + \frac{1}{2} \\
\Phi_{C_S,R}(\phi_{n,r}, \phi_{-n,s}) &= -\frac{1}{6}(n^2 - \frac{1}{4})\delta_r^s, \quad n \in \mathbb{Z} + \frac{1}{2} \\
\Phi_{C_i,R}(e_{n,r}, e_{-n,s}) &= -\frac{1}{12}(n^3 - n))\delta_r^s \delta_r^i, \quad n \in \mathbb{Z} \\
\Phi_{C_S,R}(e_{n,r}, e_{-n,s}) &= -\frac{1}{12}(n^3 - n))\delta_r^s, \quad n \in \mathbb{Z}.
\end{aligned} \tag{6.283}$$

Proof. First, that the cocycles evaluated for the vector field subalgebra are bounded respectively local is the content of Proposition 6.33. The arguments using residues extend obviously to the half-form contribution. This shows (a) and (b). The explicit expression for level zero for the vector field subalgebra is (6.127). The values for the $(-1/2)$-forms follow again by direct and straightforward residue calculus. □

In the following section we will show that the opposite is also true, meaning that every local or bounded cocycle will be equivalent to the cocycles defined above, or to a certain linear combination of them.

The following is the crucial result.

Theorem 6.91. *Let γ be a cocycle for the superalgebra $\mathcal{S}$ which is bounded from above and which vanishes on the vector field subalgebra $\mathcal{L}$, then γ vanishes in total. In other words, every bounded cocycle is uniquely given by its restriction to the vector field subalgebra.*

Before we prove this theorem in Section 6.9.1, we apply it to the question of central extension of the superalgebra.

Theorem 6.92. *Given the Lie superalgebra $\mathcal{S}$ of Krichever–Novikov type, with its induced almost-grading given by the splitting of A into I and O. Then:*
(a) *The space of bounded cohomology classes has dimension $K = \#I$. A basis is given by the classes of cocycles (6.276) integrating over the cycles C_i, $i = 1, \ldots, K$.*
(b) *The space of local cohomology classes is one-dimensional. A generator is given by the class of (6.276) integrating over a separating cycle C_S.*

(c) *Up to equivalence and rescaling there is only one nontrivial almost-graded central extension of the Lie superalgebra with even central element extending the almost-graded structure on $\mathcal{S}$.*

Proof. Let γ be a bounded cocycle for $\mathcal{S}$. After restriction to $\mathcal{L} \times \mathcal{L}$ we obtain a bounded cocycle for $\mathcal{L}$. By Theorem 6.41 it is cohomologous to a standard cocycle with a suitable projective connection R

$$\Phi := \sum_{i=1}^{K} a_i \gamma^{\mathcal{L}}_{C_i,R}, \quad a_i \in \mathbb{C}. \tag{6.284}$$

Let $\kappa : \mathcal{L} \to \mathbb{C}$ be the linear form giving the coboundary, i.e., $\gamma_{|\mathcal{L}} - \Phi = d_1(\kappa)$. We extend κ by zero for the elements of $\mathcal{F}^{-1/2}$. Then $\Phi' := \gamma - d_1(\kappa)$ is cohomologous to γ (as cocycle for $\mathcal{S}$). Moreover, $(\gamma - d_1(\kappa))_{|\mathcal{L}} = \Phi$. Next we extend Φ via (6.276) to $\mathcal{S}$ (i.e., taking the same linear combination of integration cycles and the same projective connection) and obtain another cocycle for $\mathcal{S}$, still denoted by Φ. By construction the difference $\Phi' - \Phi$ is zero on $\mathcal{L}$. Hence, by Theorem 6.91 it vanishes on $\mathcal{S}$. But Φ is exactly of the form claimed in (a). We get exactly K linearly independent cocycle classes, as they are linearly independent as cocycles for the subalgebra $\mathcal{L}$. Claim (b) follows in a completely analogous manner, now applied to local cocycles. Claim (c) is a direct consequence. □

From the proof the corollary below also follows.

Corollary 6.93. *Every almost-graded central extension of the Lie superalgebra $\mathcal{S}$ is uniquely fixed by its restriction to the Lie subalgebra of vector fields $\mathcal{L}$ (the Lie algebra of even elements in $\mathcal{S}$). Moreover, every almost-graded centrally extended vector field algebra $\widehat{\mathcal{L}}$ can be extended in a unique way to an almost-graded extension $\widehat{\mathcal{S}}$ of the Lie superalgebra $\mathcal{S}$.*

Again it has to be pointed out that it will not be case that there is only one central extension of the superalgebra $\mathcal{S}$. Only if we fix an almost-grading for $\mathcal{S}$, which means that we fix a splitting of A into I and O, will there be a unique central extension allowing us to extend the almost-grading. For another essentially different splitting (meaning that it is not only the changing of the roles of I and O), the almost-grading will be different and we will obtain a different central extension of the algebra. In fact, if the genus g of the Riemann surface is larger than zero there will be non-equivalent central extensions which are not associated to any almost-grading (neither coming from a local cocycle, nor from a bounded cocycle).

6.9.1 Proof of Theorem 6.91

We start with a bounded cocycle γ for $\mathcal{S}$ which vanishes on the subalgebra $\mathcal{L}$ and consider the cocycle condition (6.275). It remains

$$\gamma(\varphi, e\,.\,\psi) + \gamma(\psi, e\,.\,\varphi) = 0, \quad \forall e \in \mathcal{L}, \varphi, \psi \in \mathcal{F}^{-1/2}. \tag{6.285}$$

Our goal is to show that it is identically zero.

Recall that for a pair of homogeneous basis elements $(f_{m,p}, g_{n,r})$ of any combination of types the value $l = n+m$ was called the *level* of the pair. We apply the technique presented in Section 6.8. We will consider cocycle values $\gamma(f_{m,p}, g_{n,r})$ on pairs of level $l = n + m$ and will make descending induction over the level. By the boundedness from above, the cocycle values will vanish at all pairs of sufficiently high level. It will turn out that everything will be fixed by the values of the cocycle at level zero. Finally, we will show that the cocycle γ also vanishes at level zero. Hence the claim of Theorem 6.91. We will use again the symbol $\equiv$ with the meaning introduced in Section 6.8.

We consider the triplet of basis elements $(\varphi_{m,r}, \varphi_{n,s}, e_{k,p})$ for level $l = n + m + k$. Recall (3.50)

$$\begin{aligned} [e_{k,p}, \varphi_{n,s}] &\equiv \left(n - \frac{k}{2}\right) \delta_p^s\, \varphi_{n+k,p}, \\ [\varphi_{k,p}, \varphi_{n,s}] &\equiv \delta_p^s\, e_{n+k,p}. \end{aligned} \tag{6.286}$$

Hence

$$\gamma(\varphi_{m,r}, [e_{k,p}, \varphi_{n,s}]) \equiv \gamma\left(\varphi_{m,r}, \left(n - \frac{k}{2}\right) \delta_p^s\, \varphi_{n+k,p}\right) = \left(n - \frac{k}{2}\right) \delta_p^s\, \gamma(\varphi_{m,r}, \varphi_{n+k,p}). \tag{6.287}$$

If we use this we obtain from (6.285)

$$\left(n - \frac{k}{2}\right) \delta_p^s\, \gamma(\varphi_{m,r}, \varphi_{n+k,p}) + \left(m - \frac{k}{2}\right) \delta_p^r\, \gamma(\varphi_{n,s}, \varphi_{m+k,p}) \equiv 0. \tag{6.288}$$

We set $k = 0$, (now the level is $n + m$) then

$$n\, \delta_p^s\, \gamma(\varphi_{m,r}, \varphi_{n,p}) + m\, \delta_p^r\, \gamma(\varphi_{n,s}, \varphi_{m,p}) \equiv 0. \tag{6.289}$$

If $p = s$ but $p \neq r$, then we obtain

$$n \cdot \gamma(\varphi_{m,r}, \varphi_{n,p}) \equiv 0. \tag{6.290}$$

As $n \in \mathbb{Z} + \frac{1}{2}$, and hence $n \neq 0$, we have

$$\gamma(\varphi_{m,r}, \varphi_{n,p}) \equiv 0, \quad \text{for } r \neq p. \tag{6.291}$$

Next we consider $r = p = s$ and (6.289) yields

$$n \cdot c(\varphi_{m,s}, \varphi_{n,s}) + m \cdot c(\varphi_{n,s}, \varphi_{m,s}) \equiv 0. \tag{6.292}$$

As γ is symmetric on $\mathcal{F}^{-1/2}$ we get

$$(n+m)\cdot\gamma(\varphi_{m,s},\varphi_{n,s})\equiv 0. \tag{6.293}$$

This shows that as long as the level is different to zero the cocycle is given via universal cocycle values of higher level. By assumption our cocycle c is bounded from above. Hence a level M exists such that for all levels $> M$ the cocycle values will vanish. We get by induction from (6.291) and (6.293) that the cocycle values will be zero for all levels > 0. Next we will show that it will vanish also at level zero. We go back to (6.288) (for $s = p = r$) and plug in $n = m$ and $k = -2n \in \mathbb{Z}$ and obtain

$$4n\cdot\gamma(\varphi_{n,s},\varphi_{-n,s})\equiv 0,\quad \forall n\in\mathbb{Z}+\frac{1}{2}. \tag{6.294}$$

Hence, also at level zero everything will be expressed by cocycle values of higher level and consequently will be equal to zero. Continuing with (6.293), we see that also at level < 0 we get that the cocycle will vanish. Hence the claim. □.

6.9.2 The case of an odd central element

In the previous sections we considered the case of even central elements. This choice is quite natural as we want to extend the central extension of the vector field algebra to the supercase. Nevertheless, in this section we will consider the case that the central element has odd parity. Nothing new will be obtained as we will show.

Theorem 6.94. *Every bounded cocycle yielding a central extension with odd central element is a coboundary.*

This leads to the following corollary.

Corollary 6.95. *There is no nontrivial central extension of the Lie superalgebra $\mathcal{S}$ with odd central element coming from a bounded cocycle. In other words, all such central extensions will split.*

Proof of Theorem 6.94. In the odd case only the cocycle relations (6.268) for the *(even,even,odd)* and *(odd,odd,odd)* combinations will be nontrivial. We first make a cohomologous change by defining recursively a map

$$\Phi:\mathcal{F}^{-1/2}\to\mathbb{C}, \tag{6.295}$$

which will be extended by zero on $\mathcal{L}$. We consider $\gamma(e_{0,p},\varphi_{k,r})$. It is of level k. By the boundedness an M exists such that for $k > M$ all its values will vanish. Recall the almost-graded structure

$$e_{0,p}\cdot\varphi_{k,p}=k\cdot\varphi_{k,p}+y_{k,p}, \tag{6.296}$$

with $y_{k,p}$ a finite sum of elements of degree $\geq k+1$. Set

$$\Phi(\varphi_{k,p}):=0,\quad k>M,\ p=1,\dots,K \tag{6.297}$$

and then recursively for $k = M, M-1, \ldots$ (half-integer)

$$\Phi(\varphi_{k,p}) := \frac{1}{k}\left(\gamma(e_{0,p}, \varphi_{k,p}) - \Phi(y_{k,p})\right), \quad p = 1, \ldots, K. \tag{6.298}$$

For the cohomologous cocycle $\gamma' = \gamma - d_1\Phi$ we calculate for $p = 1, \ldots, K$

$$\begin{aligned}\gamma'(e_{0,p}, \varphi_{m,p}) &= \gamma(e_{0,p}, \varphi_{m,p}) - \Phi(e_{0,p} \cdot \varphi_{k,p}) \\ &= \gamma(e_{0,p}, \varphi_{m,p}) - k\Phi(\varphi_{k,p}) - \Phi(y_{k,p}) = 0.\end{aligned} \tag{6.299}$$

Next we show that the cocycle γ' will vanish identically. For simplicity we will drop the $'$. We consider the cocycle relation for *(even,even,odd)* for the elements $e_{n,p}, e_{m,r}, \varphi_{k,s}$. With the technique used in the last section we obtain

$$\begin{aligned}\left(k - \frac{m}{2}\right)\gamma(e_{n,p}, \varphi_{k+m,r})\delta_r^s - \left(k - \frac{n}{2}\right)\gamma(e_{m,r}, \varphi_{k+n,s})\delta_p^s \\ - (m-n)\gamma(e_{m+n,r}, \varphi_{k,s})\,\delta_p^r \equiv 0.\end{aligned} \tag{6.300}$$

For $n = 0$ this specializes to

$$\left(k - \frac{m}{2}\right)\gamma(e_{0,p}, \varphi_{k+m,r})\delta_r^s - k \cdot \gamma(e_{m,r}, \varphi_{k,s})\delta_p^s - m \cdot \gamma(e_{m,r}, \varphi_{k,s})\delta_p^r \equiv 0. \tag{6.301}$$

If we set $s = p \neq r$, then

$$-k \cdot \gamma(e_{m,r}, \varphi_{k,s}) \equiv 0. \tag{6.302}$$

As k is half-integer we get $\gamma(e_{m,r}, \varphi_{k,s}) \equiv 0$. For $r = s = p$ in (6.301) we obtain

$$\left(k - \frac{m}{2}\right)\gamma(e_{0,p}, \varphi_{m,p}) - (k+m)\gamma(e_{m,p}, \varphi_{k,p}) \equiv 0. \tag{6.303}$$

As $\gamma(e_{0,p}, \varphi_{m,p}) = 0$ and $k + m \neq 0$ we get $\gamma(e_{m,p}, \varphi_{k,p}) \equiv 0$ for all m and k. Hence all values are determined by values of higher level. By the boundedness it will be zero at all levels. □

6.9.3 Examples

6.9.3.1 Higher genus, two points

In this case, there is only one almost-grading for $\mathcal{S}$ because there is only one splitting possible and the separating cycle C_S coincides with the cycle C_1. In particular, integration over C_1 already gives a local cocycle. But this does not mean that every bounded cocycle will be local, it only means that the class of bounded cocycles coincides with the class of local cocycles. This case with integration over C_S was considered by Bryant [31], but he does not prove uniqueness. Also it has to be repeated that for higher genus there are other non-equivalent cocycles obtained by integration about other nontrivial cycle classes on Σ. See also Zachos [269] for $g = 1$ and two points.

6.9.3.2 Genus $g = 0$, two points

This is the classical situation. By some isomorphism of $\Sigma = S^2$, the Riemann sphere, we can assume that $I = \{0\}$ and $O = \{\infty\}$ with respect to the quasi-global coordinate z. As projective connection $R = 0$ will do. In this situation our algebras are honestly graded and the elements can be given as follows:

$$e_n = z^{n+1}\frac{d}{dz}, \quad n \in \mathbb{Z}, \qquad \varphi_m = z^{m+1/2}(dz)^{-1/2}, \quad m \in \mathbb{Z} + 1/2. \tag{6.304}$$

By calculating the cocycle values we obtain the structure equation of the centrally extended Lie superalgebra.

$$\begin{aligned} [e_n, e_m] &= (m-n)e_{m+n} + \frac{1}{12}(n^3 - n)\,\delta_n^{-m}\,t, \\ [e_n, \varphi_m] &= (m - \frac{n}{2})\,\varphi_{m+n}, \\ [\varphi_n, \varphi_m] &= e_{n+m} - \frac{1}{6}\left(n^2 - \frac{1}{4}\right)\delta_n^{-m}\,t. \end{aligned} \tag{6.305}$$

See Section 2.11.4 for the supercase without central extension. In fact, all higher order terms in the calculations above are now exact not only up to higher order. This means that there is no reference to boundedness needed and the statements are true for all cocycles. Note also, that the subspace

$$\langle e_{-1}, e_0, e_{-1}, \varphi_{-1/2}, \varphi_{1/2} \rangle \tag{6.306}$$

is a finite-dimensional sub Lie superalgebra. It consists of the global holomorphic sections of $\mathcal{F}^{-1}$ and $\mathcal{F}^{-1/2}$. Restricted to this subalgebra the cocycle vanishes.

6.9.3.3 Genus $g = 0$, more than two points

Here our algebra will not be graded anymore, but only almost-graded. Different splitting gives different separating cycles and non-equivalent central extensions. The $N = 3$ situation was studied by Kreusch [137] and the central extension was calculated. The case $N = 3$ is somehow special. If we fix the 3 points then we have 3 essentially different splittings into I and O. Hence, also 3 non-equivalent different central extensions. If we fix one splitting, then by a biholomorphic mapping of S^2 any other split can be mapped to this one. Such mapping induces an automorphism of the algebra $\mathcal{S}$ (and of course also of $\mathcal{L}$). Hence, the obtained central extensions will be isomorphic (but not equivalent).

6.10 General cohomology of Krichever–Novikov algebras

In this book we are mainly interested in the central extensions of our algebras. More precisely, even mostly in their almost-graded central extensions. In order to understand them the cohomology spaces $\mathrm{H}^2_{loc}(\mathcal{U}, \mathbb{C})$ for our algebras $\mathcal{U}$ were of fundamental importance. Essentially what we did was to determine these spaces. This in turn gave a classification of the almost-graded central extensions.

Of course, from the algebraic point of view, and for some other applications like the deformation theory of these algebras (see e.g., Chapter 12), other cohomology spaces (higher cohomology spaces and/or cohomology with values in modules different to the trivial one) are of interest.

In the infinite dimensional case one often endows the algebra and the modules with a convenient topology and considers only those algebraic cochains and coboundaries which are continuous. In this way one obtains the continuous cohomology H_{cont}. It will depend on the topology chosen. In this way sometimes one is able to get more results (but now of course only for the continuous cohomology). Sometimes even more restrictions are helpful, like differentiable cocycles [2]. Unless otherwise stated, in this book we always deal with the full algebraic cohomology.

Already for the Witt algebra there is a huge amount of mathematical literature on its cohomology. We will not even attempt to list it partially. We recommend the books by Fuks [85], Guieu and Roger [96], and Kac [118] for further reference. This field is a field of ongoing research. For example, only quite recently complete proofs of the vanishing of the second cohomology of the Witt and Virasoro algebra with values in the adjoint modules were obtained independently by Schlichenmaier [221] and Fialowski [71]. These results yield that the algebras are formally and infinitesimally rigid. The result was announced by Fialowski in 1990 [73] but no proof was supplied. Some unpublished calculations [72] exist, which have now been completed in [71].

6.10.1 Universal central extension

Recall from Section 6.2 the definition of a one-dimensional central extension and its relation to $H^2(\mathcal{L}, \mathbb{C})$. These were mainly the central extensions we were interested in. For application in one or two places in the book we have to extend the definition to central extensions of arbitrary dimension.

Definition 6.96. Given a Lie algebra L and an abelian Lie algebra K, a central extension $\widehat{L}$ of L by K is a short exact sequence of Lie algebras

$$0 \longrightarrow K \xrightarrow{\ i\ } \widehat{L} \xrightarrow{\ p\ } L \longrightarrow 0, \tag{6.307}$$

such that $i(K)$ is central in $\widehat{\mathcal{L}}$.

---

2 From this point of view we have to point out that the use of H_{loc} denoting local cocycles by Definition 6.31 which was coined by Krichever and Novikov has certain disadvantages, as it should not be confused with local cocycles in the sense of support of the arguments.

Two central extensions $\widehat{L}_1$ and $\widehat{L}_2$ of L by the same K are called equivalent if there is a Lie isomorphism $\varphi : \widehat{L}_1 \to \widehat{L}_2$, such that the diagram

$$\begin{array}{ccccccccc} 0 & \longrightarrow & K & \xrightarrow{i_1} & \widehat{L}_1 & \xrightarrow{p_1} & L & \longrightarrow & 0 \\ \| & & \| & & \varphi\downarrow & & \| & & \| \\ 0 & \longrightarrow & K & \xrightarrow{i_2} & \widehat{L}_2 & \xrightarrow{p_2} & L & \longrightarrow & 0 \end{array} \tag{6.308}$$

commutes.

In this slightly more general case, as in Section 6.2, we obtain that the equivalence classes of central extensions of L by K are in $1:1$ correspondence with the cohomology space $\mathrm{H}^2(L, K)$, where K is considered as trivial L module.

Definition 6.97. A central extension $\widehat{L}$ of L

$$0 \longrightarrow K \xrightarrow{i} \widehat{L} \xrightarrow{p} L \longrightarrow 0, \tag{6.309}$$

is called a *universal central extension* of L if for all central extensions $\widehat{L}'$ of L (with arbitrary central term K')

$$0 \longrightarrow K' \xrightarrow{i'} \widehat{L}' \xrightarrow{p'} L \longrightarrow 0, \tag{6.310}$$

unique Lie homomorphisms exist

$$\phi : \widehat{L} \to \widehat{L}', \quad \varphi : K \to K', \tag{6.311}$$

such that the diagram

$$\begin{array}{ccccccccc} 0 & \longrightarrow & K & \xrightarrow{i} & \widehat{L} & \xrightarrow{p} & L & \longrightarrow & 0 \\ \| & & \varphi\downarrow & & \phi\downarrow & & \| & & \| \\ 0 & \longrightarrow & K' & \xrightarrow{i'} & \widehat{L}' & \xrightarrow{p'} & L & \longrightarrow & 0 \end{array} \tag{6.312}$$

commutes. This says

$$p' \circ \phi = p, \quad \text{and} \quad i' \circ \varphi = \phi \circ i. \tag{6.313}$$

Recall that a Lie algebra is perfect if and only if $L = [L, L]$.

Theorem 6.98. *Every perfect Lie algebra admits a universal central extension.*

The proof can be found e.g. in [96, Theorem 2.3.3]

We already related the property of L being perfect to the fact that $\mathrm{H}^1(L, \mathbb{C}) = \{0\}$; see (6.9).

Proposition 6.99. *The Krichever–Novikov vector field algebra $\mathcal{L}$ is a simple and hence perfect Lie algebra.*

Proof. Here we will use some results from algebraic geometry. First recall that Σ can be considered a complex smooth projective curve. Furthermore, we use the fact that the open subset $\Sigma \setminus A$ will be a smooth affine curve over $\mathbb{C}$. According to Chow's Theorem [203, Theorem 6.4], all meromorphic objects on Σ correspond to rational objects on the curve. In particular, meromorphic objects (functions, vector fields, etc.) which are holomorphic outside A correspond to (algebraically-) regular objects on the affine curve $\Sigma \setminus A$. The algebra of functions $\mathcal{A} = \mathcal{A}(A)$ can be identified with the regular functions on $\Sigma \setminus A$, hence with the elements of the coordinate ring $\mathbb{C}[\Sigma \setminus A]$. The Lie algebra $\mathcal{L} = \mathcal{L}(A)$ becomes the Lie algebra of regular derivations $D : \mathcal{A} \to \mathcal{A}$, i.e.,

$$\mathcal{L}(A) = \mathrm{Der}(\mathcal{A}(A)) = \mathrm{Der}(\mathbb{C}[\Sigma \setminus A]). \tag{6.314}$$

Jordan [116] proved the result that for a smooth affine variety V, the Lie algebra $\mathrm{Der}(\mathbb{C}[V])$ will be a simple Lie algebra. But simple Lie algebras are perfect. Hence the claim. □

Moreover, also our current type algebras $\mathfrak{g} \otimes \mathcal{A}$ for $\mathfrak{g}$ semisimple will be perfect; see Proposition 9.2. Hence, at least for these two families of algebras, we have universal central extensions. In this case, the universal central extensions can be explicitly given; see (6.318) and Skryabin [252] for the vector field algebra, respectively Section 9.6 and Kassel [127] for the current algebras.

6.10.2 The full $\mathrm{H}^2(\mathcal{L}, \mathbb{C})$

Krichever and Novikov gave in [140] and [141] a sketch of a proof of the uniqueness of local cocycles for the vector field algebra for arbitrary genus in the two-point case. They assumed that every cocycle can be given by an expression of the type (6.88) with a suitable curve C on Σ. Assuming that the cocycle is local, they used quasi-periodic functions to argue that the curve C has to be cohomologous to a circle around the point P where poles are allowed.

Novikov conjectured $\mathrm{H}^2(\mathcal{L}, \mathbb{C}) \cong \mathrm{H}_1(\Sigma \setminus \{P, Q\}, \mathbb{C})$, where the latter space is the first singular homology space. Obviously, this conjecture should be extended to the multi-point case by

$$\mathrm{H}^2(\mathcal{L}, \mathbb{C}) \cong \mathrm{H}_1(\Sigma \setminus A, \mathbb{C}). \tag{6.315}$$

If $\#A = N \geq 1$, then

$$\dim \mathrm{H}_1(\Sigma \setminus A, \mathbb{C}) = 2g - N + 1. \tag{6.316}$$

A generating set is given by the classes of the $2g$ elements of the symplectic basis and the classes of the N "circles" around the points in A. One of those classes can be expressed as a combination of the others and can be left out. In this way one obtains a basis.

This conjecture was extended to the full cohomology ring by Feigin

$$\mathrm{H}^\bullet(\mathcal{L}, \mathbb{C}) \cong \Lambda(c_1, c_2, \ldots, c_{2g-N+1}, \theta). \tag{6.317}$$

Here the right-hand side is the free graded commutative algebra generated by the elements c_k in degree 2 and by θ in degree 3. This conjecture is called the *Feigin–Novikov conjecture.*

In 1997, Millionshchikov [172] was able to prove in the two-point case at least the finite-dimensionality of $\mathrm{H}^2(\mathcal{L}, \mathbb{C})$. He used spectral sequence techniques based on almost-grading.

A real breakthrough (for the algebraic cohomology) was obtained by Skryabin in 2004 [252].

He used again the fact that $X \setminus A$ is a smooth affine complex curve. Here A consists of any non-zero finite number of points. The elements of the vector field algebra $\mathcal{L}$ are the derivations $D : \mathcal{A} \to \mathcal{A}$ of the function algebra $\mathcal{A}$. Note that by Proposition 6.99 $\mathcal{L}$ is a perfect Lie algebra. Hence, there is a universal central extension. As a special case of some more general considerations he showed

$$\mathrm{H}^2(\mathcal{L}, \mathbb{C}) \cong \mathrm{H}^1_{dR}(\Sigma \setminus A)^* = H_1(\Sigma \setminus A, \mathbb{C}). \tag{6.318}$$

For more details see [252, §7].

6.10.3 Some remarks on the continuous cohomology $\mathrm{H}^\bullet_{cont}(\mathcal{L}, \mathbb{C})$

Other important contributions with respect to the Feigin-Novikov conjecture were given by Wagemann in a series of articles [258, 259, 261, 262]. His results are valid in the realm of continuous cohomology. He considers $\Sigma \setminus A$ as an open Riemann surface and studies holomorphic objects on it. Of course, meromorphic objects on Σ which have only poles at points in A (which by definition are algebraic) give subspaces of the space of holomorphic objects. In general there are more holomorphic objects on $\Sigma \setminus A$, as for them also essential singularities are allowed at A.

Let $\mathcal{L} = \mathcal{L}(A)$ be the Krichever–Novikov vector field algebra, and $Hol(\Sigma \setminus A)$ be the Lie algebra of holomorphic vector fields on $\Sigma \setminus A$. Then

$$\mathcal{L}(A) \subseteq Hol(\Sigma \setminus A). \tag{6.319}$$

Wagemann introduced for $Hol(\Sigma \setminus A)$ a certain natural Fréchet topology. With respect to this topology it becomes a Fréchet topological vector space. As subspace $\mathcal{L}(A)$ carries the induced topology but will not be complete with respect to this topology. However, he showed in [258] and [262] that $\mathcal{L}(A)$ will be dense in $Hol(\Sigma \setminus A)$. Moreover, he deduced for the full continuous cohomology

$$\mathrm{H}^\bullet_{cont}(\mathcal{L}(A), \mathbb{C}) = \mathrm{H}^\bullet_{cont}(Hol(\Sigma \setminus A), \mathbb{C}). \tag{6.320}$$

From the work of Kawazumi [126, Property 6.2] it follows that

$$\mathrm{H}^{\bullet}_{cont}(Hol(\Sigma \setminus A), \mathbb{C}) \cong \mathrm{H}^{\bullet}_{sing}(Map(\Sigma \setminus A, S^3)). \tag{6.321}$$

Here, $Map(\Sigma \setminus A, S^3)$ is the topological space of continuous maps from $\Sigma \setminus A$ to the 3-sphere S^3 in the compact-open topology, and $\mathrm{H}^{\bullet}_{sing}$ denotes its continuous topology. The later space has the claimed form and we obtain

$$\mathrm{H}^{\bullet}_{cont}(\mathcal{L}, \mathbb{C}) \cong \Lambda(c_1, c_2, \dots, c_{2g-N+1}, \theta). \tag{6.322}$$

Hence, the Feigin-Novikov conjecture is true with respect to the continuous cohomology.

Kawazumi also considers the continuous cohomology of $Hol(\Sigma \setminus A)$ with values in the modules $\mathcal{F}^{\lambda}_{hol}(\Sigma \setminus A)$ of holomorphic λ-forms of weight $\lambda \in \mathbb{Z}$. A crucial point in this consideration is the use of the Rešetnikov spectral sequence which allows splitting of the cohomology in a global and a local term involving the cohomology of the formal vector fields on the line. For the latter, results were given by Goncharova [90]; see also [85, p. 119]. In particular, using her results it turns out that

$$\mathrm{H}^{2}_{cont}(Hol(\Sigma \setminus A), \mathcal{F}^{\lambda}_{hol}(\Sigma \setminus A)) = \{0\}, \quad \text{for } \lambda \neq 0, 1, 2, 5, 7. \tag{6.323}$$

See also [260] for explicit descriptions of generators for the nonvanishing cases.

7 Semi-infinite wedge forms and fermionic Fock space representations

In this chapter we will construct representations of our differential operator algebras $\mathcal{D}^1$ which are generated from "ground states", also called vacuum states. The positive parts of the algebras will act as annihilation operators and the rest will act as creation operators. In physics, one is very often searching for exactly such representations. We will see that the procedure presented to construct such representations will only work if we pass to certain central extensions of $\mathcal{D}^1$. What has been said so far concerns of course also the subalgebras $\mathcal{L}$ and $\mathcal{A}$.

The representation of $\mathcal{D}^1$ on $\mathcal{F}^\lambda$ given by the Lie derivative (for the $\mathcal{L}$ part) and multiplication (for the $\mathcal{A}$ part) is not of the type we are looking for. There is no ground state. But from $\mathcal{F}^\lambda$ the space of semi-infinite wedge forms $\mathcal{H}^\lambda$ can be constructed as a certain type of exterior forms; see Section 7.2 below. Our goal is to extend the action of $\mathcal{D}^1$ on $\mathcal{F}^\lambda$ to $\mathcal{H}^\lambda$ via the Leibniz rule. Unfortunately, this extended action will not be well-defined for all elements of $\mathcal{D}^1$. To counteract this, the action has to be regularized. In physics language and naively speaking, we have to subtract *infinity* $\times$ *identity* in a controlled manner. In mathematics language, we modify the action to make it well-defined, with the drawback that it will not be an honest Lie representation any more, but only a projective one. This means that it is a representation up to multiples of the identity. This can be compensated by passing over to a certain central extension $\widehat{\mathcal{D}^1}$ of $\mathcal{D}^1$ where the central element acts as scalar times identity to re-establish an honest Lie action.

The representations we construct will be called *semi-infinite wedge representation*, or equivalently *fermionic Fock space representations*. With this process we also obtain Fock space representations for certain centrally extended vector field algebras $\widehat{\mathcal{L}}$ and of the abelian Lie algebra of functions $\widehat{\mathcal{A}}$ as subalgebras of $\widehat{\mathcal{D}^1}$. Moreover, we obtain a representation of a certain central extension of the full algebra of differential operators $\mathcal{D}$. We analyze the fine structure of these representation spaces. We will again find an almost-graded structure. In Chapter 8 we will use these wedge spaces to discuss $b - c$ systems which are related representations.

As a technical tool we will use infinite dimensional matrix algebras which we will study first.

7.1 The infinite matrix algebra $\overline{gl}(\infty)$

7.1.1 The algebra and its central extension

We recall the following facts about infinite-dimensional matrix algebras. They were introduced by Kac and Petersen [120], and independently by Date, Jimbo, Kashiwara and Miwa [48]. See also [122] for a lecture course treatment.

Here we will present everything necessary. Let $\mathrm{mat}(\infty)$ be the vector space of (both-sided) infinite complex matrices. An element $A \in \mathrm{mat}(\infty)$ is given as

$$A = (a_{ij})_{i,j\in\mathbb{Z}}, \quad a_{ij} \in \mathbb{C}. \tag{7.1}$$

In general it will not be possible to extend the usual definition of matrix multiplication to this space, as $\sum_{j\in\mathbb{Z}} a_{ij}b_{jk}$ will not be defined in all cases. To obtain a well-defined multiplication we have to consider certain subspaces. We will deal with the following ones.

$$gl(\infty) := \{A = (a_{ij}) \mid \exists r = r(A) \in \mathbb{N} : a_{ij} = 0, \text{ if } |i|, |j| > r\}, \tag{7.2}$$

$$\overline{gl}(\infty) := \{A = (a_{ij}) \mid \exists r = r(A) \in \mathbb{N} : a_{ij} = 0, \text{ if } |i-j| > r\}. \tag{7.3}$$

The matrices in $gl(\infty)$ have "finite support", the matrices in $\overline{gl}(\infty)$ have "finitely many diagonals". Clearly, $gl(\infty)$ is a subspace of $\overline{gl}(\infty)$.

We introduce the matrix product of two matrices $A = (a_{ij})$ and $B = (b_{kl})$ as in the finite-dimensional case by

$$C = A \cdot B, \qquad C = (c_{il}), \quad c_{il} := \sum_{j\in\mathbb{Z}} a_{ij}b_{jl}, \tag{7.4}$$

if the sum is finite. Otherwise it is not defined.

Proposition 7.1. *The matrix product is well-defined for all elements of $gl(\infty)$ and $\overline{gl}(\infty)$. Both algebras are associative algebras and Lie algebras under the matrix commutator. Moreover, $\overline{gl}(\infty)$ is a unital associative algebra. The space* $\mathrm{mat}(\infty)$ *will be a module over $\overline{gl}(\infty)$.*

Proof. The statement for $gl(\infty)$ is trivially true. Let $A \in \overline{gl}(\infty)$, and let $r \geq 0$ be such that $a_{ij} = 0$ if $|i-j| > r$, then in the summation (7.4) the non-zero part of the summation runs in the j domain only between $i - r$ and $i + r$. Hence, it is finite and C is a well-defined matrix. Now let $s \geq 0$ be such that $b_{kl} = 0$ if $|k - l| > s$. Then $c_{il} = 0$ if $|i - l| > r + s$, as in this range either a_{ij} or b_{jl} are zero. Hence, $C \in \overline{gl}(\infty)$. Consequently, the subspace $\overline{gl}(\infty)$ closes under the matrix product and becomes an associative algebra. The diagonal matrix with entry 1 for all diagonal elements lies in $\overline{gl}(\infty)$ and is the unit for associative algebra. The commutator gives now the Lie structure. We even showed that (without assuming anything about B) the product $A\cdot B$ will be well-defined matrix. Hence, the statement that all matrices constitute a module over $\overline{gl}(\infty)$ follows, as the other properties follow immediately from the definition of the matrix product. □

Note that $gl(\infty)$ will not be a unital associative algebra.[1]

As in the finite-dimensional case, it is convenient to introduce the elementary matrices

$$E_{kl} = (\delta_i^k \delta_j^l)_{i,j\in\mathbb{Z}}. \tag{7.5}$$

We calculate

$$E_{ij} \cdot E_{kl} = \delta_j^k E_{il}, \quad [E_{ij}, E_{kl}] = \delta_j^k E_{il} - \delta_l^i E_{kj}. \tag{7.6}$$

The general products of matrices are fixed by these relations.

We use for $\mu = (\dots, \mu_{-1}, \mu_0, \mu_1, \dots) \in \mathbb{C}^{\mathbb{Z}}$ and $r \in \mathbb{Z}$

$$A_r(\mu) := \sum_{i\in\mathbb{Z}} \mu_i E_{i,i+r} \tag{7.7}$$

to denote a diagonal matrix where the diagonal is shifted by r positions to the right.

The following is obvious.

Proposition 7.2.
(a) *The set of elements E_{ij} where both i and j are running through $\mathbb{Z}$ is a basis of $gl(\infty)$.*
(b) *The set of elements $A_r(\mu)$ with $r \in \mathbb{Z}$ and $\mu \in \mathbb{C}^{\mathbb{Z}}$ constitute a (vector space) generating set for $\overline{gl}(\infty)$.*

The elements $A_r(\mu)$ with different r's are linearly independent. But with the same r there are relations. Hence, they do not constitute a basis of $\overline{gl}(\infty)$.

Next we turn to central extensions of the algebras $gl(\infty)$ and $\overline{gl}(\infty)$, considered as Lie algebras. The Lie algebra $\overline{gl}(\infty)$ admits a standard 2-cocycle which we will define now; see also [85]. For $A = (a_{ij}) \in \overline{gl}(\infty)$ set $\pi(A) = (\pi(A)_{ij})$ the matrix defined by

$$\pi(A)_{ij} := \begin{cases} a_{ij}, & i \geq 0, j \geq 0 \\ 0, & \text{otherwise.} \end{cases} \tag{7.8}$$

Proposition 7.3.
(a) *The bilinear form*

$$\alpha(A, B) := \mathrm{tr}\big(\pi([A, B]) - [\pi(A), \pi(B)]\big) \tag{7.9}$$

is a cohomologically nontrivial Lie algebra 2-cocycle for $\overline{gl}(\infty)$.
(b) *The cocycle α is a multiplicative cocycle, i.e.,*

$$\alpha(A \cdot B, C) + \alpha(B \cdot C, A) + \alpha(C \cdot A, B) = 0. \tag{7.10}$$

(c) *Restricted to $gl(\infty)$ the cocycle α will be a coboundary.*

Proof. Given $X \in \overline{gl}(\infty)$ we decompose it into the four boxes X_1, X_2, X_3, X_4 by

$$X = \begin{pmatrix} X_1 & X_2 \\ X_3 & X_4 \end{pmatrix}, \qquad \text{with} \quad X_4 = \pi(X). \tag{7.11}$$

1 In the notation of Kac and Raina [122], the algebra $\overline{gl}(\infty)$ is $\bar{a}_\infty$ and $gl(\infty)$ is gl_∞.

We will not distinguish between the boxes and the matrices in $\overline{gl}(\infty)$, obtained by filling them up again by zeros to elements of $\overline{gl}(\infty)$. In particular $X = X_1+X_2+X_3+X_4$. The matrices X_2 and X_3 have finite support, the matrices X_1 and X_4 a finite number of diagonals. Let $A, B \in \overline{gl}(\infty)$, then $(AB)_4 = A_3B_2 + A_4B_4$. Hence

$$\pi([A,B]) - [\pi(A), \pi(B)] = (AB - BA)_4 - [A_4, B_4] = A_3B_2 - B_3A_2. \tag{7.12}$$

This expression has finite support. Hence, the trace

$$\alpha(A,B) = \mathrm{tr}(A_3B_2 - B_3A_2) \tag{7.13}$$

is well-defined. The antisymmetry is clear. Let us now consider $\alpha(AB, C)$. With $(AB)_3 = A_3B_1 + A_4B_3$ we calculate

$$\alpha(AB,C) = \mathrm{tr}(A_3B_1C_2) + \mathrm{tr}(A_4B_3C_2) - \mathrm{tr}(C_3A_1B_2) - \mathrm{tr}(C_3A_2B_4). \tag{7.14}$$

Because all products have finite support, all the traces make sense. Permuting (7.14) cyclically and adding the results gives (7.10). Hence we have the multiplicative property. If we change the order of all the products in (7.10) we obtain zero too. By subtracting both sums we get the cocycle conditions for the Lie algebra. Next we address the problem of whether it can be cohomologically trivial, i.e., $\alpha = d_1\phi$. We consider the matrices

$$A_{-1}(1) = \sum E_{i,i-1}, \quad A_1(1) = \sum E_{i,i+1},$$

and calculate $[A_{-1}(1), A_1(1)] = 0$. If α were a coboundary, then necessarily $\alpha(A_{-1}(1), A_1(1)) = 0$. But calculating this cocycle value using (7.13) yields

$$\alpha(\mathrm{tr}(E_{00})) = 1.$$

Hence, it is cohomologically nontrivial. This shows (a).

The multiplicativity was shown during the proof of (a). Hence, (b) is also true. If $A, B \in gl(\infty)$, the individual terms of (7.9) have a well-defined trace. As the trace of a commutator vanishes we obtain

$$\alpha(A,B) = \mathrm{tr}(\pi([A,B])) = (d_1\phi)(A,B), \tag{7.15}$$

with

$$\phi : gl(\infty) \to \mathbb{C}, \quad \phi = \mathrm{tr} \circ \pi. \tag{7.16}$$

Hence, $\alpha_{|gl(\infty)}$ is a coboundary. □

Indeed, α is the generator of the Lie algebra two cohomology of $\overline{gl}(\infty)$; see [85]. We will not need this fact in the following.

We will denote the central extension defined by α by $\widehat{gl}(\infty)$.

As we will need them later, we calculate the cocycle values for pairs of elementary matrices E_{ij}, E_{kl}. They are finite support matrices and we can use the equation (7.15)

$$\alpha(E_{ij}, E_{kl}) = \mathrm{tr}(\pi([E_{ij}, E_{kl}])) = \mathrm{tr}(\pi(\delta_j^k E_{il})) - \mathrm{tr}(\pi(\delta_i^l E_{kj})). \tag{7.17}$$

The first sum will only give a value different to zero if $i = l \geq 0$ and $j = k$. The value will be $+1$. The second sum will only give a value different to zero if $j = k \geq 0$ and $i = l$. The value will be -1. If in this condition the second indices are also ≥ 0, then the two values will cancel and the result will be zero. Hence, in total

$$\alpha(E_{ij}, E_{kl}) = \begin{cases} 1, & i = l \geq 0,\ j = k < 0 \\ -1, & j = k \geq 0,\ i = l < 0 \\ 0, & \text{otherwise.} \end{cases} \tag{7.18}$$

Proposition 7.4. *For the generators of $\overline{gl}(\infty)$ we obtain*

$$\alpha(A_r(\mu), A_{-s}(\mu')) = 0, \quad r \neq s, \tag{7.19}$$

and

$$\alpha(A_r(\mu), A_{-r}(\mu')) = \begin{cases} 0, & r = 0 \\ \sum_{k=0}^{-1-r} \mu_k \mu'_{k+r}, & r < 0 \\ -\sum_{k=0}^{r-1} \mu_{k-r}\mu'_k, & r > 0. \end{cases} \tag{7.20}$$

Proof. Let $A_r(\mu) = \sum_{i\in\mathbb{Z}} \mu_i E_{i,i+r}$ and $A_{-s}(\mu') = \sum_{j\in\mathbb{Z}} \mu'_j E_{j,j-s}$. The cocycle value calculates as

$$\alpha(A_r(\mu), A_{-s}(\mu') = \sum_{i\in\mathbb{Z}} \sum_{j\in\mathbb{Z}} \mu_i \mu'_j \alpha(E_{i,i+r}, E_{j,j-s}). \tag{7.21}$$

Using (7.18) we know that the appearing α's can be different to zero only if $i = j - s$ and $i + r = j$ (and if the signs are different). From the two equations we can already conclude that there might be a contribution only if $i = i + r - s$, meaning if $r = s$. This shows (7.19). Next we consider (7.21) for $r = s$. A non-zero contribution $\alpha(E_{i,i+r}, E_{j,j-r})$ can only occur if $i = j - r$ and as $j = i + r$, j must have a sign different to i. More precisely, if $r < 0$ then there will be only a contribution (which will be 1) if i lies between 0 and $-r-1$. If $r > 0$ then there will only be a contribution (which will be -1) if j lies between 0 and $r-1$, or equivalently i lies between $-r$ and -1. If we plug in these values to (7.21) we obtain (7.20). □

In accordance to our general treatment of central extensions, let us set

$$\widehat{E}_{ij} = (0, E_{ij}), \quad t = (1, 0), \quad \widehat{A}_r(\mu) = \sum_{i\in\mathbb{Z}} \mu_i \widehat{E}_{i,i+r} \tag{7.22}$$

with respect to the standard (linear) splitting of $\widehat{gl}(\infty)$.

For $s \in \mathbb{Z}$ we define the subspaces

$$\overline{gl}(\infty)_s := \{A_{-s}(\mu) \mid \mu \in \mathbb{C}^{\mathbb{Z}}\} \tag{7.23}$$

of $\overline{gl}(\infty)$, and for $\widehat{gl}(\infty)$

$$\begin{aligned} \widehat{gl}(\infty)_s &:= \{\widehat{A}_{-s}(\mu) \mid \mu \in \mathbb{C}^{\mathbb{Z}}\},\ s \neq 0, \\ \widehat{gl}(\infty)_0 &:= \langle\{\widehat{A}_0(\mu) \mid \mu \in \mathbb{C}^{\mathbb{Z}}\}, t\rangle_{\mathbb{C}}. \end{aligned} \tag{7.24}$$

Obviously, these are (infinite-dimensional) subspaces and we obtain the linear decomposition

$$\overline{gl}(\infty) = \bigoplus_{s\in\mathbb{Z}} \overline{gl}(\infty)_s, \quad \widehat{gl}(\infty) = \bigoplus_{s\in\mathbb{Z}} \widehat{gl}(\infty)_s. \tag{7.25}$$

We call the elements in the component s homogeneous elements of degree s. For the multiplication we obtain

$$A_{-s}(\mu)\cdot A_{-r}(\mu') = A_{-(r+s)}(\mu''), \quad \text{with } (\mu'')_i = \mu_i\mu'_{i-s}. \tag{7.26}$$

Proposition 7.5. *The Lie algebras $\overline{gl}(\infty)$ and $\widehat{gl}(\infty)$ with the above introduced degree are graded Lie algebras. The algebra $\overline{gl}(\infty)$ is also graded as associative algebra.*

Proof. The Equation (7.26) shows the gradedness for $\overline{gl}(\infty)$. For $\widehat{gl}(\infty)$ we only have to recall that the cocycle vanishes if $r \neq -s$, see (7.19). Hence, only for $r + s = 0$ will there be a term coming with the central element t, which has also degree zero. □

As usual in the graded case, the subspaces of degree zero constitute subalgebras $\overline{gl}(\infty)_0$ and $\widehat{gl}(\infty)_0$. This graded decomposition is also valid for $gl(\infty)$. The elements there correspond to "diagonals" with only a finite number of entries.

7.1.2 Semi-infinite wedge representation for $\widehat{gl}(\infty)$

Let $v_s \in \mathbb{C}^{\mathbb{Z}}$ be the infinite sequence given by $v_s = (\delta_i^s)_{i\in\mathbb{Z}}$. Set

$$V = \bigoplus_{s\in\mathbb{Z}} \mathbb{C}v_s, \tag{7.27}$$

the vector space generated by those. By defining

$$E_{ij}v_s = \delta_j^s v_i \tag{7.28}$$

we obtain an operation of $\overline{gl}(\infty)$ on V which generalizes the operation matrix × vector to infinite dimension. Indeed, each of the $A_r(\mu)$ acts in a well-defined manner as

$$A_r(\mu)v_s = \sum_{i\in\mathbb{Z}} \mu_i E_{i,i+r}v_s = \mu_{s-r}v_{s-r}. \tag{7.29}$$

Starting from V we will introduce the semi-infinite higher exterior power – also called *semi-infinite wedge space*. To achieve this goal we have to linearly order the basis elements $\{v_s \mid s \in \mathbb{Z}\}$ by increasing index s. Let H be the vector space freely generated by the formal expressions of the kind

$$\Phi = v_{i_0} \wedge v_{i_1} \cdots \wedge \cdots v_{i_k} \wedge v_{i_{k+1}} \wedge v_{i_{k+2}} \wedge \cdots, \tag{7.30}$$

where the indices i_j are in strictly increasing order, i.e., $i_0 \lneqq i_1 \lneqq i_2 \lneqq \cdots$, and they are stabilizing with a certain index. This means that starting with an index k (which depends on the element Φ), we have

$$i_{k+s} = i_k + s, \quad \forall s \in \mathbb{N}. \tag{7.31}$$

In the following, forms which do not fulfill the strictly increasing condition will occur. The wedge symbol $\wedge$ indicates how to deal with them. If two entries are the same the result will be the zero element. If two entries are not in the correct order we interchange them and change the sign. Moreover, the forms are linear in each entry. In detail, if u and w are finite neighboring pieces and ψ is an infinite neighboring piece, then

$$\begin{aligned} &u \wedge v_j \wedge w \wedge v_i \wedge \psi := -u \wedge v_i \wedge w \wedge v_j \wedge u, \quad j > i \\ &\phi \wedge v_i \wedge w \wedge v_i \wedge \psi := 0 \\ &u \wedge \left(\sum_{i=1}^{r} c_i v_i \right) \wedge \psi := \sum_{i=1}^{r} c_i \left(u \wedge v_i \wedge \psi \right). \end{aligned} \tag{7.32}$$

Next we want to extend the Lie action of $gl(\infty)$ and $\overline{gl}(\infty)$ on V to H by using the Leibniz rule. This means that if $A \in \overline{gl}(\infty)$, then A should operate on each factor in Φ separately and the results are added up, i.e.,

$$\begin{aligned} A \,.\, \Phi := (A \,.\, v_{i_0}) \wedge v_{i_1} \wedge \ldots + v_{i_0} \wedge (A \,.\, v_{i_1}) \wedge \ldots \\ + \cdots + v_{i_0} \wedge v_{i+1} \wedge \ldots \wedge (A \,.\, v_{i_k}) \wedge v_{i_{k+1}} \wedge \ldots + \ \ldots \end{aligned} \tag{7.33}$$

A priori we obtain by this definition an infinite number of summands. The (linear) action will be well-defined only if a finite number of nonvanishing summands appear. This has to be true for all Φ. First we observe that $E_{ij} \,.\, \Phi$ is always well-defined, as there can only be a contribution if the element v_j appears in Φ, which happens at least once. The result will be the Φ with v_j exchanged by v_i. Hence, $gl(\infty)$ operates on H.

Next we consider $\overline{gl}(\infty)$. For this we have to study the action of the generators $A_r(\mu)$.

Proposition 7.6. *If $r \neq 0$, then the action of $A_r(\mu)$ is well-defined.*

Proof. Let Φ be a given basis element of H and let m be an index for which the elements in Φ stabilize, i.e., $i_{m+k} = i_m + k$ for all $k \in \mathbb{N}_0$. Recall that $A_r(\mu)v_s = \mu_{s-r}v_{s-r}$. If $r < 0$, then $A_r(\mu)v_{i_m+k}$ for all $k \in \mathbb{N}$ will be annihilated by the neighboring elements to the right. If $r > 0$, then for $i_m + k$ with $k \geq r$ either the elements on the left or on the right will annihilate it. Hence, in any case only a finite number of terms will show up. □

Problems only occur for $r = 0$. In this case, any appearing v_s in Φ will be replaced by $\mu_s v_s$ and for this summand we obtain $\mu_s\Phi$. In total, $A_0(\mu)$ acts as "$\infty \times$ *identity*", which of course does not make sense.

Before we deal with this problem, the following proposition is noted.

Proposition 7.7. *The above-defined action will be a well-defined Lie action as long as the action is well-defined for all elements involved.*

Proof. Let A, B be elements of $\overline{gl}(\infty)$ such that their action and the action of $[A, B]$ is well-defined. In a first step, $B \,.\, \Phi$ will be the sum of elements in which B acts individually exactly once on an entry. If we consider now A acting on $B\Phi$, we obtain for every such summand again a sum of corresponding elements. In these, the $A \cdot B$ will act exactly at the same position and the others will have separate A and B actions. If we interchange the roles of A and B and subtract the results we find that the mixed terms will cancel. Only the $(AB - BA)$ acting on Φ will remain. Hence

$$[A, B] . \Phi = A . (B . \Phi) - B . (A . \Phi). \tag{7.34}$$

□

We return to the problem of redefining the action to make also the $A_0(\mu)$ action well-defined. The method we choose is very close to [122]. But here we use ascending forms instead of descending forms. Let r be the usual action of $gl(\infty)$ given above on H. Recall that the knowledge of $r(E_{ij})$ for all i and j fixes everything. Now we define a modified action $\hat{r}$ via

$$\hat{r}(E_{ij}) := \begin{cases} r(E_{ij}), & i \neq j \text{ or } i = j < 0 \\ r(E_{ii}) - id, & i \geq 0. \end{cases} \tag{7.35}$$

We use $\hat{r}$ to define the action of the elements $\overline{gl}(\infty)$ on H and obtain

$$\hat{r}(A_k(\mu)) = r(A_k(\mu)), \quad \text{for } k \neq 0. \tag{7.36}$$

Moreover, as the infinite tail will not contribute anymore, $\hat{r}(A_0(\mu))$ will also be well-defined, i.e.,

$$\hat{r}(A_0(\mu)) . \Phi = c(\mu, \Phi) \cdot id, \tag{7.37}$$

with a finite constant $c(\mu, \Phi)$.

Example 7.8. Let

$$\Phi_m = v_m \wedge v_{m+1} \wedge v_{m+2} \wedge \cdots, \tag{7.38}$$

then

$$\hat{r}(A_0(\mu))\Phi_m = \begin{cases} (\sum_{s=m}^{-1} \mu_s)\Phi_m, & m < 0 \\ (-\sum_{s=0}^{m-1} \mu_s)\Phi_m, & m > 0 \\ 0, & m = 0. \end{cases} \tag{7.39}$$

To check whether it is a Lie action we have to check whether the equations

$$[\hat{r}(E_{ij}), \hat{r}(E_{kl})] = \hat{r}([E_{ij}, E_{kl}]) = \delta_j^k \hat{r}(E_{il}) - \delta_l^i \hat{r}(E_{kj}) \tag{7.40}$$

are true for all indices. If we replace in (7.41) the $\hat{r}$ by r this is of course true. On the other hand, on the left-hand side all additive terms of the identity will vanish under the commutator. Hence we only have to consider the cases where on the right-hand side $\hat{r}$ is different to r. We replace r by $\hat{r}$ via (7.35) In these cases one obtains instead of (7.40) the correct expression

$$[\hat{r}(E_{ij}), \hat{r}(E_{ji})] = \hat{r}(E_{ii}) - \hat{r}(E_{jj}) + \beta(E_{ij}, E_{ji}) \cdot id \tag{7.41}$$

with

$$\beta(E_{ij}, E_{ji}) = \begin{cases} 0, & i, j < 0 \\ 0, & i, j \geq 0 \\ 1, & i \geq 0, j < 0 \\ -1, & j \geq 0, i < 0. \end{cases} \tag{7.42}$$

In particular, it is not a Lie action anymore, but only a projective Lie action. A closer looks shows that the values of β coincide exactly with the values of α used to define the central extension $\widehat{gl}(\infty)$; see (7.18). Hence, as explained also in Section 6.3, the proposition below follows.

Proposition 7.9. *By setting $\widehat{r}(\widehat{E}_{ij}) := \widehat{r}(E_{ij})$ and $\widehat{r}(t)$ we obtain an honest Lie action of $\widehat{gl}(\infty)$ on H.*

Remark 7.10 (Non-uniqueness of the regularization procedure). The procedure given by (7.35) is in no way unique. Nevertheless, we will call it the standard regularization. We could have chosen another i instead of $i = 0$ to make the cut where we modify $\widehat{r}(E_{ii})$. Moreover, more complicated procedures could be done as long as the prescription only differs by a finite number of indices from the standard regularization. We obtain again a projective action and an associated central extension (see Section 6.3) via a certain 2-cocycle. But the new cocycle will be cohomologous to the standard cocycle (7.9). This follows from the the general result about uniqueness of the cocycle class $[\alpha]$; see [85]. Another method is to calculate the difference explicitly and to verify that it is a coboundary. We will not do it here. In particular, the obtained central extension will be equivalent to the "standard" central extension. Also, it has to be pointed out that the expression (7.39) depends on the regularization prescription chosen. If we do not take 0 as cut then the summation there will be shifted.

This situation is very typical for the ambiguity of the regularization procedure. Only if we fix it will we get a well-defined central extension. If we do not fix it, the best we can hope for is that the central extensions are equivalent. This is the case in our situation.

As we will need them later we will already introduce the decomposition of the space H into the subspace of different charges and degrees.

First we consider a basis element Φ given by (7.30). The stability condition (7.31) says that there is an index k, such that $i_{k'} = k' + m$ for all $k' \geq k$, with an integer m. We call this *m charge* of Φ. The definition does not depend on the choice of the k from which we start. We define $H^{(m)}$ to be the vector space generated by the Φs of charge m. As each basis element of H has a unique charge we have the direct sum decomposition

$$H = \bigoplus_{m \in \mathbb{Z}} H^{(m)}. \tag{7.43}$$

None of these spaces is empty, as $\Phi_m \in H^{(m)}$, where Φ_m is defined by (7.38).

Proposition 7.11.
(a) *The subspace $H^{(m)}$ is an invariant subspace of H with respect to both the $gl(\infty)$ and the $\widehat{gl}(\infty)$ action.*
(b) *$H^{(m)}$ is generated by Φ_m.*

Proof. By the very definition, $E_{ij}\Phi$ will be either zero or an appearing v_j will be replaced by v_i. If v_i already appears, then the result will be zero too. Otherwise we might have to interchange elements to bring them in ascending order. If we consider the stable tail starting from an index high enough, the tail will not be affected and the number of indices before will be the same as in Φ (at least if the result is $\neq 0$). Hence, the charge of $E_{ij}\Phi$ will be equal to the charge of Φ. First we consider the $gl(\infty)$ action. In our earlier terminology this is given by $r(E_{ij})$. By the above $H^{(m)}$ is invariant. Let

$$\Phi = v_{i_0} \wedge v_{i_1} \wedge v_{i_{k-1}} \wedge v_{k+m} \wedge v_{k+1+m} \wedge \cdots \tag{7.44}$$

be an arbitrary basis element of charge m (obeying the strictly increasing order), then

$$\Phi = E_{i_0,m}((\cdots(E_{i_{k-2},k-2+m}(E_{i_{k-1},k-1+m})\Phi_m))\cdots)). \tag{7.45}$$

Hence, Φ_m is generating.

For $\widehat{gl}(\infty)$ we have to consider the $\hat{r}$-action. But $\hat{r}(E_{ij}) = r(E_{ij})$ for $j \neq i$ and for $i = j$ only for some more times the result $\hat{r}(E_{ii})\phi = 0$ will appear. The algebra $\widehat{gl}(\infty)$ is generated by $\widehat{A}_s(\mu)$ and t. The central element t acts as identity, and the operation of $\widehat{A}_s(\mu)$ can be expressed via the $\hat{r}(E_{ij})$ action. Hence also here invariance. In the proof of the generating property, only $r(E_{ij})$ with $i \neq j$ appears. Hence, Φ_m generates also $H^{(m)}$ under the $\widehat{gl}(\infty)$ action. □

Next we decompose $H^{(m)}$ further into subspaces of certain degrees. The degree measures "how far away" the basis element is from the reference element Φ_m. We define for Φ given as (7.44) its *degree*

$$\deg \Phi := \sum_{k=0}^{\infty} (i_k - (m+k)). \tag{7.46}$$

For $k \gg 0$ we have $i_k = m + k$, hence the sum is finite. The degree also calculates as

$$\deg \Phi = \sum(i_s < m \mid v_{i_s} \text{ appears in } \Phi) - \sum(i_s \geq m \mid v_{i_s} \text{ does not appears in } \Phi). \tag{7.47}$$

Proposition 7.12.

$$\deg(\Phi) \leq 0, \quad \textit{and} \quad \deg(\phi) = 0 \Leftrightarrow \Phi = \Phi_m. \tag{7.48}$$

Proof. The indices in (7.44) have to be strictly increasing and starting from a k we have $i_k = m + k$. If we assume that an i_l with $i > (m + l)$ exists, this leads to a contradiction to these requirements. Hence $\deg(\Phi) \leq 0$. Moreover, $\deg(\Phi) = 0$ is only possible if for all $i_l = (m + l)$, saying $\Phi = \Phi_m$. □

We denote the subspace of $H^{(m)}$ generated by basis elements of degree d by $H_d^{(m)}$ and call it the subspace of degree d elements. From the proposition above we know that for all subspaces of positive degree it will be zero. To describe the dimension of the others, we introduce for every positive integer n the partition function $p(n)$ to be the number of different partitions of

$$n = n_0 + n_1 + \cdots + n_l, \quad \text{with} \quad n_0 \geq n_1 \geq \cdots \geq n_l. \tag{7.49}$$

Proposition 7.13. *(See also [122])*

$$H^{(m)} = \oplus_{d \in \mathbb{Z}_-} H_d^{(m)}, \tag{7.50}$$

$$H_0^{(m)} = \mathbb{C}\Phi_m, \tag{7.51}$$

$$\dim H_d^{(m)} = p(-d). \tag{7.52}$$

Proof. Statement (7.50) is obvious, as every basis element in $H^{(m)}$ has a unique degree. Hence we can decompose it as the sum of degree d subspaces. For positive degrees they will be trivial. For degree zero we showed that Φ_m is the only basis element of this degree. Hence (7.51).

If we have a partition as in (7.49) we set

$$\Phi = v_{m-n_0} \wedge v_{m+1-n_1} \wedge \cdots \wedge v_{m+l-n_l} \wedge v_{m+l+1} \wedge \cdots . \tag{7.53}$$

This element is of charge m and its degree calculates to $-n$. Hence, it is one of the basis elements. Different partitions yield of course different basis elements, and vice versa given a basis element Φ of degree d. If we calculate for it $n_k = i_k - (m + k)$, then n_k is decreasing and $\sum_{n_k} = -d$. In particular, every such Φ can be obtained from a partition of $-d$. This shows the dimension formula (7.52). □

A priori, we used the degree definition (7.46) for basis elements which by definition fulfill the increasing index condition. In fact, given a Φ' we could use it without this restriction. This means if we have to make a finite number of transpositions to reach the required form Φ, and if this form will not vanish, then we can employ (7.46) for Φ' and this degree will coincide with the degree of Φ. It is enough to show this for 2 entries, $i_k\ i_l$, which are in reversed order. For a fixed m the degree will be a finite sum $\sum_{k=0}^{R}(i_k - (m + k))$. We might choose R big enough so that both indices are still in the summation range. The sum can be split

$$\sum_{k=0}^{R}(i_k - (m + k)) = \sum_{k=0}^{R} i_k - \sum_{k=0}^{R}(m + k).$$

Interchanging v_{i_k} with v_{i_l} will not change this value. Note that the subspaces $H_d^{(m)}$ will not be invariant under the action of $gl(\infty)$ and $\widehat{gl}(\infty)$.

Proposition 7.14.
(a)

$$r(E_{i,j+k}),\ \widehat{r}(\widehat{E}_{i,j+k})\ :\ H_d^{(m)} \to H_{d-k}^{(m)}, \tag{7.54}$$

(b)

$$\widehat{r}(A_s(\mu))\ :\ H_d^{(m)} \to H_{d-s}^{(m)}. \tag{7.55}$$

Proof. Let Φ be a basis element. Recall that $E_{i,i+k}$ replaces v_{i+k} by v_i at the same position if it appears in Φ. If $E_{i,i+k}\Phi \neq 0$ by the index change its degree calculates as

$$\deg(E_{i,i+k}\Phi) = deg(\Phi) - k. \tag{7.56}$$

This is true for the r and for the $\widehat{r}$ action. Hence, (a).

As $A_s(\mu) = \sum \mu_i E_{i,i+s}$ the statement (b) follows from (a). □

Combining Proposition 7.14 and Proposition 7.11 we arrive at the following proposition.

Proposition 7.15. *The Lie modules $H^{(m)}$ are graded modules over the algebras $gl(\infty)$ and $\widehat{gl}(\infty)$. The homogeneous subspaces of degree d elements of the modules are finite-dimensional.*

We note also that $H^{(m)}$ is also a module over the associative algebra $gl(\infty)$.

7.2 Semi-infinite wedge forms of Krichever–Novikov type elements

As usual, we will fix a splitting of the set of possible poles $A = I \cup O$ and consider the corresponding almost-grading for our objects. Also, let $\mathcal{F}^\lambda$ be the space of forms of weight $\lambda \in 1/2\ \mathbb{Z}$. As explained in the previous chapters, the Lie algebra $\mathcal{D}^1$ and its subalgebras $\mathcal{L}$ and $\mathcal{A}$ act in an almost-graded manner on $\mathcal{F}^\lambda$. We will now consider the space of semi-infinite wedge forms $\mathcal{H}^\lambda$ associated with $\mathcal{F}^\lambda$. For this we have to order the multi-indices linearly. With respect to a fixed numbering of the elements in I, we take the lexicographical order i.e., $(n, r) > (m, s)$ if $n > m$ or $(n = m$ and $r > s)$. Set $v_{Kn+r} := f^\lambda_{n,r}$.

The vector space $\mathcal{H}^\lambda$ of *semi-infinite wedge forms* is generated by basis elements which are formal expressions

$$\Phi = f^\lambda_{(i_1)} \wedge f^\lambda_{(i_2)} \wedge f^\lambda_{(i_3)} \wedge \cdots, \tag{7.57}$$

where $(i_1) = (m_1, p_1)$ is a double index indexing our basis elements. The indices are in strictly increasing lexicographical order. They are stabilizing in the sense that they will increase exactly by one step starting from a certain index, depending on Φ. The infinite rest we sometimes call ***infinite tail*** in a somewhat sloppy notation. As in the case of Section 7.1.2 (see (7.32)), the wedges should indicate how to deal with intermediate results. Mostly, we will use $f_{(m)}$ instead of $f^\lambda_{(m)}$ if the λ is clear.

Recall that we have the triangular decomposition induced by the almost-grading (see Section 3.5)

$$\mathcal{D}^1 = \mathcal{D}^1_- \oplus \mathcal{D}^1_{(0)} \oplus \mathcal{D}^1_+. \tag{7.58}$$

The algebra $\mathcal{D}^1$ operates on $\mathcal{F}^\lambda$ and we want to extend the action to $\mathcal{H}^\lambda$ via the Leibniz rule.

Before we enter into the details we would like to outline the strategy we are carrying through. For the elements of $\mathcal{D}^1_-$ and $\mathcal{D}^1_+$ the stable tail will guarantee that the action is well-defined. Problems arise for the elements of the critical strip $\mathcal{D}^1_{(0)}$ of the algebra $\mathcal{D}^1$. It is possible that they produce an infinite number of contributions. The action has to be regularized (as physicists like to call it). As already explained in Section 7.1.2, this is a well-defined mathematical procedure. In fact, we will use the results there in our context. For this we will show that by the (strong) almost-graded module structure of $\mathcal{F}^\lambda$, the algebra $\mathcal{D}^1$ can be embedded into the algebra $\overline{gl}(\infty)$ by

$$\Psi_\lambda : \mathcal{D}^1 \to \overline{gl}(\infty). \tag{7.59}$$

The embedding will depend on the weight λ. Using this embedding, the action of $\mathcal{D}^1$ on $\mathcal{F}^\lambda$ corresponds to the multiplication "matrix · vector" on V. The semi-finite forms $\mathcal{H}^\lambda$ corresponds to the forms in H. For H we discussed a well-defined modification of the action of $\overline{gl}(\infty)$ which yielded a projective action and correspondingly a well-defined action of the central extension $\widehat{gl}(\infty)$. As everything is compatible, by pulling back the defining cocycle α for $\widehat{gl}(\infty)$ via Ψ_λ we obtain a certain central extension $\widehat{\mathcal{D}^1}_\lambda$ of $\mathcal{D}^1$ and an action of it on $\mathcal{H}^\lambda$. We will show that the pulled back cocycle is a local cocycle. Hence, by the classification results of Section 6.6 we can write it as a λ depending central extension class (7.74) given by geometric cocycles. We will give the cocycle class explicitly. Up to a λ dependent rescaling, for the vector field subalgebra $\mathcal{L}$, the class will be even unique and given by (6.88) integrated over C_S.

Remark 7.16. In view of Remark 7.10, we have to stress the fact that as the defining cocycle α depends on the regularization scheme (7.35), its pullback will depend on it too. But, as explained there, by another admissible scheme we will obtain a cohomologous cocycle. Hence, the same is true for the pullback. If not otherwise stated we will take the "standard" regularization scheme (7.35).

Now we come to the details. First, recall the almost-graded structure

$$\begin{aligned} A_{n,p} \cdot f^\lambda_{m,r} &= \sum_{h=0}^{M} \sum_{h=s}^{K} D^{(h,s)}_{(n,p)(m,r)} f^\lambda_{n+m+h,s} \\ e_{n,p} \cdot f^\lambda_{m,r} &= \sum_{h=0}^{M} \sum_{h=s}^{K} C^{(h,s)}_{(n,p)(m,r)} f^\lambda_{n+m+h,s}. \end{aligned} \tag{7.60}$$

Where M is the maximum of the upper bounds for A and $\mathcal{L}$ (independent of n and m). We index the coefficients slightly differently here to other places in the book.

We will need the following values

$$D^{(0,s)}_{(n,p)(m,r)} = 1 \cdot \delta^s_r \cdot \delta^s_p, \quad e^{(0,s)}_{(n,p)(m,r)} = (m - \lambda n) \cdot \delta^s_r \cdot \delta^s_p. \tag{7.61}$$

For identification with the basis elements v_i in V (we use the notation in Section 7.1.2), we set

$$v_{Kn+r-1} := \begin{cases} f^\lambda_{n,r}, & \lambda \in \mathbb{Z} \\ f^\lambda_{n-1/2,r}, & \lambda \in \mathbb{Z} + 1/2. \end{cases} \tag{7.62}$$

Note that for $K = 1$ and $\lambda \in \mathbb{Z}$, the v_i corresponds exactly to f_i. For $\lambda \in \mathbb{Z} + 1/2$ we have to shift by $1/2$. Hence, v_0 corresponds to $f_{1/2}$ and so on. Let $\psi_\lambda : \mathcal{F}^\lambda \to V$ be the corresponding map.

After this identification we can express the action of $A_{n,p}$ and $e_{n,p}$ with the help of the structure constants given in (7.60) as multiplication with certain matrices. If we compare the definition of $A_r(\mu)$ with (7.60), we see immediately that these matrices will be finite sums of $A_r(\mu)$s. We obtain a linear map

$$\Psi_\lambda : \mathcal{D}^1 \to \overline{gl}(\infty). \tag{7.63}$$

For the basis elements we get

$$\Psi_\lambda(A_{n,r}) = \sum_{s=-K(n+M)}^{-Kn} A_s(\mu), \quad \Psi_\lambda(e_{n,r}) = \sum_{s=-K(n+M)}^{-Kn} A_s(\mu'), \tag{7.64}$$

with elements $\mu, \mu' \in \mathbb{C}^{\mathbb{Z}}$ given by the structure constants C and D respectively.

By construction we have the compatibility

$$\psi_\lambda(d \,.\, f) = \Psi_\lambda(d) \cdot \psi_\lambda(f), \quad d \in \mathcal{D}^1, f \in \mathcal{F}^\lambda. \tag{7.65}$$

From this equation it follows that ($d_1, d_2 \in \mathcal{D}^1$)

$$\psi_\lambda(d_1 \,.\, (d_2 \,.\, f)) = \Psi_\lambda(d_1) \cdot (\psi_\lambda(d_2 \,.\, f)) = (\Psi_\lambda(d_1) \cdot \Psi_\lambda(d_2)) \cdot \psi_\lambda(f). \tag{7.66}$$

As $\mathcal{F}^\lambda$ is a Lie module over $\mathcal{D}^1$ we have

$$\psi_\lambda([d_1, d_2] \,.\, f) = \psi_\lambda(d_1 \,.\, (d_2 \,.\, f) - d_2 \,.\, (d_1 \,.\, f)) = [\Psi_\lambda(d_1), \Psi_\lambda(d_2)] \; \psi_\lambda(f). \tag{7.67}$$

On the right-hand side is the bracket of matrices. Hence, Ψ_λ is a Lie homomorphism from $\mathcal{D}^1$ to $\overline{gl}(\infty)$. It is an embedding as $\mathcal{D}^1$ operates faithfully on $\mathcal{F}^\lambda$. From the compatibility it follows also that the construction of the semi-infinity wedge forms for $\overline{gl}(\infty)$ and $\mathcal{D}^1$ and the actions there correspond to each other.

In particular, we obtain by pulling back the defining cocycle α for $\widehat{gl}(\infty)$ to $\mathcal{D}^1$ also a central extension of $\mathcal{D}^1$, and correspondingly a regularized action on the semi-infinite wedge forms. The cocycle γ_λ obtained by pullback is

$$\gamma_\lambda(d_1, d_2) = \Psi^*_\lambda(\alpha)(d_1, d_2) = \alpha(\Psi_\lambda(d_1), \Psi_\lambda(d_2)). \tag{7.68}$$

Proposition 7.17.

(a) *The cocycle γ_λ obtained by pulling back α is a local cocycle which is bounded from above by zero.*

(b) *Evaluated for pairs of elements of degree zero it will vanish, i.e.,*

$$\gamma_\lambda(\mathcal{D}^1_0, \mathcal{D}^1_0) = 0. \tag{7.69}$$

(c) *The cocycle γ_λ will vanish if restricted to the subspaces corresponding to the subalgebras $\mathcal{D}^1_+$ and $\mathcal{D}^1_-$.*

(d) *As cocycle of $\mathcal{A}$ it is multiplicative.*

Proof. Let $d_1 \in \mathcal{D}^1_n, d_2 \in \mathcal{D}^1_m$ be homogeneous elements. Using (7.64) we calculate the cocycle for them

$$\gamma_\lambda(d_1, d_2) = \alpha\left(\sum_{s=-K(n+M)}^{-Kn} A_s(\mu), \sum_{t=-K(m+M)}^{-Km} A_t(\mu') \right). \tag{7.70}$$

By (7.19) there will only be a contribution if for the indices we have $t = -s$. Of course, the summation range has to be taken into account. As possible summation range which could produce a nonvanishing result we obtain the two conditions to be fulfilled simultaneously

$$-K(n+M) \le s \le -Kn, \quad Km \le s \le K(m+M). \tag{7.71}$$

The second inequality we obtain from the the summation over $t = -s$. Assume that we obtain a nonvanishing value, then there exists an s fulfilling these equations. But this is only possible if

$$Km \le -Kn \quad \text{and} \quad -K(n+M) \le K(m+M) \tag{7.72}$$

is true. If we divide these relations by K and rewrite them in terms of $n+m$, we obtain that necessarily

$$-2M \le n+m \le 0. \tag{7.73}$$

This shows locality (a).

If both n and m are zero, then by (7.71) only $s = 0$ is possible. But for $s = 0$ Equation (7.20) yields that the cocycle value will be zero. Hence, (b)

If d_1 and d_2 are both elements of $\mathcal{D}^1_\pm$, then either $n+m > 0$ or $n+m < -2M$. Hence, the cocycle vanishes for these subalgebras which shows (c).

Local cocycles of the differential operator algebra are always $\mathcal{L}$-invariant and hence by our classification also multiplicative if restricted to the abelian function algebra. Hence, (d). Indeed, a direct proof could be given by using the fact that α is a multiplicative cocycle; see (7.10). □

Theorem 7.18. *The cocycle $\gamma_\lambda = \Psi^*_\lambda(\alpha)$ can be written as the following linear combination of the separating cocycles introduced in Section 6.5:*

$$\gamma_\lambda = \Psi^*_\lambda(\alpha) = -\left(\gamma^{\mathcal{A}}_S + \frac{1-2\lambda}{2}\gamma^{(m)}_{S,T_\lambda} + 2(6\lambda^2 - 6\lambda + 1)\gamma^{\mathcal{L}}_{S,R_\lambda} + E_{V_{1/2}} \right), \tag{7.74}$$

with a suitable meromorphic affine connection T_λ, a projective connection R_λ both without poles outside A, and at most poles of order one at the points in I for T_λ and order two for R_λ, and a coboundary term $E_{V_{1/2}}$, with $V = \sum_{n\le 0}\sum_p c_{n,p}\omega^{n,p}$. The cohomology term is only needed if $\lambda = 1/2$.

Proof. By Proposition 7.17 the pullback γ_λ is a local cocycle bounded from above by zero. Hence, the existence of such a linear combination

$$\gamma_\lambda = \Psi_\lambda^*(\alpha) = \alpha^{(1)}\gamma_S^{\mathcal{A}} + \alpha^{(2)}\gamma_{S,T_\lambda}^{(m)} + \alpha^{(3)}\gamma_{S,R_\lambda}^{\mathcal{L}} + \text{local coboundary} \tag{7.75}$$

with possible coboundary terms follows from the uniqueness results of Section 6.6, in particular from Theorem 6.47. It remains to calculate the scalar factors. In the frame of the classification we gave a method to calculate them on the basis of a finite number of cocycle values for some special pairs of elements. This we will do in the following. First we remark that the lower boundary in $\mathcal{D}^1$ becomes the upper boundary in the summation in $\overline{gl}(\infty)$, and that if there is a positive n, e.g., as in $e_{n,p}$, this corresponds to a negative s in $A_s(\mu)$. Using the explicit expressions (7.20) of the cocycle α we calculate from Equation (6.135)

$$\alpha^{(1)} = -\gamma_\lambda(A_{1,r}, A_{-1,r}) = -1. \tag{7.76}$$

From (6.141) we calculate

$$\alpha^{(3)} = 2\gamma_\lambda(e_{2,r}, e_{-2,r}) - 4\gamma_\lambda(e_{1,r}, e_{-1,r}) = -2(6\lambda^2 - 6\lambda + 1), \tag{7.77}$$

and the constant

$$b_r^{(3)} = \gamma_\lambda(e_{1,r}, e_{-1,r}) = -\lambda(\lambda - 1), \tag{7.78}$$

which is of relevance for the coboundary. From (6.146) we calculate

$$\alpha^{(2)} = 1/2(\gamma_\lambda(e_{1,r}, A_{-1,r}) + \gamma_\lambda(e_{-1,r}, A_{1,r})) = \frac{2\lambda - 1}{2}, \tag{7.79}$$

with the additional constant for the coboundary

$$b_r^{(2)} = \gamma_\lambda(e_{-1,r}, A_{1,r}) = \lambda. \tag{7.80}$$

Next we want to consider the cocycle itself. The polynomial $6\lambda^2 - 6\lambda + 1$ has neither integer nor half-integer zeros, hence the vector field cocycle term will never be trivial. The corresponding coboundary D_W can be incorporated into the projective connection R_λ, depending on the weight λ. It might have poles of maximally order 2 at the points in I. Indeed, as $b_r^{(3)} = -\lambda(\lambda-1)$ it will have poles of order 2 if and only if $\lambda \neq 0$ and $\lambda \neq 1$.

As long as $\lambda \neq 1/2$, the mixing cocycle will not vanish. Hence, as above we can incorporate the coboundary E_V into the affine connection. Again from $b_r^{(2)}$ we conclude that it will have poles at I if and only if $\lambda \neq 0$. In total we get the form (7.74).

For $\lambda = 1/2$ the mixing cocycle will vanish, hence the boundary term $E_{V_{1/2}}$ will remain. For this λ value we calculate $\alpha^{(3)} = 1$. Hence,

$$\gamma_{1/2} = -\gamma_S^{\mathcal{A}} + \gamma_{S,R_{1/2}}^{\mathcal{L}} + E_{V_{1/2}}. \tag{7.81}$$

□

Remark 7.19. The values above could have been calculated also in a slightly different manner. By the $\widehat{gl}(\infty)$ techniques we know that there is a regularized action $\widehat{r}_\lambda$ on $\mathcal{H}^\lambda$. For the cocycle value we obtain

$$\gamma_\lambda(e_{2,r}, e_{-2,r}) \cdot id = [\widehat{r}_\lambda(e_{2,r}), \widehat{r}_\lambda(e_{-2,r})] - \widehat{r}_\lambda([e_{2,r}, e_{-2,r}]). \tag{7.82}$$

The cocycle value can be calculated by applying the right-hand side to an arbitrary element of $\mathcal{H}^\lambda$. In view of (7.39), Φ_0 is a convenient element as

$$\widehat{r}_\lambda([e_{2,r}, e_{-2,r}])\,\Phi_0 = \widehat{r}_\lambda(-4e_{0,r} + h.d.t.)\,\Phi_0 = 0. \tag{7.83}$$

It remains

$$\gamma_\lambda(e_{2,r}, e_{-2,r})\,\Phi_0 = \widehat{r}_\lambda(e_{2,r})(\widehat{r}_\lambda(e_{-2,r})\,\Phi_0). \tag{7.84}$$

Remark 7.20. The cocycle class $[\gamma_\lambda]$ is uniquely fixed by the values $\alpha^{(1)}$, $\alpha^{(2)}$, $\alpha^{(3)}$. In particular, the class essentially depends on λ, not only up to rescaling.

Hence, the central extensions $\widehat{\mathcal{D}}^1_\lambda$ of the differential operator algebra associated to different weights λ are not (even after rescaling the central element) equivalent. If we consider only the centrally extended $\mathcal{A}$ we see that the same central extension $\widehat{\mathcal{A}}$ will do. Clearly, the obtained central extensions $\widehat{\mathcal{L}}_\lambda$ of $\mathcal{L}$ with respect to different λ will be equivalent after rescaling of the central element. But the explicit element in the class will depend on the weight λ via the projective connection R_λ.

Remark 7.21. In the expression of the cocycle (7.74) the weight λ dependant constant

$$c_\lambda := -2(6\lambda^2 - 6\lambda + 1) \tag{7.85}$$

appears. It also has another prominent appearance in the context of Mumford's formula, relating certain natural divisors (or line bundles) on the moduli space of curves. We calculate $c_1 = -2$, hence we can deduce from (7.85)

$$c_\lambda = c_1 \cdot (6\lambda^2 - 6\lambda + 1). \tag{7.86}$$

On the moduli space of curves naturally defined line bundles λ_n exist (see [203, p. 122]) for every $n \in \mathbb{Z}$. Mumford's result is

$$\lambda_n \cong \lambda_1^{(6\lambda^2 - 6\lambda + 1)}. \tag{7.87}$$

Arbarello, deConcini, Kac, and Procesi [4] gave an explanation for this appearance in what at first sight appear to be quite different fields. The expression (7.85) has other quite interesting properties. First note that for pairs of dual weights its value is the same.

$$c_\lambda = c_{1-\lambda}. \tag{7.88}$$

Special values are

$$c_0 = c_1 = -2, \quad c_2 = c_{-1} = -26, \quad c_{1/2} = -1. \tag{7.89}$$

The number 26 appearing above is also known to be the critical space time dimension of bosonic string theory, i.e., the dimension when the conformal anomaly will be compensated by the ghost fields; see Remark 8.22. Also note that for $\lambda \in \mathbb{Q}$ the value $c_\lambda \neq 0$.

We conclude this section by formulating the following theorem.

Theorem 7.22. *The space of semi-infinite wedge forms $\mathcal{H}^\lambda$ carries a representation of centrally extended algebras $\widehat{\mathcal{A}}, \widehat{\mathcal{L}}_\lambda, \widehat{\mathcal{D}}^1_\lambda$. The defining cocycle γ_λ for the central extension is given by (7.74). The cocycles for $\widehat{\mathcal{A}}$ and $\widehat{\mathcal{L}}_\lambda$ are obtained by restricting γ to the subalgebras.*

Proposition 7.23. *With respect to the standard splitting $d \mapsto (0, d)$, the subalgebras $\mathcal{D}^1_+$ and $\mathcal{D}^1_-$ of $\mathcal{D}^1$ remain subalgebras of $\widehat{\mathcal{D}^1_\lambda}$. In particular, $\widehat{\mathcal{D}^1_{\lambda+}} \cong \mathcal{D}^1_+$ and $\widehat{\mathcal{D}^1_{\lambda-}} \cong \mathcal{D}^1_i$.*

Proof. By Proposition 7.17 the cocycle γ_λ restricted to these subalgebras vanishes. Hence the statement. □

Let us recall that the algebra $\widehat{\mathcal{A}}$ is sometimes also called (infinite-dimensional) Heisenberg algebra or oscillator algebra. Hence, we obtained with the semi-infinite wedge representations vacuum-type representations for them; see Section 7.4

7.2.1 Action of differential operators of all degrees

In Section 2.7 the algebra of differential operators $\mathcal{D}_\lambda$ on $\mathcal{F}^\lambda$ of arbitrary degree was introduced. This was done with the help of the universal enveloping algebra $U(\mathcal{D}^1)$. The embedding of $\mathcal{D}^1$ into $\overline{gl}(\infty)$ can be extended to $\mathcal{D}_\lambda$. Again, the action can be extended to $\mathcal{H}^\lambda$ if we pass to the central extension $\widehat{\mathcal{D}}_\lambda$, obtained by pulling back the cocycle α.

Theorem 7.24. *The algebra $\mathcal{D}_\lambda$ of meromorphic differential operators holomorphic outside A admits a central extension $\widehat{\mathcal{D}}_\lambda$. The restriction of the defining cocycle to the subalgebra $\mathcal{D}^1$ of differential operators of degree less than or equal to one is given by (7.74). The algebra $\widehat{\mathcal{D}}_\lambda$ can be realized as operators on the space of semi-infinite wedge forms.*

We would like to stress that the existence of this central extension is not a trivial fact. Also, the fact that exactly this linear combination of the cocycles for $\mathcal{D}^1$ can be extended is not automatic, as the extension problem is nontrivial. Our action on the semi-infinite wedge forms show both.

Let

$$u^k_{n,p} := A_{k+n,p}(e_{-1,p})^k \tag{7.90}$$

be basis elements of special type of $\mathcal{D}_\lambda$. Here the product and the power should be done in $\mathcal{D}_\lambda$. For these elements we obtain (see the calculation in [207, p. 141])

$$\gamma_\lambda(u^k_{n,p}, u^l_{m,r}) = 0, \quad n + m > 0, \tag{7.91}$$

and for $\lambda = 0$

$$\gamma_0(u^k_{n,p}, u^l_{m,r}) = 2 \cdot (-1)^{k-1} \frac{k!\,l!}{(k+l+1)!} \prod_{j=-l}^{k} (n+j)\,\delta^r_p. \tag{7.92}$$

Corresponding formulas for arbitrary λ can be obtained.

Remark 7.25. In the classical situation this extension is the extension given by the Radul cocycle [193]. Note that Radul only gave the extension for $\lambda = 0$. Again for the classical situation and also only for $\lambda = 0$, Li [160] showed that this is the only linear combination of cocycles for $\mathcal{D}^1$ which can be extended to $\mathcal{D}_0$. See also [125], where the cocycle appeared as *Schwinger term*. It is reasonable to expect that this is also the case for higher genus and many points.

Remark 7.26. In the classical situation this algebra was also called $W_{1+\infty}$ algebra. In mathematics it appeared for the first time in [120]. In this sense, the algebra $\widehat{\mathcal{D}}_\lambda$ constructed here by the action on the semi-infinite wedge forms is the higher genus multipoint $W_{1+\infty}$ algebra. For more details about the classical $W_{1+\infty}$ see [213].

7.2.2 Fine structure of the representation space

The representation space $\mathcal{H}^\lambda$ has an infinite number of invariant subspaces. In fact, we can decompose it with respect to the charge (up to a possible shift) introduced in Section 7.1.2. For $K = 1$ and $\lambda \in \mathbb{Z}$ a basis element

$$\Phi = f_{i_0} \wedge f_{i_1} \wedge \cdots \wedge f_{i_k} \wedge \cdots \tag{7.93}$$

will be of *charge* m, if for $k \gg 0$ we have $i_k = k + m$. Note that in our normalization for half-integer λ, the charge will also be a half-integer. For $K > 1$ we have to take into account the rewriting of the indices. Otherwise the conditions stay exactly the same.

We denote by $\mathcal{H}^{\lambda,(m)}$ the subspace generated by basis elements of charge m. From Proposition 7.13 follows the proposition below.

Proposition 7.27. *$\mathcal{H}^\lambda$ decomposes as a direct sum of invariant subspaces $\mathcal{H}^{\lambda,(m)}$*

$$\mathcal{H}^\lambda = \bigoplus_{m \in \mathbb{J}_\lambda} \mathcal{H}^{\lambda,(m)} \tag{7.94}$$

with $\mathbb{J}_\lambda = \mathbb{Z}$ if $\lambda \in \mathbb{Z}$, and $\mathbb{J}_\lambda = \mathbb{Z} + 1/2$ if $\lambda \in \mathbb{Z} + 1/2$.

In a more intuitive way, we can say that $\mathcal{H}^{\lambda,(m)}$ is generated by basis elements which differ from a fixed element which is stable already from the beginning only for a finite number of indices. Such an element is uniquely fixed. For $K = 1$ this element is

$$\Phi_T = f_T \wedge f_{T+1} \wedge \cdots \wedge f_{T+k} \wedge \cdots . \tag{7.95}$$

For $K > 1$ these elements are

$$\begin{aligned}
\Phi_{T,1} &= f_{T,1} \wedge f_{T,2} \wedge \cdots \wedge f_{T,K} \wedge f_{T+1,1} \wedge \cdots \\
\Phi_{T,2} &= f_{T,2} \wedge f_{T,3} \cdots \wedge f_{T,K} \wedge f_{T+1,1} \wedge \cdots \\
\cdots \quad & \quad \cdots \\
\Phi_{T,K} &= f_{T,K} \wedge f_{T+1,1} \wedge \cdots
\end{aligned} \tag{7.96}$$

For simplicity, we denote $\Phi_{T,1}$ also by Φ_T

These special basis elements are called *ground states* or *vacua states*. Respectively vacuum vector of weight λ and level T. We will denote them also by Φ_{vac}.

For the applications it is more convenient to denote the charge m subspace $\mathcal{H}^{\lambda,(m)}$ by the vacuum element $\Phi_{T,p}$ it contains. For this we use the notation $\mathcal{H}^{\lambda,(T,p)}$. Recall that $\mathcal{H}^{\lambda,(T,p)}$ is generated by all semi-infinite wedge forms which differ from $\Phi_{T,p}$ only for a finite number of indices.

Proposition 7.28.

(a) *The subalgebra* $\mathcal{D}^1_+ \cong \widehat{\mathcal{D}^1}_+$ *of* $\widehat{\mathcal{D}^1}$ *annihilates the vacuum elements* Φ_{vac}, *i.e.,* $\mathcal{D}^1_+\Phi_{vac} = 0$.

(b) *The central element and the other elements of degree zero act by multiplication with a constant on the vacuum.*

(c) *The whole representation state is generated by* $\mathcal{D}^1_- \oplus \widehat{\mathcal{D}^1}_{(0)}$ *from the vacuum.*

(d) *The same statement is true for the representation of the subalgebras* $\widehat{\mathcal{A}}$ *and* $\widehat{\mathcal{L}}$.

Proof. Recall that we can identify these subalgebras of $\widehat{\mathcal{D}^1}$ with those of $\mathcal{D}^1$. By the almost-graded structure for $d \in \mathcal{D}^1_n$ with $n > 0$ and $f \in \mathcal{F}^\lambda_m$ we get

$$\deg(d\,.\,f) = \deg(f) + n > \deg f. \tag{7.97}$$

Hence, if f is present in Φ_{vac}, then $d\,.\,f$ will be annihilated by neighboring elements to the right. This shows (a).

The central element operates as scalar by definition. Let now d be of degree zero. Then in $d\,.\,f$ one might have elements of degree zero and higher. The higher degree elements will again be annihilated. If there is a component of the same degree then it has to be a multiple of f (by the form of the lowest degree contribution). Hence, after regularization of the action it will remain just a multiple of Φ_{vac}. This shows (b).

To show (c), let V be the subspace generated by successively applying the complementary subspace $\mathcal{D}^1_- \oplus \widehat{\mathcal{D}^1}_0$ to the vacuum vector Φ_{vac}. These vectors are linear combinations of elements of the type

$$\psi_k = d_k\,.\,(d_{k-1}\,.\,\ldots\ldots(d_1\,.\,\Phi_{vac})\cdots), \quad d_i \in \mathcal{D}^1_- \oplus \widehat{\mathcal{D}^1}_0,\ 0 \le i \le k. \tag{7.98}$$

Here $k \in \mathbb{N}_0$. We have to show that V is invariant under $\widehat{\mathcal{D}^1}$. Meaning that every $e\,.\,\psi_k$ is again a linear combination of elements of the type (7.98) (of course, the k might be different). This we only have to show for $\mathcal{D}^1_+$. We make induction with respect to the number k in (7.98). Let e be an arbitrary element of $\mathcal{D}^1_+$. If $k = 0$, then $e\,.\,\Phi_{vac} = 0$ for all elements e of positive degree. Now

$$\begin{aligned} e\,.\,\psi_k &= e\,.\,(d_k\,.\,(d_{k-1}\,.\,\ldots\ldots\,(d_1\,.\,\Phi_{vac})\cdots)) \\ &= ([e,d_k]\psi_{k-1} + \gamma_\lambda(e,d_k)\psi_{k-1} - d_k\,.\,(e\,.\,(d_{k-1}\,.\,\ldots\ldots(d_1\,.\,\Phi_{vac})\cdots)) \end{aligned} \tag{7.99}$$

By induction the last term is of the required type, the second term is always of this type. For the first term, depending on the elements in $[e, d_k]$, we either get that they are directly in V or by induction.

Clearly, the arguments work also for the subalgebras $\widehat{\mathcal{A}}$ and $\widehat{\mathcal{L}}$. □

Proposition 7.29. *Let V be the representation space generated by the vacuum Φ_{vac}, then*

$$\mathcal{D}^1_+\,.\,\Phi_{vac} = 0\,, \tag{7.100}$$

$$t\,.\,\Phi_{vac} = \Phi_{vac}\,, \tag{7.101}$$

$$A_{0,p}\,.\,\Phi_{T,s} = \begin{cases} -T\ \Phi_{T,s} & p \ge s \\ -(T-1)\ \Phi_{T,s} & p < s \end{cases} \tag{7.102}$$

$$e_{0,p}\,.\,\Phi_{T,s} = \begin{cases} -\frac{T(T-1)}{2}\ \Phi_{T,s} & p \ge s \\ -\frac{T(T-3)}{2}\ \Phi_{T,s} & p < s. \end{cases} \tag{7.103}$$

Proof. The relations (7.100) and (7.101) have been shown in Proposition 7.28. To demonstrate the other claims we use the embedding of $\mathcal{H}^\lambda$ into H and the standard regularization (7.35) and its consequence (7.39). First, let $s = 1$. As $A_{0,p}f_{m,r} \equiv f_{m,r}\delta^p_r$ modulo higher degree terms in $\mathcal{F}^\lambda$, in the range $r = 1, \ldots, K$, only maximally one $f_{m,r}$ could produce a coefficient 1. To check let $m \ge 0$. Hence the summation (7.39) yields $\hat{r}(A_{0,p})\Phi_{T,1} = (-T)\ \Phi_{T,1}$. Similarly, $e_{0,p}\cdot f_{m,r} \equiv m f_{m,r}\delta^p_r$, and $\hat{r}(e_{0,p})\Phi_{T,1} = -T(T-1)/2\ \Phi_{T,1}$.

If $s \ne 1$ there might also be a contribution in the summation over m in (7.39) appearing in "level" T which we have to take into consideration. But this is only needed if $p < s$. In particular, in this case the first alternative always appears. Now if $s > 1$, then for $p \ge s$ nothing is changed. For $p < s$ the coefficient related to $A_{0,p}F_{T,p}$ respectively $e_{0,p}F_{T,p}$ will not appear. Hence, following (7.39) once 1 or T will not be subtracted. This yields the expression for the second alternative in (7.102) and (7.103). □

We note that the values of the coefficients on the right of (7.102) and (7.103) are based on the choice of the standard regularization. A different regularization $\hat{r}'$ corresponds to the fact that

$$\hat{r}'(A_{0,p}) = \hat{r}(A_{0,p}) + \alpha\, id, \tag{7.104}$$

with a constant α. Hence the values there can be adjusted by adding an (T,s) independent constant.

Also $t\,.\,\Phi_{vac} = \Phi_{vac}$ can be adjusted to any non-zero number c, i.e., $t\,.\,\Phi_{vac} = c\,\Phi_{vac}$ by rescaling the central element. A constant sometimes chosen is $c = c_\lambda = -2(6\lambda^2 - 6\lambda + 1)$.

Clearly the spaces $\mathcal{H}^{\lambda,(m)}$ are infinite-dimensional. With the help of the $b - c$ systems discussed in Chapter 8 we will see that the spaces for different m are naturally isomorphic.

Next we decompose each $\mathcal{H}^{\lambda,(m)}$ into subspaces of fixed degree. Again, we use exactly the degree definition for $\widehat{gl}(\infty)$. For $K = 1$ we get

$$\deg(\Phi) = \sum_{k=0}^{\infty} (i_k - (m + k). \tag{7.105}$$

For $K > 1$ we have to take the rescaling into account. Let $\mathcal{H}_d^{\lambda,(m)}$ be the degree d subspace. Proposition 7.12 and Proposition 7.13 yield the following proposition.

Proposition 7.30.

(a)

$$\mathcal{H}^{\lambda,(m)} = \bigoplus_{d \in \mathbb{Z}_-} \mathcal{H}_d^{\lambda,(m)}. \tag{7.106}$$

(b) *The degree subspaces d are finite-dimensional of dimension $p(-d)$, where $p(-d)$ is the number of partitions of $-p$ as introduced in (7.49).*

(c) $\dim \mathcal{H}_0^{\lambda,(m)} = 1$.

For our algebra $\widehat{\mathcal{D}^1}$ we consider the rescaled degree

$$\deg_K(e_{n,r}) = \deg_K(A_{n,r}) = K \cdot n \quad \text{and} \quad \deg(t) = 0. \tag{7.107}$$

Of course, for $K = 1$ this is the original degree. Our algebra will remain almost-graded with respect to the new degree. We showed in Proposition 7.14 that $\hat{r}(A_s(\mu))$ maps the degree d space to the degree $d - s$ space. Using (7.64) we see that for the matrices corresponding to $A_{n,r}$ and $e_{n,r}$ the s range is between $-K(n + M), \ldots.. - Kn$. Hence,

$$e_{n,p} \cdot \; : \mathcal{H}_d^{\lambda,(m)} \quad \to \quad \bigoplus_{h=d+\deg_K(e_{n,p})}^{d+\deg_K(e_{n,p})+M} \mathcal{H}_h^{\lambda,(m)}. \tag{7.108}$$

The same is true for $A_{n,r}$. The central element leaves the spaces invariant. Hence, if there are no elements of positive degree the proposition below.

Proposition 7.31. *With respect to the degree decomposition, the module $\mathcal{H}^{\lambda,(m)}$ is an almost-graded module over $\widehat{\mathcal{D}^1}$, $\widehat{\mathcal{L}}$, and $\widehat{\mathcal{A}}$. The algebras $\widehat{\mathcal{A}}_+$, $\widehat{\mathcal{L}}_+$, and $\widehat{\mathcal{D}^1}_+$ operate trivially on the elements of degree zero. For $D \in \widehat{\mathcal{D}^1}_-$ we have $\deg(D \,.\, \Phi) < 0$. Here for $K > 1$ the degrees in the algebras have to be rescaled.*

The dimension of the homogeneous subspaces is finite-dimensional but not bounded by a constant. Hence, it is not strongly almost-graded.

For the centrally extended current algebras $\widehat{\mathfrak{g}}$, the affine algebra of Krichever–Novikov type has been constructed similar to fermionic Fock space representations. We will return to this in Section 9.10.

In the following section we will also consider the subspace of $\mathcal{H}^{\lambda,(T,s)}$ generated by the action of $\widehat{\mathcal{D}^1}$ on $\Phi_{T,s}$. Clearly, we can restrict our attention also to the smaller subalgebras $\widehat{\mathcal{A}}$ and $\widehat{\mathcal{L}}$ and talk about the subspace generated by them from $\Phi_{T,s}$.

7.3 Highest weight representations and Verma modules

The semi-infinite wedge representations constructed in this chapter are examples of highest weight representations for our algebras. We will give the general definition now.

7.3.1 Highest weight representations

In the following section all central extensions of the function algebra $\mathcal{A}$, the vector field algebra $\mathcal{L}$, and of the differential operator algebra $\mathcal{D}^1$ are given with respect to local cocycles, i.e., obtained by integrating over a separating cycle C_S. We simply denote the central extensions by $\widehat{\mathcal{A}}$, $\widehat{\mathcal{L}}$, and $\widehat{\mathcal{D}^1}$ respectively.

We have the triangular decomposition

$$\widehat{\mathcal{A}} = \widehat{\mathcal{A}}_- \oplus \widehat{\mathcal{A}}_{(0)} \oplus \widehat{\mathcal{A}}_+, \quad \widehat{\mathcal{L}} = \widehat{\mathcal{L}}_- \oplus \widehat{\mathcal{L}}_{(0)} \oplus \widehat{\mathcal{L}}_+, \quad \mathcal{D}^1 = \mathcal{D}^1_- \oplus \mathcal{D}^1_{(0)} \oplus \mathcal{D}^1_+. \tag{7.109}$$

We recall that $\widehat{\mathcal{A}}_+ \cong \mathcal{A}_+$ and $\widehat{\mathcal{A}}_- \cong \mathcal{A}_-$ and similar for the other algebras. The lift of the basis elements $A_{n,p}$ of the function with respect to the representation of the local cocycle will be denoted by $\widehat{A}_{n,p}$. Correspondingly, the lift of the basis elements $e_{n,p}$ of the vector fields by $E_{n,p}$. The central element is called t. The universal enveloping algebra for a Lie algebra B is denoted by $U(B)$.

Definition 7.32. Let V be an $\widehat{\mathcal{A}}$ module with a weight decomposition $V = \bigoplus_{n\in\mathbb{Z}} V_n$ as vector space. The module V is called *highest weight module* of $\widehat{\mathcal{A}}$ with *highest weight vector* $v = v_{h,c}$, *highest weight* $h = (h_1, h_2, \ldots, h_K) \in \mathbb{C}^K$, and *central charge* $c \in \mathbb{C}$ if the following conditions are fulfilled.

(1) $V = U(\widehat{\mathcal{A}})\, v$.

(2) The element v is homogeneous and

$$V = \bigoplus_{n=-\infty}^{\mathrm{weight}(v)} V_n \quad \text{with} \quad V_{\mathrm{weight}(v)} = \mathbb{C} \cdot v.$$

(3) $\widehat{\mathcal{A}}_+ v = 0$ and $\mathrm{weight}(\widehat{\mathcal{A}}_- v) < \mathrm{weight}(v)$.

(4) The central element t operates as $c \cdot$ *identity* on V.

(5)

$$\widehat{A}_{0,p} \,.\, v = h_p \cdot v, \quad p = 1, \ldots, K. \tag{7.110}$$

We can define highest weight modules for the vector field algebra and the differential operator algebra in a completely analogous way.

Definition 7.33. Let V be an $\widehat{\mathcal{L}}$ module with a weight decomposition $V = \bigoplus_{n\in\mathbb{Z}} V_n$ as vector space. The module V is called *highest weight module* of $\widehat{\mathcal{L}}$ with *highest weight vector* $v = v_{h,c}$, *highest weight* $h = (h_1, h_2, \ldots, h_K) \in \mathbb{C}^K$, and *central charge* $c \in \mathbb{C}$ if the following conditions are fulfilled.

(1) $V = U(\widehat{\mathcal{L}})\, v$.

(2) The element v is homogeneous and

$$V = \bigoplus_{n=-\infty}^{\text{weight}(v)} V_n \quad \text{with} \quad V_{\text{weight}(v)} = \mathbb{C}\cdot v .$$

(3) $\widehat{\mathcal{L}}_+ v = 0$ and $\text{weight}(\widehat{\mathcal{L}}\_v) < \text{weight}(v)$.

(4) The central element t operates as $c \cdot$ *identity* on V.

(5)

$$E_{0,p} \,.\, v = h_p \cdot v, \quad p = 1, \ldots, K. \tag{7.111}$$

Definition 7.34. Let V be an $\widehat{\mathcal{D}^1}$ module with a weight decomposition $V = \bigoplus_{n\in\mathbb{Z}} V_n$ as vector space. The module V is called *highest weight module* of $\widehat{\mathcal{D}^1}$ with *highest weight vector* $v = v_{h,c} \in V$, *highest weight* $(h, h') = ((h_1, h_2, \ldots, h_K), ((h'_1, h'_2, \ldots, h'_K) \in \mathbb{C}^K \times \mathbb{C}^K$, and *central charge* $c \in \mathbb{C}$ if the following conditions are fulfilled.

(1) $V = U(\widehat{\mathcal{D}^1})\, v$.

(2) The element v is homogeneous and

$$V = \bigoplus_{n=-\infty}^{\text{weight}(v)} V_n \quad \text{with} \quad V_{\text{weight}(v)} = \mathbb{C}\cdot v .$$

(3) $\widehat{\mathcal{D}^1}_+ v = 0$ and $\text{weight}(\widehat{\mathcal{D}^1}\_v) < \text{weight}(v)$.

(4) The central element t operates as $c \cdot$ *identity* on V.

(5)

$$\widehat{A}_{0,p} \,.\, v = h_p \cdot v, \quad E_{0,p} \,.\, v = h'_p \cdot v, \qquad p = 1, \ldots, K. \tag{7.112}$$

Without restriction after a shift we might always assume that the vacuum has weight zero and we will do so in the following.

It is by intention that we do not put the requirement that V is an almost-graded module in the definition, as there are examples for which this is not the case.

Proposition 7.35. *In the case of the semi-infinite wedge representations $\mathcal{H}^\lambda$, the submodules $U(\widehat{\mathcal{A}})\Phi_{T,s}$, $U(\widehat{\mathcal{L}})\Phi_{T,s}$, and $U(\widehat{\mathcal{D}^1})\Phi_{T,s}$ of $\mathcal{H}^{\lambda,(T,s)}$ are highest weight modules for the corresponding algebras $\widehat{\mathcal{A}}$, $\widehat{\mathcal{L}}$, and $\widehat{\mathcal{D}^1}$ with highest weight vector $\Phi_{T,s}$, central charge 1, and highest weights given by* (7.102), *respectively* (7.103). *The weight decomposition is given by the degree* (7.105).

Proof. If we restrict the situation to the corresponding submodules depending on the algebra under consideration, then Condition 1 in the definition of a highest weight module is automatic. Furthermore, by Proposition 7.31 they are almost-graded modules with degree bounded by zero. Condition 2 comes from the degree decomposition, and $v = \Phi_{T,s}$ will have degree zero. Condition 3 is part of Proposition 7.31. Conditions 4 and 5 are parts of Proposition 7.29. □

7.3.2 Verma modules

For the sake of clarity we will now concentrate on the vector field algebra case $\widehat{\mathcal{L}}$ with its almost-grading. The other cases work completely analogously. Let V be a highest weight module with highest weight vector v. By its very definition, V equals $U(\widehat{\mathcal{L}})v$. Let $\widehat{\mathcal{L}}^- = \bigoplus_{n<0} \widehat{\mathcal{L}}_n$ be the subspace with basis $E_{n,p}$, $n \in \mathbb{Z}$, $n < 0$, $p = 1, \ldots, K$.

Proposition 7.36. *The module V is generated as vector space by the elements*

$$w = E_{n_1,p_1}(\,.\,E_{n_2,p_2} \cdots (\,.\,(E_{n_r,p_r}\,.\,v)\ldots)), \tag{7.113}$$

with

$$(n_1, p_1) \le (n_2, p_2) \le \cdots \le (n_r, p_r) < (0, 1). \tag{7.114}$$

Proof. The proof is based on the almost-graded structure of the algebra $\widehat{\mathcal{L}}$. The statement is nearly a $1 : 1$ copy of part of the proof of Proposition 7.28. We will not reproduce it here. The relation corresponding to (7.99) can of course be used to realize the decreasing condition in the indices. □

Definition 7.37. A highest weight representation is called a *Verma module* or *Verma representation* of $\widehat{\mathcal{L}}$ if the generating elements (7.113) are linearly independent, i.e., if they constitute a basis of V.

We want to give another definition and will discuss the equivalence between the definitions further down. It is done with the help of universal constructions. The subspace $\widehat{\mathcal{L}}_0 \oplus \widehat{\mathcal{L}}_+$ of $\widehat{\mathcal{L}}$ is by the almost-graded structure a subalgebra. Let $V = V(c, h)$, with $c \in \mathbb{C}$, $h \in \mathbb{C}^K$, be the one-dimensional vector space with basis element v on which $\widehat{\mathcal{L}}_+$ acts as zero and

$$E_{0,p}\,.\,v = h_p \cdot v,\ p = 1, \ldots, K \qquad t\,.\,v = c \cdot v. \tag{7.115}$$

By the almost-graded structure $V(c, h)$ is Lie module over $\widehat{\mathcal{L}}_0 \oplus \widehat{\mathcal{L}}_+$.

Definition 7.38. The Verma module of central charge c and highest weight h is the $\widehat{\mathcal{L}}$ module

$$M(c, h) = U(\widehat{\mathcal{L}}) \otimes_{U(\widehat{\mathcal{L}}_0 \oplus \widehat{\mathcal{L}}_+)} V(c, h). \tag{7.116}$$

By the action of $\widehat{\mathcal{L}}$ on $U(\widehat{\mathcal{L}})$ it is indeed an $\widehat{\mathcal{L}}$ module. The vector v will be the vacuum vector, which leads to the following proposition.

Proposition 7.39. *For every charge c and weight $h = (h_1, h_2, \ldots, h_K)$, a Verma module $M(c, h)$ of $\widehat{\mathcal{L}}$ exists.*

Remark 7.40. An equivalent construction is given as follows. Moreover, it shows that both definitions are indeed equivalent. Let $U := U(\widehat{\mathcal{L}})$ be the universal enveloping algebra of $\widehat{\mathcal{L}}$, $\mathbf{1}$ its unit element and J the left ideal generated by

$$E_{n,p},\ n \geq 1,\ p = 1, \ldots, K, \quad E_{0,p} - h_p \cdot \mathbf{1},\ p = 1, \ldots K, \quad t - c \cdot \mathbf{1}. \tag{7.117}$$

We take the quotient

$$M := M(c, h) = M(c, h_1, h_2, \ldots, h_K) := U/J. \tag{7.118}$$

The algebra $\widehat{\mathcal{L}}$ operates on M by left multiplication and gives a Lie algebra representation. If we set $v = \mathbf{1} \bmod J$, then the conditions on the actions of the degree zero elements are clearly fulfilled. Moreover, the module will be generated by v. Next we show that the elements of the type (7.113) are indeed a basis of V. For this we order the basis elements in $\widehat{\mathcal{L}}$ in the usual lexicographical order and give the central element t the order $(0, K) < \operatorname{ord}(t) < (1, 1)$. By the Poincaré-Birkhoff-Witt Theorem (see Section 1.7) a basis of U is given by increasing chains of these elements. A basis of the ideal J is given by chains which end with elements $E_{n,p}$ with $n \geq 1$ and the differences

$$w \cdot E_{0,p} - h_p\, w, \quad w \cdot t - c\, w. \tag{7.119}$$

Here w should denote a chain for which the largest index is still $\leq (0, 1)$, respectively $\leq \operatorname{ord}(t)$. Consequently, a basis of the complement will be given by increasing chains for which the largest index $< (0, 1)$. In particular, these elements constitute a basis of V. To see that these chains are a basis we have to again use the same kinds of arguments as in the proof of Proposition 7.28 to move via the almost-grading the higher degrees to the right. □

In the Verma module case, when the elements of the type in (7.113) constitute a basis, we can assign to the element w given like there

$$\operatorname{weight}(w) = \sum_{i=1}^{r} n_i. \tag{7.120}$$

Note that $\operatorname{weight}(w) < 0$ if $w \neq v$, the vacuum. Also,

$$\operatorname{weight}(E\,.\,w) \geq \operatorname{weight}(E) + \operatorname{weight}(w). \tag{7.121}$$

Proposition 7.41. *The two definitions of a Verma module coincide.*

Proof. By the previous remark, a Verma module in the new definition is equipped with a weight decomposition. Only the vacuum vector has weight zero. Condition 3 is automatic from the definition of the weight. Conditions 4 and 5 are true by the construction of the Verma modules, and vice versa, the basis elements of a Verma module in the first definition can be identified with the basis elements in $M(c, h)$. □

Proposition 7.42. *We set*

$$E_0 := \sum_{p=1}^{K} E_{(0,p)}, \quad \underline{h} = \sum_{p=1}^{K} h_p. \tag{7.122}$$

Let

$$w = E_{n_1,p_1} \,.\, E_{n_2,p_2} \,.\, \dots \,.\, E_{n_r,p_r} \,.\, v$$

with the increasing condition as in (7.113), *then we obtain*

$$E_0 \,.\, w \equiv (deg(w) + \underline{h}) \cdot w. \tag{7.123}$$

Here $\equiv$ *means up to terms of higher weight.*

Proof. Let $E_{0,p}$ act on $w \neq 0$. If w is the vacuum vector, then weight$(w) = 0$ and $E_0 \,.\, w = (deg(w) + \underline{h})$. Now let w be different to the vacuum vector. Hence $w = E_{n_1,p_1} \,.\, w'$ with $n_1 < 0$. Moreover, weight$(w') >$ weight(w).

$$E_{0,p} \,.\, w = E_{0,p} \,.\, (E_{n_1,p_1} \,.\, w').$$

This can be rewritten as

$$E_{0,p} \,.\, w = ([E_{0,p}, E_{n_1,p_1}])w' + E_{n_1,p_1} E_{0,p} w' \equiv n_1 \delta_p^{p_1} E_{n_1,p_1} w' + E_{n_1,p_1} E_{0,p} w'.$$

We can replace the last summand by induction till we reach the vacuum vector and obtain

$$E_{0,p} \,.\, w \equiv \left(\sum_{k=1}^{r} \delta_p^{p_k} n_k + h_p \right) w. \tag{7.124}$$

By summing this up over $p = 1, \dots, K$ we obtain

$$E_0 \,.\, w \equiv \left(\sum_{k=1}^{r} n_k + \underline{h} \right) \cdot w. \tag{7.125}$$

□

In this respect, the element E_0 might be considered a shifted weight operator.

Proposition 7.43. *The Verma module* $M(c, h)$ *is universal with respect to any highest weight representations* W *of* $\widehat{\mathfrak{L}}$ *with the same charge and weight. This means that a unique surjective* $\widehat{\mathfrak{L}}$*-module morphism exists*

$$\phi : M(c, h) \to W. \tag{7.126}$$

Consequently, W *is a quotient of* $M(c, h)$.

Proof. As V is a highest weight representation, we have a $v \in V$ which generates the whole space V, $t \,.\, v = c \cdot v$, $E_{0,p} \,.\, v = h_p \cdot v$, and $E_{n,p} \,.\, v = 0$, for $n > 0$. By the definition of the universal enveloping algebra we have a map

$$\tilde{\phi} : U(\widehat{\mathfrak{L}}) \otimes_{\mathbb{C}} V(c, h) \to W, \quad x \otimes 1 \mapsto \tilde{\phi}(x \otimes 1) = x \,.\, v. \tag{7.127}$$

This map factors via $M(c, h)$ by the very definition of W. Take as ϕ the quotient map of $\tilde{\phi}$. By the universality condition of the universal enveloping algebra this map is unique.

□

Remark 7.44. The semi-infinite wedge representations defined as submodules of $\mathcal{H}^{\lambda,(T,s)}$ generated by $\widehat{\mathcal{L}}$ from the vacuum $\Phi_{T,s}$ are quotients of Verma modules $M(c,h)$, where $c = 1$ and

$$h_p = \begin{cases} -\frac{T(T-1)}{2}, & p \geq s \\ -\frac{T(T-3)}{2}, & p < s. \end{cases} \tag{7.128}$$

Here the central extension $\widehat{\mathcal{L}}$ is defined by the cocycle

$$(-2)(6\lambda^2 - 6\lambda + 1)\,\gamma_{S,R}^{\mathcal{L}}. \tag{7.129}$$

7.4 Some remarks on the Heisenberg algebra representations

This section is a digression on the relationship between the fermionic and bosonic representation of the Heisenberg algebra in the classical genus zero and $N = 2$ situation.

Recall that via Theorem 7.18 (by the action on the semi-infinite wedge forms) we get a representation of the almost-graded centrally extended function algebra $\widehat{\mathcal{A}}$, the Heisenberg or oscillator algebra (see also Remark 6.40). The defining cocycle is given by

$$\gamma(f,g) = -\gamma_S^{\mathcal{A}}(f,g) = -\frac{1}{2\pi \mathrm{i}} \int_{C_S} f df. \tag{7.130}$$

In our normalization it is independent of the weight λ. In the classical case, the algebra has as basis elements $A_n = z^n$, $n \in \mathbb{Z}$ and a central element t. With the help of the cocycle we calculate

$$[A_n, A_m] = -n\,\delta_m^{-n}\, t. \tag{7.131}$$

If we consider the subspace of the semi-infinite wedge forms generated by the vacuum (λ arbitrary)

$$\Phi_T := f_T \wedge f_{T+1} \wedge \ldots, \tag{7.132}$$

then this representation is called *fermionic Fock space representation of charge* T and denoted by F_T. We have

(1) The set $\{A_n \,.\, \Phi_T \mid n < 0\} \cup \{\Phi_T\}$ will be a basis of the representation space.
(2) The algebra $\widehat{\mathcal{A}}_+$ annihilates the vacuum, i.e., $\widehat{\mathcal{A}}_+ \,.\, \Phi_T = 0$.
(3) $A_0 \,.\, \Phi_T = (-T)\cdot \phi_T$ (see (7.102)).
(4) The central element t operates as $-id$.
(5) With respect to the degree decomposition (7.106), the representation module is a graded module and the representation is irreducible.

For the Heisenberg algebra another representation exists, the *bosonic Fock space representation,* which is defined in the following way. Let

$$B = \mathbb{C}[x_1, x_2, x_3, \ldots, x_n, \ldots] \tag{7.133}$$

be the algebra of polynomials in infinitely many variables $\{x_i \mid i \in \mathbb{N}\}$. The vector space B is generated by a basis given as monomials consisting individually of a finite number of variables x_i. We define the following action of $\widehat{\mathcal{A}}$ on B. Let $\mu \in \mathbb{C}$ be a fixed constant and $f \in B$, then

$$\begin{aligned} b(A_0) . f &:= \mu \cdot f \\ b(A_{-n}) . f &:= n \cdot x_n \cdot f, \quad n \in \mathbb{N} \\ b(A_n) . f &:= \frac{1}{\partial x_n} f, \quad n \in \mathbb{N} \\ b(t) . f &= -f. \end{aligned} \tag{7.134}$$

Obviously, as operators

$$\begin{aligned} [b(A_n), b(A_m)] &= 0, \quad m \neq -n, \\ [b(A_n), b(A_{-n})] &= n \cdot f = (-n) b(t) = b([A_n, A_{-n}]), \\ [b(A_n), b(t)] &= 0. \end{aligned} \tag{7.135}$$

Hence, the linearly extended map b is a representation of $\widehat{\mathcal{A}}$ on the space B. We denote this representation space and also the representation by $B(\mu)$. It is clear that it is irreducible and it is generated by the constant polynomial 1. The uniqueness result for irreducible representations of the Heisenberg algebra (see, e.g., [122, Proposition 2.1]) says that two such representations generated by one vector (the vacuum), are equivalent if t operates as scalar, A_0 operates as scalar on the vacuum, and the individual scalars are the same. If we compare these values we see that both representations $B(-T)$ and F_T are equivalent[2]. In particular, an intertwining map exists

$$\Psi : \quad B(-T) \to F_T. \tag{7.136}$$

This map, which carries polynomials f to semi-infinite wedge forms, is also called *boson-fermion correspondence*. The introduced standard basis elements of the semi-infinite wedge forms will not correspond to a monomial basis of B. They are related to *Schur polynomials*. For more details see [122, Lectures 5 and 6].

Remark 7.45. The question is whether a similar correspondence for the higher genus and multi-point case also exists. Via vertex algebra type construction, Linde [161, 162] was able to give an analogue at least in the two-point case. This should extend also to the multi-point case. See Section 14.1 for a few more remarks.

2 Of course, a direct proof could also be given.

7.5 Left semi-infinite forms

Corresponding to the (right) semi-infinite wedge forms we can also introduce such semi-infinite wedge forms running to the left. For this we make $\mathcal{F}^\lambda$ to a right module via

$$f^\lambda . e := -e . f^\lambda, \quad f \in \mathcal{F}^\lambda,\ e \in \mathcal{L}. \tag{7.137}$$

The left semi-infinite forms are now linear combinations of basis elements

$$\dots f_{m-1,1} \wedge f_{m,K} \dots \wedge f_{m,2} \wedge f_{m,1} \wedge \dots \wedge f_{(i_2)} \wedge f_{(i_1)}.$$

Here (i_k) denotes a double index and the elements are lexicographically ordered, where the order in the second index is the inverted one (this means $(m,p) < (m,p')$ for $p > p'$). Up to a certain index all forms will appear. To distinguish them we will denote by $\mathcal{H}^\lambda{}_r$ the right forms already introduced and the left forms by $\mathcal{H}^\lambda{}_l$. If we do not state left or right, we assume by convention right forms.

Again the right action of $\mathcal{L}_+$ and $\mathcal{L}_-$ on $\mathcal{H}^\lambda{}_l$ is well-defined. It can be extended to an action of a central extension $\widehat{\mathcal{L}}$. The defining cocycle is local and by the uniqueness (up to equivalence and rescaling), up to an identification only involving the critical subspace, the same algebra as on the right forms will act on the left forms too.

Below we consider pairs of weights $(\lambda, 1-\lambda)$ and can be more precise.

Lemma 7.46. *Let $C_{(k,r)(m,p)}^{(n,s)}(\lambda)$ be the structure constants of the module $\mathcal{F}^\lambda$ over $\mathcal{L}$ defined by*

$$e_{k,r} . f_{m,p}^\lambda = \sum_{n=k+m}^{k+m+S} \sum_{p=1}^{K} C_{(k,r)(m,p)}^{(n,s)}(\lambda)\, f_{n,s}^\lambda, \tag{7.138}$$

then

$$C_{(k,r)(m,p)}^{(n,s)}(\lambda) = -C_{(k,r)(-n,s)}^{(-m,p)}(1-\lambda). \tag{7.139}$$

Proof. Lemma 6.17 says for pairs $f \in \mathcal{F}^\lambda$ and $h \in \mathcal{F}^{1-\lambda}$

$$\frac{1}{2\pi i}\int_{C_S} (e . f) \cdot h = -\frac{1}{2\pi i}\int_{C_S} (f \cdot (e . h). \tag{7.140}$$

Using the method of calculating the structure constants with the help of the Krichever–Novikov duality we obtain

$$\begin{aligned} C_{(k,r)(m,p)}^{(n,s)}(\lambda) &= \frac{1}{2\pi i}\int_{C_S} (e_{k,r} . f_{m,p}^\lambda) \cdot f_{-n,s}^{1-\lambda} = -\frac{1}{2\pi i}\int_{C_S} f_{m,p}^\lambda \cdot (e_{k,r} . f_{-n,s}^{1-\lambda}) \\ &= -C_{(k,r)(-n,s)}^{(-m,p)}(1-\lambda). \end{aligned}$$

□

Using (7.139) the pulled back cocycle fulfills $\Psi_\lambda = -\Psi_{1-\lambda}$. There is an isomorphism

$$h : \widehat{\mathcal{L}}_r \to \widehat{\mathcal{L}}_l \tag{7.141}$$

(r (l) denote the extension defined via the action on the right (left) forms). On $\widehat{\mathcal{L}}_+$ and $\widehat{\mathcal{L}}_-$ it is the identity. Of course, everything makes sense for the other algebras $\widehat{\mathcal{A}}$ and $\widehat{\mathcal{D}}^1$ too.

In physics, but of course not only there, one needs a scalar product or at least a pairing between dual spaces. A natural pairing is given between the spaces $\mathcal{H}_l^{1-\lambda}$ and $\mathcal{H}_r^{\lambda}$. If we take the respective basis elements

$$\begin{aligned}\psi &= f_{(j_1)}^{\lambda} \wedge f_{(j_2)}^{\lambda} \cdots \wedge f_{(j_k)}^{\lambda} \cdots,\\ \phi &= \ldots \wedge f_{(i_k)}^{1-\lambda} \wedge \ldots f_{(i_2)}^{1-\lambda} \wedge f_{(i_1)}^{1-\lambda},\end{aligned} \tag{7.142}$$

we set

$$\langle \phi, \psi \rangle = \prod_{k \in \mathbb{N}} \frac{1}{2\pi \mathrm{i}} \int_{C_S} f_{(i_k)}^{1-\lambda} \cdot f_{(j_k)}^{\lambda}, \tag{7.143}$$

and extend it bilinearly. By the Krichever–Novikov duality only factors 0 or 1 will appear in the product. The dual basis element of

$$\phi_T = \phi_{T,1} = f_{T,1}^{\lambda} \wedge f_{T,2}^{\lambda} \wedge \ldots$$

is

$$\phi_T^* = \phi_{T,1}^* = \cdots \wedge f_{-T,2}^{1-\lambda} \wedge f_{-T,1}^{1-\lambda}.$$

In particular now we have

$$\phi_T^* \,.\, \widehat{\mathcal{L}}_- = 0. \tag{7.144}$$

Proposition 7.47. *For $e \in \mathcal{L}$ we denote by $\hat{e}$ the standard lift to $\widehat{\mathcal{L}}$. Then*

$$\langle \psi, \hat{e} \,.\, \phi \rangle = \langle \psi \,.\, h(\hat{e}), \phi \rangle \quad \langle \psi, t \,.\, \phi \rangle = \langle \psi \,.\, h(t), \phi \rangle. \tag{7.145}$$

Proof. First assume that our element e is from $\widehat{\mathcal{L}}_+$ or $\widehat{\mathcal{L}}_-$. By Lemma 6.17 we have the relation

$$\frac{1}{2\pi \mathrm{i}} \int_{C_S} (e \,.\, f_{i_k}^{1-\lambda}) \cdot f_{j_k}^{\lambda} = -\frac{1}{2\pi \mathrm{i}} \int_{C_S} f_{i_k}^{1-\lambda} \cdot (e \,.\, f_{j_k}^{\lambda}). \tag{7.146}$$

Taking the minus from the right module action into account, the individual terms in (7.143) remain invariant. Hence, for them we get the statement (7.145).
For the elements from the critical strip we need to have a closer look at the regularization procedure again. We will skip the details here. We only indicate that we need to show (see notation in Section 7.1.2)

$$\langle \psi_r \check{r}(A_0^*(\mu^*)), \phi_t \rangle = \langle \psi_r, \hat{r}(A_0(\mu)), \phi_t \rangle. \tag{7.147}$$

Here $\check{r}$ corresponds to the regularization with respect to left forms, $\hat{r}$ to the right forms treated there. Both operators are scalar operators. In the sequence μ the coefficients of the $\mathcal{F}^{\lambda}$-module show up, in μ^* the coefficients of $\mathcal{F}^{1-\lambda}$. Using Lemma 7.46 we will obtain the result.[3] □

3 Those readers wanting more details are referred to [207, p. 121].

Remark 7.48. It is also possible to introduce other pairings.

(a) One could also simply take the basis given by the basis semi-infinite wedge forms of $\mathcal{H}^\lambda = \mathcal{H}^\lambda{}_r$ and declare them to be an orthonormal basis of the space. In this case it seems that there is no concept of self-adjoint action.

(b) Let ϕ be a basis semi-infinite wedge form from $\mathcal{H}^\lambda_r$, and ψ a basis semi-infinite wedge form from $\mathcal{H}^\lambda_l$ (note in this case we take the same weight λ). Now we take the both-sides formal concatenation $\psi \wedge \phi$. If in the result not all basis elements $f^\lambda_{m,r}$ appear, or if a least one appears twice, we set $\langle \phi, \psi \rangle = 0$. Otherwise we set

$$\langle \phi, \psi \rangle = \operatorname{sign}(\sigma), \tag{7.148}$$

where σ is the permutation needed to establish the complete natural order and sign is the signature of σ. The complete definition follows from bilinear extension. Krichever and Novikov [141, 142, 144] showed that in the two-point case, the action of $\widehat{\mathcal{L}}$ is again self-adjoint as above. Their proof should also extend to the multi-point situation.

8 $b-c$ systems

In the quantization of conformal field theory, string theory, etc., so-called (Polyakov–Faddeev–Popov) ghost fields show up. These are pairs of operator fields of conformal weight, -1 and 2 respectively. They play an important role in the context of anomalies of the quantized string theory. They are called ghost fields as they do not have a correspondence to objects in non-quantized physics. These fields make sense for every pair of complementary weights $(\lambda, 1-\lambda)$ and are called $b-c$ fields in physics. In the language of physics they are fermionic fields.

From the mathematical point of view, the b and c operators are defined via anti-commutators, obey a Clifford algebra like structure, and are mathematically interesting objects.

In this chapter we will show how the Krichever–Novikov objects will give a global treatment for $b-c$ systems. We start by realizing these operators as operators on the semi-infinite wedge forms of weight λ. One type act as insertion, the other type by contraction. More generally we will consider arbitrary representation spaces of these field operators and will define the energy momentum tensor. We show that its "modes" will define a representation of an almost-graded central extension of the vector field algebra $\widehat{\mathcal{L}}$. In the two-point case in the Krichever–Novikov setting, $b-c$ systems were discussed by Bonora, Lugo, Matone, and Russo [22]. For an example in the torus case see also Ruffing, Deck, and Schlichenmaier [197]. The multi-point situation is given in [207] and [210]. These results will now be reproduced.

In a similar way, on the basis of the product of the $b-c$ fields, we obtain a representation of the (higher genus) Heisenberg algebra.

Furthermore, in Section 8.2 we will study general global conformal operator fields in the Krichever–Novikov setting.

8.1 The Clifford algebra like structure

Recall that given a ring $(R, \circ)$, the anti-commutator is defined as

$$\{D, E\}_+ = D \circ E + E \circ D \quad D, E \in R. \tag{8.1}$$

The ring we will consider here is the ring of (linear) endomorphisms of a vector space V. More specifically, we consider in this section $V = \mathcal{H}^\lambda$ the space of semi-infinite wedge forms. As we want to also have a pairing, we will consider left and right forms. Let $\mathcal{H}_r^\lambda$ be the space of right forms of λ, and $\mathcal{H}_l^{1-\lambda}$ the dual space of left forms of weight $1-\lambda$. To simplify the notation the weight will only be mentioned if necessary, and $f_{n,p} = f_{n,p}^\lambda$, and correspondingly $h_{n,p} = f_{n,p}^{1-\lambda}$, will be used. Also, let (j_k) denote a double index.

The vector spaces $\mathcal{F}^\lambda$ and $\mathcal{F}^{1-\lambda}$ act on the semi-infinite wedge forms as follows. Let $f \in \mathcal{F}^\lambda$ and $h \in \mathcal{F}^{1-\lambda}$. We define

$$c_f \,.\, w = f \wedge w, \quad b_h \,.\, w = i_h(w), \quad w \in \mathcal{H}_r^\lambda \tag{8.2}$$

$$w \,.\, c_f = (w) i_f, \quad w \,.\, b_h = w \wedge h, \quad w \in \mathcal{H}_l^{1-\lambda}. \tag{8.3}$$

Here i_h is the contraction. On every factor $f_{m,r}$ it is defined as

$$i_h(f_{m,r}) = \frac{1}{2\pi i} \int_{C_S} h \cdot f_{m,r}, \tag{8.4}$$

and on

$$w = f_{(j_1)} \wedge f_{(j_2)} \wedge \ldots \wedge f_{(j_l)} \wedge \ldots$$

by the modified Leibniz rule

$$i_h(w) = \sum_{l=1}^{\infty} (-1)^{l-1} i_h(f_{(j_l)}) \cdot f_{(j_1)} \wedge \ldots \wedge \check{f}_{(j_l)} \wedge \ldots. \tag{8.5}$$

Here $\check{f}_{(j_l)}$ denotes as usual that this factor is skipped. From the almost-graded structure and the duality (3.10) it follows that the contraction is well-defined, as in the sum (8.5) only a finite number of terms remain. The definition on the left forms is accordingly.

Obviously the maps are linear. They are also linear in the index, this says

$$c_{e+f} = c_e + c_f, \quad b_{g+h} = b_g + b_h. \tag{8.6}$$

For the operators corresponding to the basis elements $f_{n,p}$, respectively $h_{n,p}$, we will use

$$c_{n,p} := c_{f_{n,p}}, \quad \text{and} \quad b_{n,p} := b_{h_{n,p}}. \tag{8.7}$$

Intuitively the operation of $c_{n,p}$ corresponds (on the right forms) to the insertion of $f_{n,p}$ and the operation of $b_{n,p}$ to removing $f_{-n,p}$. On the left forms it is just the opposite.

Proposition 8.1. *Both for the operation on the left and on the right forms we have*

$$\begin{aligned} \{b_g, b_h\}_+ &= 0, \quad g, h \in \mathcal{F}^{1-\lambda}, \\ \{c_e, c_f\}_+ &= 0, \quad e, f \in \mathcal{F}^\lambda, \\ \{b_{n,p}, c_{m,r}\}_+ &= \delta_m^{-n} \delta_p^r \, \mathrm{id}\,. \end{aligned} \tag{8.8}$$

Proof. We will only consider the right forms. The proof for the left forms works accordingly. As

$$e \wedge f \wedge w = -f \wedge e \wedge w$$

we directly get $\{c_e, c_f\} = 0$.

For $k \neq l$ we use the abbreviation

$$y_{kl} = i_h(f_{(j_k)}) \cdot i_g(f_{(j_l)}) \cdot f_{(j_1)} \wedge \ldots \check{f}_{(i_l)} \wedge \ldots \check{f}_{(i_k)} \wedge \ldots.$$

Note that for a given g and h of course only a finite number of these y_{kl} will be non-vanishing. With a semi-infinite from w we obtain

$$
\begin{aligned}
i_g(i_h(w)) &= \sum_{k=1}^{\infty}\sum_{l=1}^{k-1}(-1)^{k-1}(-1)^{l-1}\, y_{kl} + \sum_{k=1}^{\infty}\sum_{l=k+1}^{\infty}(-1)^{k-1}(-1)^{l-2}\, y_{kl} \\
&= \sum_{\substack{k,l\\ l<k}}(-1)^{k+l}\, y_{kl} + \sum_{\substack{k,l\\ l>k}}(-1)^{k+l-1}\, y_{kl}.
\end{aligned}
$$

Again all the sums above are finite. If we change the order of h and g this corresponds to the fact that the roles of k and l are switched. Hence, we obtain the negative of the expression above. Which shows $\{b_g, b_h\} = 0$.

This leaves the last relation.

$$
\begin{aligned}
b_{n,p}\cdot(c_{m,r}\,.\,w) &= b_{n,p}\cdot(f_{m,r}\wedge w) = i_{h_{n,p}}(f_{m,r})\cdot w - f_{m,r}\wedge i_{h_{n,p}}(w), \\
c_{m,r}\cdot(b_{n,p}\,.\,w) &= c_{m,r}\cdot(i_{h_{n,p}}(w)) = f_{m,r}\wedge i_{h_{n,p}}(w).
\end{aligned}
$$

Hence,

$$
\{b_{n,p}, c_{m,r}\} = i_{h_{n,p}}(f_{m,r})\cdot w = \delta_m^{-n}\delta_p^r\cdot w. \qquad \square
$$

Proposition 8.2.

(a) *The operators $b_{n,p}$ and $c_{n,p}$ on the left and right forms are self-adjoint with respect to the duality pairing (7.142), this says for $\phi \in \mathcal{H}_l^{1-\lambda}$, $\psi \in \mathcal{H}_r^{\lambda}$*

$$
\langle \phi\,.\,c_{n,p}, \psi\rangle = \langle \phi, c_{n,p}\,.\,\psi\rangle, \tag{8.9}
$$

$$
\langle \phi\,.\,b_{n,p}, \psi\rangle = \langle \phi, b_{n,p}\,.\,\psi\rangle. \tag{8.10}
$$

(b) *Let Φ_T be the vacuum vector of weight λ and level T (see (7.96)), and Φ_T^* the dual vacuum vector. Then*

$$
c_{n,p}\,.\,\Phi_T = 0,\quad n \ge T, \qquad b_{n,p}\,.\,\Phi_T = 0,\quad n \ge 1-T, \tag{8.11}
$$

$$
\Phi_T^*\,.\,c_{n,p} = 0,\quad n \le T-1, \qquad \Phi_T^*\,.\,b_{n,p} = 0,\quad n \le -T. \tag{8.12}
$$

(c) *Let $w \in \mathcal{H}_r^{\lambda}$, then $n_0, m_0 \in \mathbb{Z}$ (both depending on w) exist, such that*

$$
c_{n,p}\,.\,w = 0,\quad b_{m,p}\,.\,w = 0,\quad \forall n \ge n_0,\quad \forall m \ge m_0. \tag{8.13}
$$

The corresponding with $n \le n_0$ and $m \le m_0$ is true for the action on the left forms.

Proof. (a) Let $c_{n,p}$ be given. Case 1: the element $f_{n,p}$ appears in ψ. Then $c_{n,p}\,.\,\psi = 0$. On the other side $c_{n,p}$ applied to ϕ removes $h_{-n,p}$ if there. In this case $f_{n,p}$ does not have a partner anymore. This says that the pairing yields zero. Case 2: the element $f_{n,p}$ does not appear in ψ. First assume that (n, p) is smaller than the first appearing index. The pairing will be non-zero if and only if $h_{-n,p}$ appears as the first element in ϕ and all other elements are pairwise dual to the elements in ψ. The same value is given by the operation on ϕ. In the case that (n, p) is larger than the first appearing index, then by interchanging and taking up signs $f_{m,p}$ has to be brought to the correct position. This corresponds exactly to the sign factor for the contraction. Hence (8.9). The same argument works for (8.10).

(b) On the right forms the action of $c_{n,p}$ is the insertion of $f_{n,p}$. If $n \geq T$ it is already there and the result will be zero. The element $b_{n,p}$ is the removing of $f_{-n,p}$. Again for $n \geq 1 - T$ it is not there. The argument is similar for the left forms.

(c) Recall that in $w \in \mathcal{H}_r^\lambda$ starting from a certain index all elements appear and only a finite number of elements with negative index will appear. The same kinds of arguments used for (b) works for an arbitrary w if n is big enough. These n will depend on w. □

By linearity we can generalize the results (8.9), (8.10) also to

$$\langle \phi \,.\, c_f, \psi \rangle = \langle \phi, c_f \,.\, \psi \rangle, \quad \langle \phi \,.\, b_h, \psi \rangle = \langle \phi, b_h \,.\, \psi \rangle. \tag{8.14}$$

The results obtained can be interpreted as follows. The relations (8.8) say that the subalgebra of the endomorphism of $\mathcal{H}_r^\lambda$ spanned by

$$\langle \, \{b_{n,p}, c_{n,p} \mid n \in \mathbb{Z}, p = 1, \ldots, K\} \, , 1 \rangle_{\mathbb{C}} \tag{8.15}$$

with the anti-commutator carries a Clifford algebra like structure. If we interpret Φ_T as ground state, then relations (8.11) and (8.13) say that the operators $c_{n,p}$ and $b_{n,p}$ can be considered annihilation operators on the right forms. These results are of the type physicists expect from $b-c$ systems.

If we consider the subspaces $\mathcal{H}_{T,p}^\lambda$ of fixed charge given by (T, p) as they were introduced in (7.96), then the operators b and c do not leave them invariant. On the contrary,

$$\mathcal{H}_{T,K-1}^\lambda \;\underset{c}{\overset{b}{\rightleftarrows}}\; \mathcal{H}_{T,K}^\lambda \;\underset{c}{\overset{b}{\rightleftarrows}}\; \mathcal{H}_{T+1,1}^\lambda \;\underset{c}{\overset{b}{\rightleftarrows}}\; \cdots . \tag{8.16}$$

This shows in particular that the spaces of different charges are isomorphic.

On $\mathcal{H}_r^\lambda$ and $\mathcal{H}_l^{1-\lambda}$, besides $\mathcal{F}^\lambda$ and $\mathcal{F}^{1-\lambda}$, also $\widehat{\mathcal{A}}$, $\widehat{\mathcal{L}}$, and $\widehat{\mathcal{D}}^1$ operate. We have the following interesting commutator relations.

Proposition 8.3. *Let $\hat{e} \in \widehat{\mathcal{D}}^1$ be a lift of a vector field, $e \in \mathcal{L}$, $\hat{a} \in \widehat{\mathcal{D}}^1$ a lift of a function $a \in \mathcal{A}$, and t the central element in $\widehat{\mathcal{D}}^1$, $f \in \mathcal{F}^\lambda$ and $h \in \mathcal{F}^{1-\lambda}$. Then for the operators on $\mathcal{H}_r^\lambda$ we obtain*

$$\begin{aligned} [\,\hat{e}, c_f\,] &= c_{e\,.\,f}, \quad [\,\hat{e}, b_h\,] = b_{e\,.\,h}, \\ [\,\hat{a}, c_f\,] &= c_{a\cdot f}, \quad [\,\hat{a}, b_h\,] = b_{-a\cdot h}, \\ [\,t, c_f\,] &= [\,t, b_h\,] = 0. \end{aligned} \tag{8.17}$$

In particular, the subspace of End $\mathcal{H}_r^\lambda$ spanned by $\widehat{\mathcal{D}}^1$, $c(\mathcal{F}^\lambda)$, and $b(\mathcal{F}^{1-\lambda})$ is a Lie subalgebra.

Proof. Let $\phi \in \mathcal{H}_r^\lambda$ be a basis element. We might write it as $\phi = w \wedge \psi$ with a finite part w and an infinite part ψ. We make the starting chain w long enough so that neither ψ nor $\hat{e}\,.\,\psi$ "has something to do" with the indices on which i_h and $i_{e\,.\,h}$ operate nontrivially. Given an arbitrary lift $\hat{e}$ of $e \in \mathcal{L}$, we split the action

$$\hat{e}\,.\,(w \wedge \psi) = (e\,.\,w) \wedge \psi + w \wedge (\hat{e}\,.\,\psi). \tag{8.18}$$

Note that on the finite part the $\hat{e}$-action coincides with the e-action. We obtain

$$\hat{e}\,.\,(f \wedge \phi) = (e\,.\,(f \wedge w)) \wedge \psi + f \wedge w \wedge (\hat{e}\,.\,\psi)$$

and

$$f \wedge (\hat{e}\,.\,\phi) = f \wedge (e\,.\,w) \wedge \psi + f \wedge w \wedge (\hat{e}\,.\,\psi),$$

which yields

$$[\hat{e}, f\wedge]\,\phi = ([e, f\wedge]\,.\,w) \wedge \psi. \tag{8.19}$$

Accordingly,

$$\hat{e}\,.\,(i_h(\phi)) = \hat{e}\,.\,(i_h(w) \wedge \psi) = (e\,.\,i_h(w)) \wedge \psi + i_h(w) \wedge (\hat{e}\,.\,\psi),$$
$$i_h(\hat{e}(\phi)) = i_h(e\,.\,w \wedge \psi + w \wedge (\hat{e}\,.\,\psi)) = i_h(e\,.\,w) \wedge \psi + i_h(w) \wedge (\hat{e}\,.\,\psi),$$

which yields

$$[\hat{e}, i_h]\,\phi = ([e, i_h]\,w) \wedge \psi. \tag{8.20}$$

From (8.19) and (8.20) it follows that if the results are true for finite starting chains, then

$$[\hat{e}, f\wedge]\,\phi = (e\,.\,f) \wedge w \wedge \psi = (e\,.\,f) \wedge \phi,$$
$$[\hat{e}, i_h]\,\phi = i_{e\,.\,h}(w) \wedge \psi = i_{e\,.\,h}(\phi).$$

Hence, it is enough to show the results for finite starting chains. For the first expression

$$e\,.\,(f \wedge w) = e\,.\,f \wedge w + f \wedge (e\,.\,w), \quad \text{and hence} \quad [e, f\wedge]\,w = (e\,.\,f) \wedge w.$$

This was to be shown.

For the second expression we calculate (with n the length of the finite piece w)

$$\begin{aligned} e\,.\,(i_h(w)) &= e\,.\left(\sum_{k=0}^{n}(-1)^{k-1} i_h(f_{(j_k)}) \cdot f_{(j_1)} \wedge \ldots \check{f}_{(j_k)} \wedge \ldots f_{(j_n)}\right) \\ &= \sum_{\substack{k,l=0\\ l\neq k}}^{n} (-1)^{k-1} i_h(f_{(j_k)}) \cdot f_{(j_1)} \wedge \ldots (e\,.\,f_{(i_l)}) \ldots \check{f}_{(j_k)} \wedge \ldots f_{(j_n)}. \end{aligned}$$

In the reverse order we obtain

$$\begin{aligned} i_h(e\,.\,w) = &\sum_{\substack{k,l=0\\ l\neq k}}^{n} (-1)^{k-1} i_h(f_{(j_k)}) \cdot f_{(j_1)} \wedge \ldots \check{f}_{(j_k)} \wedge (e\,.\,f_{(i_l)}) \ldots f_{(j_n)} \\ &+ \sum_{l=0}^{n}(-1)^{l-1} i_h(e\,.\,f_{(j_l)}) \cdot f_{(j_1)} \wedge \ldots \check{f}_{(j_l)} \wedge \ldots f_{(j_n)}. \end{aligned}$$

Hence, as difference

$$[e, i_h](w) = -\sum_{l=0}^{n}(-1)^{l-1} i_h(e \,.\, f_{(j_l)}) \cdot f_{(j_1)} \wedge \dots \check{f}_{(j_l)} \wedge \dots f_{(j_n)}.$$

By Lemma 6.17

$$i_h(e \,.\, f_{(i_l)}) = \frac{1}{2\pi\mathrm{i}} \int_{C_S} h \cdot (e \,.\, f_{(i_l)}) = -\frac{1}{2\pi\mathrm{i}} \int_{C_S} (e \,.\, h) \cdot f_{(i_l)}, \tag{8.21}$$

and hence

$$[e, i_h](w) = \sum_{l=0}^{n}(-1)^{l-1} i_{e.h}(f_{(j_l)}) \cdot f_{(j_1)} \wedge \dots \check{f}_{(j_l)} \wedge \dots f_{(j_n)} = i_{e.h}(w).$$

This shows the claim for the action associated to the vector fields. The considerations above work (nearly) unchanged for $\hat{a} \in \widehat{\mathcal{A}} \subset \widehat{\mathcal{D}}^1(A)$. In this case $e \,.\, f$ has to be replaced by $a \,.\, f$, etc.,. Instead of (8.21) one has to use

$$i_h(a \cdot f_{(i_l)}) = \frac{1}{2\pi\mathrm{i}} \int_{C_S} h \cdot a \cdot f_{(i_l)} = i_{a \cdot h}(f_{(i_l)})$$

without a change of sign. Hence the changed sign in the final result. The results involving the central element are clear. □

For the importance of this algebra in the classical $g = 0$ and $N = 2$ case see [125, Equation 3.32].

8.2 Operator valued fields in conformal field theory

The way physicists use quantum fields in conformal field theory (CFT) can be formulated mathematically in the method now described. We will formulate it via global operator fields on Riemann surfaces of arbitrary genus and a finite number of "insertion" points. The local description usually found in the mathematical literature [119, 123], etc., will show up as genus zero and two-point situation.

Let V be a vector space called the *state space* . In applications V is quite often an infinite-dimensional $\mathbb{Z}$ (or $\mathbb{J}_\lambda$) graded vector space

$$V = \bigoplus_{n \in \mathbb{Z}} V_n, \tag{8.22}$$

such that the homogeneous spaces for $n \gg 0$ will become trivial (or alternatively for $n \ll 0$).

Example 8.4. An example is given by the space $\mathcal{H}^\lambda$ of semi-infinite wedge forms of weight λ, or the subspace $\mathcal{H}^{\lambda,(m)}$ of fixed charge m. These spaces are graded with the help of the (charge dependant) degree. Accordingly, we take

$$V' = \mathcal{H}^{\lambda,(m)} = \bigoplus_{d\le 0} \mathcal{H}^{\lambda,(m)}_{(d)}, \tag{8.23}$$

and for the full space $\mathcal{H}^{\lambda}$

$$V = \mathcal{H}^{\lambda} = \bigoplus_{d\le 0} \mathcal{H}^{\lambda}_{(d)}, \quad \mathcal{H}^{\lambda}_{(d)} = \bigoplus_{m\in\mathbb{Z}} \mathcal{H}^{\lambda,(m)}_{(d)}. \tag{8.24}$$

Note that by Proposition 7.30 the degree is always ≤ 0. In the first example the homogeneous subspaces are finite-dimensional, in the second the subspaces of degree ≤ 0 will be infinite-dimensional. We stress the fact that if we consider the operators b and c introduced in the last section we have to take the larger space as they act between different charges (8.16).

For our general consideration here it is not necessary to assume such a grading of V. Let

$$\{f^{\lambda}_{n,p} \mid n \in \mathbb{J}_{\lambda},\ p = 1\dots, K\} \tag{8.25}$$

be the Krichever–Novikov basis elements for $\mathcal{F}^{\lambda}$. Also, despite the fact that we use the symbols $b_{n,r}$ and $c_{m,r}$ for the operators, we do not assume that these fields are the above operators, in particular we do not assume that we have fermionic fields.

Definition 8.5. The formal series

$$b(Q) = \sum_{n\in\mathbb{J}_{\lambda}} \sum_{p=1}^{K} b_{n,p} f^{\lambda}_{-n,p}(Q), \tag{8.26}$$

with $b_{n,p} \in \mathrm{End}(V)$ (this means they are operators on the state space) is called *(conformal) operator field of (conformal) weight* λ if

$$\forall v \in V,\ \exists n_0, \quad \text{such that} \quad b_{n,p} \,.\, v = 0,\ \forall n \ge n_0. \tag{8.27}$$

Equivalently, (8.27) can be described as that for all $v \in V$ an $M = M(v)$ exists, such that

$$\begin{aligned} b(Q)\,.\,v &= \sum_{n\le M(v)} \sum_{p} (a_{n,p} \,.\, v) f^{\lambda}_{-n,p}(Q) \\ &= \sum_{n\ge -M(v)} \sum_{p} (a_{-n,p} \,.\, v) f^{\lambda}_{n,p}(Q). \end{aligned} \tag{8.28}$$

Recall that

$$\mathrm{ord}_{P_i}(f^{\lambda}_{n,p}) = n - \lambda + (1 - \delta^{p}_{i}), \quad P_i \in I. \tag{8.29}$$

Hence, (8.28) can also be interpreted as the condition that the formal series "at the point P_i" is a Laurent series.

The operators $b_{n,p} : V \to V$ are sometimes called *modes* of the operator field $b(Q)$. Using the Krichever–Novikov duality (3.10), we obtain from (8.26) the expression for the modes

$$b_{n,p} = \frac{1}{2\pi \mathrm{i}} \int_{C_S} b(Q) f^{1-\lambda}_{n,p}(Q). \tag{8.30}$$

Example 8.6. In the classical $g = 0$ and $N = 2$ situation we have

$$f_{-n}^{\lambda} = z^{-n-\lambda}(dz)^{\lambda}. \tag{8.31}$$

Hence a field $b(z)$ of conformal weight λ is given as formal series[1]

$$b(z) = \sum_{n\in\mathbb{J}_\lambda} b_n z^{-n-\lambda}(dz)^{\lambda}, \quad b_n : V \to V. \tag{8.32}$$

In this case (8.30) yields

$$b_n = \frac{1}{2\pi \mathrm{i}} \oint b(z) z^{n+\lambda-1}(dz)^{1-\lambda}. \tag{8.33}$$

Here the integration is over a circle around $z = 0$.

Let $c(Q)$ be another field now of weight μ, i.e.,

$$c(Q) = \sum_{n\in\mathbb{J}_\lambda} \sum_{p=1}^{K} c_{n,p} f_{-n,p}^{\mu}(Q), \tag{8.34}$$

with $c_{n,p} \in \mathrm{End}(V)$ operators on the same state space. By the definition of a field, given $v \in V$ there exists $M'(v)$ such that if $n \geq M'(v)$, then $c_{n,p} . v = 0$. If we multiply the fields $b(Q)$ and $c(Q')$ at different points, Q and Q', we obtain formal series in two "variables". But we have to consider the product also at the same point. And here problems will occur. For the product we obtain another field $a(Q)$, given by

$$a(Q) =:b(Q)c(Q):= \sum_{n,p}\sum_{m,r} :b_{n,p}c_{m,r}:\, f_{-n,p}^{\lambda}(Q) \cdot f_{-m,r}^{\mu}(Q). \tag{8.35}$$

To ease reading we do not write down the range of the summation if it is clear from the context. Moreover, we use the symbol :...:. This should denote normal ordering to be discussed further down. For the beginning we assume that it is not there, respectively it just says that it is the usual product. We will show that this will create problems and define :...: accordingly to avoid them.

Assume that the product is a well-defined field, then it will be a field of conformal weight $\lambda + \mu$. Hence it can be written as

$$a(Q) =:b(Q)c(Q):= \sum_{k,s} a_{k,s} f_{-k,s}^{\lambda+\mu}. \tag{8.36}$$

The coefficient operators are calculated

$$a_{k,s} = \sum_{n,p}\sum_{m,r} :b_{n,p}c_{m,r}:\, \frac{1}{2\pi \mathrm{i}} \int_{C_S} f_{-n,p}^{\lambda} f_{-m,r}^{\mu} f_{k,s}^{1-(\lambda+\mu)}. \tag{8.37}$$

1 Often the $(dz)^{\lambda}$ is suppressed.

A priori the expression for $a_{k,s}$ is a formal sum of products of elements from End(V). Hence, it is not clear whether it will be an element of End(V) at all. Furthermore, if this is the case, it will not be automatically clear that given $v \in V$ for $k \gg 0$ we have the relation $a_{k,s} . v = 0$. Only if both conditions are fulfilled will $a(Q)$ be a field.

Let γ be the expression given by the integral in (8.37). The value of the expression will be zero if the product of the integrands is either holomorphic at the points in I or holomorphic at the points in O. From the definition of the orders in Section 4.2 and Section 4.3 we know that there will be an $S \geq 0$ independent of n and m, such that the value of γ will be zero outside of the range

$$0 \leq k - (n + m) \leq S. \tag{8.38}$$

Here k is fixed. Given k and n there will only be a finite number of m yielding a non-zero value. More precisely, the corresponding m is bounded by

$$k - n - S \leq m \leq k - n. \tag{8.39}$$

The infinite sum with respect to m disappears, but there is still the infinite sum with respect to n left. If we apply this sum to a vector $v \in V$ in general it will be an infinite sum of vectors.

Now we specify what we mean by :...:. It should take care that all other criteria for a field are satisfied. Recall that for m and n big enough, the vectors $b_{n,p} . v$ and $c_{m,r} . v$ will vanish. Hence, if we move them to the right in the product they will behave better. Exactly this is done with the *normal ordering*. An example of a normal ordering, called the standard normal ordering, is

$$:b_{n,p}c_{m,r}: := \begin{cases} \epsilon \cdot c_{m,r} \cdot b_{n,p}, & \text{if } n > m \\ b_{n,p} \cdot c_{m,r}, & \text{otherwise.} \end{cases} \tag{8.40}$$

Here ϵ is a sign factor depending on the parity of the field. Our $b - c$ fields which we study in the next section will be fermionic fields. Hence for them $\epsilon = -1$.

For $n \gg 0$ in (8.35), by (8.39) $m < n$, and the operator $b_{n,p}$ will be moved to the right and annihilates v. Hence the summation over n will be bounded from above. If $n \ll 0$, then $m \gg 0$ and the product is already normally ordered and $c_{m,r}$ will annihilate v. Consequently, the summation is also bounded from below. In total we showed that $a_{k,s} \in \text{End}(V)$.

Proposition 8.7. *Let* a =: $b(Q)c(Q)$: *be the normal ordered product of the fields* $b(Q)$, *and* $c(Q)$ *of weight* λ *and* μ *respectively, then* $a(Q)$ *is a field of weight* $\lambda + \mu$. *Its modes are calculated via* (8.37).

Proof. It remains to show that for all $v \in V$ a $k_0 = k_0(v)$ exists such that for all $k \geq k_0$ we have $a_{k,s} . v = 0$. Given v, then set $n_0 = M(v)$ and $m_0 = M'(v)$ the corresponding bounds with respect to the field $b(Q)$ and $c(Q)$ respectively. Furthermore, let $S \geq 0$ the constant introduced above. We set $k_0 := 2\max(n_0, m_0) + S$ and will show that this bound will do.

In dropping all irrelevant (for this problem) parts of the expression we obtain that the original sum is obtained as sum over $l = 0, 1, \dots, S$ of sums of the kind

$$\sum_{n,p}\sum_{r} :b_{n,p}c_{(k-l)-n,r}: .$$

For this the order has to be changed if $n > (k-l) - n$, which is the case if and only if $2n > (k-l)$. If $k \geq k_0$, then $n \geq n_0$ and the $b_{n,p}$ annihilates the vector. If the order has not been changed, then $n \leq \frac{1}{2}(k-l)$ and c_{k-l-n} acts first on v. As

$$(k-l) - n \geq \frac{1}{2}(k-l) \geq \frac{1}{2}(k-S) \geq m_0,$$

the operator $c_{k-l-n,r}$ annihilates v. □

Example 8.8. In the classical case ($g = 0$, $N = 2$) we have $S = 0$ and the integrals will be non-zero if and only if $m = k - n$. The value will then be equal to 1. Hence we get

$$a_k = \sum_{n\in\mathbb{Z}} :b_n c_{k-n}: . \tag{8.41}$$

What was been done here with respect to the product of fields can be done with the Lie algebra structure of the forms given by the Poisson bracket; see Section 2.4. Recall that our $\mathcal{F}$ carries a Lie algebra structure

$$\mathcal{F}^{\lambda} \times \mathcal{F}^{\mu} \to \mathcal{F}^{\lambda+\mu+1} \tag{8.42}$$

given by

$$(h, f) \mapsto P(h, f) := [h, f] := (1-\lambda)\, h\frac{df}{dz} - \lambda\frac{dh}{dz} f, \tag{8.43}$$

where the h and f are identified with their local representing functions. Instead of the Lie bracket of the forms, we prefer to use here the symbol $P(h, f)$ to avoid a conflict with the bracket of the operators.

We obtain the following proposition.

Proposition 8.9. *Given two fields $b(Q)$ and $c(Q)$ of weights λ and μ respectively, then*

$$a(Q) =: P(b(Q), c(Q)) := \sum_{n,p}\sum_{m,r} :b(Q)c(Q):\ P(f^{\lambda}_{-n,p}, f^{\mu}_{-m,r}) \tag{8.44}$$

will be a field of weight $\lambda + \mu + 1$, i.e.,

$$a(Q) = \sum_{k,s} a_{k,s} f^{\lambda+\mu+1}_{-k,s} . \tag{8.45}$$

Its modes are given by

$$a_{k,s} = \sum_{n,p}\sum_{m,r} :b_{n,p}c_{m,r}: \frac{1}{2\pi \mathrm{i}}\int_{C_S} P(f^{\lambda}_{-n,p}, f^{\mu}_{-m,r}) \cdot f^{\lambda+\mu}_{k,s} . \tag{8.46}$$

Such an example will be provided in Section 8.4.

Remark 8.10. The normal ordering is in a certain sense a regularization procedure which yields quite often the necessity of a central extension of the algebraic structures involved. The calculations might yield a two-cocycle for it.

Remark 8.11. The recipe (8.40) is not really canonical. It is just in our case a convenient one. Others are possible as long as the basic property of moving the annihilators to the right is kept. For example, we could allow a normal ordering which differs from the above standard ordering only for pairs $(n, p), (m, r)$ in a finite number of n and m components. We could also replace the $n > m$ by $n \geq m$. Hence, we have to keep the ambiguity in mind. With respect to the remark on an associated cocycle above, quite often a different normal ordering will yield a cohomologous cocycle.

8.3 $b - c$ fields

Let b and c be fields of weight λ and $1 - \lambda$ respectively. Furthermore, we assume that they are now fermionic fields. Our state space $\mathcal{V}$ can be arbitrary. We can write these fields as

$$b(Q) = \sum_{n \in \mathbb{J}_\lambda} \sum_{p=1}^{K} b_{n,p} f^{\lambda}_{-n,p}(Q), \tag{8.47}$$

$$c(Q) = \sum_{n \in \mathbb{J}_\lambda} \sum_{p=1}^{K} c_{n,p} f^{1-\lambda}_{-n,p}(Q) \tag{8.48}$$

with $b_{n,p}, c_{m,r} \in \mathrm{End}(V)$.

A particular example is given by $V = \mathcal{H}^\lambda$, and $b_{n,c}$, and $c_{n,p}$, the operators introduced in Section 8.1. Proposition 8.2 (c) shows that they are indeed fields.

We return to the general case with arbitrary endomorphism $b_{n,p}$ and $c_{n,p}$ fulfilling the annihilation conditions.

Let C_τ, $\tau \in \mathbb{R}$ be a level line as introduced in Section 3.9. Recall that τ might be considered as proper time of the string in the interpretation of the Riemann surface, Σ as world-sheet in the frame of string theory. For the global fields $b(Q)$ and $c(Q)$ the following anti-commutator relations are postulated for $Q, Q' \in C_\tau$ (this says at the same proper time τ)

$$\{b(Q), c(Q')\}_+ = \Delta_\lambda(Q, Q'), \tag{8.49}$$

$$\{b(Q), b(Q')\}_+ = \{c(Q), c(Q')\}_+ = 0. \tag{8.50}$$

Here $\Delta_\lambda(Q, Q')$ is the delta distribution for $(\lambda, 1 - \lambda)$ systems on C_τ introduced in Section 3.10. The delta distribution is given as (see (3.92))

$$\Delta_\lambda(Q, Q') = \sum_{n \in \mathbb{J}_\lambda} \sum_{p=1}^{K} f^{\lambda}_{n,p}(Q) \cdot f^{1-\lambda}_{-n,r}(Q'). \tag{8.51}$$

For every field h of weight λ we have

$$h(Q) = \frac{1}{2\pi i} \int_{C_\tau} h(Q') \cdot \Delta_\lambda(Q, Q'), \tag{8.52}$$

where the integration is taken over the variable Q'.

Proposition 8.12. *The modes of the fields $b(Q)$ and $c(Q)$ which fulfill* (8.49) *and* (8.50) *fulfill the Clifford algebra like relations*

$$\begin{aligned} \{b_{n,p}, b_{m,r}\}_+ &= 0, \\ \{c_{n,p}, c_{m,r}\}_+ &= 0, \\ \{b_{n,p}, c_{m,r}\}_+ &= \delta_m^{-n}\, \delta_p^r \,\mathrm{id}\,. \end{aligned} \tag{8.53}$$

Vice versa, fields $b(Q)$ and $c(Q)$ based on a system of operators fulfilling (8.53) *will satisfy* (8.49) *and* (8.50).

Proof. If we express the anti-commutator as multiple sum

$$\{b(Q), c(Q')\}_+ = \sum_{n,m,p,r} \{b_{n,p}, c_{m,r}\}_+ \, f^\lambda_{-n,p}(Q) \cdot f^{1-\lambda}_{-m,r}(Q')$$

and compare it with (8.51), we see that the relation (8.49) is equivalent to the last relation of (8.53). In the same way (8.50) is equivalent to the other relations in (8.53). □

Remark 8.13. From this proposition it follows that the representation defined via (8.2) on $\mathcal{H}_r^\lambda$ realizes the anti-commutator relations (8.49) and (8.50) of the associated fields.

Remark 8.14. In physics, two special cases are of particular importance. The first case is if the b-field is of conformal weight 2, and hence the complementary field c of conformal weight -1. This system is the usual ghost field system. The system for $\lambda = 1 - \lambda = 1/2$ has maximal duality and can be realized on semi-infinite wedge forms of weight $1/2$. See [22] for a discussion of partition functions etc., in the two-point case for such $b - c$ systems.

8.4 Energy-momentum tensor

In this section we consider arbitrary b–c fields, not necessarily their explicit realization via semi-infinite forms. But of course the basic relations (8.53) are required. In the last section we obtained that they are equivalent to (8.49) and (8.50).

In physics, the energy-momentum tensor for $b - c$ systems is defined as

$$T(Q) =:P(b(Q), c(Q)):\,. \tag{8.54}$$

In local coordinates it can be written as

$$T(z) = :(1-\lambda)\, c(z) \frac{db}{dz}(z) - \lambda \frac{dc}{dz}(z) b(z):\,. \tag{8.55}$$

Here $c(z)$ and $b(z)$ are local representing functions of the c and b fields in a local coordinate z. The coefficient operators are considered to be constant with respect to $\frac{d}{dz}$. Only the local representing functions for the forms will be differentiated. The symbol :....: means normal ordering of the operator part; see (8.40).

By Proposition 8.9 the field $T(Q)$ has conformal weight two and can be written as

$$T = \sum_{k,s} L_{k,s}\Omega^{k,s}, \tag{8.56}$$

with modes given by the operators $L_{k,s}$. If we plug the expansions (8.47) and (8.48) into the definition (8.55) we obtain

$$T = \sum_{n,m\in\mathbb{J}_\lambda} \sum_{r,p=1}^{K} :c_{n,p}b_{-m,r}:\ P(f^{1-\lambda}_{-n,p}, f^{\lambda}_{m,r}). \tag{8.57}$$

Before we can continue we have to discuss the normal ordering in this context. Normal ordering should mean that in the product of two operators the annihilation operator should be on the right.

As we deal with fermionic fields and anti-commutators we have to change the sign if we change the order to avoid a contradiction. By the relation (8.53) we have for the products of operators

$$\begin{aligned} c_{n,p}\cdot c_{m,r} &= -c_{m,r}\cdot c_{n,p}\,, \\ b_{n,p}\cdot b_{m,r} &= -b_{m,r}\cdot b_{n,p}\,, \\ c_{n,p}\cdot b_{m,r} &= -b_{m,r}\cdot c_{n,p} + \delta_m^{-n}\delta_p^r. \end{aligned} \tag{8.58}$$

As for $n, m < 0$ the elements anti-commute, we can rewrite the standard normal ordering also as

$$:c_{n,p}b_{m,r}: := \begin{cases} c_{n,p}b_{m,r}, & m \ge 0 \\ -b_{m,r}c_{n,p}, & m < 0. \end{cases} \tag{8.59}$$

Or equivalently,

$$:c_{n,p}b_{m,r}: := \begin{cases} c_{n,p}\cdot b_{m,r}, & m \ne -n \\ c_{n,p}\cdot b_{-n,r}, & m = -n,\ r \ne p \\ c_{n,p}\cdot b_{-n,p}, & m = -n,\ r = p,\ n \le 0 \\ c_{n,p}\cdot b_{-n,p} - 1, & m = -n,\ r = p,\ n > 0 \end{cases}. \tag{8.60}$$

By Proposition 8.9 the modes of the operator T are well-defined and can be calculated by

$$\begin{aligned} L_{k,s} &= \frac{1}{2\pi\mathrm{i}} \int_{C_S} T\cdot e_{k,s} \\ &= \sum_{n,m\in\mathbb{J}_\lambda} \sum_{r,p=1}^{K} :c_{n,p}b_{-m,r}:\ \frac{1}{2\pi\mathrm{i}} \int_{C_S} (P(f^{1-\lambda}_{-n,p}, f^{\lambda}_{m,r}))\cdot e_{k,s}. \end{aligned} \tag{8.61}$$

We will re-express this with the the help of the following lemma.

Lemma 8.15.

$$\frac{1}{2\pi i}\int_{C_S} P(h,f)e = -\frac{1}{2\pi i}\int_{C_S}(e\,.\,h)f = \frac{1}{2\pi i}\int_{C_S}(e\,.\,f)h. \tag{8.62}$$

Proof. Let $w = P(h,f)\cdot e + (e\,.\,h)\cdot f$. In a local chart we get

$$w = (1-\lambda)eh\frac{df}{dz} - \lambda ef\frac{dh}{dz} + ef\frac{dh}{dz} + (1-\lambda)fh\frac{de}{dz} = (1-\lambda)\frac{d}{dz}(ehf).$$

Now by Lemma 6.17 (a) we get the first relation, and by Lemma 6.17 (e) the second. □

Proposition 8.16.

$$L_{k,s} = \frac{1}{2\pi i}\int_{C_\tau} T\cdot e_{k,s} = \sum_{n,m,p,r} C^{(n,p)}_{(k,s),(m,r)}(\lambda)\,:c_{n,p}b_{-m,r}:\,. \tag{8.63}$$

Proof. Recall that the structure equations of the module $\mathcal{F}^\lambda$ over $\mathcal{L}$ are defined by

$$e_{k,s}\,.\,f^\lambda_{m,r} = \sum_{h=k+m}^{k+m+S}\sum_{t=1}^{K} C^{(h,t)}_{(k,s),(m,p)}(\lambda)f^\lambda_{h,t}. \tag{8.64}$$

Now we calculate the coefficients in (8.63)

$$\begin{aligned}\frac{1}{2\pi i}\int_{C_S} P(f^{1-\lambda}_{-n,p}, f^\lambda_{m,r})\cdot e_{k,s} &= \frac{1}{2\pi i}\int_{C_S}(e_{k,s}\,.\,f^\lambda_{m,r})\cdot f^{1-\lambda}_{-n,p}\\ &= \sum_{h,t} C^{(h,t)}_{(k,s),(m,p)}(\lambda)\frac{1}{2\pi i}\int_{C_S} f^\lambda_{(h,t)}\cdot f^{1-\lambda}_{-n,p}\\ &= \sum_{h,t} C^{(h,t)}_{(k,s),(m,p)}(\lambda)\delta^t_p\,\delta^n_h.\\ &= C^{(n,p)}_{(k,s),(m,p)}(\lambda).\end{aligned} \tag{8.65}$$

From (8.61) the relation (8.63) follows. □

Theorem 8.17. *The modes $L_{k,r}$ of the energy momentum tensor of the b–c system define a representation of an almost-graded central extension $\widehat{\mathcal{L}}$ of the vector field algebra $\mathcal{L}$ defined by the cocycle $\gamma^{\mathcal{L}}_{S,R}$ (6.140) with central charge*

$$c_\lambda = -2(6\lambda^2 - 6\lambda + 1). \tag{8.66}$$

The calculation will yield the following proposition as a result aside.

Proposition 8.18. *With respect to the standard normal ordering, the cocycle is given as*

$$\chi(e_{k,r}, e_{l,s}) = \Bigg(\sum_{\substack{n\le 0\\ m>0}} - \sum_{\substack{m\le 0\\ n>0}}\Bigg)\sum_{p,t} C^{(n,t)}_{(l,s)(m,p)}(\lambda)\cdot C^{(m,p)}_{(k,r)(n,t)}(\lambda). \tag{8.67}$$

Remark 8.19. Again the famous number c_λ shows up. Recall that $c_\lambda = c_{1-\lambda}$. This is in complete accordance with the fact that from the general setting in this section the operators of the b and c fields cannot be distinguished by their individual weights.

Before we give the proof of Theorem 8.17 we define the operator

$$T[e] := \frac{1}{2\pi i} \int_{C_S} T(Q)e(Q), \quad e \in \mathcal{L}. \tag{8.68}$$

Theorem 8.17 says that

$$T : \mathcal{L} \to \mathrm{End}(V), \quad e \mapsto T[e] \tag{8.69}$$

gives a projective representation. Moreover,

$$T[[e, f]] = [T[e], T[f]] - 2(6\lambda^2 - 6\lambda + 1)\,\gamma_{S,R}^{\mathcal{L}} \cdot id. \tag{8.70}$$

Now we start preparing the proof of the main theorem. The proof resembles the proof of the Sugawara construction for representations of the affine Lie algebras which will be presented in Chapter 10. There the proofs will be much more involved. In the definition of the $L_{k,r}$ formal infinite series appear. Applied to individual elements in the given representation this has to define an operator. To keep this under control we use the cut-off function as it was used by Kac and Raina in [122]. Let ψ be the $\mathbb{R}$-valued function given by

$$\psi(x) = 1 \quad \text{if } |x| \le 1 \quad \text{and} \quad \psi(x) = 0 \quad \text{if } |x| > 1. \tag{8.71}$$

Proposition 8.20.

(a)

$$[L_{k,r}, c_{l,s}] = \sum_{n,p} C_{(k,r)(l,s)}^{(n,p)}(\lambda)c_{n,p}\,, \tag{8.72}$$

(b)

$$\begin{aligned} [L_{k,r}, b_{l,s}] &= -\sum_{n,p} C_{(k,r)(-n,p)}^{(-l,s)}(\lambda)b_{n,p} \\ &= \sum_{n,p} C_{(k,r)(l,s)}^{(n,p)}(1-\lambda)b_{n,p}. \end{aligned} \tag{8.73}$$

Proof. With the help of $\psi(\epsilon n)$ we define

$$L_{k,r}(\epsilon) = \sum_{n,m,p,t} C_{(k,r)(m,p)}^{(n,t)}(\lambda) :c_{n,t}b_{-m,p}: \psi(\epsilon n). \tag{8.74}$$

Let k be fixed, For every n there are only a finite number of m, such that $C_{(k,r)(m,p)}^{(n,t)}(\lambda) \neq 0$. Hence, for $\epsilon > 0$ the sum is reduced to a finite number of summands. For $v \in V$, based on the normal ordering, only a finite number of operators act nontrivially on v. Hence, if we choose $\epsilon > 0$ small enough, we get $L_{k,r}(\epsilon) . v = L_{k,r} . v$. This we understand by

$$\lim_{\epsilon \to 0} L_{k,r}(\epsilon) = L_{k,r}. \tag{8.75}$$

The expression (8.74) without normal ordering is denoted by $\tilde{L}_{k,r}(\epsilon)$. As long as $\epsilon \neq 0$ it is well-defined. The difference between the normal ordered product and the usual product is a scalar operator, and hence will disappear in the commutator. This means $[\tilde{L}_{k,r}(\epsilon), u] = [L_{k,r}(\epsilon), u]$, and as long as $\epsilon \neq 0$ we can ignore the normal ordering signs. We have

$$\begin{aligned}[c_{n,t} \cdot b_{-m,p}, c_{l,s}] &= c_{n,t} \cdot b_{-m,p} \cdot c_{l,s} - c_{l,s} \cdot c_{n,t} \cdot b_{-m,p} \\ &= c_{n,t} \left(b_{-m,p} \cdot c_{l,s} + c_{l,s} \cdot b_{-m,p} \right) = c_{n,t} \cdot \delta^l_m \delta^p_s, \end{aligned} \tag{8.76}$$

and hence with (8.63)

$$\begin{aligned}[L_{k,r}(\epsilon), c_{l,s}] &= \sum_{m,n,p,t} C^{(n,t)}_{(k,r)(m,p)}(\lambda)\, c_{n,t}\, \delta^l_m \delta^p_s \psi(\epsilon n) \\ &= \sum_{n,t} C^{(n,t)}_{(k,r)(l,s)}(\lambda)\, c_{n,t} \psi(\epsilon n). \end{aligned} \tag{8.77}$$

Here we can pass without problems to $\epsilon = 0$. This shows part (a).
To show (b) we calculate accordingly

$$[c_{n,t} b_{-m,p}, b_{l,s}] = -\delta^s_t \delta^{-n}_l b_{-m,p}. \tag{8.78}$$

Hence,

$$\begin{aligned}[L_{k,r}(\epsilon), b_{l,s}] &= - \sum_{n,m,p,t} C^{(n,t)}_{(k,r)(m,p)}(\lambda) \cdot \delta^s_t \delta^{-n}_l\, b_{-m,r} \\ &= - \sum_{n,p} C^{(-l,s)}_{(k,r)(-n,p)}(\lambda)\, b_{n,p} = \sum_{n,p} C^{(n,p)}_{(k,r)(l,s)}(1-\lambda)\, b_{n,p}. \end{aligned} \tag{8.79}$$

For the last equality we used (7.139). □

Proof of Theorem 8.17. We consider $[L_{k,r}, L_{l,s}(\epsilon)]$ first for $\epsilon \neq 0$. For $\epsilon \to 0$ we obtain the operator which we are looking for.

$$[L_{k,r}, L_{l,s}(\epsilon)] = \sum_{n,m,t,p} C^{(n,t)}_{(l,s)(m,p)}(\lambda)\ [L_{k,r}\,,\ :c_{n,t} b_{-m,p}:] \psi(\epsilon n). \tag{8.80}$$

As long as $\epsilon \neq 0$ we are allowed to ignore the normal ordering. With Proposition 8.20 we obtain

$$\begin{aligned}[L_{k,r}, c_{n,t} b_{-m,p}] &= [L_{k,r}, c_{n,t}] b_{-m,p} + c_{n,t} [L_{k,r}, b_{-m,p}] \\ &= \sum_{h,u} \left(C^{(h,u)}_{(k,r)(n,t)}(\lambda) c_{h,u} b_{-m,p} - C^{(m,p)}_{(k,r)(h,u)}(\lambda) c_{n,t} b_{-h,u} \right). \end{aligned} \tag{8.81}$$

If we plug this into (8.80) and split the sum we obtain

$$\begin{aligned}[L_{k,r}, L_{l,s}(\epsilon)] = &\sum_{\substack{m,n,h \\ p,t,u}} C^{(n,t)}_{(l,s)(m,p)}(\lambda) C^{(h,u)}_{(k,r)(n,t)}(\lambda) c_{h,u} b_{-m,p} \psi(\epsilon n) \\ &- \sum_{\substack{m,n,h \\ p,t,u}} C^{(n,t)}_{(l,s)(m,p)}(\lambda) C^{(m,p)}_{(k,r)(h,u)}(\lambda) c_{n,t} b_{-h,u} \psi(\epsilon n). \end{aligned} \tag{8.82}$$

For $\epsilon \to 0$ both sums are only defined if we pass to normal ordering. Hence, let the sum be normally ordered. By this we pick up terms which we will discuss in a minute. First, we let $\epsilon \to 0$ for the normal ordered parts and change the summation variables

$$(n,t) \to (h,u) \to (m,p) \to (n,t).$$

The normally ordered parts can be grouped together as

$$\sum_{\substack{m,n,h\\p,t,u}} (C^{(n,t)}_{(l,s)(m,p)}(\lambda)C^{(h,u)}_{(k,r)(n,t)}(\lambda) - C^{(h,u)}_{(l,s)(m,p)}(\lambda)C^{(n,t)}_{(k,r)(m,p)}(\lambda)) \,{:}c_{h,u}b_{-m,p}{:}\,. \tag{8.83}$$

As $\mathcal{F}^\lambda$ is a $\mathcal{L}$ module, we can rewrite this as

$$\sum_{n,t} C^{(n,t)}_{(k,r)(l,s)}(-1)\Bigg(\sum_{m,n,p,u} C^{(h,u)}_{(n,t)(m,p)}(\lambda)\,{:}c_{h,u}b_{-m,p}{:}\Bigg) = \sum_{n,t} C^{(n,t)}_{(k,r)(l,s)}(-1)\,L_{n,t}. \tag{8.84}$$

Hence, without taking the additional terms due to normal ordering into account, we get that that the $L_{k,r}$ gives a representation of the vector field algebra $\mathcal{L}$.

But the splitting only makes sense if the additional terms also exist in the limit $\epsilon \to 0$. Assume that they will be well-defined (which we will show in the following). The difference between the product of the operators with or without normal ordering are scalar operators. More precisely, we have (8.84)

$$[T[e], T[f]] = T[[e,f]] + \chi(e,f)\cdot id. \tag{8.85}$$

As operators on V the $T[e]$ fulfill the Jacobi identity. Also, for the coefficients for $[e,f]$ the Jacobi identity is true. Hence χ will be necessarily a Lie algebra two-cocycle. We will show that it is a local cocycle.

Based on the relations (8.60), normal ordering terms in the first sum of (8.82) only appear for $h = m > 0$ and $u = p$, and in the second sum for $n = h > 0$ and $u = t$. Hence, the additional terms add up to

$$\sum_{\substack{n\\m>0\\p,t}} C^{(n,t)}_{(l,s)(m,p)}(\lambda)C^{(m,p)}_{(k,r)(n,t)}(\lambda)\psi(\epsilon n) - \sum_{\substack{m\\n>0\\p,t}} C^{(n,t)}_{(l,s)(m,p)}(\lambda)C^{(m,p)}_{(k,r)(n,t)}(\lambda)\psi(\epsilon n).$$

If we split the sum into $(n, m > 0) = (n \le 0, m > 0) + (n > 0, m > 0)$ and $(n > 0, m) = (n > 0, m > 0) + (n > 0, m \le 0)$ respectively, then the corresponding parts cancel and it remains

$$\sum_{\substack{n\le 0\\m>0\\p,t}} C^{(n,t)}_{(l,s)(m,p)}(\lambda)C^{(m,p)}_{(k,r)(n,t)}(\lambda)\psi(\epsilon n) - \sum_{\substack{m\le 0\\n>0\\p,t}} C^{(n,t)}_{(l,s)(m,p)}(\lambda)C^{(m,p)}_{(k,r)(n,t)}(\lambda)\psi(\epsilon n). \tag{8.86}$$

By the almost-graded structure of the module $\mathcal{F}^\lambda$ we have

$$\begin{aligned} C^{(n,t)}_{(l,s)(m,p)}(\lambda) \neq 0 &\implies l+m \le n \le l+m+R, \\ C^{(m,p)}_{(k,r)(n,t)}(\lambda) \neq 0 &\implies k+n \le m \le k+n+R \end{aligned} \tag{8.87}$$

with a constant R, which is independent of n and m. This means that the individual sums in (8.86) are well-defined and we can forget about ϵ. On the other hand, we get the claimed form of the cocycle as expressed in Proposition 8.18.

Moreover, if we add the conditions for the nonvanishing of the cocycle it will be zero if we are outside the range $-2R \le l+k \le 0$. Hence, the cocycle is local. In fact it is bounded by zero. As it is a local cocycle it has to be a multiple of the separating cocycle $\gamma_{S,R}^{\mathcal{L}}$ (6.140), with a suitable projective connection R. Theorem 6.41(d) gives the explicit form on level zero. We determine it by calculating

$$\chi_{(1,r),(-1,r)} \quad \text{and} \quad \chi_{(2,r),(-2,r)}.$$

On the upper bound for the cocycle given for k and $l=-k$ there will only be a contribution if the second indices are the same, hence we obtain

$$\chi_{(k,r),(-k,r)} = \sum_{\substack{n\le 0\\ m>0\\ p,t}} C^{(n,t)}_{(-k,r)(m,p)}(\lambda)C^{(m,p)}_{(k,r)(n,t)}(\lambda) - \sum_{\substack{m\le 0\\ n>0\\ p,t}} C^{(n,t)}_{(-k,r)(m,p)}(\lambda)C^{(m,p)}_{(k,r)(n,t)}(\lambda). \tag{8.88}$$

If $k>0$ the second sum will disappear. Recall that for the structure constants of the modules we have

$$C^{(n+m,r)}_{(n,r)(m,r)}(\lambda) = (m+\lambda n). \tag{8.89}$$

We calculate (dropping (λ) in the notation for the structure constants)

$$\chi_{(1,r),(-1,r)} = C^{(0,r)}_{(1,r)(-1,r)}\cdot C^{(1,r)}_{(1,r)(0,r)} - C^{(-1,r)}_{(-1,r)(0,r)}\cdot C^{(0,r)}_{(1,r)(-1,r)} \tag{8.90}$$

$$= -\lambda^2+\lambda, \tag{8.91}$$

$$\chi_{(2,r),(-2,r)} = C^{(0,r)}_{(-2,r)(2,r)}\cdot C^{(2,r)}_{(2,r)(0,r)} - C^{(1,r)}_{(-2,r)(1,r)}\cdot C^{(1,r)}_{(2,r)(-1,r)} \tag{8.92}$$

$$= -8\lambda^2+8\lambda-1. \tag{8.93}$$

Hence via (6.141) and in the notation there

$$\begin{aligned} \alpha &= 2\chi_{(2,r),(-2,r)} - 4\chi_{(1,r),(-1,r)} = (-2)(6\lambda^2 = 6\lambda+1)\\ \beta &= -\lambda^2+\lambda. \end{aligned} \tag{8.94}$$

With these values we obtain

$$\chi_{(n,r),(-n,r)} = \frac{(n+1)n(n-1)}{12}\cdot\alpha + n\cdot\beta. \tag{8.95}$$

The term $n\beta$ is a coboundary, which could be brought to zero by a change of basis. The value $(-2)(6\lambda^2+6\lambda+1)$ is the central charge. In this way, Theorem 8.17 is proven. □

Remark 8.21. We stress the fact that for the above result our representation space V was arbitrary. Hence it is a very general result. In the case that our representation space is the space $\mathcal{H}^\lambda$ and the operators are the operators defined via insertion or contraction, then Proposition 8.20 can be reformulated as

$$[T[e],c_g] = c_{e.g}, \qquad [T[e],b_f] = b_{e.f}. \tag{8.96}$$

Furthermore, by the very definition of the operators $L_{k,r}$ as sums of terms which individually are a product of one wedging and one contracting operator, they will leave the subspaces of $\mathcal{H}^\lambda$ of a fixed charge m invariant. On the charge m subspace we have another representation of the centrally extended vector field $\widehat{\mathcal{L}}$ given by the cocycle (7.74) restricted to the vector fields. It has the same central charge $-2(6\lambda^2-6\lambda+1)$. We will show in Section 8.6 that indeed both representations coincide. As for an abstract representation space V there are no semi-infinite wedge forms, this only makes sense if $V=\mathcal{H}^{\lambda,(m)}$.

Remark 8.22. The following is only intended as a heuristic interpretation of these results in the frame of string theory. By the quantization condition for the position and momentum variables on the world-sheet of the bosonic string in D-dimensional space-time, one obtains a representation of a D–dimensional Heisenberg algebra (see [141]). The energy-momentum tensor is given by the Sugawara operator. We will define it in Chapter 10. As we will show there, the modes will define a representation of the almost-graded central extension of the vector field algebra. The central charge will be $c=\dim\mathfrak{g}=D$; see (10.32). For a complete theory one has to add the ghost fields. This means a $b-c$ system of weight 2 has to be incorporated. The complete representation space of bosonic string theory will be the tensor product of both representations. As shown above, the almost-graded centrally extended vector field algebra operates with central charge $c=-2(6\lambda^2-6\lambda+1)=-26$ on the ghost-field part. To be free of (conformal) anomaly, on the complete theory the vector fields should act without central extension. More precisely, the central term should act with central charge zero. In the tensor product the central charges add up. Hence, by the requirement of being anomaly-free the space-time dimension of the bosonic string theory has to be 26.

8.5 Representation of the Heisenberg algebra via $b-c$ systems

Let the situation be as in Section 8.4. In particular, the $b-c$ fields are not necessarily realized on the semi-infinite wedge forms. We consider their normally ordered product

$$a(Q)=:c(Q)b(Q):\,. \tag{8.97}$$

By Proposition 8.7 it is a well-defined field of conformal weight one. It can be expanded in the operator modes $a_{k,s}$ with

$$a(Q)=\sum_{k,s}a_{k,s}\omega^{k,s}, \tag{8.98}$$

and the modes calculated as

$$a_{k,s}=\sum_{n,p}\sum_{m,r}:c_{n,p}b_{-m,r}:\frac{1}{2\pi\mathrm{i}}\int_{C_S}f^{1-\lambda}_{-n,p}\cdot f^{\lambda}_{m,r}\cdot A_{k,s}. \tag{8.99}$$

The spaces $\mathcal{F}^\lambda$ are modules over the associative algebra $\mathcal{A}$. The structure constants with respect to the Krichever–Novikov basis elements are $\alpha_{(k,s)(m,r)}^{(n,p)}(\lambda)$ given by

$$A_{k,s} \cdot f_{m,r}^\lambda = \sum_{n,p} \alpha_{(k,s)(m,r)}^{(n,p)}(\lambda) f_{n,p}^\lambda. \tag{8.100}$$

By the Krichever–Novikov duality they calculate as

$$\alpha_{(k,s)(m,r)}^{(n,p)}(\lambda) = \frac{1}{2\pi \mathrm{i}} \int_{C_S} A_{k,s} \cdot f_{m,r}^\lambda \cdot f_{-n,p}^{1-\lambda}, \tag{8.101}$$

which yields exactly the same expression as in (8.99). Hence

$$a_{k,s} = \sum_{n,p} \sum_{m,r} \alpha_{(k,s)(m,r)}^{(n,p)}(\lambda) :c_{n,p} b_{-m,r}: . \tag{8.102}$$

Theorem 8.23. *The modes $a_{k,s}$ of the normally ordered product $:c(Q)b(Q):$ define a representation of an almost-graded central extension $\widehat{\mathcal{A}}$ of the function algebra $\mathcal{A}$ (also called Heisenberg algebra) defined by the cocycle $\gamma_S^{\mathcal{A}}(f,g) = \frac{1}{2\pi \mathrm{i}} \int_{C_S} f dg$ with central charge* (-1).

Proof. For the proof we can partly use exactly the same proof as for the Sugawara case for $b-c$ fields. We just have to replace the coefficients $C_{\cdot\cdot,\cdot\cdot}^{\cdot\cdot}(\lambda)$ by $\alpha_{\cdot\cdot,\cdot\cdot}^{\cdot\cdot}(\lambda)$. In this case (8.63) is already fulfilled. The proof of Proposition 8.20 involves only operations with the b and c operators and remains true in the current setting. The expressions of the coefficients in (8.83) will be zero. This follows from the explicit expressions of the coefficients given in (8.101). Hence, up to additional terms due to normal ordering the modes commute. The additional terms define a two-cocycle for $\mathcal{A}$. In particular, the representation will be a projective representation of $\mathcal{A}$, given by this cocycle. The corresponding expression of the cocycle is (like (8.67))

$$\chi(A_{k,r}, A_{l,s}) = \Bigg(\sum_{\substack{n\le 0 \\ m>0}} - \sum_{\substack{m\le 0 \\ n>0}} \Bigg) \sum_{p,t} \alpha_{(l,s)(m,p)}^{(n,t)}(\lambda) \cdot \alpha_{(k,r)(n,t)}^{(m,p)}(\lambda). \tag{8.103}$$

The cocycle is local and bounded by zero by estimates of the type (8.87). Unfortunately, we cannot immediately conclude using our general classification results regarding local cocycles for $\mathcal{A}$, as we have not yet established $\mathcal{L}$ invariance. Instead of doing this we will use the explicit form (8.103). For this sum we use Lemma 10.33 which we will show in the context of the Sugawara construction for the affine algebra. By this lemma our cocycle will be

$$\chi(A_{k,r}, A_{l,s}) = -\gamma_{(k,r)(l,s)} = -\frac{1}{2\pi \mathrm{i}} \int_{C_S} A_{k,r} dA_{l,s}. \tag{8.104}$$

This is the form of the cocycle to be shown. □

In case our $b - c$ fields are defined with the b and c operators operating on the semi-infinite wedge forms, we again obtain that the constructed representation equals the function part of the differential operator algebra action on the forms. In particular the central charges also coincide (see Theorem 7.18). The proof for the equality will be given in Section 8.6.

Remark 8.24. In field theoretical terms the Heisenberg operators are bosonic operators, the $b - c$ operators fermionic. The presented representation realizes the bosonic field $a(Q)$ as a product of two fermionic fields.

Remark 8.25. In the classical case ($g = 0$, $N = 2$) we have

$$\alpha^n_{k,m} = 1 \cdot \delta^n_{k+m}. \tag{8.105}$$

Hence for the operators

$$a_k = \sum_{n \in \mathbb{J}_\lambda} :c_n b_{-(n+k)}: . \tag{8.106}$$

As long as $k \neq 0$ normal ordering will not change anything. We only have to take it into account for $k = 0$ and get

$$\begin{aligned} a_k &= \sum_{n \in \mathbb{J}_\lambda} c_n b_{-(n+k)}, \quad k \in \mathbb{Z} \\ a_0 &= \sum_{n \le 0} c_n b_{-n} - \sum_{n > 0} b_{-n} c_n. \end{aligned} \tag{8.107}$$

Recall that b_{-n} "removes" the element f^λ_n from the semi-infinite wedge form and c_n "adds" it.

8.6 $b - c$ systems and the algebra $\widehat{gl}(\infty)$

In Section 7.1.2 we introduced the semi-infinite wedge representation of the algebra $\widehat{gl}(\infty)$. It was the basic model as we proved the results on the semi-infinite wedge representations for our Krichever–Novikov algebras by embedding them into $\overline{gl}(\infty)$. The almost-graded structure of the modules was important for the embedding.

We can also define a system of b and c operators for the $\widehat{gl}(\infty)$ case. Applied to the form

$$\Phi = v_{i_0} \wedge v_{i_1} \cdots \wedge \cdots v_{i_k} \wedge v_{i_{k+1}} \wedge v_{i_{k+2}} \wedge \cdots , \tag{8.108}$$

the operator c_m wedges it with v_m and b_{-m} removes v_m, if it appears, with a sign factor, depending on its position it appears in complete accordance with (8.2) and (8.5).

We recall the definition of the generating elements for $\overline{gl}(\infty)$

$$A_k(\mu) := \sum_{i \in \mathbb{Z}} \mu_i E_{i,i+k}. \tag{8.109}$$

As long as $k \neq 0$ the action of $A_k(\mu)$ on the wedge space H is well-defined. The action of $A_0(\mu)$ has to be regularized. The regularization will be induced by setting

$$\hat{r}(E_{ij}) := \begin{cases} r(E_{ij}), & i \neq j \text{ or } i = j < 0 \\ r(E_{ii}) - id, & i \geq 0. \end{cases} \tag{8.110}$$

The (regularized) action on H is denoted by $\hat{r}$. For example, we have

$$\hat{r}(A_0(\mu))\Phi_m = \begin{cases} \left(\sum_{s=m}^{-1} \mu_s\right)\Phi_m, & m < 0 \\ \left(-\sum_{s=0}^{m-1} \mu_s\right)\Phi_m, & m > 0 \\ 0, & m = 0. \end{cases} \tag{8.111}$$

Now we consider the operators

$$D_k = \sum_{n\in\mathbb{Z}} :c_n b_{-(n+k)}: \quad k \in \mathbb{Z}. \tag{8.112}$$

First, note that the product of the operators $c_n b_{-(n+k)}$ operates as $E_{n,n+k}$. Also for $k \neq 0$ the normal ordering in(8.112) does not have an effect. Hence

$$D_k = \sum_{n\in\mathbb{Z}} c_n b_{-(n+k)} = \hat{r}(A_k(1)), \quad k \neq 0. \tag{8.113}$$

Here 1 denotes the constant sequence 1. For $k = 0$ we obtain[2]

$$D_0 = \sum_{n\in\mathbb{Z}} :c_n b_{-n}:= \sum_{n<0} c_n b_{-n} - \sum_{n\geq 0} b_{-n} c_n = \hat{r}(A_k(0)). \tag{8.114}$$

An easy calculation in $\widehat{gl}(\infty)$ shows

$$[\hat{r}(A_k(1)), \hat{r}(A_l(1))] = k\,\delta_l^{-k}. \tag{8.115}$$

Hence the charge m subspace is a representation of the classical Heisenberg algebra with central charge -1. For the vacuum vector Φ_m (see (7.38)) we have

$$\hat{r}(A_k(1))\Phi_m = -m\Phi_m. \tag{8.116}$$

The operators $\hat{r}(A_k(\mu))$, by definition and after identification with the $b-c$ operators, can be given as

$$\hat{r}(A_k(\mu)) = \sum_n \mu_n :c_n b_{n+k}: . \tag{8.117}$$

We will sketch now why, in the case that our fields are defined for the state space given by the semi-infinite wedge forms, the representations which we constructed in Section 8.4 and Section 8.5 coincide with the representation of $\widehat{\mathcal{L}}$ and $\widehat{\mathcal{A}}$ constructed in Chapter 7.

---

2 We adapted the normal ordering for the $b-c$ system by changing the $n > m$ to $n \geq m$, to obtain exactly the same form as for the regularized action (8.110).

For $K > 1$ we passed over to a linearized order. Hence, for simplicity we might assume from the very beginning that there is no second index, i.e., that $K = 1$. Our operators (8.102) for the Heisenberg algebra case are

$$a_k = \sum_{m,n} :c_n b_{-m}: \alpha^n_{k,m}(\lambda). \tag{8.118}$$

By the almost-graded structure of the module $\mathcal{F}^\lambda$ over $\mathcal{A}$ we have a constant R independent of n, m, and k such that

$$\alpha^n_{k,m}(\lambda) \neq 0 \implies k + m \leq n \leq k + m + R, \tag{8.119}$$

which in terms of m reads as

$$n - k - R \leq m \leq n - k. \tag{8.120}$$

Hence we write $m = (n-k) - l$ with $0 \leq l \leq R$ and the sum (8.118) rewrites

$$a_k = \sum_{n\in\mathbb{Z}} \sum_{l=0}^{R} :c_n b_{(k+l)-n}: \alpha^n_{k,n-(k+l)}(\lambda). \tag{8.121}$$

Under the identification with the elements from $\widehat{gl}(\infty)$ we obtain

$$\hat{r}(a_k) = \sum_{l=0}^{R} \hat{r}(A_{-(k+l)})(\mu^{(k+l)}), \quad \mu_n^{(k+l)} = \alpha^n_{k,n-(k+l)}(\lambda). \tag{8.122}$$

But these are exactly the expressions obtained via the embedding of $\mathcal{A}$ into $\overline{gl}(\infty)$ based on the action on the module $\mathcal{F}^\lambda$ and the regularization procedure applied; see (7.64). Hence, both representations on $\mathcal{H}^\lambda$ are identical.

For the $b-c$ Sugawara operators $L_{k,s}$ we obtained the description (8.63) in terms of the structure constants for the $\mathcal{L}$-module $\mathcal{F}^\lambda$, which has exactly the same structure as (8.118). Consequently, both representations of the centrally extended vector field algebra $\widehat{\mathcal{L}}$ on $\mathcal{H}^\lambda$ coincide too.

Finally, we point out that this identification only makes sense if our representation space, the state space, is the space of semi-infinite wedge forms.

9 Affine algebras

We have already introduced higher genus current type algebras in Section 2.9, and given examples of central extensions in Section 6.5.5. In this chapter we will recall their definition and study them in more detail.

The *current algebras* come with an almost-grading. One of the main goals of this chapter is to classify their almost-graded central extensions; or equivalently, classify the local cocycle classes. This is achieved for the case if the finite-dimensional Lie algebra is reductive. In particular, it will turn out that if $\mathfrak{g}$ is simple, then there is up to equivalence and rescaling only one nontrivial almost-graded central extension $\widehat{\mathfrak{g}}$ of $\mathfrak{g}$.

Additionally, we will discuss extended current algebras obtained as semidirect products with the vector field algebra $\mathcal{L}$ and their central extensions.

Finally, we will introduce as examples of representations the Verma modules and the fermionic Fock space representations.

9.1 Higher genus current algebras

Let $\mathfrak{g}$ be an arbitrary finite-dimensional Lie algebra and $\mathcal{A}$ the function algebra. The *multi-point higher genus current algebra* (or *multi-point higher genus loop algebra*) was defined as

$$\overline{\mathfrak{g}} := \mathfrak{g} \otimes \mathcal{A}, \quad \text{with Lie product} \quad [x \otimes f, y \otimes g] := [x, y] \otimes (f \cdot g). \tag{9.1}$$

Indeed, such a construction works for every commutative and associative algebra $\mathcal{A}$ with unit. See Section 2.9 for the proofs.

The almost-grading of $\mathcal{A} = \bigoplus_{n\in\mathbb{Z}}(\mathcal{A})_n$ induced by the splitting $A = I \cup O$ gives an almost-grading

$$\overline{\mathfrak{g}} = \bigoplus_{n\in\mathbb{Z}}(\overline{\mathfrak{g}})_n, \quad \overline{\mathfrak{g}}_n = \mathfrak{g} \otimes \mathcal{A}_n \tag{9.2}$$

via $\deg(\overline{\mathfrak{g}}_n) := n$. Or

$$\deg(x \otimes A_{n,p}) := n, \quad n \in \mathbb{Z},\ p = 1, \ldots, K. \tag{9.3}$$

Here $A_{n,p}$ are the basis elements (3.50) of the algebra $\mathcal{A}$. It follows directly that $L \in \mathbb{Z}$ exists such that for all $n, m \in \mathbb{Z}$

$$[\overline{\mathfrak{g}}_n, \overline{\mathfrak{g}}_m] \subseteq \bigoplus_{h=n+m}^{n+m+L} \overline{\mathfrak{g}}_h. \tag{9.4}$$

Here L is the bound given by the almost-grading of $\mathcal{A}$. This makes $\overline{\mathfrak{g}}$ a (strongly) almost-graded Lie algebra, $\overline{\mathfrak{g}} = \oplus_{n\in\mathbb{Z}}(\overline{\mathfrak{g}})_n$, with homogeneous subspaces $(\overline{\mathfrak{g}})_n = \mathfrak{g} \otimes \mathcal{A}_n$. To simplify the notation we will sometimes use $x(g)$ to denote the element $x \otimes g$.

Recall that we have shown in Proposition 3.14

$$1 = \sum_{p=1}^{K} A_{0,p}, \tag{9.5}$$

where $A_{0,p}$ are the basis elements generating $\mathcal{A}_0$.

Proposition 9.1. *Via the embedding $x \mapsto x \otimes 1$, the algebra $\mathfrak{g}$ becomes a subalgebra of $\overline{\mathfrak{g}}$ lying in the degree zero subspace $\overline{\mathfrak{g}}_0$.*

Proof. As $1 \in \mathcal{A}_0$ we obtain an injective map $\psi : \mathfrak{g} \to \overline{\mathfrak{g}}$, $x \mapsto \psi(x) = x \otimes 1$. By (9.1)

$$\psi([x, y]) = [x, y] \otimes 1 = [x \otimes 1, y \otimes 1] = [\psi(x), \psi(y)].$$

Hence, it is a Lie homomorphism. Using (9.5) we see that $1 \in \mathcal{A}_0$ and consequently $\psi(\mathfrak{g}) \in \overline{\mathfrak{g}}_0$. □

Recall that a Lie algebra $\mathfrak{g}$ fulfilling $\mathfrak{g}' := [\mathfrak{g}, \mathfrak{g}] = \mathfrak{g}$ is called *perfect*. Examples of perfect Lie algebras are simple or semisimple Lie algebras. Later we will need the following observation.

Proposition 9.2. *If the finite-dimensional Lie algebra $\mathfrak{g}$ is perfect, then the current algebra is $\overline{\mathfrak{g}}$ also perfect. In particular, $\overline{\mathfrak{g}}$ for $\mathfrak{g}$ semisimple is perfect.*

Proof. The elements of the type $x \otimes f$, with $x \in \mathfrak{g}$ and $f \in \mathcal{A}$, generate $\overline{\mathfrak{g}}$. Take any $x \otimes g$, then by the perfectness of $\mathfrak{g}$ there exists $z, y \in \mathfrak{g}$ with $x = [z, y]$. Hence, $x \otimes f = [z \otimes f, y \otimes 1]$. □

9.2 Central extensions

Next we consider central extensions. As explained in Chapter 6, central extensions are given via Lie algebra 2-cocycles with values in the trivial module. For short they will just be called cocycles.

Examples of central extensions are obtained as follows. Let α be a fixed invariant, symmetric bilinear form for the Lie algebra $\mathfrak{g}$. Invariance means that it obeys

$$\alpha([x, y], z) = \alpha(x, [y, z]). \tag{9.6}$$

We do not require it to be non-degenerate. If $\mathfrak{g}$ is abelian, every symmetric bilinear form will be invariant and will do. If $\mathfrak{g}$ is a simple Lie algebra, any scalar multiple of the Cartan–Killing form will do. Moreover, in this case only such scalar multiples will be of the required type. Choose a multiplicative cocycle[1] γ for $\mathcal{A}$. We recall from (6.86) that this says

$$\gamma(fg, h) + \gamma(gh, f) + \gamma(hf, g) = 0, \quad \forall f, g, h \in \mathcal{A}. \tag{9.7}$$

1 Which is the same as a 1-cocycle in cyclic cohomology.

We set $\widehat{\overline{\mathfrak{g}}}_{\alpha,\gamma} = \mathbb{C} \oplus \overline{\mathfrak{g}}$ as vector space and introduce a Lie product

$$[\widehat{x \otimes f}, \widehat{y \otimes g}] = \widehat{[x,y] \otimes (fg)} + \alpha(x,y)\gamma(f,g) \cdot t, \quad [t, \widehat{\overline{\mathfrak{g}}}_{\alpha,\gamma}] = 0. \tag{9.8}$$

As usual, we set $\widehat{x \otimes f} := (0, x \otimes f)$ and $t := (1, 0)$.

Proposition 9.3. *The vector space $\widehat{\overline{\mathfrak{g}}}_{\alpha,\gamma}$ with structure (9.8) is a Lie algebra and a central extension of $\overline{\mathfrak{g}}$.*

Proof. By general constructions of central extensions (see Section 6.2) it is enough to show that $\psi(x \otimes g, y \otimes h) := \alpha(x,y)\gamma(g,h)$ is a cocycle for $\overline{\mathfrak{g}}$. The antisymmetry is clear. Now

$$\psi([x \otimes f, y \otimes g], z \otimes h) = \psi([x,y] \otimes fg, z \otimes h) = \alpha([x,y],z)\gamma(fg,h). \tag{9.9}$$

By cyclically permuting and adding up the three terms obtained we get

$$\begin{aligned}\alpha([x,y],z)\gamma(fg,h) + \alpha([y,z],x)\gamma(gh,f) + \alpha([z,x],y)\gamma(hf,g) \\ = \alpha([x,y],z) \cdot \big(\gamma(fg,h) + \gamma(gh,f) + \gamma(hf,g)\big) = 0.\end{aligned} \tag{9.10}$$

Here we used the invariance of α and the multiplicative property of γ. □

Let C be any non-singular curve on Σ, then $\gamma_C^{\mathcal{A}}(f,g) = \frac{1}{2\pi\mathrm{i}}\int_C fdg$ is a multiplicative cocycle for $\mathcal{A}$; see Section 6.5.1. Hence,

$$\gamma_{\alpha,C}(x \otimes f, y \otimes g) = \alpha(x,y)\frac{1}{2\pi\mathrm{i}}\int_C fdg \tag{9.11}$$

gives, via the expression (9.8), a central extension $\widehat{\overline{\mathfrak{g}}}_{\alpha,C}$ of $\overline{\mathfrak{g}}$. Examples of such curves are given by the separating cycles C_S.

9.3 Local cocycles

Cocycles which are local (see Definition 6.31) are of special importance. As explained there, in this case the almost-grading of $\overline{\mathfrak{g}}$ can be extended to the central extension, defined via this cocycle by setting $\deg t = 0$ for the central element t. From Theorem 6.38 we immediately get the following proposition.

Proposition 9.4. *Assume that $\alpha \not\equiv 0$, then the cocycle $\alpha(x,y)\gamma_C^{\mathcal{A}}(f,g)$ is local if and only if the integration cycle C is a separating cycle C_S.*

Recall that

$$\gamma_S^{\mathcal{A}}(f,g) = \gamma_{C_S}^{\mathcal{A}}(f,g) = \frac{1}{2\pi\mathrm{i}}\int_{C_S} fdg. \tag{9.12}$$

In this case, $\widehat{\overline{\mathfrak{g}}}_{\alpha,S} := \widehat{\overline{\mathfrak{g}}}_{\alpha,C_S}$ is an almost-graded central extension of $\overline{\mathfrak{g}}$.

It should be remarked that everything depends on the bilinear form α. If $\mathfrak{g}$ is simple, there is up to multiplication with a scalar only one such form, the Cartan–Killing form. In this case it is even a non-degenerate form.

Definition 9.5. The algebras obtained via a cocycle γ_{C_S} (or more generally via γ_C) are called the *higher genus (multi-point) affine Lie algebras (or Krichever–Novikov algebras of affine type).*

Remark 9.6. In the classical situation (and $\mathfrak{g}$ simple), these are nothing other than the usual affine Lie algebras (i.e., the untwisted affine Kac-Moody algebras) [117, 118]. In physics they made their first appearance as symmetries of two-dimensional conformal field theory models employed for the description of quarks; see Bardakçi and Halpern [6]. See also, e.g., [84] for further physics references.
For higher genus and for the two-point situations, such algebras were introduced by Krichever and Novikov [140–142] and studied in more detail by Sheinman [233, 235]. The multi-point case was treated by the author in [209, 210].

Remark 9.7. Everything which will be done in this chapter for local cocycles can easily be extended to the bounded cocycles[2]. In particular, Proposition 9.4 has as its counterpart the following proposition.

Proposition 9.8. *Assume that $\alpha \not\equiv 0$, then the cocycle $\alpha(x,y)\gamma_C^{\mathcal{A}}(f,g)$ is bounded (from above) if and only if the cohomology class $[C]$ of the integration cycle C is a linear combination*

$$[C] = \sum_{i=1}^{K} \beta_i [C_i], \quad \beta_i \in \mathbb{C}, \tag{9.13}$$

where the C_i are deformed circles around the points $P_i \in I$, $i = 1, \dots, K$.

Again, this follows from our classification results for bounded cocycles for $\mathcal{A}$.

The question which we now want to address is in which sense Proposition 9.4 is also true in the opposite direction. In other words, if we have a local cocycle for $\overline{\mathfrak{g}}$, are we able to write it as product $\gamma(x \otimes f, y \otimes g) = \alpha(x,y)\gamma_S^{\mathcal{A}}(f,g)$ with an invariant symmetric bilinear form α on $\mathfrak{g}$? Of course, this should always be understood as "up to coboundary". Indeed we will show that this is the case for the current algebra associated with semisimple Lie algebras. In the reductive case we again require an additional $\mathcal{L}$-invariance condition.

Proposition 9.9. *Let $\mathfrak{g}$ be a finite-dimensional Lie algebra which fulfills the condition $\mathfrak{g}' := [\mathfrak{g}, \mathfrak{g}] \neq 0$. Let γ be a cocycle of $\overline{\mathfrak{g}}$. Assume that there is an invariant symmetric bilinear form α on γ fulfilling $\alpha(\mathfrak{g}', \mathfrak{g}) \neq 0$ (e.g., α non-degenerate), and a bilinear form*

2 Recall that if not otherwise stated, "bounded" means bounded from above.

$\gamma^{\mathcal{A}}$ on $\mathcal{A}$ such that γ can be written as

$$\gamma(x \otimes f, y \otimes z) = \alpha(x, y)\gamma^{\mathcal{A}}(f, g). \tag{9.14}$$

(a) *If γ is a local cocycle, then $\gamma^{\mathcal{A}}$ is a multiple of the local cocycle $\gamma_S^{\mathcal{A}}$ for the function algebra.*

(b) *If γ is a bounded cocycle, then $\gamma^{\mathcal{A}}$ is a linear combination of the cocycles $\gamma_{C_i}^{\mathcal{A}}$, $i = 1, \dots, K$.*

Proof. First, $\gamma^{\mathcal{A}}$ is obviously antisymmetric and hence a cocycle for $\mathcal{A}$. We calculate

$$\gamma([x \otimes f, y \otimes g], z \otimes h) = \gamma([x, y] \otimes f \cdot g, z \otimes h) = \alpha([x, y], z)\gamma^{\mathcal{A}}(f \cdot g, h). \tag{9.15}$$

For the cocycle condition for the elements $x \otimes f$, $y \otimes g$ and $z \otimes h$ we have to permute this cyclically and add up the results to yield 0. We obtain (using the invariance of α)

$$\alpha([x, y], z)\left(\gamma^{\mathcal{A}}(f \cdot g, h) + \gamma^{\mathcal{A}}(g \cdot h, f) + \gamma^{\mathcal{A}}(h \cdot f, g)\right) = 0. \tag{9.16}$$

By the condition $[\mathfrak{g}, \mathfrak{g}] \neq 0$, and by $\alpha(\mathfrak{g}', \mathfrak{g}) \neq 0$ it follows that we can find $x, y, z \in \mathfrak{g}$, such that $\alpha([x, y], z) \neq 0$. This implies that $\gamma^{\mathcal{A}}$ is a multiplicative cocycle. If γ is local or bounded the same will be true for $\gamma^{\mathcal{A}}$, and Theorem 6.38 yields the claim. Note that for α non-degenerate, $\alpha(\mathfrak{g}', \mathfrak{g}) \neq 0$ is always true. □

Proposition 9.10. *Let γ be a cocycle of $\overline{\mathfrak{g}}$ which can be written as product*

$$\gamma(x \otimes f, y \otimes z) = \alpha(x, y) \cdot \gamma^{\mathcal{A}}(f, g), \tag{9.17}$$

with $\gamma^{\mathcal{A}}$ a nonvanishing multiplicative cocycle for $\mathcal{A}$ and α a bilinear form on $\mathfrak{g}$, then α is a symmetric invariant bilinear form.

Proof. First we apply the multiplicative property for $\gamma^{\mathcal{A}}$ to the triple $(1, 1, f)$, and obtain $\gamma^{\mathcal{A}}(1 \cdot 1, f) + \gamma^{\mathcal{A}}(1 \cdot f, 1) + \gamma^{\mathcal{A}}(f \cdot 1, 1) = 0$. This implies that $\gamma^{\mathcal{A}}(1, f) = 0$ for all $f \in \mathcal{A}$. From the antisymmetry of γ and $\gamma^{\mathcal{A}}$ it follows that α is symmetric. Now take $f, h \in \mathcal{A}$ such that $\gamma^{\mathcal{A}}(f, h) \neq 0$. We obtain

$$\gamma([x \otimes f, y \otimes 1], z \otimes h) = \alpha([x, y], z) \cdot \gamma^{\mathcal{A}}(f \cdot 1, h). \tag{9.18}$$

Adding this cyclically permuted up, yields

$$0 = \alpha([x, y], z) \cdot \gamma^{\mathcal{A}}(f, h) + \alpha([y, z], x) \cdot \gamma^{\mathcal{A}}(h, f) + \alpha([z, x], y) \cdot \gamma^{\mathcal{A}}(hf, 1). \tag{9.19}$$

Hence, $(\alpha([x, y], z) - \alpha([y, z], x))\,\gamma^{\mathcal{A}}(f, h) = 0$. From $\gamma^{\mathcal{A}}(f, h) \neq 0$ and the symmetry of α, the invariance $\alpha([x, y], z) = \alpha(x, [y, z])$ follows. □

Recall that we called a cohomology class *local* if it contained a representing local cocycle. Note that not all elements in a local cohomology class will be local. But the sum of two local cocycles will be local again. The subspace of $\mathrm{H}^2(\overline{\mathfrak{g}}, \mathbb{C})$, consisting of local cohomology classes, is denoted by $\mathrm{H}^2_{loc}(\overline{\mathfrak{g}}, \mathbb{C})$. The bounded cocycles yield $\mathrm{H}^2_b(\overline{\mathfrak{g}}, \mathbb{C})$. Of course, $\mathrm{H}^2_{loc}(\overline{\mathfrak{g}}, \mathbb{C}) \subseteq \mathrm{H}^2_b(\overline{\mathfrak{g}}, \mathbb{C})$, as every local cocycle is bounded.

9.4 $\mathcal{L}$-invariant cocycles

The following definition generalizes the definition of $\mathcal{L}$-invariance for cocycles of $\mathcal{A}$ to cocycles of arbitrary current algebras $\overline{\mathfrak{g}}$.

Definition 9.11. A cocycle γ of $\overline{\mathfrak{g}}$ is called $\mathcal{L}$-*invariant* if

$$\forall x, y \in \mathfrak{g}, \quad e \in \mathcal{L},\ g, h \in \mathcal{A} : \gamma(x(e\,.\,g), y(h)) + \gamma(x(g), y(e\,.\,h)) = 0. \tag{9.20}$$

Proposition 9.12. *The cocycle* (9.11) *is* $\mathcal{L}$*-invariant.*

Proof.

$$\gamma_{\alpha,C}(x(e.g), y(f)) + \gamma_{\alpha,C}(x(e.g), y(f)) = \alpha(x, y)\left(\gamma_C^{\mathcal{A}}(e.g, f) + \gamma_C^{\mathcal{A}}(g, e.f)\right) = 0, \tag{9.21}$$

as $\gamma_C^{\mathcal{A}}$ is $\mathcal{L}$-invariant. □

Proposition 9.13. *Let* $\psi = d\phi$ *be a coboundary of* $\overline{\mathfrak{g}}$ *which is bounded (e.g., local) and* $\mathcal{L}$*-invariant, then* $\psi = 0$.

Proof. The coboundary ψ is given as $\psi(x(g), y(h)) = \phi([x, y](gh))$, with a linear form $\phi : \overline{\mathfrak{g}} \to \mathbb{C}$. By the $\mathcal{L}$-invariance

$$0 = \psi(x(e\,.\,g), y(h)) + \psi(x(g), y(e\,.\,h)) = \phi([x, y](e\,.\,(gh)).$$

We set $h = 1$ and obtain

$$\phi([x, y](e\,.\,g)) = 0, \quad \forall x, y \in \mathfrak{g}, \quad e \in \mathcal{L},\ g \in \mathcal{A}. \tag{9.22}$$

Assume that $\psi \neq 0$, then $x, y \in \mathfrak{g}$ and $f \in \mathcal{A}$ must exist such that $\phi([x, y](f)) \neq 0$. Hence, a basis element $A_{k,r}$ also exists such that $\phi([x, y](A_{k,r})) \neq 0$. Assume first that $k > 0$, then using the almost-graded structure for the module $\mathcal{A}$ over $\mathcal{L}$ employed for the pairs of elements $(e_{0,r}, A_{k,r})$, we can write (with a certain $S' \in \mathbb{N}$ and coefficients $a'_{h,s} \in \mathbb{C}$)

$$e_{0,r}\,.\,A_{k,r} = kA_{k,r} + \sum_{h=k+1}^{k+S'} \sum_{s=1}^{K} a'_{h,s} A_{h,s}. \tag{9.23}$$

The $A_{h,s}$ appearing in the sum can be replaced by $(e_{0,s}\,.\,A_{h,s})/h + h.d.t.$ until we can write

$$e_{0,r}\,.\,A_{k,r} = kA_{k,r} + \sum_{h=k+1}^{S} \sum_{s=1}^{K} a_{h,s}(e_{0,s}\,.\,A_{h,s}) + D, \quad \text{with } D \in \bigoplus_{h>S}^{\infty} \mathcal{A}_h,$$

and S chosen such that

$$\psi((\overline{\mathfrak{g}})_n, (\overline{\mathfrak{g}})_m) = 0 \quad \text{for } n + m > S. \tag{9.24}$$

By the assumed boundedness of ψ such an S exists. Now we calculate

$$0 = \phi([x, y](e_{0,r}\,.\,A_{k,r})) = k\phi([x, y](A_{k,r})) + \sum_{h=k+1}^{S} \sum_{s=1}^{K} a_{h,s}\phi([x, y](e_{0,s}\,.\,A_{h,s})) + \phi([x, y](D)).$$

The first equation follows from (9.22). On the right, the second term will vanish by (9.22) too. The third term can be written as $\phi([x,y](D)) = \psi(x \otimes D, y \otimes 1)$. Using (9.24) we obtain that the third term also vanishes. Hence $k \cdot \phi([x,y](A_{k,r})) = 0$, in contradiction to the assumption.

If $k = 0$, for the first step we take $e_{-1,r} \,.\, A_{1,r} = 1 \cdot A_{0,r}$ + higher terms. For the higher terms we can apply $e_{0,s}$ as above. For $k < 0$ we start as above and use, instead of $e_{0,s} \,.\, A_{0,s}$, the element $e_{-1,s} \,.\, A_{1,s}$ to obtain $A_{0,s}$ in the corresponding intermediate step. Altogether this leads to a contradiction. □

Proposition 9.14. *Let γ be an $\mathcal{L}$-invariant and bounded (e.g., local) cocycle for the current algebra $\overline{\mathfrak{g}}$, then*

$$\gamma(x(1), y(h)) = 0, \quad \forall x, y \in \mathfrak{g},\ h \in \mathcal{A}. \tag{9.25}$$

Proof. We use the $\mathcal{L}$-invariance (9.20) for $e \in \mathcal{L}$, $g = 1$, and $h \in \mathcal{A}$. It follows that $\gamma(x(1), y(e\,.\,h)) = 0$ for all $e \in \mathcal{L}$ and $h \in \mathcal{A}$. Using the same kind of argument as in the proof of Proposition 9.13, from the boundedness of γ and the almost-graded structure, the claim follows. □

We call a cohomology class which has an $\mathcal{L}$-invariant cocycle as a representing element an *$\mathcal{L}$-invariant cohomology class*. By $\mathrm{H}^2_{loc,\mathcal{L}}(\overline{\mathfrak{g}}, \mathbb{C})$ we denote the subspace of $\mathcal{L}$-invariant and local cocycle classes. In the same way, $\mathrm{H}^2_{b,\mathcal{L}}(\overline{\mathfrak{g}}, \mathbb{C})$ denotes the subspace of $\mathcal{L}$-invariant and bounded cocycle classes.

Proposition 9.15. *Each class in $\mathrm{H}^2_{loc,\mathcal{L}}(\overline{\mathfrak{g}}, \mathbb{C})$ contains a unique representing cocycle which is local and $\mathcal{L}$-invariant. Each class in $\mathrm{H}^2_{b,\mathcal{L}}(\overline{\mathfrak{g}}, \mathbb{C})$ contains a unique representing cocycle which is bounded and $\mathcal{L}$-invariant.*

Proof. Let γ_1 and γ_2 be two bounded and $\mathcal{L}$-invariant cocycles representing the same class. The difference $\gamma_1 - \gamma_2$ is a coboundary which is also bounded and $\mathcal{L}$-invariant. Proposition 9.13 shows that it has to vanish. □

9.5 Current algebras of reductive Lie algebras

Let $\mathfrak{g}$ be a finite-dimensional reductive Lie algebra. Recall that $\mathfrak{g}$ is reductive if and only if $\mathfrak{g}$ is the direct sum of Lie algebras $\mathfrak{g}_0, \mathfrak{g}_1, \ldots, \mathfrak{g}_M$, with $\mathfrak{g}_0$ abelian and $\mathfrak{g}_1, \ldots, \mathfrak{g}_M$ simple Lie algebras, i.e.,

$$\mathfrak{g} = \mathfrak{g}_0 \oplus \mathfrak{g}_1 \oplus \cdots \oplus \mathfrak{g}_M. \tag{9.26}$$

In particular, $\mathfrak{g}_0$ is the center of $\mathfrak{g}$, for the derived algebra we obtain $\mathfrak{g}' = [\mathfrak{g}, \mathfrak{g}] = \mathfrak{g}_1 \oplus \cdots \oplus \mathfrak{g}_M$, and we have $[\mathfrak{g}_i, \mathfrak{g}_i] = \mathfrak{g}_i$, for $i \neq 0$, and $[\mathfrak{g}_i, \mathfrak{g}_j] = 0$ for $i \neq j$.

For the current algebra $\overline{\mathfrak{g}} = \mathfrak{g} \otimes \mathcal{A}$ we obtain the direct decomposition

$$\overline{\mathfrak{g}} = \overline{\mathfrak{g}}_0 \oplus \overline{\mathfrak{g}}_1 \oplus \cdots \oplus \overline{\mathfrak{g}}_M, \quad \text{with } \overline{\mathfrak{g}}_i = \mathfrak{g}_i \otimes \mathcal{A}. \tag{9.27}$$

Furthermore, note that every simple Lie algebra is perfect. Hence, using Proposition 9.2 we obtain

$$\begin{aligned} [\overline{\mathfrak{g}}_i, \overline{\mathfrak{g}}_i] = \overline{\mathfrak{g}}_i,\ i = 1, \ldots, M \quad [\overline{\mathfrak{g}}_0, \overline{\mathfrak{g}}_0] = 0,\ [\overline{\mathfrak{g}}_i, \overline{\mathfrak{g}}_j] = 0,\ i \neq j, \\ \overline{\mathfrak{g}}' = [\overline{\mathfrak{g}}, \overline{\mathfrak{g}}] = \overline{\mathfrak{g}'} = \overline{\mathfrak{g}}_1 \oplus \cdots \oplus \overline{\mathfrak{g}}_M. \end{aligned} \tag{9.28}$$

In particular, if the abelian part $\mathfrak{g}_0$ vanishes, then $\overline{\mathfrak{g}}$ will be a perfect Lie algebra.

Let γ be a cocycle for $\overline{\mathfrak{g}}$, then by restriction $\gamma_i := \gamma_{|\overline{\mathfrak{g}}_i \times \overline{\mathfrak{g}}_i}$ defines a cocycle for the subalgebra and direct summand $\overline{\mathfrak{g}}_i$ for $i = 0, 1, \ldots, M$. Vice versa, given a cocycle γ_i on $\overline{\mathfrak{g}}_i$, we define the extension $\tilde{\gamma}_i$ to $\overline{\mathfrak{g}} \times \overline{\mathfrak{g}}$ as γ_i on $\overline{\mathfrak{g}}_i \times \overline{\mathfrak{g}}_i$ and zero otherwise.

Lemma 9.16.
(a) *$\tilde{\gamma}_i$ is a cocycle for $\overline{\mathfrak{g}}$ if and only if γ_i is a cocycle for $\overline{\mathfrak{g}}_i$.*
(b) *$\tilde{\gamma}_i$ is a coboundary for $\overline{\mathfrak{g}}$ if and only if γ_i is a coboundary for $\overline{\mathfrak{g}}_i$.*
(c) *Let γ be a cocycle for $\overline{\mathfrak{g}}$, then*

$$\gamma(x_i, y_j) = 0, \quad \textit{for } x_i \in \overline{\mathfrak{g}}_i,\ y_j \in \overline{\mathfrak{g}}_j,\ \textit{if } i \neq j. \tag{9.29}$$

Proof. (a) One direction has already been explained above. Assume that γ_i is a cocycle for $\overline{\mathfrak{g}}_i$. Clearly, $\tilde{\gamma}_i$ is an antisymmetric bilinear form for $\overline{\mathfrak{g}}$. We have to show the cocycle condition. Let $x = \sum_{j=0}^M x_j$, $y = \sum_{k=0}^M y_k$, and $z = \sum_{l=0}^M z_l$ be three elements of $\overline{\mathfrak{g}}$ with their unique decompositions, with $x_m, y_m, z_m \in \overline{\mathfrak{g}}_m$ for $m = 0, \ldots, M$. Using (9.28) we obtain

$$\begin{aligned} \tilde{\gamma}_i([x, y], z) &= \tilde{\gamma}_i\left(\left[\sum_{j=0}^M x_j, \sum_{k=0}^M y_k\right], \sum_{l=0}^M z_l\right) \\ &= \sum_{j,k,l} \tilde{\gamma}_i([x_j, y_k], z_l) = \sum_{j,l} \tilde{\gamma}_i([x_j, y_j], z_l) = \gamma_i([x_i, y_i], z_i). \end{aligned} \tag{9.30}$$

Hence, from the cocycle condition of γ_i the cocycle condition for $\tilde{\gamma}_i$ follows.

(b) Let γ be a coboundary for $\overline{\mathfrak{g}}$, i.e., $\gamma(x, y) = d\phi(x, y) = \phi([x, y])$, with $\phi : \overline{\mathfrak{g}} \to \mathbb{C}$ a linear form. Set $\phi_i = \phi_{|\overline{\mathfrak{g}}_i}$. For $x_i, y_i \in \overline{\mathfrak{g}}_i$ we obtain

$$\gamma_i(x_i, y_i) = \gamma(x_i, y_i) = d\phi(x_i, y_i) = \phi([x_i, y_i]) = \phi_i([x_i, y_i]). \tag{9.31}$$

Hence, γ_i will be a coboundary for $\overline{\mathfrak{g}}_i$. If γ_i is a coboundary for $\overline{\mathfrak{g}}_i$, then

$$\gamma_i(x_i, y_i) = d\phi_i(x_i, y_i) = \phi_i([x_i, y_i]), \tag{9.32}$$

with a linear form ϕ_i on $\overline{\mathfrak{g}}_i$. We use $\tilde{\phi}_i$ to denote ϕ_i extended by zero on the other components of $\overline{\mathfrak{g}}$. But

$$\tilde{\gamma}_i(x, y) = \gamma_i(x_i, y_i) = \phi_i([x_i, y_i]) = \tilde{\phi}_i([x, y]). \tag{9.33}$$

Hence the claim.

(c) Because $i \neq j$ at least one of the indices is $\neq 0$. Assume $i \neq 0$. As $[\overline{\mathfrak{g}}_i, \overline{\mathfrak{g}}_i] = \overline{\mathfrak{g}}_i$, we can write $x_i = [x_i^{(1)}, x_i^{(2)}]$ with $x_i^{(1)}, x_i^{(2)} \in \overline{\mathfrak{g}}$. The cocycle condition for $x_i^{(1)}, x_i^{(2)}$ and y_j is

$$\gamma([x_i^{(1)}, x_i^{(2)}], y_j) + \gamma([x_i^{(2)}, y_j], x_i^{(1)}) + \gamma([y_j, x_i^{(1)}], x_i^{(2)}) = 0. \tag{9.34}$$

The last two summands are zero because the commutators are zero. Hence, the first summand which equals $\gamma(x_i, y_j)$ is also zero. □

Proposition 9.17.

(a) *Let γ be a cocycle for $\overline{\mathfrak{g}}$, $\gamma_i = \gamma_{|\overline{\mathfrak{g}}_i \times \overline{\mathfrak{g}}_i}$ its restriction to the i-th summand, and $\tilde{\gamma}_i$ the extension of γ_i by zero to $\overline{\mathfrak{g}}$ again as above, then*

$$\gamma = \sum_{i=0}^{M} \tilde{\gamma}_i =: \bigoplus_{i=0}^{M} \gamma_i. \tag{9.35}$$

(b) *The above decomposition of γ induces*

$$\mathrm{Z}^2(\overline{\mathfrak{g}}, \mathbb{C}) = \bigoplus_{i=0}^{M} \mathrm{Z}^2(\overline{\mathfrak{g}}_i, \mathbb{C}), \quad \mathrm{B}^2(\overline{\mathfrak{g}}, \mathbb{C}) = \bigoplus_{i=0}^{M} \mathrm{B}^2(\overline{\mathfrak{g}}_i, \mathbb{C}), \quad \mathrm{H}^2(\overline{\mathfrak{g}}, \mathbb{C}) = \bigoplus_{i=0}^{M} \mathrm{H}^2(\overline{\mathfrak{g}}_i, \mathbb{C}), \tag{9.36}$$

where Z^2 denotes the vector space of 2-cocycles and B^2 denotes the vector space of 2-coboundaries.

(c) *The cocycle γ is a local (respectively bounded, respectively $\mathcal{L}$-invariant) cocycle if and only if all γ_i, $i = 0, 1, \ldots, M$ are local (respectively bounded, respectively $\mathcal{L}$-invariant). In particular,*

$$\mathrm{H}^2_{loc}(\overline{\mathfrak{g}}, \mathbb{C}) = \bigoplus_{i=0}^{M} \mathrm{H}^2_{loc}(\overline{\mathfrak{g}}_i, \mathbb{C}), \quad \mathrm{H}^2_{loc,\mathcal{L}}(\overline{\mathfrak{g}}, \mathbb{C}) = \bigoplus_{i=0}^{M} \mathrm{H}^2_{loc,\mathcal{L}}(\overline{\mathfrak{g}}_i, \mathbb{C}). \tag{9.37}$$

$$\mathrm{H}^2_{b}(\overline{\mathfrak{g}}, \mathbb{C}) = \bigoplus_{i=0}^{M} \mathrm{H}^2_{b}(\overline{\mathfrak{g}}_i, \mathbb{C}), \quad \mathrm{H}^2_{b,\mathcal{L}}(\overline{\mathfrak{g}}, \mathbb{C}) = \bigoplus_{i=0}^{M} \mathrm{H}^2_{b,\mathcal{L}}(\overline{\mathfrak{g}}_i, \mathbb{C}). \tag{9.38}$$

Proof. Per construction γ coincides with $\sum_i \tilde{\gamma}_i$ on $\overline{\mathfrak{g}}_j \times \overline{\mathfrak{g}}_j$ for all j. By Lemma 9.16 both also coincide on $\overline{\mathfrak{g}}_j \times \overline{\mathfrak{g}}_k$ for all j, k with $j \neq k$. Hence (a). Part (b) is a reformulation of Lemma 9.16 (a) and (b). Clearly locality, boundedness, and $\mathcal{L}$-invariance are conditions which have to be checked on each component $\overline{\mathfrak{g}}_i$ separately. Hence (c) follows. □

One way to obtain central extensions of $\overline{\mathfrak{g}}$ is to choose a symmetric invariant bilinear form α for γ and use the defining equation (9.8). With the same type of argument as above any such α for reductive γ can be decomposed as

$$\alpha = \oplus_{i=0}^{M} \alpha_i, \quad \text{with } \alpha_i = \alpha_{|\mathfrak{g}_i \times \mathfrak{g}_i} \tag{9.39}$$

symmetric invariant bilinear forms on $\mathfrak{g}_i$. Conversely, every such sum of α_i gives a symmetric invariant bilinear form α on $\mathfrak{g}$. Note also that $\alpha(\mathfrak{g}_i, \mathfrak{g}_j) = 0$ for $i \neq j$. Recall that for a simple Lie algebra $\mathfrak{g}_i$ over $\mathbb{C}$ there is up to multiplication with a scalar r_i only one such invariant form, the Cartan–Killing form β_i of $\mathfrak{g}_i$. The form β_i is non-degenerate. We obtain $\alpha = \alpha_0 + \sum_{i=1}^{M} r_i\beta_i$ with $r_i \in \mathbb{C}$. For the abelian Lie algebra $\mathfrak{g}_0$ every symmetric bilinear form is invariant. If $\dim \mathfrak{g}_0 = n$, then the space of symmetric bilinear forms on $\mathfrak{g}_0$ is $\frac{n(n+1)}{2}$-dimensional. Altogether we obtain that the space of symmetric invariant bilinear forms on $\mathfrak{g}$ has dimension $\frac{n(n+1)}{2} + M$.

Remark 9.18. For general results on the universal central extensions and the second homology of current algebras, now with function algebra the holomorphic functions on $\Sigma \setminus A$, see also the work of Neeb and Wagemann [178–180].

9.6 Classification results

Let $\mathfrak{g}$ be a reductive Lie algebra, and $\mathfrak{g} = \mathfrak{g}_0 \oplus \mathfrak{g}_1 \oplus \cdots \oplus \mathfrak{g}_M$ its decomposition into its abelian $\mathfrak{g}_0$ summand and M simple summands $\mathfrak{g}_i$, $i = 1, \ldots M$. Assume that the abelian summand has the dimension n. We now give the classification of local and $\mathfrak{L}$-invariant cocycle classes for the associated current algebra $\overline{\mathfrak{g}}$.

Theorem 9.19.

(a) *Given a cocycle γ for $\overline{\mathfrak{g}}$ which is local, and whose restriction γ_0 to the abelian summand $\overline{\mathfrak{g}}_0$ is $\mathfrak{L}$-invariant, then a symmetric invariant bilinear form α for $\mathfrak{g}$ exists such that γ is cohomologous to*

$$\gamma'_\alpha(x \otimes f, y \otimes g) = \frac{\alpha(x, y)}{2\pi i} \int_{C_S} f dg. \tag{9.40}$$

Conversely, every such α determines a local and $\mathfrak{L}$-invariant cocycle.

(b) *If the cocycle γ is already local and $\mathfrak{L}$-invariant for the whole $\overline{\mathfrak{g}}$, then it coincides with the cocycle γ'_α.*

(c)

$$\dim \mathrm{H}^2_{loc,\mathfrak{L}}(\overline{\mathfrak{g}}, \mathbb{C}) = \frac{n(n+1)}{2} + M. \tag{9.41}$$

For the proof we will consider the simple, semisimple and abelian cases separately in the following sections and then put everything together.

Corollary 9.20. *Those equivalence classes of almost-graded central extensions of the current algebra $\overline{\mathfrak{g}}$ of a reductive Lie algebra $\mathfrak{g}$, whose corresponding cocycles can be given by $\mathfrak{L}$-invariant cocycles if restricted to $\overline{\mathfrak{g}}_0$, correspond 1 : 1 to the space of symmetric invariant bilinear forms α on $\mathfrak{g}$. This space has the dimension* $\dfrac{n(n+1)}{2} + M$.

Remark 9.21. The corresponding classification results exists for bounded cocycles. Instead of (9.40) we obtain (with $K = \#I$)

$$\gamma'_\alpha(x \otimes f, y \otimes g) = \alpha(x, y) \cdot \sum_{i=1}^{K} \beta_i \frac{1}{2\pi\mathrm{i}} \int_{C_i} f dg, \quad \beta_i \in \mathbb{C}. \tag{9.42}$$

The dimension formula for the bounded cocycles is now

$$\dim \mathrm{H}^2_{b,\mathfrak{L}}(\overline{\mathfrak{g}}, \mathbb{C}) = \left(\frac{n(n+1)}{2} + M\right) \cdot K. \tag{9.43}$$

9.6.1 Cocycles for the simple case

Theorem 9.22.

(a) *Let $\mathfrak{g}$ be a finite-dimensional simple Lie algebra, then every local cocycle of the current algebra $\overline{\mathfrak{g}} = \mathfrak{g} \otimes \mathcal{A}$ is cohomologous to a cocycle given by*

$$\gamma(x \otimes f, y \otimes g) = r \cdot \frac{\beta(x, y)}{2\pi\mathrm{i}} \int_{C_S} f dg, \quad \text{with } r \in \mathbb{C}, \tag{9.44}$$

and β the Cartan–Killing form of $\mathfrak{g}$. In particular, γ is cohomologous to a local and $\mathfrak{L}$-invariant cocycle.

(b) *If the cocycle is already local and $\mathfrak{L}$-invariant, then it coincides with the cocycle (9.44) with $r \in \mathbb{C}$ suitably chosen.*

Proof. Kassel [127] proved that the algebra $\overline{\mathfrak{g}} = \mathfrak{g} \otimes \mathcal{A}$ for any commutative algebra $\mathcal{A}$ over $\mathbb{C}$ and any simple Lie algebra $\mathfrak{g}$ admits a universal central extension which is given by

$$\widehat{\mathfrak{g}}^{univ} = (\Omega^1_{\mathcal{A}}/d\mathcal{A}) \oplus \overline{\mathfrak{g}} \tag{9.45}$$

with Lie structure

$$[x \otimes f, y \otimes g] = [x, y] \otimes fg + \beta(x, y)\overline{fdg}, \quad [\Omega^1_{\mathcal{A}}/d\mathcal{A}, \widehat{\mathfrak{g}}^{univ}] = 0. \tag{9.46}$$

Here $\Omega^1_{\mathcal{A}}/d\mathcal{A}$ denotes the vector space of Kähler differentials of the algebra $\mathcal{A}$, and β is the Cartan–Killing form. The elements in $\Omega^1_{\mathcal{A}}$ can be given as fdg with $f, g \in \mathcal{A}$, and $\overline{fdg}$ denotes its class modulo $d\mathcal{A}$. This universal extension is not necessarily one-dimensional. Let $\widehat{\mathfrak{g}}$ be any one-dimensional central extension of $\overline{\mathfrak{g}}$. By the very definition of a universal central extension, $\widehat{\mathfrak{g}}$ (see Section 6.10.1) will be given as a quotient of $\widehat{\mathfrak{g}}^{univ}$. Up to equivalence we have a Lie homomorphism Φ

$$\widehat{\mathfrak{g}}^{univ} = \Omega^1_{\mathcal{A}}/d\mathcal{A} \oplus \overline{\mathfrak{g}} \xrightarrow{\Phi=(\varphi, id)} \widehat{\mathfrak{g}} = \mathbb{C} \oplus \overline{\mathfrak{g}} \tag{9.47}$$

with a linear form φ on $\Omega^1_{\mathcal{A}}/d\mathcal{A}$. The structure of $\widehat{\mathfrak{g}}$ is then equal to

$$[x \otimes f, y \otimes g] = [x, y] \otimes fg + \beta(x, y)\, \varphi(\overline{fdg})\, t, \quad [t, \widehat{\mathfrak{g}}] = 0. \tag{9.48}$$

In our situation $\Sigma \setminus A$ is an affine curve and $\Omega^1_{\mathcal{A}}/d\mathcal{A}$ is the first algebraic deRham cohomology group of the complex of meromorphic functions on Σ which are holomorphic on $\Sigma \setminus A$ (see also Bremner [28]). By Grothendieck's algebraic deRham theorem [92, p. 453], the cohomology of the complex is isomorphic to the singular cohomology of $\Sigma \setminus A$, which is finite-dimensional. A linear form φ as above is given by choosing a linear combination of cycle classes in $\Sigma \setminus A$ and integrating the differential class $\overline{fdg}$ over this combination. As γ is assumed to be local, Theorem 6.38 implies that the combination is a multiple of the separating cocycle. This shows that the given cocycle is cohomologous to (9.44). But this cocycle is local and $\mathcal{L}$-invariant. Hence, (a). By Proposition 9.15 it is the only local and $\mathcal{L}$-invariant cocycle in this class. Hence, (b). □

Corollary 9.23. *For $\mathfrak{g}$ simple, up to equivalence of extensions and rescaling of the central element there is a unique nontrivial almost-graded central extension $\widehat{\mathfrak{g}}$ of its higher genus multi-point current algebra $\overline{\mathfrak{g}}$. It is given by the cocycle* (9.44).

Corollary 9.24. *For $\mathfrak{g}$ simple*

$$\mathrm{H}^2_{loc}(\overline{\mathfrak{g}}, \mathbb{C}) = \mathrm{H}^2_{loc,\mathcal{L}}(\overline{\mathfrak{g}}, \mathbb{C}). \tag{9.49}$$

This space is one-dimensional.

Remark 9.25. It is possible to avoid the use of Kassel's Theorem about the structure of the universal central extension by a direct proof using the internal structure (i.e., roots systems) of the finite-dimensional Lie algebras. A corresponding proof was given by the author together with O.K. Sheinman in the context of Lax operator algebras [224, 228]. The proof presented there is valid also for Krichever–Novikov current algebras associated with finite simple Lie algebras. This technique is related to the work of Garland [87], in which he proves the existence and uniqueness of central extensions in the simple Lie algebra case in the classical ($g = 0, N = 2$) situation.

9.6.2 Cocycles for the semisimple case

Let $\mathfrak{g}$ be a semisimple Lie algebra. It can be written as the direct sum of simple Lie algebras $\mathfrak{g}_i$, i.e., $\mathfrak{g} = \mathfrak{g}_1 \oplus \cdots \oplus \mathfrak{g}_M$.

Theorem 9.26.

(a) *For every local cocycle γ for the current algebra $\overline{\mathfrak{g}}$ of a semisimple Lie algebra $\mathfrak{g}$ a symmetric invariant bilinear form α for $\mathfrak{g}$ exists such that γ is cohomologous to*

$$\gamma'_{\alpha,S}(x \otimes f, y \otimes g) = \frac{\alpha(x,y)}{2\pi \mathrm{i}} \int_{C_S} fdg. \tag{9.50}$$

In particular, γ is cohomologous to a local and $\mathcal{L}$-invariant cocycle.

(b) *If the cocycle is already local and $\mathcal{L}$-invariant, then it coincides with the cocycle* (9.50).

(c)
$$\dim \mathrm{H}^2_{loc}(\overline{\mathfrak{g}}, \mathbb{C}) = \dim \mathrm{H}^2_{loc,\mathcal{L}}(\overline{\mathfrak{g}}, \mathbb{C}) = M, \tag{9.51}$$

where M is the number of simple summands of γ.

Proof. As explained in Section 9.5, every local cocycle for $\overline{\mathfrak{g}}$ can be decomposed as $\gamma = \oplus_{i=1}^{M} \gamma_i$ with γ_i local cocycles on the simple summands $\mathfrak{g}_i$. Denote by β_i the Cartan–Killing form of $\mathfrak{g}_i$. By Theorem 9.22 the γ_i is cohomologous to

$$\gamma_i'(x_i \otimes f, y_i \otimes g) = r_i \cdot \frac{\beta_i(x_i, y_i)}{2\pi \mathrm{i}} \int_{C_S} f dg, \quad \text{with } r_i \in \mathbb{C}. \tag{9.52}$$

We set $\alpha := \sum_{i=1}^{M} r_i \beta_i$, which is defined for $x = \sum_j x_j$, $y = \sum_k y_k$ by $a(x, y) := \sum_{i=1}^{M} r_i \beta_i(x_i, y_i)$. The form α is a symmetric invariant bilinear form and

$$\gamma_{\alpha,S}(x \otimes f, y \otimes g) = \frac{\alpha(x, y)}{2\pi \mathrm{i}} \int_{C_S} f dg \tag{9.53}$$

is a cocycle which is cohomologous to γ (use Proposition 9.17). This shows (a). Part (b) follows from Proposition 9.15.

All linear combinations of the Cartan–Killing forms β_i give the whole (local) cohomology space. Hence (c). This follows also from Proposition 9.17 (b). □

Corollary 9.27. *The space of equivalence classes of almost-graded central extensions of the current algebra $\overline{\mathfrak{g}}$ of a semisimple Lie algebra $\mathfrak{g}$ is in 1 : 1 correspondence with the space of symmetric invariant bilinear forms for $\mathfrak{g}$. Its dimension is the number of simple summands of $\mathfrak{g}$.*

Remark 9.28. In general, the claim of the above theorem is not true for $\mathfrak{g}$ not semisimple. As a nontrivial example, take $\mathfrak{g} = \mathfrak{gl}(n)$. Let α be any nonvanishing symmetric invariant bilinear form for $\mathfrak{gl}(n)$. For any antisymmetric bilinear form ψ on $\mathcal{A}$ the form

$$\gamma(x \otimes f, y \otimes g) := \alpha(x, y)\psi(f, g)$$

defines a cocycle for $\overline{\mathfrak{gl}(n)}$. But ψ can be chosen to be local without being a geometric cocycle. We require the cocycle to be $\mathcal{L}$-invariant to obtain a statement corresponding to the above theorem at least in the reductive case.

9.6.3 Cocycles for the abelian case

Let $\mathfrak{g}$ be a finite-dimensional abelian Lie algebra. Note that in the abelian case no nontrivial coboundaries exist. Hence two different cocycles will never be cohomologous. Already in the one-dimensional abelian case, i.e., $\mathfrak{g} = \mathbb{C}$, $\overline{\mathfrak{g}} = \mathbb{C} \otimes \mathcal{A} = \mathcal{A}$, if one wants to obtain uniqueness results, it is necessary to allow only $\mathcal{L}$-invariant cocycles.

If $\dim \mathfrak{g} \geq 2$, an additional phenomena will show up. In contrast to the semisimple case, the Lie algebra $\mathfrak{g}$ will have nontrivial central extensions. These are given by alternating bilinear forms on $\mathfrak{g}$. If we were to allow arbitrary central extensions of $\overline{\mathfrak{g}}$, then also central extensions induced by extensions of $\mathfrak{g}$ would show up. It will turn out that the required $\mathcal{L}$-invariance will exclude cocycles coming from $\mathfrak{g}$.

Lemma 9.29. *Let γ be a local and $\mathcal{L}$-invariant cocycle for $\overline{\mathfrak{g}}$.*
Set $\gamma_{xy}(f,g) := \gamma(x \otimes f, y \otimes g)$ for fixed $x, y \in \mathfrak{g}$. Then $\alpha_{xy} \in \mathbb{C}$ exists with

$$\gamma_{xy}(f,g) = \alpha_{xy}\gamma_S^{\mathcal{A}}(f,g). \tag{9.54}$$

Proof. For fixed x, y the form γ_{xy} will be a bilinear form on $\mathcal{A}$ which is $\mathcal{L}$-invariant. Proposition 6.60 shows the claim. □

Theorem 9.30. *Let $\mathfrak{g}$ be an abelian Lie algebra of dimension n. Then*

$$\dim \mathrm{H}^2_{loc,\mathcal{L}}(\overline{\mathfrak{g}}, \mathbb{C}) = \frac{n(n+1)}{2}, \tag{9.55}$$

and the cocycles are given by

$$\gamma_{\alpha,S}(x \otimes f, y \otimes g) = \frac{\alpha(x,y)}{2\pi \mathrm{i}} \int_{C_S} f dg, \tag{9.56}$$

where α is an arbitrary symmetric bilinear form for $\mathfrak{g}$.

Proof. For $n = 1$ this is the result for $\overline{\mathfrak{g}} = \mathcal{A}$ given in Theorem 6.38. Clearly, given any such α we obtain via (9.56) a cocycle which is local and $\mathcal{L}$-invariant. Now let γ be a local and $\mathcal{L}$-invariant cocycle. From Lemma 9.29 we know that it writes as $\gamma(x \otimes f, y \otimes g) = \alpha_{xy} \cdot \gamma_S^{\mathcal{A}}$. The map $\alpha : (x,y) \to \alpha_{x,y}$ is bilinear. By the antisymmetry of γ, and $\gamma_S^{\mathcal{A}}$, and the nonvanishing of the latter, α has to be a symmetric form on $\mathfrak{g}$. Hence, γ is indeed of the form (9.56). But there are exactly $\frac{n(n+1)}{2}$ symmetric bilinear forms which are linearly independent. The corresponding $\gamma_{\alpha,S}$s will stay linearly independent. This shows the formula for the dimension. □

Let us repeat that if $\mathfrak{g}_0$ is one-dimensional, our current algebra will be $\mathcal{A}$, and hence $\widehat{\mathcal{A}}$ defined via (9.56) will be the infinite-dimensional higher genus Heisenberg algebra (oscillator algebra).

Now we collect the individual results to prove Theorem 9.19 for the reductive case.

Proof. The statement is obtained by restricting the cocycle to the semisimple and abelian parts. Using Theorem 9.26 and Theorem 9.30 together with the decomposition result of Proposition 9.17 gives the result including the dimension formula. □

9.7 Algebras of $\overline{\mathfrak{g}}$-valued differential operators

In Section 2.6, the algebra of differential operators of degree ≤ 1 associated with the function algebra $\mathcal{A}$ was introduced as semidirect sum of $\mathcal{A}$ with the vector field algebra $\mathcal{L}$. In this section, the construction will be extended to the case when $\mathcal{A}$ is replaced by the current algebra $\overline{\mathfrak{g}}$ of a general finite-dimensional Lie algebra $\mathfrak{g}$. Again, central extensions are studied and classification results for local cocycles are given. The algebras play an important role in the context of the Sugawara construction (see Chapter 10) and fermionic representations for $\widehat{\mathfrak{g}}$. See also Section 9.10 later in this chapter.

9.7.1 $\overline{\mathfrak{g}}$-valued differential operators

Let $\mathfrak{g}$ be an arbitrary finite-dimensional Lie algebra and $\overline{\mathfrak{g}}$ the associated current algebra. Set $\mathcal{D}^1_{\mathfrak{g}} := \overline{\mathfrak{g}} \oplus \mathcal{L}$ as vector space. By direct verification it follows that

$$e\,.\,(x \otimes g) := x \otimes (e\,.\,g), \qquad e \in \mathcal{L},\ x \in \mathfrak{g},\ g \in \mathcal{A}, \tag{9.57}$$

defines an $\mathcal{L}$-module structure on the space $\overline{\mathfrak{g}}$.

Proposition 9.31. *$\mathcal{L}$ operates as derivation on $\overline{\mathfrak{g}}$, i.e.,*

$$e\,.\,[x \otimes f, y \otimes g] = [e\,.\,(x \otimes f), y \otimes g] + [x \otimes f, e\,.\,(y \otimes g)]. \tag{9.58}$$

We define the semidirect sum $\overline{\mathfrak{g}} \rtimes \mathcal{L}$ on the basis of this action as explained in Section 1.6.

Definition 9.32. The Lie algebra $\mathcal{D}^1_{\mathfrak{g}}$ is the semidirect sum structure on $\overline{\mathfrak{g}} \oplus \mathcal{L}$ induced by $[e, x \otimes g] := x \otimes (e\,.\,g)$.

In more detail, the product is given as (for $e, f \in \mathcal{L}$, $x, y \in \mathfrak{g}$, $g, h \in \mathcal{A}$)

$$[(x \otimes g, e), (y \otimes h, f)] = (\ [x, y] \otimes (g \cdot h) + y \otimes (e\,.\,h) - x \otimes (f\,.\,g)\ ,\ [e, f]\ ). \tag{9.59}$$

The subspace $\overline{\mathfrak{g}}$ is a Lie-ideal and $\mathcal{L}$ is a subalgebra. We have the short exact sequence of Lie algebras

$$0 \longrightarrow \overline{\mathfrak{g}} \xrightarrow{\ i_1\ } \mathcal{D}^1_{\mathfrak{g}} \xrightarrow{\ p_2\ } \mathcal{L} \longrightarrow 0. \tag{9.60}$$

Using the almost-gradings of $\overline{\mathfrak{g}}$ and $\mathcal{L}$ we obtain an almost-grading for $\mathcal{D}^1_{\mathfrak{g}}$ by taking $(\mathcal{D}^1_{\mathfrak{g}})_m := (\overline{\mathfrak{g}})_m \oplus \mathcal{L}_m$ for $m \in \mathbb{Z}$ as homogeneous subspaces. From (9.59) one immediately verifies that this defines an almost-graded structure. For $\mathfrak{g} = \mathbb{C}$ (and hence $\overline{\mathfrak{g}} = \mathcal{A}$) we recover the algebra $\mathcal{D}^1$ as a special case.

Remark 9.33. If we fix an element $e \in \mathcal{L}$, then the subspace $\overline{\mathfrak{g}}_e := \overline{\mathfrak{g}} \oplus \mathbb{C}e$ is a Lie subalgebra of $\mathcal{D}^1_{\mathfrak{g}}$. This follows immediately from (9.59). In the classical case this construction is of certain relevance. There one adjoins the derivation $e = z\frac{d}{dz}$ to $\overline{\mathfrak{g}}$. This element works as a grading element, i.e., we have

$$e\,.\,(x \otimes z^n) = n(x \otimes z^n), \quad \forall n \in \mathbb{Z}. \tag{9.61}$$

In our only almost-graded case we do not have such an element.

Proposition 9.34. *Set $e := \sum_{r=1}^{K} e_{0,r}$, then for a homogeneous element $X \in \overline{\mathfrak{g}}_m$ we have*

$$e\,.\,X = mX + h.d.t. \tag{9.62}$$

Proof. As $X \in \overline{\mathfrak{g}}_m$, $x_i \in \mathfrak{g}$ exists such that $X = \sum_{i=1}^{K} x_i \otimes A_{m,i}$. Hence,

$$\begin{aligned} e\,.\,X &= \sum_{r=1}^{K}\sum_{i=1}^{K} x_i \otimes (e_{0,r}\,.\,A_{m,i}) = \sum_{r,i} x_i \otimes (m \cdot A_{m,i}\delta_i^r + h.d.t.) \\ &= m\sum_{i} x_i \otimes A_{m,i} + h.d.t. = m \cdot X + h.d.t. \end{aligned}$$

□

9.7.2 Cocycles

Next we consider cocycles and central extensions of $\mathcal{D}^1_{\mathfrak{g}}$. By bilinearity the cocycle condition (6.11) can be equivalently formulated by considering cocycle conditions for triples of elements which are individually of "pure types", i.e., elements which are either vector fields or elements from $\overline{\mathfrak{g}}$.

(1) If all three are vector fields, the condition is that the cocycle defines by restriction a cocycle for $\mathcal{L}$.
(2) If all three elements are from $\overline{\mathfrak{g}}$, the condition is that the cocycle defines by restriction a cocycle for $\overline{\mathfrak{g}}$.
(3) Now let $e, f \in \mathcal{L}$, and $x(g) \in \overline{\mathfrak{g}}$, then using the Lie structure we rewrite the cocycle condition as
$$\gamma([e, f], x(g)) - \gamma(e, x(f\,.\,g)) + \gamma(f, x(e\,.\,g)) = 0. \tag{9.63}$$
(4) If $e \in \mathcal{L}$ and $x(g), y(h) \in \overline{\mathfrak{g}}$, then the cocycle condition rewrites as
$$\gamma(x(e\,.\,g), y(h)) - \gamma(e, [x, y](gh)) + \gamma(x(g), y(e\,.\,h)) = 0. \tag{9.64}$$

An antisymmetric form on $\mathcal{D}^1_{\mathfrak{g}}$ will be a cocycle if and only if all 4 conditions are fulfilled.

Let V be one of the subalgebras $\mathcal{L}$ or $\overline{\mathfrak{g}}$ of $\mathcal{D}^1_{\mathfrak{g}}$. For a cocycle of V we will use the expression "extension by zero on the complementary space" to denote the extended bilinear form which coincides with the cocycle on pairs of elements from V and will be set to zero if either of the two entries in the bilinear form is from the complementary subspace. The rest follows from bilinear extension. Note that such an extension is not necessarily a cocycle for $\mathcal{D}^1_{\mathfrak{g}}$.

Proposition 9.35.
(a) *Every cocycle of $\mathcal{L}$ can be extended by zero on $\overline{\mathfrak{g}}$ to a cocycle of $\mathcal{D}^1_{\mathfrak{g}}$.*
(b) *A cocycle γ of $\overline{\mathfrak{g}}$ can be extended by zero on $\mathcal{L}$ to a cocycle of $\mathcal{D}^1_{\mathfrak{g}}$ if and only if the cocycle γ is $\mathcal{L}$-invariant (see Definition 9.11).*
(c) *Let α be an invariant symmetric bilinear form for $\mathfrak{g}$. The geometric cocycles*

$$\gamma_{\alpha,C}(x(g),y(h)) = \alpha(x,y)\cdot\gamma_C^{\mathcal{A}}(g,h) \tag{9.65}$$

of $\overline{\mathfrak{g}}$ can be extended by zero to a cocycle for $\mathcal{D}^1_{\mathfrak{g}}$.

Proof. From the above 4 separate cocycle conditions, only the first one is of relevance for a cocycle of $\mathcal{L}$ which is extended by zero. This is exactly the condition that it is a cocycle for $\mathcal{L}$. This follows also from the sequence (9.60) because the extended cocycle is nothing but the pull-back by p_2. This shows (a).

The situation is different for cocycles of $\overline{\mathfrak{g}}$. If we set the cocycle to zero outside $\overline{\mathfrak{g}}$, then (1), (2), and (3) are automatically fulfilled. From (9.64) it follows that it will define a cocycle if and only if

$$\gamma(x(e\,.\,g),y(h)) + \gamma(x(g),y(e\,.\,h)) = 0. \tag{9.66}$$

But this is Definition 9.11 of $\mathcal{L}$-invariance for cocycles of $\overline{\mathfrak{g}}$. Hence, (b). Part (c) follows from the fact that $\gamma_C^{\mathcal{A}}$ is $\mathcal{L}$-invariant, hence the same is true for $\gamma_{\alpha,C}(x(g),y(h))$. □

Now we consider cocycles γ of pure mixing type. By a cocycle of pure mixing type we understand a cocycle which might be non-zero only if the arguments are of opposite pure type. From (9.64) it follows that

$$\gamma(e,[x,y](g)) = 0, \quad \forall x,y\in\mathfrak{g},\ e\in\mathcal{L},\ g\in\mathcal{A}, \tag{9.67}$$

is a necessary condition for γ to be a cocycle of pure mixing type.

Proposition 9.36. *For a perfect Lie algebra $\mathfrak{g}$ (i.e., $\mathfrak{g}=[\mathfrak{g},\mathfrak{g}]$) no nonvanishing cocycles of $\mathcal{D}^1_{\mathfrak{g}}$ exist of pure mixing type.*

Proof. By assumption, we can express every $z\in\mathfrak{g}$ as $z=[x,y]$. Hence, by (9.67) $\gamma(e,z(g))=0$ for all $e\in\mathcal{L}$, $z\in\mathfrak{g}$, $g\in\mathcal{A}$ and the claim follows. □

Let γ be a cocycle of pure mixing type which is of the form $\gamma(e,x(g)) = \phi(x)\gamma^{(m)}(e,g)$, with $\phi\in\mathfrak{g}^*$ a linear form on $\mathfrak{g}$, and $\gamma^{(m)}$ bilinear on $\mathcal{L}\times\mathcal{A}$. Condition (9.67) implies that either $\gamma^{(m)}\equiv 0$ or that $\phi([x,y])=0$ for all $x,y\in\mathfrak{g}$. By Condition (9.63) for $x\in\mathfrak{g}$, $e,f\in\mathcal{L}$ and $g\in\mathcal{A}$ the relation

$$\phi(x)\left(\gamma^{(m)}([e,f],g) - \gamma^{(m)}(e,f\,.\,g) + \gamma^{(m)}(f,e\,.\,g)\right) = 0 \tag{9.68}$$

follows. Excluding the trivial case $\phi=0$, this implies that $\gamma^{(m)}$ is a mixing cocycle for $\mathcal{D}^1$. Conversely, every bilinear form $\gamma(e,x(g)) = \phi(x)\gamma^{(m)}(e,g)$ with $\phi\in\mathfrak{g}^*$ such that $\phi_{|[\gamma,\gamma]}=0$, and $\gamma^{(m)}$ a mixing cocycle of $\mathcal{D}^1$ defines a cocycle of $\mathcal{D}^1_{\mathfrak{g}}$.

Let ϕ be a linear form ϕ of $\mathfrak{g}$ which vanishes on $\mathfrak{g}' = [\mathfrak{g}, \mathfrak{g}]$. Let T be a meromorphic connection, holomorphic outside A. We define

$$\gamma_{S,\phi,T}(e, x(g)) := \frac{\phi(x)}{2\pi \mathrm{i}} \int_{C_S} \left(e \cdot g'' + T \cdot (e \cdot g')\right) dz. \tag{9.69}$$

This will be antisymmetrically extended and put zero on the pairs of elements of the same pure type. Above we demonstrated the following proposition.

Proposition 9.37. *The extended bilinear form $\gamma_{\phi,T}$ defines a local cocycle for $\mathcal{D}^1_{\mathfrak{g}}$.*

9.7.3 The classification result for reductive Lie algebras

We first give the result on local cocycle classes for $\mathcal{D}^1_{\mathfrak{g}}$. This means we classify almost-graded central extensions. Recall the usual decomposition $\mathfrak{g} = \mathfrak{g}_0 \oplus \mathfrak{g}_1 \oplus \cdots \oplus \mathfrak{g}_M$ with $\mathfrak{g}_0$ abelian of dimension n and the $\mathfrak{g}_1, \ldots, \mathfrak{g}_M$ simple. Let $\mathfrak{g}' = [\mathfrak{g}, \mathfrak{g}] = \mathfrak{g}_1 \oplus \cdots \oplus \mathfrak{g}_M$ be the derived subalgebra, which is now semisimple. The algebras $\mathcal{D}^1_{\mathfrak{g}_0}$ and $\mathcal{D}^1_{\mathfrak{g}'}$ are in a natural way subalgebras of $\mathcal{D}^1_{\mathfrak{g}}$. They are even ideals, but not complementary ones.

Recall that we have the current algebra cocycle $\gamma_{\alpha,S}$ defined via (9.50), and the vector field cocycle $\gamma^{\mathcal{L}}_{C_S,R}$ (6.140) with a projective connection R. Both can be extended by zero on the complementary space to yield a cocycle for $\mathcal{D}^1_{\mathfrak{g}}$. Furthermore, we have the mixing cocycle (9.69).

Theorem 9.38.

(a) *For every local cocycle γ for $\mathcal{D}^1_{\mathfrak{g}}$ a symmetric invariant bilinear form α of $\mathfrak{g}$, and a linear form ϕ of $\mathfrak{g}$ which vanishes on $\mathfrak{g}' = [\mathfrak{g}, \mathfrak{g}]$ exists, such that γ is cohomologous to*

$$\gamma' = r_1 \gamma_{\alpha,S} + r_2 \gamma_{S,\phi,T} + r_3 \gamma^{\mathcal{L}}_{S,R}. \tag{9.70}$$

With $r_1, r_2, r_3 \in \mathbb{C}$, a current algebra cocycle $\gamma_{\alpha,T}$ given by (9.50), a mixing cocycle $\gamma_{S,\phi,T}$ given by (9.69), and the vector field cocycle $\gamma^{\mathcal{L}}_{S,R}$ given by (6.140). Conversely, any such α, ϕ, $r_1, r_2, r_3 \in \mathbb{C}$ determine a local cocycle.

(b) *The space of local cocycles $\mathrm{H}^2_{loc}(\mathcal{D}^1_{\mathfrak{g}}, \mathbb{C})$ is $\frac{n(n+1)}{2} + n + M + 1$ dimensional.*

Recall that in the semisimple case there is no mixing cocycle. In particular, the cohomology space of local cohomology classes is of dimension $M + 1$.

The proof of the theorem will be given in the following section, which might be skipped on first reading.

9.7.4 The proof

Later we will need the following result.

Lemma 9.39. *Let γ be a cocycle of pure mixing type which is a coboundary. Then a linear form $\psi : \overline{\mathfrak{g}} \oplus \mathcal{L} \to \mathbb{C}$ exists which vanishes on $\mathcal{L}$ with $\gamma = d_1\psi$.*

Proof. As a coboundary $\gamma = d_1\psi'$ with $\psi' : \overline{\mathfrak{g}} \oplus \mathcal{L} \to \mathbb{C}$ a linear form. We decompose $\psi' = \psi'_{|\overline{\mathfrak{g}}} \oplus \psi'_{|\mathcal{L}}$. For all $e, f \in \mathcal{L}$ we obtain

$$0 = \gamma(e, f) = d_1\psi(e, f) = \psi([e, f]), \tag{9.71}$$

by the requirement of pure mixing type. Hence, $\psi'_{|[\mathcal{L},\mathcal{L}]} = 0$. We extend $\psi'_{|\overline{\mathfrak{g}}}$ by zero on $\mathcal{L}$ to a linear form ψ. Note that $\psi([e, x(g)]) = \psi(x(e\,.\,g)) = \psi'([e, x(g)])$, and $\psi([x(g), y(h)]) = \psi'([x(g), y(h)]) = 0$. Hence we get $\gamma = \delta\psi$ with a linear form ψ of the required kind. □

Without assuming further properties of the finite-dimensional Lie algebra $\mathfrak{g}$ we obtain the following theorem.

Theorem 9.40. *Let $\mathfrak{g}$ be a finite-dimensional Lie algebra.*

(a) *Let $\gamma^{\mathcal{A}}$ be any $\mathcal{L}$-invariant cocycle of the function algebra, $\gamma^{\mathcal{L}}$ any cocycle of the vector field algebra $\mathcal{L}$, and $\gamma^{(m)}$ any mixing cocycle of $\mathcal{D}^1$. Furthermore, let α be any symmetric invariant bilinear form on $\mathfrak{g}$, and ϕ any linear form on $\mathfrak{g}$ which vanishes on the derived subalgebra $\mathfrak{g}' = [\mathfrak{g}, \mathfrak{g}]$, then*

$$\begin{aligned}\gamma((x(g), e), (y(h), f)) &= r_1\alpha(x, y)\gamma^{\mathcal{A}}(g, h)\\ &\quad + r_2(\phi(y)\gamma^{(m)}(e, h) - \phi(x)\gamma^{(m)}(f, g)) + r_3\gamma^{\mathcal{L}}(e, f)\end{aligned} \tag{9.72}$$

for arbitrary $r_1, r_2, r_3 \in \mathbb{C}$ defines a cocycle of $\mathcal{D}^1_{\mathfrak{g}}$.

(b) *Let γ be a local cocycle of $\mathcal{D}^1_{\mathfrak{g}}$. Assume that it can be written as*

$$\begin{aligned}\gamma((x(g), e), (y(h), f)) &= r_1\alpha(x, y)\tilde{\gamma}^{\mathcal{A}}(g, h)\\ &\quad + r_2(\phi(y)(\tilde{\gamma}^{(m)}(e, h) - \phi(x)\tilde{\gamma}^{(m)}(f, g)) + r_3\tilde{\gamma}^{\mathcal{L}}(e, f)\end{aligned} \tag{9.73}$$

with a symmetric invariant bilinear form α for $\mathfrak{g}$ fulfilling $\alpha(\mathfrak{g}', \mathfrak{g}) \neq 0$, ϕ a linear form on $\mathfrak{g}$, $\tilde{\gamma}^{\mathcal{A}}$ a bilinear form on $\mathcal{A}$, $\tilde{\gamma}^{\mathcal{L}}$ a bilinear form on $\mathcal{L}$ and $\tilde{\gamma}^{(m)}$ a bilinear form on $\mathcal{L} \times \mathcal{A}$, and $r_1, r_2, r_3 \in \mathbb{C}$. Assume either $[\mathfrak{g}, \mathfrak{g}] \neq 0$ or that $\tilde{\gamma}^{\mathcal{A}}$ is $\mathcal{L}$-invariant. Then, ignoring all terms which are identically zero, the remaining forms are multiples of the corresponding unique local cocycles introduced for $\mathcal{D}^1$ in Section 6.5.3, or respectively coboundaries, and the linear form ϕ vanishes on $\mathfrak{g}' = [\mathfrak{g}, \mathfrak{g}]$.

Proof. Part (a) follows from the previous discussions. To see (b) we note that the restriction of γ to $\mathcal{L}$ defines a local cocycle, hence for this part Theorem 6.38 shows the correct form. The restriction of the cocycle to $\overline{\mathfrak{g}}$ also defines a cocycle. Hence, if $[\mathfrak{g}, \mathfrak{g}] \neq 0$ we can use Proposition 9.9 and obtain the claim for this part. In case $[\mathfrak{g}, \mathfrak{g}] = 0$

we assume $\mathcal{L}$-invariance of $\tilde{\gamma}^{\mathcal{A}}$, and by locality of the cocycle Theorem 6.38 shows the claim directly. Hence, the first and last terms define cocycles of $\mathcal{D}^1_{\mathfrak{g}}$. This implies that the second term is also a cocycle. In particular, it is a pure mixing cocycle. For the mixing cocycle the relation (9.68) shows that either ϕ is identical to zero (and the term does not appear at all), or $\tilde{\gamma}^{(m)}$ defines a local cocycle for $\mathcal{D}^1$. Again, Theorem 6.38 shows the claim. Finally we saw above that for these kinds of cocycles the linear form ϕ has to vanish on $[\mathfrak{g}, \mathfrak{g}]$. □

9.7.4.1 Cocycles for the semisimple case

Theorem 9.41.

(a) *Let $\mathfrak{g}$ be a semisimple Lie algebra and γ a local cocycle of $\mathcal{D}^1_{\mathfrak{g}}$. Then a symmetric invariant bilinear form α for $\mathfrak{g}$ exists, such that γ is cohomologous to a linear combination of the local cocycle $\gamma_{\alpha,S}$ given by (9.50) and the local cocycle $\gamma^{\mathcal{L}}_{S,R}$ (6.140) for $C = C_S$ of the vector field algebra $\mathcal{L}$.*

(b) *If $\mathfrak{g}$ is a simple Lie algebra, then $\gamma_{\alpha,S}$ is a multiple of the standard cocycle (9.44) for $\mathfrak{g}$.*

(c) $\dim \mathrm{H}^2_{loc}(\mathcal{D}^1_{\mathfrak{g}}, \mathbb{C}) = M + 1$, *where M is the number of simple summands of $\mathfrak{g}$.*

Proof. Let γ be a local cocycle. By restricting it to $\mathcal{L}$ and $\overline{\mathfrak{g}}$, we obtain local cocycles γ_1 and γ_2. By Theorem 9.26, γ_2 is cohomologous to the cocycle $\gamma_{\alpha,S}$ on $\overline{\mathfrak{g}}$ with a suitable α. The coboundary $d_1\psi$ can be extended to the whole algebra by setting the linear form ψ zero on $\mathcal{L}$. Hence, by replacing γ by a cohomologous cocycle we might even assume that γ_2 is already $\gamma_{\alpha,S}$. By the $\mathcal{L}$-invariance this cocycle can be extended by zero to the whole algebra. Hence, $\gamma - (\gamma_1 + \gamma_2)$ is again a cocycle which is now of pure mixing type. By Proposition 9.36 for $\mathfrak{g}$ semisimple they are vanishing. This shows (a). Statement (b) follows from the uniqueness of the invariant symmetric bilinear forms for simple Lie algebras. Statement (c) follows from Theorem 9.26. □

If the cocycle γ_2 appearing in the proof of Theorem 9.41 restricted to $\overline{\mathfrak{g}}$ is $\mathcal{L}$-invariant, then by Theorem 9.26 (b) already the cocycle γ_2 equals $\gamma_{\alpha,S}$.

9.7.4.2 Cocycles for the abelian case

Let $\mathfrak{g}$ be an abelian Lie algebra and γ a local cocycle for $\mathcal{D}^1_{\mathfrak{g}}$. From (9.64) it follows that

$$\gamma(x(e\,.\,g), y(h)) + \gamma(x(g), y(e\,.\,h)) = 0. \tag{9.74}$$

By definition this implies that $\gamma_2 := \gamma_{|\overline{\mathfrak{g}}}$ is $\mathcal{L}$-invariant. Also $\gamma_1 := \gamma_{|\mathcal{L}}$ is a vector field cocycle. Both cocycles can be extended by zero on the complement to $\mathcal{D}^1_{\mathfrak{g}}$. Hence, $\gamma - (\gamma_1 + \gamma_2)$ will be a local cocycle for $\mathcal{D}^1_{\mathfrak{g}}$ which is of pure mixing type.

Proposition 9.42. *Every local cocycle γ of pure mixing type is cohomologous to a scalar multiple of $\gamma_{S,\phi,T}$, with a suitable linear form $\phi \in \mathfrak{g}^*$ and meromorphic connection T.*

Proof. If $\dim \mathfrak{g} = 1$, then by Theorem 6.38 γ is up to coboundary a scalar multiple of $\gamma_{S,T}^{(m)}$, the standard mixing cocycle, see Equation (6.108). Let $x_1, \ldots, x_n$ be a basis of $\mathfrak{g}$. Set $\mathcal{A}_i := x_i \otimes \mathcal{A} \cong \mathcal{A}$ and $\mathcal{D}_i := \mathcal{A}_i \oplus \mathcal{L}$. The space $\mathcal{D}_i$ is a subalgebra of $\mathcal{D}^1_{\mathfrak{g}}$ isomorphic to $\mathcal{D}^1$. Restricting the cocycle to $\mathcal{D}_i$, we obtain a cocycle γ_i cohomologous to $r_i\gamma_{S,T}^{(m)}$ with suitable $r_i \in \mathbb{C}$. We set $\phi(x_i) := r_i$ and extend ϕ linearly to $\mathfrak{g}$. If $x = \sum_i s_i x_i$, then

$$\gamma(e, x(g)) = \sum_i s_i \gamma(e, x_i(g)) \sim \left(\sum_i s_i r_i\right)\gamma_{S,T}^{(m)}(e, g) = \phi(x)\gamma_{S,T}^{(m)}(e, g).$$

Here $\sim$ denotes cohomologous. For this step the following remark has to be taken into account. The individual cocycles γ_i are cohomologous to a scalar multiple of the standard cocycle. The individual coboundary is determined by a linear form on $\mathcal{A}_i \oplus \mathcal{L}$. By the pure mixing type using Lemma 9.39, the corresponding form can be chosen to vanish on the summand $\mathcal{L}$. Hence, after such a modification they stick together to a linear form on $\overline{\mathfrak{g}}$ and define a coboundary for $\mathcal{D}^1_{\mathfrak{g}}$. This shows that indeed γ is a multiple of (9.69). □

Theorem 9.43. *Let $\mathfrak{g}$ be an abelian finite-dimensional Lie algebra of dimension n. The subspace of local cohomology classes $\mathrm{H}^2_{loc}(\mathcal{D}^1_{\mathfrak{g}}, \mathbb{C})$ of $\mathrm{H}^2(\mathcal{D}^1_{\mathfrak{g}}, \mathbb{C})$ is $\frac{n(n+1)}{2} + n + 1$-dimensional. Up to coboundary, every local cocycle is a linear combination of*

$$\gamma_{\alpha,S}(x(f), y(g)) = \frac{\alpha(x, y)}{2\pi \mathrm{i}} \int_{C_S} f dg, \tag{9.75}$$

with an arbitrary symmetric bilinear form α, and of $\gamma_{S,\phi,T}$ with an arbitrary linear form ϕ (see (9.69)), and of $\gamma_{S,R}^{\mathcal{L}}$ (see (6.140)).

Proof. Let γ be a local cocycle for $\mathcal{D}^1_{\mathfrak{g}}$. As explained above, such γ can be written by restriction as $\gamma = \gamma_{|\mathcal{L}} + \gamma_{|\overline{\mathfrak{g}}} + \gamma_3$, with γ_3 a local cocycle of pure mixing type. By Proposition 9.42, γ_3 can be given up to coboundary as $\gamma_{\phi,T}$ with a suitable ϕ. The cocycle $\gamma_{|\mathcal{L}}$ is a local vector field cocycle. Hence, $\gamma_{|\mathcal{L}}$ is a multiple of $\gamma_{S,R}^{\mathcal{L}}$. By Theorem 9.30, $\gamma_{|\overline{\mathfrak{g}}} = \gamma_\alpha$ with a suitable symmetric bilinear form α. Hence every local cocycle is cohomologous to a linear combination of cocycles of the required type. Vice versa, every such linear combination is a local cocycle and all the basis cocycles remain linearly independent. □

9.7.4.3 Cocycles for the reductive case

Here we have to combine the results on the semisimple and the abelian case.

Proposition 9.44. *Let γ be a cocycle of $\mathcal{D}^1_{\mathfrak{g}}$, then γ restricted to $\overline{\mathfrak{g}}$ is $\mathcal{L}$-invariant if and only if γ restricted to $\overline{\mathfrak{g}}'$ is $\mathcal{L}$-invariant.*

Proof. We only have to show that a cocycle γ of $\mathcal{D}^1_{\mathfrak{g}}$, which restricted to $\overline{\mathfrak{g}}'$ is $\mathcal{L}$-invariant, is also $\mathcal{L}$-invariant if restricted to $\overline{\mathfrak{g}}$. By the cocycle condition (9.64), the

$\mathcal{L}$-invariance on $\overline{\mathfrak{g}}$ is true if and only if $\gamma(e, [x, y](gh)) = 0$ for $x, y \in \mathfrak{g}$. $\mathfrak{g}$ is reductive, hence $\mathfrak{g}' = [\mathfrak{g}', \mathfrak{g}']$. This implies $[x, y] = [x', y']$, with suitable $x', y' \in \mathfrak{g}'$, and $\gamma(e, [x', y'](gh)) = 0$ by assumption. □

Proposition 9.45.

(a) *Every local cocycle for $\mathcal{D}^1_{\mathfrak{g}}$ restricted to $\overline{\mathfrak{g}}_0$ is $\mathcal{L}$-invariant.*

(b) *Every local cocycle for $\mathcal{D}^1_{\mathfrak{g}}$ which vanishes on pairs of mixed types will be $\mathcal{L}$-invariant if restricted to $\overline{\mathfrak{g}}$. In case $\mathfrak{g}$ is semisimple (or more generally perfect), vanishing on mixed types is also necessary for $\mathcal{L}$-invariance.*

(c) *Every local cocycle for $\mathcal{D}^1_{\mathfrak{g}}$ is cohomologous to a local cocycle which, restricted to $\overline{\mathfrak{g}}$, is an $\mathcal{L}$-invariant cocycle.*

Proof. (a) Let γ be a local cocycle for $\mathcal{D}^1_{\mathfrak{g}}$. It defines by restriction to $\mathcal{D}^1_{\mathfrak{g}_0}$ a local cocycle. Using (9.64) for $\mathcal{D}^1_{\mathfrak{g}_0}$ we obtain, using $[x_0, y_0] = 0$, that the restriction to $\overline{\mathfrak{g}}_0$ is $\mathcal{L}$-invariant.

(b) Recall that by (9.63) the restricted cocycle is $\mathcal{L}$-invariant if and only if $\gamma(e, [x, y](gh)) = 0$. Hence, the vanishing of γ on mixed types is sufficient for $\mathcal{L}$-invariance. If $\mathfrak{g}$ is perfect, then $\overline{\mathfrak{g}}$ is also perfect. Hence, for $e \in \mathcal{L}$ fixed, $\gamma(e, [x, y](gh)) = 0$ for all x, y and g, h if and only if $\gamma(e, z(f)) = 0$. Hence the claim.

(c) Let γ be a local cocycle. By Theorem 9.41, γ restricted to $\mathcal{D}^1_{\mathfrak{g}'}$ is cohomologous to a linear combination of the $\mathcal{L}$-invariant cocycle (9.50) and the separating vector field cocycle. Let $\psi = \delta\varphi$ be the coboundary which appears as difference. The linear form φ can be extended to $\tilde{\varphi}$ by setting it zero on $\mathfrak{g}_0$. We obtain a coboundary $\tilde{\psi} = d_1\tilde{\varphi}$ on $\mathcal{D}^1_{\mathfrak{g}}$. The cocycle $\gamma' = \gamma + \tilde{\psi}$ is cohomologous to the one we started with, and restricted to $\overline{\mathfrak{g}'}$ it is $\mathcal{L}$-invariant. The claim (b) follows now from Proposition 9.44. □

Proof of Theorem 9.38. According to Proposition 9.45 we may replace γ by a cohomologous cocycle which, restricted to $\overline{\mathfrak{g}}$, is $\mathcal{L}$-invariant. We denote it by the same symbol. Hence, by Theorem 9.19, γ restricted to $\overline{\mathfrak{g}}$ is given as $\gamma_{\alpha,S}$ with α a suitable symmetric invariant bilinear form. Also by restriction to $\mathcal{L}$ we obtain $\gamma_{|\mathcal{L}} = r_3\gamma^{\mathcal{L}}_{S,R} + \psi$, with ψ a coboundary of $\mathcal{L}$. By the $\mathcal{L}$-invariance of $\gamma_{\alpha,S}$, all these cocycles can be extended by zero to the whole of $\mathcal{D}^1_{\mathfrak{g}}$ and we can consider the cocycle

$$\gamma' = \gamma - (\gamma_{\alpha,S} + r_3\gamma^{\mathcal{L}}_{S,R} + \psi). \tag{9.76}$$

It will vanish on pairs of elements of the same type. From (9.64) follows that $\gamma'(e, [x, y](g)) = 0$. Hence by decomposing $x \in \mathfrak{g}$ as $x = x_0 + x'$ with $x_0 \in \mathfrak{g}_0$ and $x' \in \mathfrak{g}'$, we get $\gamma'(e, x(g)) = \gamma'(e, x_0(g))$. It follows that γ' is a local cocycle of $\mathcal{D}^1_{\mathfrak{g}_0}$ of pure mixing type which is extended by zero on $\mathfrak{g}'$. In particular, using Theorem 9.43, up to coboundary it can be given as $\gamma_{S,\phi,T}$ with $\phi \in \mathfrak{g}^*$ satisfying $\phi_{|\mathfrak{g}'} \equiv 0$. This shows the first claim. Clearly, we obtain a local cocycle by any such linear combination. Altogether (using the results of Theorem 9.41 and Theorem 9.43), we obtain the statement (b) about the dimension of the cohomology space. □

The following proposition is a further consequence.

Proposition 9.46.

(a) *Let γ be a local cocycle of $\mathcal{D}^1_{\mathfrak{g}}$. A linear form ϕ on $\mathfrak{g}$ which vanishes on $\mathfrak{g}'$ and a linear form ψ on $\overline{\mathfrak{g}}$ exist such that*

$$\gamma(e, x(g)) = \phi(x)\gamma^{(m)}(e, g) + \psi(x(e.g)), \quad \forall e \in \mathcal{L}, g \in \mathcal{A}, x \in \mathfrak{g}, \tag{9.77}$$

with $\gamma^{(m)}$ a mixing cocycle of $\mathcal{D}^1$.

(b) *Assume in addition that the cocycle γ is $\mathcal{L}$-invariant if restricted to $\overline{\mathfrak{g}}$ (or equivalently to $\overline{\mathfrak{g}}'$), then the linear form ψ on $\overline{\mathfrak{g}}$ can be chosen such that it vanishes on $\overline{\mathfrak{g}}'$.*

Proof. (a) We use Theorem 9.38. If we evaluate γ for pairs $(e, x(g))$ we obtain from (9.70) the expression (9.77). Note that every coboundary will be given by a linear form ψ on $\mathcal{D}^1_{\mathfrak{g}}$, evaluated at the Lie bracket of the two elements. But this reduces exactly to the expression above.

(b) Now assume γ to be $\mathcal{L}$-invariant if restricted to $\overline{\mathfrak{g}}$. Let γ' be the corresponding cohomologous cocycle of type (9.70). The difference will be a coboundary which, evaluated at the relevant pairs, will be $d_1\psi$. In particular it will be an $\mathcal{L}$-invariant cocycle. Assume that ψ does not vanish on some element of the type $x(e\,.\,g)$ with $x \in \mathfrak{g}'$, then choose $y, z \in \mathfrak{g}'$ such that $x = [y, z]$. Hence, $\psi([y, z](e\,.\,g)) \neq 0$ in contradiction to the assumed $\mathcal{L}$-invariance; see (9.64). Hence, by setting $\tilde{\psi}(x(h)) = 0$ for all $x \in \mathfrak{g}', h \in \mathcal{A}$, we obtain $\psi(x(h)) = \tilde{\psi}(x(h))$ and get the required form in (9.77). □

9.8 Examples: $\mathfrak{sl}(n)$ and $\mathfrak{gl}(n)$

As examples we deal with the important special case $\mathfrak{sl}(n)$ (which is a simple algebra) and the case of $\mathfrak{gl}(n)$ (which is reductive but not semisimple). We will study the affine algebras, the differential operator algebras, and their central extensions.

9.8.1 $\mathfrak{sl}(n)$

First, we consider $\mathfrak{sl}(n)$, the Lie algebra of trace-less complex $n \times n$ matrices. Up to multiplication with a scalar the Cartan–Killing form $\beta(x, y) = \operatorname{tr}(xy)$ is the unique symmetric invariant bilinear form. It is non-degenerate.

Due to the fact that $\mathfrak{sl}(n)$ is simple, from Theorem 9.22 and Theorem 9.41 the proposition below follows.

Proposition 9.47.

(a) *Every local cocycle for the current algebra $\overline{\mathfrak{sl}}(n)$ is cohomologous to*

$$\gamma(x(g), y(h)) = r\frac{\operatorname{tr}(xy)}{2\pi \mathrm{i}} \int_{C_S} g\,dh, \quad r \in \mathbb{C}. \tag{9.78}$$

(b) *Every $\mathcal{L}$-invariant local cocycle equals the cocycle (9.78) with a suitable r.*

(c) *Every local cocycle for the differential operator algebra $\mathcal{D}^1_{\mathfrak{sl}(n)}$ is cohomologous to a linear combination of (9.78) and the standard local cocycle $\gamma^{\mathcal{L}}_{S,R}$ for the vector field algebra. In particular, no cocycles of pure mixing type exist.*

9.8.2 $\mathfrak{gl}(n)$

Next, we deal with $\mathfrak{gl}(n)$, the Lie algebra of all complex $n \times n$-matrices. Recall that $\mathfrak{gl}(n)$ can be written as Lie algebra direct sum

$$\mathfrak{gl}(n) = \mathfrak{s}(n) \oplus \mathfrak{sl}(n) \cong \mathbb{C} \oplus \mathfrak{sl}(n). \tag{9.79}$$

Here $\mathfrak{s}(n)$ denotes the $n \times n$ scalar matrices. This decomposition is the decomposition as reductive Lie algebra into its abelian and semisimple summands.

After tensoring with $\mathcal{A}$ we obtain

$$\overline{\mathfrak{gl}}(n) = \overline{\mathfrak{s}}(n) \oplus \overline{\mathfrak{sl}}(n) \cong \mathcal{A} \oplus \overline{\mathfrak{sl}}(n). \tag{9.80}$$

From Theorem 9.19 it follows that we need to determine all symmetric invariant bilinear forms for $\overline{\mathfrak{gl}}(n)$. From the above decomposition it follows that the space of such forms is two-dimensional. An adapted basis is

$$\alpha_1(x, y) = \operatorname{tr}(xy), \quad \text{and} \quad \alpha_2(x, y) = \operatorname{tr}(x)\operatorname{tr}(y). \tag{9.81}$$

The form α_1 is the "natural" extension of the Cartan–Killing form for $\mathfrak{sl}(n)$ to $\mathfrak{gl}(n)$. It is also $\mathfrak{gl}(n)$ invariant. The form α_2 vanishes on $\mathfrak{sl}(n)$ and is non-zero on $\mathfrak{s}(n)$. Note that $\alpha_{1|\mathfrak{s}(n)} \neq 0$. The $\tilde{\alpha}_1$ on $\mathfrak{gl}(n)$ obtained by extending the Cartan–Killing form for $\mathfrak{sl}(n)$ by zero would be

$$\tilde{\alpha}_1(x, y) = \operatorname{tr}(xy) - \frac{1}{n}\operatorname{tr}(x)\operatorname{tr}(y). \tag{9.82}$$

From Theorem 9.19 we conclude the following proposition.

Proposition 9.48.

(a) *A cocycle γ for $\overline{\mathfrak{gl}}(n)$ is local and restricted to $\overline{\mathfrak{s}}$ is $\mathcal{L}$-invariant if and only if it is cohomologous to a linear combination γ' of the following two cocycles*

$$\begin{aligned} \gamma_1(x(g), y(h)) &= \frac{\operatorname{tr}(xy)}{2\pi \mathrm{i}} \int_{C_S} g\,dh, \\ \gamma_2(x(g), y(h)) &= \frac{\operatorname{tr}(x)\operatorname{tr}(y)}{2\pi \mathrm{i}} \int_{C_S} g\,dh. \end{aligned} \tag{9.83}$$

(b) *If the cocycle γ of (a) is $\mathcal{L}$-invariant, then γ is equal to the linear combination $r_1\gamma_1 + r_2\gamma_2$ of the cocycles (9.83).*

(c) $\dim \mathrm{H}^2_{loc}(\mathfrak{gl}(n), \mathbb{C}) = 2.$

Proposition 9.49.

(a) *Every local cocycle γ for $\mathcal{D}^1_{\mathfrak{gl}(n)}$ is cohomologous to a linear combination of the cocycles γ_1 and γ_2 of (9.83), of the mixing cocycle*

$$\gamma^{(m)}_{S,tr,T}(e, x(g)) = \frac{\operatorname{tr}(x)}{2\pi \mathrm{i}} \int_{C_S} (\tilde{e}g'' + T\tilde{e}g')dz, \tag{9.84}$$

and of the standard local cocycle $\gamma^{\mathcal{L}}_{S,R}$ for the vector field algebra, i.e.,

$$\gamma = r_1\gamma_1 + r_2\gamma_2 + r_3\gamma^{(m)}_{S,tr,T} + r_4\gamma^{\mathcal{L}}_{S,R} + \text{coboundary}, \tag{9.85}$$

with suitable $r_1, r_2, r_3, r_4 \in \mathbb{C}$.

(b) *If the cocycle γ is local and restricted to $\overline{\mathfrak{gl}}(n)$, is $\mathcal{L}$-invariant, and $r_3, r_4 \neq 0$, then an affine connection T' and a projective connection R', holomorphic outside A, exist such that $\gamma = r_1\gamma_1 + r_2\gamma_2 + r_3\gamma_{S,tr,T'} + r_4\gamma^{\mathcal{L}}_{S,R'}$.*

(c) $\dim \mathrm{H}^2_{loc}(\mathcal{D}^1_{\mathfrak{gl}(n)}, \mathbb{C}) = 4$.

Proof. For $\mathfrak{gl}(n) = \mathfrak{s}(n) \oplus \mathfrak{sl}(n)$ up to multiplication with a scalar, the unique linear form vanishing on $\mathfrak{sl}(n)$ is $x \mapsto \operatorname{tr}(x)$. Hence, from Proposition 9.48 and Theorem 9.38 the claim (a) follows. To prove (b) we note that by the $\mathcal{L}$-invariance using Proposition 9.48(b) it follows that the restriction to $\overline{\mathfrak{gl}}(n)$ is, without changing the element in the cohomology class, already of the required type. Also, by the nonvanishing of the coefficients r_3 and r_4 the coboundaries can be included in the definition of $\gamma^{\mathcal{L}}_{S,R'}$ and $\gamma_{S,tr,T'}$, by choosing suitable projective and affine connections. Part (c) follows from Theorem 9.38, Part (b). □

9.9 Verma modules

In this section, let $\mathfrak{g}$ be either a simple or an abelian finite-dimensional Lie algebra. In fact, everything can be extended to the reductive case by writing the reductive algebra as a direct sum of its simple and abelian summands. By $\mathfrak{h}$ we denote a Cartan subalgebra of $\mathfrak{g}$ which we fix, $\mathfrak{n}_+$ the upper nilpotent subalgebra, $\mathfrak{n}_-$ the lower nilpotent subalgebra, and $\mathfrak{b} = \mathfrak{h} \oplus \mathfrak{n}_+$ the corresponding Borel subalgebra. We have $\mathfrak{g} = \mathfrak{n}_+ \oplus \mathfrak{h} \oplus \mathfrak{n}_-$. In the abelian case $\mathfrak{g} = \mathfrak{h}$ and there will be no nilpotent subalgebras.

Let $\overline{\mathfrak{g}} = \mathfrak{g} \otimes \mathcal{A}$ be the associated current algebra of KN-type. We fix α a symmetric, invariant form for $\mathfrak{g}$ and consider the affine Lie algebra $\widehat{\mathfrak{g}} = \widehat{\mathfrak{g}}_\alpha$ given by the local cocycle

$$\gamma_{S,\alpha}(x \otimes f, y \otimes g) = \alpha(x, y)\frac{1}{2\pi \mathrm{i}} \int_{C_S} f dg. \tag{9.86}$$

Recall that the (strong) almost-grading of $\mathcal{A}$ induces a (strong) almost-graded structure, both for $\overline{\mathfrak{g}}$ and $\widehat{\mathfrak{g}}$. We have (as vector subspaces)

$$\overline{\mathfrak{g}}_n = \widehat{\mathfrak{g}}_n = \mathfrak{g} \otimes \mathcal{A}_n, \ (n \neq 0), \text{ and } \ \widehat{\mathfrak{g}}_0 = \overline{\mathfrak{g}}_0 \oplus \mathbb{C} \cdot t = \mathfrak{g} \otimes \mathcal{A}_0 \oplus \mathbb{C} \cdot t. \tag{9.87}$$

We will denote the center $\mathbb{C} \cdot t$ also by Z. Furthermore, we will consider the following subalgebras

$$\widehat{\mathfrak{g}}_+ = \overline{\mathfrak{g}}_+ = \bigoplus_{n>0} \overline{\mathfrak{g}}_n, \quad \widehat{\mathfrak{g}}'_+ = \overline{\mathfrak{g}}_0 \oplus Z \oplus \overline{\mathfrak{g}}_+, \quad \widehat{\mathfrak{b}} := \overline{\mathfrak{b}}_0 \oplus Z \oplus \widehat{\mathfrak{g}}_+. \tag{9.88}$$

By the almost-grading they are indeed subalgebras of $\widehat{\mathfrak{g}}$.

By Proposition 9.1 the map

$$\mathfrak{g} \to \overline{\mathfrak{g}}, \quad x \mapsto x \otimes 1 = \sum_{p=1}^{K} x \otimes A_{0,p} \tag{9.89}$$

embeds the finite-dimensional Lie algebras $\mathfrak{g}, \mathfrak{h}, \mathfrak{n}_+, \mathfrak{n}_-, \mathfrak{b}$ as subalgebras into the degree zero part of $\overline{\mathfrak{g}}$ (and $\widehat{\mathfrak{g}}$).

Obviously, the subspaces $\overline{\mathfrak{g}}_0, \overline{\mathfrak{b}}_0, \overline{\mathfrak{n}_{+0}}$ are (vector space) direct sums of the subspaces of the type $\mathfrak{g} \otimes A_{0,p}$, $p = 1, \dots, K$, i.e.,

$$\overline{\mathfrak{g}}_0 = \oplus_{p=1}^{K} \mathfrak{g} \otimes A_{0,p}, \quad \overline{\mathfrak{b}}_0 = \oplus_{p=1}^{K} \mathfrak{b} \otimes A_{0,p}, \quad \overline{\mathfrak{n}_{+0}} = \oplus_{p=1}^{K} \mathfrak{n}_+ \otimes A_{0,p}. \tag{9.90}$$

We consider the (external) direct sum $\mathfrak{g}_{(K)}$ of K copies of the Lie algebra $\mathfrak{g}$:

$$\mathfrak{g}_{(K)} := \mathfrak{g}_1 \oplus \ldots \oplus \mathfrak{g}_K. \tag{9.91}$$

Here $\mathfrak{g}_i = \mathfrak{g}$. We could also incorporate certain twists in the identification. We will not do this here, but see [226]. Correspondingly, we denote

$$\mathfrak{b}_{(K)} := \mathfrak{b}_1 \oplus \ldots \oplus \mathfrak{b}_K, \quad \mathfrak{n}_{+(K)} := \mathfrak{n}_{+1} \oplus \ldots \oplus \mathfrak{n}_{+K}, \tag{9.92}$$

with $\mathfrak{b}_p = \mathfrak{b}$ (and $\mathfrak{n}_{+p} = \mathfrak{n}_+$) at the position p. Of course, as vector spaces

$$\mathfrak{g}_{(K)} \cong \overline{\mathfrak{g}}_0, \quad \mathfrak{b}_{(K)} \cong \overline{\mathfrak{b}}_0, \quad \mathfrak{n}_{+(K)} \cong \overline{\mathfrak{n}_{+0}}. \tag{9.93}$$

This should be regarded in an intuitive way as assigning to every point in I a copy of $\mathfrak{g}$.

We define a linear map by

$$\phi : \widehat{\mathfrak{g}}_0 \oplus \widehat{\mathfrak{g}}_+ \to \mathfrak{g}_{(K)}, \quad \text{with} \quad \phi_{|Z} = 0, \quad \phi_{|\widehat{\mathfrak{g}}_+} = 0, \tag{9.94}$$

and on the subspace $\overline{\mathfrak{g}}_0$

$$x = \sum_{p=1}^{K} x_p \otimes A_{0,p} \in \overline{\mathfrak{g}}_0 \mapsto (x_1, x_2, \dots, x_K). \tag{9.95}$$

Proposition 9.50. *The map ϕ is a Lie homomorphism with kernel $Z \oplus \widehat{\mathfrak{g}}_+$. Moreover,*

$$\widehat{\mathfrak{b}} \bmod Z \oplus \widehat{\mathfrak{g}}_+ \cong \overline{\mathfrak{b}}_0 \cong \mathfrak{b}_{(K)} \quad \widehat{\mathfrak{g}} \bmod Z \oplus \widehat{\mathfrak{g}}_+ \cong \overline{\mathfrak{g}}_0 \cong \mathfrak{g}_{(K)}. \tag{9.96}$$

Proof. We write

$$x = \sum_i x_i \otimes A_{0,i} + R(x), \quad y = \sum_j y_j \otimes A_{0,j} + R(y),$$

with $R(x), R(Y) \in Z \oplus \widehat{\mathfrak{g}}_+$. By the almost-grading

$$[x, y] = \sum_{i,j} [x_i \otimes A_{0,i}, y_j \otimes A_{0,j}] = \sum_i [x_i, y_i] \otimes A_{0,i} + u, \tag{9.97}$$

with $u \in Z \oplus \widehat{\mathfrak{g}}_+$. Recall that

$$A_{0,i} \cdot A_{0,j} = A_{0,i}\delta_i^j + h.d.t.$$

The relation (9.97) says that ϕ is a Lie homomorphism. Its kernel is $Z \oplus \widehat{\mathfrak{g}}_+$. Restricted to the degree zero parts we get the claimed isomorphisms. □

Via ϕ, the degree zero part of the algebras (ignoring the degree zero central term) will be identified with K copies of the corresponding finite-dimensional Lie algebras.

Let $\chi := (\chi_1, \dots, \chi_K) \in (\mathfrak{h}^*)^K$. Furthermore, let V_p be a one-dimensional linear space over $\mathbb{C}$ with a fixed basis vector v_p $(p = 1, \dots, K)$. We set $V := \bigotimes_{p=1}^{K} V_p$. For the basis of V we take $v = v_1 \otimes v_2 \otimes \cdots \otimes v_K$.

Next, we define for $p = 1, \dots, K$ a one-dimensional representation of $\mathfrak{b}_p$ on V_p by

$$h_p v_p = \chi_p(h_p)v_p, \quad n_p v_p = 0, \quad \text{for} \quad h_p \in \mathfrak{h}_p, \quad n_p \in \mathfrak{n}_+. \tag{9.98}$$

We extend this to a one-dimensional representation of the Lie algebra $\mathfrak{b}_{(K)}$ on the linear space V as follows. We decompose $x_{(K)} \in \mathfrak{b}_{(K)}$ as $x_{(K)} = x_1 \oplus \dots \oplus x_K$ with $x_1 \in \mathfrak{b}_1, \dots, x_K \in \mathfrak{b}_K$, and set

$$\begin{aligned} x_{(K)}(v_1 \otimes \dots \otimes v_K) &= (x_1 v_1) \otimes v_2 \otimes \dots \otimes v_K \\ &\quad + v_1 \otimes (x_2 v_2) \otimes \dots \otimes v_K + \dots + v_1 \otimes v_2 \otimes \dots \otimes (x_K v_K). \end{aligned} \tag{9.99}$$

Let us denote this representation of the Lie algebra $\mathfrak{b}_{(K)}$ on the linear space V by τ_χ.

Let $c \in \mathbb{C}$ be a number. Now we define a representation $\tau_{\chi,c}$ of $\widehat{\mathfrak{b}}$ on the one-dimensional vector space V by defining τ_χ via

$$\tau_{\chi,c}(\phi(x_0)) = \tau_\chi(\phi(x_0)), \; x_0 \in \widehat{\mathfrak{b}}_{(0)} \quad \tau_{\chi,c}(t) = c \cdot id, \quad \tau_{\chi,c|\widehat{\mathfrak{g}}_+} = 0. \tag{9.100}$$

As ϕ is a Lie homomorphism, this defines a representation of $\widehat{\mathfrak{b}}$.

We collect the data as follows:

$$\begin{aligned} \tau_{\chi,c}(u) &= 0, & & u \in \widehat{\mathfrak{g}}_+, \\ \tau_{\chi,c}(t) &= c \cdot id, & & \\ \tau_{\chi,c}(h(0,p))v &= \chi_p(h)v, & & \forall h \in \mathfrak{h}, \; \forall p = 1, \dots, K, \\ \tau_{\chi,c}(n(0,p)) &= 0, & & \forall n \in \mathfrak{n}_+, \; \forall p = 1, \dots, K. \end{aligned} \tag{9.101}$$

Recall that for a Lie algebra $\mathfrak{g}$, the universal enveloping algebra is denoted by $U(\mathfrak{g})$; see Section 1.7.

Definition 9.51. The vector space

$$\widehat{V}_{\chi,c} = U(\widehat{\mathfrak{g}}) \otimes_{U(\widehat{\mathfrak{b}})} V, \tag{9.102}$$

with its natural structure of a $\widehat{\mathfrak{g}}$-module is called *Verma module* of the Lie algebra $\widehat{\mathfrak{g}}$, corresponding to the weight χ and level c. The action of $U(\widehat{\mathfrak{b}})$ on V is given by the representation $\tau_{\chi,c}$. The vector $v = v_{\chi,c}$ is called *highest weight vector*, or vacuum of the module $\widehat{V}_{\chi,c}$.

Proposition 9.52. *The $\widehat{\mathfrak{g}}$ module $\widehat{V}_{\chi,c}$ is also a $\mathfrak{g}$-module. It contains the module*

$$V_\chi = V_{\chi_1} \otimes \cdots V_{\chi_K} \tag{9.103}$$

as submodule generated from the vacuum $v_{\chi,c}$. Here, each V_{χ_p} is the highest weight module of $\mathfrak{g}$ of weight χ_p.

Proof. Recall that $\mathfrak{g}$ is embedded into $\widehat{\mathfrak{g}}$ via (9.89). Hence, $\widehat{V}_{\chi,c}$ is also a $\mathfrak{g}$-module. Let V_χ be the tensor product of the representation spaces V_{χ_p} of the finite-dimensional algebra. The $\mathfrak{g}$-submodule of $\widehat{V}_{\chi,c}$ generated from the vector $v_{\chi,c}$ by $\mathfrak{g}$ is exactly the module V_χ. Note that there will be no action of elements of negative degree involved. Moreover, the action of $g \in \mathfrak{g}$ on decomposable tensors $w_1 \otimes w_2 \otimes w_K$ is exactly given by

$$g\,.\,w = g\,.\,w_1 \otimes w_2 \cdots \otimes w_K + \cdots + w_1 \otimes w_2 \cdots \otimes g\,.\,w_K. \tag{9.104}$$

□

Remark 9.53. This proposition can also regarded in the inverse sense. We assign to every point $P_p \in I$ a highest weight representation V_p of the finite Lie algebra $\mathfrak{g}$. These highest weight representations are uniquely given by their weights. They do not see the geometry. We take their tensor product. After choosing a level c the data is fixed for the construction of $\widehat{V}_{\chi,c}$, which is now a representation of the geometric algebra $\widehat{\mathfrak{g}}$. In fact, the geometry induces the representation from the individual V_{χ_p}.

Recall the definition (9.88) of the subalgebra $\widehat{\mathfrak{g}}'_+$. It contains all elements of (Krichever–Novikov-)degree greater than or equal to zero. Consequently, the module $\widehat{V}_{\chi,c}$ contains a $\widehat{\mathfrak{g}}'_+$-submodule $\widehat{V}_{(\chi,c),0}$ generated by v. It is called the *subspace of degree zero elements.* Obviously,

$$V_{\chi,c} \subseteq \widehat{V}_{(\chi,c),0}. \tag{9.105}$$

Next, we take the subalgebra

$$\overline{\mathfrak{g}}^r = \mathfrak{g} \otimes \mathcal{A}^r, \tag{9.106}$$

where $\mathcal{A}^r$ is the subalgebra of meromorphic functions which might have poles at I, but vanish at O. Of course, $\overline{\mathfrak{g}}^r$ is a subalgebra of $\widehat{\mathfrak{g}}$ (we always use the fact that the cocycle values will vanish for pairs of elements coming from these kind of subalgebras).

Denote by $\overline{\mathfrak{g}}^r\widehat{V}_{\chi,c}$ the subspace of $\widehat{V}_{\chi,c}$ generated by the vectors of this set.

Definition 9.54. The quotient space

$$C(\widehat{V}_{\chi,c}) := \widehat{V}_{\chi,c}/\widehat{\mathfrak{g}}^r\widehat{V}_{\chi,c} \tag{9.107}$$

is called the space of coinvariants, or *conformal blocks* of weight χ and level c.

Remark 9.55. This definition plays an important role in the context of the *Wess–Zumino–Novikov–Witten models* of Conformal Field Theory. As described in Remark 9.53, at the "insertion points" $P_p \in I$, representations are assigned. The geometry induces the space of conformal blocks. Roughly speaking, varying the geometric data (meaning the complex moduli of the Riemann surface, respectively moving the points), the conformal blocks define a vector bundle over the moduli space of the data. This bundle is called the Verlinde bundle. The *Knizhnik–Zamolodchikov connection* will single out a subspace of horizontal sections of the Verlinde bundle. We will return to this in Chapter 11.

Proposition 9.56. *The module $\widehat{V}_{\chi,c}$ will be generated as vector space by elements given as*

$$x_k \otimes A_{i_k,p_k}(\cdots(x_2 \otimes A_{i_2,p_2}(x_1 \otimes A_{i_1,p_1} . v_{\chi,c}))\cdots), \tag{9.108}$$

with $x_i \in \mathfrak{g}$ fulfilling the conditions

$$i_k \le i_{k-1} \le \cdots \le i_2 \le i_1 \le 0. \tag{9.109}$$

Proof. By definition, the module $\widehat{V}_{\chi,c}$ will be generated by chains of the form (9.108) without necessarily decreasing the conditions on the indices. We induce the length k. If $k = 0$ then nothing has to be shown. If $k = 1$, then $(x_1 \otimes A_{i_1,p_1}) . v_{\chi,c} \neq 0$, then $i_1 \le 0$. Let w be a chain which is not in the correct order. This means it does not fulfill (9.109). If the wrong order is not at the left-most place we can, by induction, arrange to write it as a finite sum of chains which are in the correct order. Hence, the only crucial place is the left-most, i.e., if $i_k > i_{k-1}$. We have

$$x_k \otimes A_{i_k,p_k} . ((x_{k-1} \otimes A_{i_{k-1},p_{k-1}}) . w'). \tag{9.110}$$

Using the structure of $\widehat{\mathfrak{g}}$ we get

$$\begin{aligned}([x_k, x_{k-1}]A_{i_k,p_k}A_{i_{k-1}p_{k-1}}) . w' + \gamma_\alpha(x_k \otimes A_{i_k,p_k}, x_{k-1} \otimes A_{i_{k-1},p_{k-1}})c \cdot w' \\ + x_{k-1} \otimes A_{i_{k-1},p_{k-1}} . (x_k \otimes A_{i_k,p_k} . w').\end{aligned} \tag{9.111}$$

The first two terms are of shorter length, also $x_k \otimes A_{i_k,p_k} . w'$ is of shorter length, and by induction we get the result. □

Definition 9.57. A representation V of $\widehat{\mathfrak{g}}$ is called *admissible* if
(1) the central element t operates as $c \cdot id$, with $c \in \mathbb{C}$;
(2) for all $v \in V$ an $n_0 \in \mathbb{Z}$ exists such that for all $n \ge n_0$ we have $\widehat{\mathfrak{g}}_n . v = 0$.

Proposition 9.58. *The Verma module $\widehat{V}_{\chi,c}$ is an admissible representation.*

Proof. The central element t operates as scalar by construction. Let w be an arbitrary element of $\widehat{V}_{\chi,c}$. By Proposition 9.56, the element w will be a finite sum of elements of the type (9.108). If we consider each such chain separately and take the maximum of n, such that $\widehat{\mathfrak{g}}_n$ annihilates the chain, we get the claim. If the chain is a multiple of the vacuum vector v, we take $m = 1$, otherwise we choose an

$$m \in \mathbb{N} \quad \text{such that} \quad m > -\sum_{l=1}^{k} i_l. \tag{9.112}$$

Note that $-\sum_{l=1}^{k} i_l > 0$. Furthermore, $m + i_k = -\sum_{l=1}^{k-1} > 0$. Let $y \in \mathfrak{g}$ and consider $y \otimes A_{m,p}$. We claim that $y \otimes A_{m,p} \,.\, w = 0$. We make induction over the length of the chain. For w the vacuum vector this is by definition. We replace

$$(y \otimes A_{m,p})(x_k \otimes A_{i_k,p_k}) = [y \otimes A_{m,p}, x_k \otimes A_{i_k,p_k}] + (x_k \otimes A_{i_k,p_k})(y \otimes A_{m,p}). \tag{9.113}$$

As $m + i_k > 0$ by the locality of the cocycle, no central term will appear. Also in the product

$$[y \otimes A_{m,p}, x_k \otimes A_{i_k,p_k}] = [y, x_k] \otimes (A_{m,p} \cdot A_{i_k,p_k})$$

the range of the degree will $\geq m + i_k$. This is still a positive number. For the first part of the result in (9.113) we get a finite sum of chains of shorter length and we can apply induction. For the second part, we move the $A_{m,p}$ to the right and by induction the result of the remaining chain will also be annihilated. Hence, the proof. □

9.10 Fermionic representations

There is another type of representation of the affine algebra $\widehat{\mathfrak{g}}$ which we will now discuss. We begin rather generally. Let W be an almost-graded $\mathcal{A}$ module, and

$$\tau : \mathfrak{g} \to V_\tau \tag{9.114}$$

a finite-dimensional representation with representation space V_τ of the finite dimensional Lie algebra $\mathfrak{g}$. For definiteness let us assume that the module is $\mathbb{Z}$-graded. Obviously, also half-integer grading will work in the following. We set

$$W_\tau := W \otimes V_\tau, \tag{9.115}$$

and define a $\overline{\mathfrak{g}}$-action on W_τ as follows:

$$(x \otimes g)(s \otimes v) = (g \cdot s) \otimes \tau(x)v \quad \text{for all} \quad x \in \mathfrak{g},\ g \in \mathcal{A},\ s \in W,\ v \in V_\tau. \tag{9.116}$$

One verifies directly that this defines a representation of $\overline{\mathfrak{g}}$.

Of course, W is an $\mathcal{A}$-module via

$$g \,.\, (s \otimes v) := (g \cdot s) \otimes v. \tag{9.117}$$

In case W is more generally an almost-graded $\mathcal{D}^1$-module, we can extend this action by the action of the vector field algebra $\mathcal{L}$ given as

$$e\,.\,(s \otimes v) = (e\,.\,s) \otimes v \quad \text{for all} \quad e \in \mathcal{L},\ s \in W,\ v \in V_\tau. \tag{9.118}$$

It is straightforward to verify directly that (9.116) and (9.118) give a representation of $\mathcal{D}^1_{\mathfrak{g}}$ on W_τ.

As W is an almost-graded $\mathcal{A}$-module we decompose it

$$W = \bigoplus_{n \in \mathbb{Z}} W_n, \tag{9.119}$$

and define the following degree decomposition for W_τ

$$W_\tau = \bigoplus_{n \in \mathbb{Z}} W_{\tau,n}, \qquad W_{\tau,n} = W_n \otimes V_\tau. \tag{9.120}$$

Proposition 9.59. *W_τ with the above grading is an almost-graded $\overline{\mathfrak{g}}$ (respectively $\mathcal{D}^1_{\mathfrak{g}}$) module. If the module W is strongly almost-graded, then W_τ is also strongly almost-graded.*

Proof. Let $d = \dim V_\tau$ and $v_1, v_2, \ldots, v_d$ be a basis of V_τ. The elements of the space W_τ are given as linear combinations of $w_{i,n} \otimes v_i$, with $w_{i,n}$ of degree n. If we apply $g \in \mathcal{A}_m$ or $e \in \mathcal{L}_m$, then $\deg((g\,.\,w_{i,n}) \otimes v_i)$ (respectively $\deg((e\,.\,w_{i,n}) \otimes v_i)$) lies by the almost-graded structure of the module in a fixed degree range between $n + m - S_1$ and $n + m + S_2$. The same is then true for the $\overline{\mathfrak{g}}$ action, via (9.116). In case W is strongly almost-graded, by definition the dimensions of the homogeneous subspace is bounded by a constant. For the module W_τ, this bound for the dimension is just multiplied by d. □

Now we assume that our module W is strongly almost-graded. We fix an (ordered) basis inside the homogeneous subspaces W_n and fix an (ordered) basis of V_τ. In this way we obtain a basis of W_τ, where the individual basis element is given by the index

$$(\,n, \quad \text{“index of } W_n \text{ basis part”}, \quad \text{“index of } V_\tau \text{ basis part”}\,). \tag{9.121}$$

The first part, n, runs through the integer (or half-integer depending on the situation), the second and third part have only a finite range. By choosing a lexicographical order we obtain a linearly ordered sequence of indices.

Now we are exactly in the set-up of the construction of the semi-infinite wedge products as explained in Chapter 7. Let us denote the space of semi-infinite wedge products by $\mathcal{H}(W_\tau)$. Again we define the charge $m \in \mathbb{Z}$ subspace $\mathcal{H}^{(m)}(W_\tau)$. Recall from there that these subspaces are generated by basis elements which differ from a standard element in a finite number of positions. This element is also called the vacuum, and it starts with the element of (linearized) index m and increases at every position exactly the index by one. For a basis element Φ of charge m we define again its *degree* as follows:

$$\deg \Phi = \sum_{k=0}^{\infty} (n_k - (k + m)). \tag{9.122}$$

See also (7.46) and (7.44) for the structure. Here n_k denotes the (linearized) index of the element at the position k in ϕ. We decompose further

$$\mathcal{H}^{(m)}(W_\tau) = \bigoplus_{d\in\mathbb{Z}} (\mathcal{H}^{(m)}(W_\tau))_d, \quad \text{with} \quad (\mathcal{H}^{(m)}(W_\tau))_d := \langle \Phi \mid \deg \Phi = d\rangle_{\mathbb{C}}. \tag{9.123}$$

Proposition 9.60. *The subspaces $(\mathcal{H}^{(m)}(W_\tau))_d$ of degree d are finite-dimensional. If $d > 0$, then $(\mathcal{H}^{(m)}_{F,\tau})_k = 0$.*

Proof. From the very definition of the degree it follows that the degree d is always ≤ 0. Notice that all summands in (9.122) are ≤ 0. Hence for a given d, only a finite number of semi-infinite wedges of charge m can realize this k. □

By (7.52) we know that its dimension is $p(-d)$ with p the partition function.

We employ the regularization procedure given in Chapter 7. First, by the action on W_τ our algebras can be embedded in the algebra into $\overline{gl}(\infty)$, the algebra of both-sided infinite matrices with a finite number of diagonals (Section 7.1). Now the procedure described there works in exactly the same way in this setting. By the modified action we obtain a projective action of our algebras, or equivalently a Lie action of centrally extended algebras.

Theorem 9.61. *Let W be either an $\mathcal{A}$, or more generally a $\mathcal{D}^1$, strongly graded almost-graded module, and V_τ a finite-dimensional representation of $\mathfrak{g}$, then:*
(a) *the space $\mathcal{H}(W_\tau)$ is a $\widehat{\mathfrak{g}}$-module, or $\widehat{\mathcal{D}^1_{\mathfrak{g}}}$-module, of suitable central extensions of $\overline{\mathfrak{g}}$ and $\mathcal{D}^1_{\mathfrak{g}}$;*
(b) *the defining cocycles for the central extensions are local, hence the central extensions are almost-graded;*
(c) *the subspaces $\mathcal{H}^{(m)}(W_\tau)$ are invariant subspaces;*
(d) *with respect to the degree definition (9.122), the modules $\mathcal{H}^{(m)}(W_\tau)$ are almost-graded modules (but not strongly almost-graded).*

Proof. Using the embedding into $\overline{gl}(\infty)$, the claim (a) is obvious. By the strongly almost-gradedness as in the proof of Proposition 7.17, the pull-back of the cocycle defining $\widehat{\mathfrak{g}}$ will be local. Hence (b). By Proposition 7.15, the subspaces of a fixed charge m are submodules, hence the same is true with respect to the action of our algebras. This shows (c). The proof of Proposition 7.31 works word by word as far as the statement of the almost-grading is concerned also in our case. Hence we obtain (d). □

Similar to the other part of Proposition 7.31, the following is also true.

Proposition 9.62. *Let the situation be as in Theorem 9.61. Assume in addition; that for the module W we have*

$$\mathcal{A}_m \,.\, W_n \subseteq \bigoplus_{h=n+m}^{n+m+S}, \qquad \mathcal{L}_m \,.\, W_n \subseteq \bigoplus_{h=n+m}^{n+m+S}, \tag{9.124}$$

then the algebras $\widehat{\mathcal{A}}_+$, $\widehat{\mathfrak{g}}_+$; and $\widehat{\mathcal{D}^1_{\mathfrak{g}+}}$ will operate trivially on the elements of degree zero (in particular on the vacuum) in $\mathcal{H}^{(m)}(W_\tau)$.

Recall the Definition 9.57 of an admissible representation.

Proposition 9.63. *$\mathcal{H}^{(m)}(W_\tau)$ carries an admissible representation of $\widehat{\mathfrak{g}}$.*

Proof. The central element operates by construction by a scalar. Let $v \in \mathcal{H}^{(m)}(W_\tau)$ be a homogeneous element of degree k, $k \le 0$. Due to the almost-graded structure; if we apply elements from $\widehat{\mathfrak{g}}_n$ with n high enough; $\widehat{\mathfrak{g}}_n \,.\, v = 0$. □

Above we were rather general with respect to the Lie algebra $\mathfrak{g}$. We obtained that the defining cocycle by the representation is a local cocycle. In case $\mathfrak{g}$ is semisimple, we had given in this chapter their complete classification. In the reductive case we needed for the classification the cocycle to be $\mathcal{L}$-invariant if restricted to the abelian summand. From Theorem 9.61 we deduce the following proposition.

Proposition 9.64.

(a) *If $\mathfrak{g}$ is semisimple and W is an $\mathcal{A}$-module, then the defining cocycle for the central extension $\widehat{\mathfrak{g}}$ acting on $\mathcal{H}^{(m)}(W_\tau)$ is cohomologous to (9.50), i.e., to*

$$\gamma_{\alpha,S}(x \otimes f, y \otimes g) = \frac{\alpha(x,y)}{2\pi \mathrm{i}} \int_{C_S} f dg, \tag{9.125}$$

with α a symmetric invariant bilinear for $\mathfrak{g}$.

(b) *In case the defining cocycle is $\mathcal{L}$-invariant it will be equal to (9.125).*

(c) *For $\mathfrak{g}$ reductive and W an $\mathcal{A}$-module such that the cocycle is $\mathcal{L}$-invariant, then also in this case the expression of the cocycle will be given by (9.125).*

(d) *In particular, if $\mathfrak{g}$ is simple then the cocycle class, or the cocycle itself in the case of $\mathcal{L}$-invariance, will be unique.*

We can extend this to the proposition below.

Proposition 9.65.

(a) *If $\mathfrak{g}$ is semisimple and W is a $\mathcal{D}^1$-module, then the defining cocycle for the central extension $\widehat{\mathfrak{g}}$ acting on $\mathcal{H}^{(m)}(W_\tau)$ is cohomologous to a sum of (9.125) and the vector field algebra cocycle $\gamma^{\mathcal{L}}_{S,R}$ (6.140). In case the cocycle is $\mathcal{L}$-invariant it is equal to such an expression.*

(b) *For $\mathfrak{g}$ reductive and W a $\mathcal{D}^1$-module, then the cocycle is cohomologous to a cocycle expressed in (9.70). In particular in contrast to (a) there could be an additional term given by a mixing cocycle (9.69).*

For the reductive case see also Proposition 9.45 for more information related to $\mathcal{L}$-invariance.

The above introduced representations are called *semi-infinite wedge representations* or *fermionic representations*.

Until now we have neither specified the module W nor the representation space V_τ.

Example 9.66. As one extreme we might take $W = \mathcal{F}^\lambda$, $\mathfrak{g} = \mathbb{C}$, and on $V_\tau = \mathbb{C}$ the identity representation. Here W is a $\mathcal{D}^1$-module. Obviously, this gives back the representations we studied in Chapter 7. There the cocycle is given as a sum of the three types.

Example 9.67. A related, only slightly more general, example is given by $W = \mathcal{F}^\lambda$ and $\mathfrak{g}$ arbitrary, and τ an arbitrary representation of $\mathfrak{g}$. Again, W is a $\mathcal{D}^1$-module and we obtain a $\widehat{\mathcal{D}^1_\mathfrak{g}}$ representation.

Example 9.68. Let E be a trivial rank n vector bundle over Σ. Let W be its space of global meromorphic sections, which are holomorphic outside A. The algebra A operates on W. After fixing a basis of E, the space of meromorphic sections can be identified with the space of meromorphic vector valued functions with n components. Sheinman introduced for this space a Krichever–Novikov like basis and showed that W with respect to the splitting $A = I \cup O$ is a strongly almost-graded $\mathcal{A}$-module. See [227, 240, 241, 245]. As the bundle is trivial we can even introduce the action of the vector fields by taking the derivative in the components (or any other convenient connection). In this way, W becomes a $\mathcal{D}^1$-module. Consequently, the above deduced results are true.

Example 9.69. In the last example our vector bundle was trivial. We could extend the above to nontrivial bundles E. For this, let E be a rank n generic semi-stable vector bundle of degree $g \cdot n$ (as usual, g denotes the genus of our Riemann surface Σ). See Section 13.2 for an explanation of the notion. Also, as shown there, the space of holomorphic sections is n-dimensional. After the choice of a basis, a framing, the space of holomorphic sections can be identified with the space of vector-valued functions with n-components. The fact that the bundle is nontrivial corresponds to the fact that these functions are not a basis of the fiber at every point. In fact, we have $g \cdot n$ points where they only generate a one-codimensional subspace of the vector space lying above the point. (We obtain the Tyurin parameter defined by (13.29).)

We take as W the space of meromorphic sections. Obviously, W is again an $\mathcal{A}$-module. After framing we identify it with meromorphic vector-valued functions with n components, which are holomorphic outside the set of points A. The construction of Sheinman in the last example extends to this case too, and we obtain with respect to the splitting $A = I \cup O$ an almost-graded $\mathcal{A}$-module.

There is not a naturally defined action of $\mathcal{L}$ on W. For this we need to introduce first a connection. In Section 13.3.2, such connections are introduced. With respect to the $\mathcal{L}$ action, W is a strongly almost-graded $\mathcal{D}^1$-module. These results are recalled in Chapter 13. For the proofs see [228]. As far as the wedge space representation is concerned, see further results of Sheinman [240]. See in particular also [250]. There he also calculated the cocycle explicitly in the case of $\mathfrak{gl}(n)$. It turns out to be a multiple of

$$\alpha(x \otimes g, y \otimes h) = \frac{\operatorname{tr}(x \cdot y)}{2\pi \mathrm{i}} \int_{C_S} g dh, \tag{9.126}$$

which is a more detailed statement than Proposition 9.48. Moreover, he showed that for $\mathfrak{g}$ reductive the cocycle is $\mathfrak{L}$-invariant [250, Lemma 2.10.].

Remark 9.70. The objects introduced in Example 9.69 have a long history. See e.g. [138, 139, 145–147, 256] for related holomorphic vector bundles, integrable systems, and non-linear equations. They will also show up in Chapter 13.

10 The Sugawara construction

In this chapter we will present the Sugawara construction for arbitrary genus and in the multi-point situation.

Given an admissible representation of a centrally extended current algebra $\widehat{\mathfrak{g}}$, we construct the so-called *Sugawara operators*. Here, admissible means that the central element operates as constant × identity, and that every element v in the representation space will be annihilated by the elements in $\widehat{\mathfrak{g}}$ of sufficiently high degree (the degree might depend on the element v).

The Sugawara operator field is an infinite formal sum of operators and is constructed as the product of the current operator fields, which are again formal infinite sum of operators. To make the product well-defined, a normal ordering has to be fixed which moves the annihilation operators to the right to act first. It turns out, in the case of $\mathfrak{g}$ simple or abelian that, after rescaling, the operators appearing in the formal sum of the Sugawara operator field give a representation of a centrally extended vector field algebra $\widehat{\mathcal{L}}$. The result can be extended to case $\mathfrak{g}$, a reductive Lie algebra.

The central extension is due to the unavoidable appearance of normal ordering. It will turn out that the cocycle defining the central extension is local. Hence, by our classification results of Chapter 6, the central extension is (up to equivalence and rescaling) given by a cocycle $\gamma_{S,R}^{\mathcal{L}}$.

In physics terms, the affine algebra corresponds to "gauge symmetry", the vector field algebra corresponds to "conformal symmetry". The central term corresponds to *conformal anomaly*. The family of operators show up as modes of the "energy-momentum tensor".

The above-mentioned rescaling is done by $1/(c + \kappa)$, where c is the constant with which the central element t operates as $c \cdot id$ (also called the *level*), and κ is the dual Coxeter number for $\mathfrak{g}$ simple if the invariant form used to define the affine algebra $\widehat{\mathfrak{g}}$ is suitably normalized. In the abelian case $\kappa = 0$. We obtain the representations mentioned only if our level is non-critical, i.e., $(c + \kappa) \neq 0$.

First, we will recall the classical situation $g = 0, N = 2$. Then we will formulate the general results. In a separate section we will present the rather technical proofs. The results presented were obtained by the author in joint work with Oleg Sheinman [225].

10.1 The classical Sugawara construction

This section is mainly intended for the reader aware of the classical genus zero Sugawara construction of Conformal Field Theory. As the classical construction will follow from our general results, its knowledge is not necessary.

From a mathematical point of view, the classical Sugawara construction is a method of obtaining from a highest weight representation of the classical affine Lie algebra a representation of the Viraroso algebra. Let V be a highest weight representation of the affine algebra $\widehat{\mathfrak{g}} = \mathfrak{g} \otimes \mathbb{C}[z^{-1}, z]$ associated with the finite-dimensional simple Lie algebra $\mathfrak{g}$. Highest weight representations are generated starting from an element ψ. We have $\overline{\mathfrak{g}}_{+}\psi = 0$, the Cartan and positive nilpotent algebra of $\widehat{\mathfrak{g}}_{(0)}$ operates in a certain manner on ψ, and the whole representation space will be generated by $\widehat{\mathfrak{g}}$ from ψ. In particular, for every $v \in V$ and every $x \in \mathfrak{g}$ with n big enough, we have $x(n)v = 0$.

Let $\{u_i,\ i = 1, \ldots, \dim \mathfrak{g}\}$ be a basis of $\mathfrak{g}$ and $\{u^i\}$ the dual basis (with respect to the normalized invariant symmetric bilinear form). Furthermore, let κ be the dual Coxeter number (the definition and its relation to the adjoint representation will be given later). The central element t operates as $c \cdot id$ on V. The scalar c is called level of the representation. Assume that $\kappa + c \neq 0$. In this case the operator

$$S_k := -\frac{1}{2(c+\kappa)} \sum_{n \in \mathbb{Z}} \sum_{i=1}^{\dim \mathfrak{g}} :u_i(-n)u^i(n+k): \tag{10.1}$$

is well-defined. Here $u_i(n) = u_i \otimes z^n$ is considered as operator on V and $:\ldots.:$ denotes the normal ordering

$$:x(n)y(m): := \begin{cases} x(n)y(m), & n \le m \\ y(m)x(n), & n > m. \end{cases} \tag{10.2}$$

By normal ordering, operators of high degree will be brought to the right. Given an arbitrary vector v, then operators of high enough degree will annihilate this vector. Hence, by normal ordering the operators S_k will be well-defined operators on V.

The result is that by $L_k \mapsto S_k$ and $t \mapsto id$ a representation of the Virasoro algebra with central charge

$$\frac{c \cdot \dim \mathfrak{g}}{c + \kappa} \tag{10.3}$$

is defined. (The central charge of the representation of the Virasoro algebra is the scalar by which the central element of the Virasoro algebra operates on this representation.) A proof is given for example in [122]. Or, alternatively, it will be special case of the general case which we present now.

Remark 10.1. The original constructions of Sugawara [253], with the title "A field theory of currents", appeared in 1968. It was done in the context of a four-dimensional field theory related to strong interaction. Sugawara himself did not consider the two-dimensional conformal situation which we are examining here. The author does not feel competent to give reliable historical references to its appearance in physics (and later in mathematics). It seems to go back to work of Bardakçi and Halpern [6] in 1971. Kac and Raina [122] state that the first case, when the central term for $su(n)$ was calculated correctly, was done by Dashen and Freshman [47] in 1975. A lot of other names should be mentioned too, for example Knizhnik and Zamolodchikov [131], Segal, Kac, Feigin and Fuks, Todorov, Wakimoto. For a complete mathematically correct treatment of the classical case see [118, 122, 124, 263]. As far as the naming is concerned, we would like to mention that in addition to Sugawara construction (which we will use for simplicity too) also *affine Sugawara construction*, and *Segal–Sugawara construction* are used.

10.2 General Sugawara construction

Following the general philosophy, it is necessary to formulate the classical algebraic objects with the help of the geometric Krichever–Novikov objects. Let $\mathfrak{g}$ be a finite-dimensional reductive Lie algebra with a fixed symmetric, invariant, and non-degenerate bilinear form $\alpha(.,.) = (..|..)$. In contrast to Chapter 9, we need the form to be non-degenerate. In case $\mathfrak{g}$ simple this form is necessarily a nonvanishing multiple of the Cartan–Killing form. Let $\widehat{\mathfrak{g}}$ be the associated affine Lie algebra to $\mathfrak{g}$, where the central extension is obtained by defining cocycle $-\gamma_{\alpha,C}$ (see (9.11)), obtained by the integration over a separating cycle C_S:

$$[\widehat{x \otimes f}, \widehat{y \otimes g}] = \widehat{[x, y] \otimes (fg)} - \left((x|y) \cdot \frac{1}{2\pi \mathrm{i}} \int_{C_S} f.dg \right) \cdot t. \qquad (10.4)$$

Here $x, y \in \mathfrak{g}$ and $f, g \in \mathcal{A}$. The algebra $\widehat{\mathfrak{g}}$ is equipped with an almost-grading, induced by the almost-grading of $\mathcal{A}$; see Section 9.1.

Definition 10.2. A module V over the Lie algebra $\widehat{\mathfrak{g}}$ (or a representation) is called *admissible module*, or *admissible representation* if for every $v \in V$ an $n_0 \in \mathbb{Z}$ exists such that $\widehat{\mathfrak{g}}_n . v = 0$ for $n \geq n_0$, and the central element t operates as a scalar multiple of the identity. This multiple (with respect to the presentation (10.4)) is called *level* of the representation. Sometimes also *central charge* of the affine algebra is used.

Let V be a fixed admissible module with level c. For the element $x \otimes g$ (more precisely $\widehat{x \otimes g}$) of $\widehat{\mathfrak{g}}$, we often use the notation $x(g)$ to denote the corresponding operator on V. For $x \otimes A_{n,p}$, with the special basis elements $A_{n,p}$, we also use $x(n, p)$, and $x(n)$ if $K = 1$, for $x(A_{n,p})$.

Examples of admissible representations are Verma modules (see Section 9.9) and fermionic modules (see Section 9.10).

We choose a basis u_i, $i = 1, \dots, \dim \mathfrak{g}$ of $\mathfrak{g}$. Let u^i, $i = 1, \dots, \dim \mathfrak{g}$ be the dual basis with respect to the bilinear form (..|..). By definition we have $(u_i|u^j) = \delta_i^j$. The *Casimir element* of $\mathfrak{g}$ is given by

$$\Omega = \sum_{i=1}^{\dim \mathfrak{g}} u_i u^i. \qquad (10.5)$$

It is an element of the universal enveloping algebra $U(\mathfrak{g})$. To simplify the notation, a summation over the index i, i.e., over the basis elements of $\mathfrak{g}$, should always run over the range $1, \dots, \dim \mathfrak{g}$.

Lemma 10.3.
(a) *Ω is independent of the choice of a basis;*
(b) $[\Omega, \mathfrak{g}] = 0$;
(c) $\sum_i [u_i, u^i] = 0$;
(d) $\sum_i [u_i \otimes f, u^i \otimes g] = -\dim \mathfrak{g} \cdot \gamma_S^{\mathcal{A}}(f, g) \cdot t$;
(e) *For abelian or simple $\mathfrak{g}$ there is a constant κ with*

$$\sum_i ad_{u_i} \circ ad_{u^i} = 2\kappa. \tag{10.6}$$

This means that 2κ is the eigenvalue of the Casimir operator in the adjoint representation.

Proof. Statement (a): This is linear algebra. For completeness we give the proof. Let $\{v_k\}$ and $\{v^k\}$ be another system of dual basis elements. If $x, y \in \mathfrak{g}$, then

$$(x|y) = \sum_i (x|u_i)(y|u^i) = \sum_k (x|v_k)(y|v^k). \tag{10.7}$$

Moreover,

$$v_k = \sum_i (v_k|u^i)u_i, \quad v^k = \sum_j (v^k|u_j)u^j, \tag{10.8}$$

and we calculate

$$\sum_k v_k v^k = \sum_{i,j} \sum_k (v_k|u^i)(v^k|u_j)u_i u^j = \sum_{i,j} (u_i|u^j)u_i u^j = \sum_i u_i u^i. \tag{10.9}$$

Statement (b): In $U(\mathfrak{g})$ we have $[a, bc] = [a, b]c + b[a, c]$. Hence,

$$[x, \sum_i u_i u^i] = \sum_i [x, u_i u^i] = \sum_i [x, u_i]u^i + \sum_i u_i [x, u^i]. \tag{10.10}$$

By the dual basis property we decompose

$$[x, u^i] = \sum_j ([x, u^i]|u_j)u^j = -\sum_j ([x, u_j]|u^i)u^j.$$

Here the invariance and the symmetry of (..|..) were used. Applied to the second sum in (10.10) we get

$$\sum_i u_i [x, u^i] = -\sum_{i,j} u_i ([x, u_j]|u^i)u^j = -\sum_j [x, u_j]u^j,$$

which cancels the first sum.

Statement (c):

$$\sum_i [u_i, u^i] = \sum_i u_i u^i - \sum_i u^i u_i = \Omega - \Omega = 0\,,$$

as the Casimir element was expressed once for the basis u_i and once for the basis u^i. Statement (d): Recall the structure equation for the centrally extended algebra

$$[u_i \otimes f, u^i \otimes g] = [u_i, u^i] \otimes (fg) - \gamma_S^{\mathcal{A}}(f, g) \cdot (u_i|u^i) \cdot t. \tag{10.11}$$

By (c), after summation over i the first term vanishes and $\sum_i (u_i|u^i) = \dim \mathfrak{g}$ gives the claim.

Statement (e): If $\mathfrak{g}$ is abelian, then

$$ad_{u_i}(x) = [u_i, x] = 0, \quad \forall x \in \mathfrak{g}.$$

Hence, $\kappa = 0$ fulfills the condition. If $\mathfrak{g}$ is simple then the adjoint representation is irreducible. The operator $\sum_i ad_{u_i} \circ ad_{u^i}$ is nothing other than the operator on the adjoint representation which corresponds to the Casimir element. It commutes with all $x \in \mathfrak{g}$, and hence by irreducibility it operates as a scalar. This scalar we denote by 2κ. □

Remark 10.4. Our invariant form $(..|..)$ has been fixed. If we rescale it by a non-zero factor r, then the the central charge will be rescaled by $1/r$ too. Moreover, the dual basis $\{u^i\}$ will be rescaled also by $1/r$. In case $\mathfrak{g}$ is simple we can normalize our $(..|..)$ such that the long roots have square length 2. In this case, κ is the *dual Coxeter number* of $\mathfrak{g}$ associated with the root system of $\mathfrak{g}$. It can be expressed with the help of the positive roots of the Lie algebra and will be a positive integer.

For $\mathfrak{g} = \mathfrak{sl}(n)$, the normalized bilinear form is given by the trace form

$$(X|Y) = \operatorname{tr}(XY), \tag{10.12}$$

and we have $\kappa = n$. A complete list of dual Coxeter numbers for the simple Lie algebras is given for example in [122, Table 10.1]. We will not need the dual Coxeter number interpretation in the following.

Example 10.5. For illustration we give the example of $\mathfrak{sl}(2)$. A basis is given by the matrices

$$E = \begin{pmatrix} 0 & 1 \\ 0 & 0 \end{pmatrix}, \quad H = \begin{pmatrix} 1 & 0 \\ 0 & -1 \end{pmatrix}, \quad F = \begin{pmatrix} 0 & 0 \\ 1 & 0 \end{pmatrix}. \tag{10.13}$$

The trace form (10.12) is the normalised Cartan–Killing form. The basis dual to the basis given above is (in this order)

$$F, \quad \frac{1}{2}H, \quad E. \tag{10.14}$$

Hence, the Casimir element is

$$C = E \cdot F + \frac{1}{2}H \cdot H + F \cdot E = \frac{1}{2}H \cdot H + H + 2F \cdot E. \tag{10.15}$$

Here we used $[E, F] = H$. We calculate

$$[C, H] = 2[F \cdot E, H] = 2H.$$

Hence, 2 is the eigenvalue of the Casimir operator, as it has to be.

We suggest to those readers who have not yet done it to first look at Section 8.2, where we give an account of the mathematical notion of an operator field.

First, we will consider here *formal currents*, which are fields of conformal weight one. Given $x \in \mathfrak{g}$ and a point $Q \in \Sigma \setminus A$, we assign to these data the formal sum (sometimes called "generating function" or formal distribution)

$$\widehat{x}(Q) = \sum_n \sum_p x(n,p) \cdot \omega^{n,p}(Q). \tag{10.16}$$

Here the summation over n denotes a summation over $\mathbb{Z}$, and the summation over p a summation over $1, \ldots, K$. In cases where the summation ranges are obvious we will drop them from the notation. In the sum the operators $x(n,p)$ are the operators on the representation space corresponding to the elements $x \otimes A_{n,p}$, and the $\omega^{n,p}$ are the basis elements in the space of differentials dual to the $A_{n,p}$. Obviously, this is a formal series as described by (8.26). We only have to take into account that by convention $f^1_{-n,p} = \omega^{n,p}$.

If we apply such a formal current $\widehat{x}(Q)$ to a vector $v \in V$ from the representation space we obtain

$$\widehat{x}(Q)\,.\,v = \sum_{n \in \mathbb{Z}} \sum_p (x(n,p)\,.\,v) \cdot \omega^{n,p}(Q)\,, \tag{10.17}$$

a formal sum of vectors from the representation space. By the definition of admissibility given a $v \in V$, a constant $M(v)$ exists such that $x(n,p)\,.\,v = 0$ if $n \geq M(v)$. Hence (8.27) is fulfilled and $\widehat{x}(Q)$ is a field in the sense of Definition 8.5. In particular, the outer summation in (10.17) is bounded from above.

Of special importance will be in the following the current fields $\widehat{u_i}(Q)$ and $\widehat{u^i}(Q)$. To define the Sugawara operators we have to deal with products of the currents fields. As explained in Section 8.2, such products are well-defined if we use instead of the product the normal ordered product. By Proposition 8.7, this product will then be a field of conformal weight two. Hence, it will be a formal sum with respect to the quadratic differentials $\Omega^{n,p}$. Let us repeat the *standard normal ordering*.

$$:x(n,p)y(m,r): \quad := \begin{cases} x(n,p)y(m,r), & (n,p) \leq (m,r) \\ y(m,r)x(n,p), & (n,p) > (m,r) \end{cases} \tag{10.18}$$

Here we again order our indices (n,p) lexicographically. Recall that the intention of the modification was to bring the annihilation operator, i.e., the operator of positive degree, as much as possible to the right, so that they act first. Other normal orderings are possible.

Remark 10.6. As there is a certain arbitrariness in the description it is necessary to keep an eye on the way the normal ordering will play a role in the proofs. Certain types of result will not depend at all on the normal ordering. But we will have other results which will do so. A typical example is the following. We will obtain by the Sugawara construction a projective representation of the vector field algebra $\mathcal{L}$. To make it into an honest representation, we have to take the cocycle defined by the projective representation into account. The choice of an admissible normal ordering (different to the standard ordering) will change this cocycle. But the difference will only be a coboundary. Hence, the equivalence class of the central extension will not depend on the chosen normal ordering.

Definition 10.7. Let :.....: be a normal ordering. The *Sugawara operator field* is defined as

$$T(Q) := \frac{1}{2}\sum_i :\widehat{u_i}(Q)\widehat{u^i}(Q) := \frac{1}{2}\sum_{n,m}\sum_{p,r}\sum_i :u_i(n,p)u^i(m,r):\ \omega^{n,p}(Q)\omega^{m,r}(Q). \quad (10.19)$$

If we consider $T(Q)$ as "generating function", then it is a quadratic differential. Hence, we write

$$T(Q) = \sum_{k\in\mathbb{Z}}\sum_{p=1}^{K} L_{k,p}\cdot\Omega^{k,p}(Q) \quad (10.20)$$

with certain operators $L_{k,p}$. The $L_{k,p}$ are called *modes of the Sugawara field* T, or just Sugawara operators. With the help of the duality (3.10) we get

$$L_{k,p} = \frac{1}{2\pi\mathrm{i}}\int_{C_S} T(Q)e_{k,p}(Q) = \frac{1}{2}\sum_{n,m}\sum_{r,t}\sum_i :u_i(n,r)u^i(m,t):\ l_{k,p}^{(n,r)(m,t)}, \quad (10.21)$$

with

$$l_{k,p}^{(n,r)(m,t)} = \frac{1}{2\pi\mathrm{i}}\int_{C_S} w^{n,r}(Q)w^{m,t}(Q)e_{k,p}(Q).$$

By Proposition 8.7 we know that $L_{k,p} \in gl(V)$. The necessary calculation is given in the passage after (8.37).

For $g = 0$ and two points the coefficients calculate directly to $l_k^{nm} = \delta_k^{m+n}$, and we obtain the usual definition of the Sugawara operator discussed in Section 10.1; see also [122] and the references given there.

For an element $e \in \mathcal{L}$ we define an action on the formal sum $\widehat{x}(Q) = \sum_{n,r} x(n,r)\omega^{n,r}(Q)$ by extending the Lie derivative action on the one-forms as

$$e\,.\,\widehat{x}(Q) := \sum_{n,r} x(n,r)(e\,.\,\omega^n)(Q). \quad (10.22)$$

The following is the key result of our construction.

Proposition 10.8. *Let* $\mathfrak{g}$ *be either an abelian or a simple Lie algebra. Then*

$$[L_{k,p}, x(g)] = -(c+\kappa)\cdot x(e_{k,p}\cdot g). \quad (10.23)$$

$$[L_k, \widehat{x}(Q)] = (c+\kappa)\cdot(e_k\,.\,\widehat{x}(Q)). \quad (10.24)$$

In the classical situation we obtain $[L_k, x(n)] = -(c+\kappa)\cdot n\cdot x(n+k)$. Recall that $x(n) = x(z^n)$. The proof of the above proposition will be given in Section 10.4

Remark 10.9. In Conformal Field Theory in genus zero, fields $\Phi(z)$ in the representation under consideration are called *primary fields* of conformal weight h if

$$[L_n, \Phi(z)] = \left(z^{n+1}\frac{d}{dz} + h(n+1)z^n\right)\Phi(z). \tag{10.25}$$

If one interprets $\Phi(z)$ as operator valued section of $\mathcal{K}^h$ ($\mathcal{K}$ the canonical bundle), then the right-hand side of (10.25) can be written as $e_n \,.\, \Phi(z)$. The similarity with (10.24) is obvious. Note that in this case we have $h = 1$.

Proposition 10.10. *The operators $L_{k,p} \in gl(V)$ and $id = 1 \in gl(V)$ close to a Lie subalgebra of $gl(V)$ with the commutator relations*

$$[L_{k,p}, L_{l,r}] = -(c+\kappa)\sum_n C_{(k,p)(l,r)}^{(n,s)} L_{n,s} - \frac{1}{2}c(c+\kappa)\dim\mathfrak{g}\cdot\chi_{(k,p)(l,r)}\cdot id \tag{10.26}$$

with the structure constants $C_{(k,p)(l,r)}^{(n,s)}$ of the vector field algebra $\mathcal{L}$ in respect to the basis elements $e_{n,p}$. The constants $\chi_{(k,p)(l,r)}$ are given as

$$\chi_{(k,p)(l,r)} = (-2)\gamma_{C_S,R}^{\mathcal{L}}(e_{k,p}, e_{l,r}), \tag{10.27}$$

where $\gamma_{C_S,R}^{\mathcal{L}}$ is the unique local separating cocycle for the vector field algebra (up to equivalence and rescaling) given by (6.140) with a suitable projective connection R.

Moreover, for the standard normal ordering, this cocycle is bounded from above by zero, and on the upper bound we have

$$\chi_{(k,p),(-k,r)} = -\frac{1}{6}(k^3 - k)\delta_p^r. \tag{10.28}$$

The proof will be given in Section 10.4

It is clear that the case $c = -\kappa$ plays a special role. In this case, Proposition 10.10 shows that the $L_{k,p}$, $k \in \mathbb{Z}$, $p = 1\dots,K$ define an abelian subalgebra. By Proposition 10.8 this subalgebra commutes with all operators $x(n,r)$. The case $c + \kappa = 0$ is called *critical level.*

If $c + \kappa \neq 0$, then we take the rescaled elements $L_{k,p}^* = \frac{-1}{c+\kappa}L_{k,p}$ and obtain

$$[L_{k,p}^*, L_{l,r}^*] = \sum_{n,s} C_{(k,p)(l,r)}^{(n,s)} L_{n,s}^* + \frac{c}{(c+\kappa)}\dim\mathfrak{g}\cdot\gamma_{C_S,R}^{\mathcal{L}}(e_{k,p}, e_{l,r})\cdot id. \tag{10.29}$$

Which leads to the following theorem.

Theorem 10.11. *Let $\mathfrak{g}$ be a finite-dimensional abelian or simple Lie algebra, 2κ the eigenvalue of the Casimir operator in the adjoint representation, and $\widehat{\mathfrak{g}}$ the affine Lie algebra (of arbitrary genus and arbitrary number of points A). Let V be an admissible representation, such that the central element of $\widehat{\mathfrak{g}}$ operates as $c\cdot$ identity. If $c + \kappa \neq 0$, then the rescaled modes of the Sugawara operator*

$$L_{k,p}^* = \frac{-1}{2(c+\kappa)}\sum_{n,m}\sum_{r,s}\sum_i {:}u_i(n,r)u^i(m,s){:}\; l_{k,p}^{(n,r)(m,p)} = \frac{-1}{2\pi \mathrm{i}(c+\kappa)}\int_{C_S} T(Q)e_{k,p}(Q) \tag{10.30}$$

give a representation of a centrally extended Krichever–Novikov vector field algebra. The central extension is given by a geometric cocycle

$$\gamma_{C_S,R}^{\mathcal{L}}(e,f) = \frac{1}{24\pi i}\int_{C_S}\left(\frac{1}{2}(e'''f - ef''') - R\cdot(e'f - ef')\right)dz\,, \tag{10.31}$$

with a certain meromorphic connection R holomorphic outside A. The central charge of the representation is

$$c_{\mathcal{L}} = \frac{c\cdot\dim\mathfrak{g}}{(c+\kappa)} \tag{10.32}$$

The central charge $c_{\mathcal{L}}$ is sometimes also called *conformal anomaly.*

Remark 10.12. The choice of the normal ordering fixes the projective connection R. A different rule for the normal ordering will only yield a cohomologous cocycle.

Remark 10.13. As the cocycle is bounded from above by zero (at least in the standard normal ordering), we conclude that if our projective connection has poles at the points in I their order is at most two. In fact we will see that even order two is excluded.

Let us denote by $T[e]$ the *Sugawara representation* at non-critical level. This means

$$T[e_{k,p}] = \mathcal{L}_{k,p}^{*} = \frac{-1}{c+\kappa}\frac{1}{2\pi i}\int_{C_S} T(Q)e_{k,p}(Q). \tag{10.33}$$

This integral expression makes complete sense also for $e \in \mathcal{L}$ arbitrary.

$$T[e] = \frac{-1}{c+\kappa}\frac{1}{2\pi i}\int_{C_S} T(Q)e(Q). \tag{10.34}$$

Theorem 10.11 says that the linear map

$$T : \mathcal{L} \to gl(V) \tag{10.35}$$

defines a projective representation, i.e.,

$$T[[e,f]] = [T[e],T[f]] + \frac{c\dim\mathfrak{g}}{c+\kappa}\gamma_{C_S,R}^{\mathcal{L}}(e,f)id. \tag{10.36}$$

Note that the factor in front of the central term does not depend on the normalization of the invariant form α. A possible rescaling factor will cancel. By setting

$$T[\hat{e}] := T[e], \quad T[t] := \frac{c\dim\mathfrak{g}}{c+\kappa}id, \tag{10.37}$$

we obtain a Lie representation of the unique almost-graded centrally extended vector field algebra $\widehat{\mathcal{L}}$.

In the purely abelian case we obtain

$$T[[e,f]] = [T[e],T[f]] + \dim\mathfrak{g}\,\gamma_{C_S,R}^{\mathcal{L}}(e,f)\,id. \tag{10.38}$$

With Proposition 10.8 we obtain the fundamental property in the non-critical case

Proposition 10.14. *For $e \in \mathcal{L}$, $g \in \mathcal{A}$, $x \in \mathfrak{g}$ we obtain*

$$[T[e], x(g)] = x(e\,.\,g), \quad [T[e], \widehat{x}(Q)] = -(e\,.\,\widehat{x}(Q)). \tag{10.39}$$

Equations (10.36) and (10.39) supplemented by the operation of the affine algebra $\widehat{\mathfrak{g}}$ on V

$$[x(f), y(g)] = [x, y](fg) - c \cdot (x|y)\gamma_S^{\mathcal{A}}(f, g)\, id \tag{10.40}$$

yield the following proposition.

Proposition 10.15. *The assignment*

$$e \mapsto T[e], \quad \text{and} \quad x \otimes g \mapsto x(g), \tag{10.41}$$

defines a projective Lie representation of $\mathcal{D}^1_{\mathfrak{g}}$ on V. In particular a Lie homomorphism from a central extension $\widehat{\mathcal{D}^1_{\mathfrak{g}}}$ to $gl(V)$ exists for which the defining cocycle is given by the sum of the vector field cocycle $\frac{c \dim \mathfrak{g}}{c+\kappa}\gamma^{\mathcal{L}}_{C_S,R}$ and the current algebra cocycle $-c \cdot (..|..)\gamma^{\mathcal{A}}_{C_S}$. There is no mixing cocycle.

As a separate result aside let us also note the proposition below.

Proposition 10.16. *Let E be an arbitrary operator from $gl(V)$. If there is a basis element $e_{l,s}$ of $\mathcal{L}$ with*

$$[E, x(g)] = -(c + \kappa)\, x(e_{l,s} \cdot g) \tag{10.42}$$

for all $x \in \mathfrak{g}$ and $g \in \mathcal{A}$, then for all k, p

$$[L_{k,p}, E] = [L_{k,p}, L_{l,s}]\,,$$

with $L_{l,s} = T[e_{l,s}]$.

This will be a corollary to the proof of Proposition 10.10. We will show it in Section 10.4

Remark 10.17. If $\mathfrak{g}$ is an abelian Lie algebra, then the central charge is $\dim \mathfrak{g}$. In case $\mathfrak{g} = \mathbb{C}$, the one-dimensional Lie algebra, then $\widehat{\mathfrak{g}} = \widehat{\mathcal{A}}$ and starting from a representation of $\widehat{\mathcal{A}}$ (the Heisenberg algebra, or oscillator algebra), by the Sugawara construction a representation of the centrally extended vector field algebra with central charge $c_{\widehat{\mathcal{L}}} = 1$ is obtained.

10.2.1 The reductive case

Let $\mathfrak{g}$ be a reductive Lie algebra and

$$\mathfrak{g} = \mathfrak{g}_0 \oplus \mathfrak{g}_1 \oplus \cdots \oplus \mathfrak{g}_M \tag{10.43}$$

be its decomposition into the abelian summand g_0 and the simple summands g_i, $i = 1, \ldots, M$. Let α be a non-degenerate symmetric invariant bilinear form for $\mathfrak{g}$. With respect to this α, we consider the affine Lie algebra $\widehat{\mathfrak{g}}_{\alpha,S}$ obtained via the cocycle

$$-\gamma_{\alpha,S}(x \otimes g, y \otimes g) = -\alpha(x, y)\frac{1}{2\pi \mathrm{i}} \int_{C_S} f dg. \tag{10.44}$$

Recall from Chapter 9 that the decomposition (10.43) is also an orthogonal decomposition with respect to α. Denote by α_i the restriction of α to the component $\mathfrak{g}_i$. We choose a basis $\{u_i\}$ of $\mathfrak{g}$ which is a collection of basis elements for the components. In particular, the dual basis also respects this decomposition. We build the Casimir element C_i for the component $\mathfrak{g}_i$ as in (10.5) and obtain the results of Lemma 10.3 individually. In particular, let $2\kappa_i$ be the eigenvalue of the Casimir operator C_i on the adjoint representation of $\mathfrak{g}_i$. Of course, the properties (1) to (4) of Lemma 10.3 are also valid for the full Casimir operator $C = \sum_i C_i$, but not (5).

Let $x = \sum_i x_i$ and $y = \sum_j y_j$ be the decomposition of the elements $x, y \in \mathfrak{g}$. Then $x \otimes f = \sum_i x_i \otimes g$ and $y \otimes f = \sum_j y_j \otimes g$. We have (dropping the ˆ for the elements of the central extension)

$$[x_i \otimes f, y_j \otimes g] = [x_i, y_j] \otimes (fg) - \alpha(x_i, y_j)\frac{1}{2\pi \mathrm{i}} \int_{C_S} f dg \cdot t. \tag{10.45}$$

If $i \neq j$, then both summands on the right-hand side will vanish and the elements left will commute. If $i = j$, then $\alpha(x_i, y_i) = \alpha_i(x_i, y_i)$. Let $\widehat{\mathfrak{g}}_i$ be the subalgebra of $\widehat{\mathfrak{g}}$ generated by $\mathfrak{g}_i$ and the central element t. The central element lies in all $\widehat{\mathfrak{g}}_i$. Let V be an admissible representation for $\widehat{\mathfrak{g}}$. The V is also an admissible representation for $\widehat{\mathfrak{g}}_i$. Assume that it is non-critical for every summand of g. This means that $c + \kappa_i \neq 0$ for all $i = 0, \ldots, M$. Let T_i be the Sugawara representation for the summand $\mathfrak{g}_i$ defined via the action of $\widehat{\mathfrak{g}}_i$ on V. We set

$$T[e] := \sum_{i=0}^{M} T_i[e]. \tag{10.46}$$

The individual operators T_i are built from elements $x_i(f)$. Elements with different indices commute as discussed above, and consequently $[T_i[e], T_j[f]] = 0$, and $[T_i[e], x_j(g)] = 0$ for $i \neq j$. Hence,

$$[T[e], T[f]] = \sum_{i=0}^{M} [T_i[e], T_i[f]] = T[e, f] + \left(\sum_{i=0}^{M} \frac{c \dim \mathfrak{g}_i}{c + \kappa_i} \gamma_{S,R_i}^{\mathcal{L}}(e, f) \right) \cdot id. \tag{10.47}$$

In view of (10.39) we calculate

$$[T[e], x(g)] = \left[\sum_{i=0}^{M} T_i[e], \sum_{j=0}^{M} x_j(g) \right] = \sum_{i=0}^{M} x_i(e.g) = x(e.g). \tag{10.48}$$

From the above the following is clear.

Proposition 10.18. *Let $\mathfrak{g}$ be a reductive Lie algebra with a chosen bilinear form α, and $T[e]$ for every $e \in \mathcal{L}$ the operator as defined above. Then T defines a representation of a centrally extended vector field algebra $\widehat{\mathcal{L}}$ given by the local cocycle*

$$\left(\sum_{i=0}^{M} \frac{c \dim \mathfrak{g}_i}{c + \kappa_i} \gamma^{\mathcal{L}}_{C_S,R_i}\right). \tag{10.49}$$

Moreover, together with the action of $\widehat{\mathfrak{g}}$ on the basis of the relation (10.48), the representation extends to a representation of the differential operator algebra $\widehat{\mathcal{D}^1_{\mathfrak{g}}}$.

Note that up to coboundary (10.49) can also be written as

$$\left(\sum_{i=0}^{M} \frac{c \dim \mathfrak{g}_i}{c + \kappa_i}\right) \gamma^{\mathcal{L}}_{C_S,R}, \tag{10.50}$$

with a single R.

10.2.2 Almost-graded structure

Until now our $\widehat{\mathfrak{g}}$-module V was only required to be an admissible module. In cases where V is also an almost-graded module we show the following proposition.

Proposition 10.19. *The almost-graded $\widehat{\mathfrak{g}}$-module V is an almost-graded $\widehat{\mathcal{L}}$-module under the Sugawara action, and hence also an almost-graded $\widehat{\mathcal{D}^1_{\mathfrak{g}}}$ module.*

Proof. We have to show that constants M_1, M_2 exist, such that for every given homogeneous element $\psi_l \in V$ of degree l, and every k we have

$$k + l + M_1 \leq \deg(L^*_{k,r} \cdot \psi_l) \leq k + l + M_2. \tag{10.51}$$

Starting from the description (10.20) we consider first the coefficients $l^{(n,p)(m,s)}_{(k,r)}$, which are given as integrals (10.21). The integral could only be non-vanishing if the integrand has poles both at the points in I and at the point in O. We note

$$\operatorname{ord}_{P_i}(\omega^{n,r}) = -n - 1 + (1 - \delta^r_i), \quad \operatorname{ord}_{P_i}(e_{l,p}) \geq k + 1 + (1 - \delta^p_i).$$

Hence,

$$\operatorname{ord}_{P_i}(w^{n,r} w^{m,t} e_{k,p}) = -(n + m) + k - 1 + (3 - \delta^r_i + \delta^t_i + \delta^t_i). \tag{10.52}$$

If $-(n+m)+k > 0$, then there will be no residue at the points in I. In the same way we get estimates of the type that there will be no residue at the points in O if $-(n+m)+k < -S$ with a suitable S. Hence the integral could only be nonvanishing if

$$k \leq n + m \leq k + S. \tag{10.53}$$

By the almost-gradedness of V as $\widehat{\mathfrak{g}}$-module, the constants C_1 and C_2 exist such that for all m, r, s

$$m + l + C_1 \leq \deg(u(m,r)\psi_l) \leq m + l + C_2, \tag{10.54}$$

if the element $u(m,r)\psi_l \neq 0$. Hence,

$$n + m + l + 2C_1 \leq \deg(:u(n,p)u(m,r): \psi_l) \leq n + m + l + 2C_2, \tag{10.55}$$

if the element $:u(n,p)u(m,r): \psi_l \neq 0$. Using (10.53) we obtain Equation (10.51) if we set $M_1 = 2C_1$ and $M_2 = 2C_2 + S$.

The action of $\widehat{\mathfrak{g}}$ is by assumption almost-graded. Hence, the action of $\widehat{\mathcal{D}^1_{\mathfrak{g}}}$ is almost-graded too. □

10.3 Verma module representations

Let $\mathfrak{g}$ be a reductive finite-dimensional Lie algebra. In fact, let us assume that $\mathfrak{g}$ is either simple or abelian, the general reductive case will then follow. Let V be now a Verma module of $\widehat{\mathfrak{g}}$ as introduced in Section 9.9, which is generated by the vacuum vector $\psi \in V$. By construction we require

(1) $\widehat{\mathfrak{g}}_n \cdot \psi = 0$ for $n \geq 1$;
(2) for the elements $x \in \mathfrak{n}_+$, the upper nilpotent algebra, $x(0,p)\psi = 0$;
(3) for the elements h of the Cartan subalgebra $\mathfrak{h}$, the operators $h(0,p)$ operate as scalar on ψ. More precisely, an element $\chi \in \mathfrak{h}^{*K}$ exists such that $h(0,p)\psi = \chi_p(h)\psi$.
(4) The central element t operates as a scalar c on the whole representation space.

We recall that χ is called the weight and c the level of the Verma module V. In fact, the following works also for highest weight representations, i.e., quotients of the Verma modules.

In case $\mathfrak{g}$ is abelian, $\mathfrak{h}$ is the full algebra and there is no nilpotent subalgebra. If $\mathfrak{g}$ is reductive, we take its decomposition into summands and pose the conditions for each summand separately.

Let $(.|.)$ be an invariant non-degenerate symmetric bilinear form for $\mathfrak{g}$ and consider the central extension $\widehat{\mathfrak{g}}$ with respect to this form. Assume that the Verma module for $\widehat{\mathfrak{g}}$ is non-critical for every summand of $\mathfrak{g}$. Recall that in this case, the Sugawara operator for $\widehat{\mathfrak{g}}$ can be defined as a balanced sum of Sugawara operators of the summands.

Theorem 10.20. *Let V be a Verma module for $\widehat{\mathfrak{g}}$ which is non-critical for every summand of $\mathfrak{g}$. Assume that normal ordering is given by the standard normal ordering. Denote by $L^*_{k,r}$ the rescaled Sugawara operators* (10.30), *and the sum* (10.46) *in the reductive case. Then:*

(a) *the Sugawara operators $L^*_{k,r}$ for $k \geq 1$ operate trivially on ψ, i.e., $L^*_{k,r} \cdot \psi = 0$.*
(b) *The Sugawara operators $L^*_{0,r}$ operate as scalars on ψ. This means $\lambda_r \in \mathbb{C}$ exists, such that*

$$L^*_{0,r} \cdot \psi = \lambda_r \psi. \tag{10.56}$$

(c) *The central element* $t_{\mathcal{L}} \in \widehat{\mathcal{L}}$ *operates as*

$$\frac{c \dim \mathfrak{g}}{c+\kappa}\, id. \tag{10.57}$$

(d) *The subspace generated by* $T[\mathcal{L}]$ *from the vacuum vector is a highest weight representation of the almost-graded centrally extended vector field algebra* $\widehat{\mathcal{L}}$ *given by* $\gamma^{\mathcal{L}}_{C_S,R}$.

Proof. In the definition of the operators summands of the the type

$$:u_i(n,r)u^i(m,s):\, l^{(n,r)(m,s)}_{(k,p)} \tag{10.58}$$

appear; see (10.30). For the coefficients we have the expressions of the orders (10.52). Hence there can only be a residue if $n+m \geq k$, and for $n+m=k$ there will only be a residue if $p=r=s$ and it appears at the point P_p. If $k \geq 1$, then either n or m has to be ≥ 1. Hence either directly or by normal ordering the corresponding operator will act first and will annihilate the vacuum ψ. This shows (a).

If $k = 0$ and n or m is positive, then we have again annihilation. It remains to consider $n=m=k=0$. The corresponding factor will be 1. The remaining part of the operator will be

$$L^*_{0,p}\psi = \frac{-1}{2(c+\kappa)} \sum_i u_i(0,p)u^i(0,p)\psi. \tag{10.59}$$

We have to show that the result will be a scalar multiple of ψ. To see this we make a convenient choice of a dual system of basis elements. For this we assume that $\mathfrak{g}$ is simple. In this case we can decompose

$$\mathfrak{g} = \mathfrak{n}_- \oplus \mathfrak{h} \oplus \mathfrak{n}_+, \tag{10.60}$$

where $\mathfrak{n}_+$ is the upper nilpotent subalgebra spanned by the positive root vectors, $\mathfrak{n}_-$ the lower nilpotent subalgebra spanned by the negative root vectors, and $\mathfrak{h}$ is the Cartan matrix (i.e., in the terminology of matrix groups the diagonal matrices). Its dimension is called the rank of $\mathfrak{g}$, rk $\mathfrak{g}$. Let $\{e_\alpha\}$ be the system of positive root vectors (which is a basis of $\mathfrak{n}_+$), $\{h_i\}$ a basis of $\mathfrak{h}$, and $\{f_{-\alpha}\}$ a system of negative root vectors (a basis of $\mathfrak{n}_-$). This collection of vectors we now take as basis for our $\mathfrak{g}$. With respect to the invariant non-degenerate symmetric bilinear form $(.|.)$, the dual basis is given by $\{f_{-\alpha}\}, \{h^i\}, \{e_\alpha\}$. Here the system $\{h^i\}$ is a dual basis of the to $\{h_i\}$, with respect to the (non-degenerate) restriction of the form to $\mathfrak{h}$. The sum in (10.59) decomposes as

$$\sum_\alpha e_\alpha(0,p)f_{-\alpha}(0,p) + \sum_{i=1}^{\mathrm{rk}\,\mathfrak{g}} h_i(0,p)h^i(0,p) + \sum_\alpha f_{-\alpha}(0,p)e_\alpha(0,p). \tag{10.61}$$

As V is a Verma module for $\widehat{\mathfrak{g}}$, by definition the elements $e_\alpha(0,p)$ annihilate ψ. The elements in the first sum we can express as

$$e_\alpha(0,p)f_{-\alpha}(0,p) = [e_\alpha(0,p), f_{-\alpha}(0,p)] + f_{-\alpha}(0,p)e_\alpha(0,p). \tag{10.62}$$

Applied to ψ, the second term on the right will vanish. The first term we rewrite as

$$[e_\alpha(0,p), f_{-\alpha}(0,p)] = [e_\alpha, f_{-\alpha}](0,p) + R = h_\alpha(0,p) + R, \quad h_\alpha \in \mathfrak{h} \quad R \in \widehat{\mathfrak{g}}_+. \tag{10.63}$$

Hence, using the fact that for a Verma module it is required that the elements of $\mathfrak{h}$ operate as scalars on ψ we get

$$L^*_{0,p}\psi = \frac{-1}{2(c+\kappa)}\left(\sum_\alpha h_\alpha(0,p) + \sum_{i=1}^{\operatorname{rk}\mathfrak{g}} h_i(0,p)h^i(0,p)\right)\psi = \lambda_p\psi. \tag{10.64}$$

If we evaluate this scalar we get exactly the form (10.65). The statement that it is a representation of $\widehat{\mathfrak{L}}$ is the result of Theorem 10.11. Above we showed that it is a highest weight representation as defined in Section 7.3. For the abelian case, we just ignore the part of the positive roots. For the reductive case we take the above-noted combination of the Sugawara operators and obtain the same statement. □

Within the above proof we also showed the following proposition.

Proposition 10.21. *In the case of $\mathfrak{g}$ simple or abelian, the highest weight representation of $\widehat{\mathfrak{L}}$ has weight*

$$\lambda_p = \frac{-1}{2(c+\kappa)}\left(\sum_\alpha \chi_p(h_\alpha) + \sum_{i=1}^{\operatorname{rk}\mathfrak{g}} \chi_p(h_i)\chi_p(h^i)\right). \tag{10.65}$$

i.e., $\mathfrak{L}^*_{0,p}\psi = \lambda_p\psi$, where ψ is the highest weight vector.

Remark 10.22.

(a) In general, the full representation space V of $\widehat{\mathfrak{g}}$ might be reducible under the $\widehat{\mathfrak{L}}$ action even if irreducible under the $\widehat{\mathfrak{g}}$ action.

(b) The Sugawara construction provides a useful method for constructing highest weight representation of the centrally extended vector field algebra. See for example [122] for the classical Virasoro case.

(c) It is also possible to introduce Casimir operators for the algebra $\widehat{\mathfrak{g}}$, and to extend the concept of highest weight. For this see [225, §5 and 6].

10.4 The proofs

The reader only interested in the results might skip this rather technical section. Quite often we will not mention the explicit range of the summation. The range should always be clear from the positions of the indices.

Lemma 10.23. *Let*

$$K^{(v,s)}_{(k,p)(n,r)} := \frac{1}{2\pi \mathrm{i}}\int_{C_S} \omega^{v,s}(Q)e_{k,p}(Q)dA_{n,r}(Q). \tag{10.66}$$

Then

$$K^{(v,s)}_{(k,p)(n,r)} = \sum_{m,t} l^{(v,s)(m,t)}_{(k,p)} \gamma_{(m,t)(n,r)}, \tag{10.67}$$

with

$$\gamma_{(m,t)(n,r)} := \gamma^{\mathcal{A}}_{C_S}(A_{m,t}, A_{n,r}) = \frac{1}{2\pi \mathrm{i}} \int_{C_S} A_{m,t}(Q) dA_{n,r}(Q). \tag{10.68}$$

Proof. With the "delta distribution" (3.92) we get

$$\begin{aligned}
\sum_m l^{(v,s)(m,t)}_{(k,p)} \gamma_{(m,t)(n,r)} &= \sum_{m,t} \frac{1}{2\pi \mathrm{i}} \int_{C_S} \omega^{v,s}(Q)\omega^{m,t}(Q) e_{k,p}(Q) \frac{1}{2\pi \mathrm{i}} \int_{C_S}' A_{m,t}(Q') dA_{n,r}(Q') \\
&= \frac{1}{(2\pi \mathrm{i})^2} \iint_{C_S\, C_S}' \omega^{v,s}(Q) e_{k,p}(Q) dA_{n,r}(Q') \Delta(Q', Q) \\
&= \frac{1}{2\pi \mathrm{i}} \int_{C_S}' \omega^{v,s}(Q') e_{k,p}(Q') dA_{n,r}(Q') = K^{(v,s)}_{(k,p)(n,r)}.
\end{aligned}$$

□

We now give a more technical version of Proposition 10.8 which we will finally prove.

Proposition 10.24. *Let $\mathfrak{g}$ be either an abelian or a simple Lie algebra. Then*

$$[L_{k,p}, x(n,r)] = -(c+\kappa) \sum_{v,s} K^{v,s}_{(k,p)(n,r)} x(v,s). \tag{10.69}$$

The result does not depend on the chosen normal ordering.

Proof of Proposition 10.8. In local coordinates $e_{k,p} . A_{n,r}$ writes as $e_{k,p}(z)\frac{dA_{n,r}(z)}{dz}$. Hence,

$$(e_{k,p} . A_{n,r})(Q) = e_{k,p}(Q) \cdot dA_{n,r}(Q) = \sum_{v,s} \beta^{(v,s)}_{(k,p)(n,r)} A_{v,s}(Q).$$

By duality the coefficient $\beta^{(v,s)}_{(k,p)(n,r)}$ calculates as

$$\beta^{(v,s)}_{(k,p)(n,r)} = \frac{1}{2\pi \mathrm{i}} \int_{C_S} e_{k,p}(Q) dA_{n,r}(Q) \omega^{v,s}(Q) = K^{(v,s)}_{(k,p)(n,r)}.$$

Hence, (10.69) yields (10.23) for $g = A_{n,r}$. By linearity the general result follows.

To show (10.24) we express $e_{k,p} . \omega^{v,s} = \sum_{n,r} \zeta^{(v,s)}_{(k,p)(n,r)} \omega^{n,r}$ with

$$\begin{aligned}
\zeta^{(v,s)}_{(k,p)(n,r)} &= \frac{1}{2\pi \mathrm{i}} \int_{C_S} (e_{k,p} . \omega^{v,s}) A_{n,r} \\
&= \frac{1}{2\pi \mathrm{i}} \int_{C_S} e_{k,p} . (\omega^{v,s} A_{n,r}) - \frac{1}{2\pi \mathrm{i}} \int_{C_S} \omega^{v,s} e_{k,p} . A_{n,r} = -K^{(v,s)}_{(k,p)(n,r)}.
\end{aligned}$$

Here we used Lemma 6.17. □

In the definition of the $L_{k,p}$ formal infinite series appear. Applied to individual elements in the given representation this has to define an operator. To keep this under control in Section 8.4 we used the cut-off function, as used by Kac and Raina in [122]. As we do not assume that all readers have studied this section we will reproduce the method here.

Let ψ be the $\mathbb{R}$-valued function given by

$$\psi(x) = 1 \quad \text{if } |x| \le 1 \quad \text{and} \quad \psi(x) = 0 \quad \text{if } |x| > 1. \tag{10.70}$$

For $\epsilon \in \mathbb{R}$ we define

$$L_{k,p}(\epsilon) = \frac{1}{2} \sum_{n,m,r,s} \sum_i :u_i(n,r)u^i(m,s):\, l_{(k,p)}^{(n,r)(m,s)} \psi(\epsilon n). \tag{10.71}$$

Let k be fixed, For every n there are only a finite number of m, such that $l_{(k,p)}^{(n,r)(m,s)} \neq 0$. Hence, for $\epsilon > 0$ the sum is reduced to a finite number of summands. For $v \in V$ based on the normal ordering, only a finite number of operators $l_{(k,p)}^{(n,r)(m,s)} :u_i(n,r)u^i(m,s):$ act non-trivially on v. Hence, if we choose $\epsilon > 0$ small enough we get $L_{k,p}(\epsilon)v = L_{k,p}v$. This we understand by

$$\lim_{\epsilon \to 0} L_{k,p}(\epsilon) = L_{k,p}. \tag{10.72}$$

If we drop the normal ordering symbols in (10.71) we get the expression $\widetilde{L}_{k,p}(\epsilon)$. As long as $\epsilon \neq 0$ it is well-defined. We compare it to $L_{k,p}(\epsilon)$. For every pair (n,r), (m,s), which is not normally ordered we take up the sum of commutators $\sum_i [u_i(n,r), u_i(m,s)]$. The commutator is a scalar by Lemma 10.3 (4). Hence, $L_{k,p}(\epsilon) = \widetilde{L}_{k,p}(\epsilon) + \alpha \cdot t$ with a scalar α as long as $\epsilon \neq 0$. In particular in the calculations of commutators we ignore the normal ordering as long as we stay with $\epsilon \neq 0$.

In the following section we will concentrate first on the situation that $K = 1$ (which is for example true in the two-point case). In Section 10.4.3 we will discuss the extension to $K > 1$. The main difference between these two cases is the notational complexity. Within some of the proofs we already use arguments which are adapted to the general situation (like for example for order estimates we will use an arbitrary finite set I of in-points).

Remark 10.25. The two-point case was also considered by the physicists Bonora, Rinaldi, Rosso and Wu [24]. Not all parts of their proof fulfill the requirements of a rigorous mathematical proof. In particular, in the context of the subtle points regarding normal ordering. The proof presented here for the multi-point situation is in some places clearly inspired by [24]. It is a minor simplification of a proof given by the author in joint work with Sheinman [225]. The Heisenberg two-point case was also treated by Krichever and Novikov in [141].

10.4.1 Proof of Proposition 10.24

We start with the expression (10.71) of $L_{k,p}(\epsilon)$. After writing out the commutator and regrouping the elements we obtain

$$R_\epsilon := 2\,[\tilde{L}_k(\epsilon), x(r)\,] = \sum_{n,m}\sum_i [u_i(n)u^i(m), x(r)]\, l_k^{nm}\psi(\epsilon n)$$
$$= \sum_{n,m}\sum_i \left(u_i(n)[u^i(m), x(r)] + [u_i(n), x(r)]u^i(m)\right) l_k^{nm}\psi(\epsilon n). \quad (10.73)$$

Each commutator can be written as

$$[u^i(m), x(r)] = [u^i, x](A_m A_r) - (u^i|x)\gamma_{mr}\cdot c$$

(note $t.v = c\cdot v$). We decompose $R_\epsilon = A_\epsilon + B_\epsilon - (C_\epsilon + D_\epsilon)$ as follows:

$$\begin{aligned} A_\epsilon &= \sum_{n,m}\sum_i u_i(n)[u^i, x](A_m A_r) l_k^{nm}\psi(\epsilon n), \\ B_\epsilon &= \sum_{n,m}\sum_i [u_i, x](A_n A_r) u^i(m) l_k^{nm}\psi(\epsilon n), \\ C_\epsilon &= \sum_{n,m}\sum_i u_i(n)(u^i|x)\gamma_{mr} l_k^{nm} c\psi(\epsilon n), \\ D_\epsilon &= \sum_{n,m}\sum_i (u_i|x)u^i(m)\gamma_{nr} l_k^{nm} c\psi(\epsilon n). \end{aligned} \quad (10.74)$$

If we use $\sum_i u_i \otimes A_n(u^i|x) = \left(\sum_i (u^i|x)u_i\right)\otimes A_n = x\otimes A_n = x(n)$, we obtain $C_\epsilon = \sum_{n,m} x(n)\gamma_{mr} l_k^{nm} c\psi(\epsilon n)$ and $D_\epsilon = \sum_{n,m} x(m)\gamma_{nr} l_k^{nm} c\psi(\epsilon n)$. For fixed r and k only a finite number of terms appear. Consequently, for $\epsilon = 0$

$$\lim_{\epsilon\to 0}(C_\epsilon + D_\epsilon) = 2c\cdot\sum_n\left(\sum_m l_k^{nm}\gamma_{mr}\right)x(n) = 2c\sum_n K_{k,r}^n x(n). \quad (10.75)$$

Here we used (10.67). The sums A_ϵ and B_ϵ separately do not make sense for $\epsilon \to 0$. To make them well-defined we have to pass to normal ordering. For this we write $A_m A_r = \sum_s \alpha_{mr}^s A_s$ with $\alpha_{mr}^s = \frac{1}{2\pi \mathrm{i}}\int_{C_S} A_m A_r \omega^s$. We obtain

$$\begin{aligned} A_\epsilon &= \sum_{n,m,s}\sum_i u_i(n)[u^i, x](s)\alpha_{mr}^s l_k^{nm}\psi(\epsilon n), \\ B_\epsilon &= \sum_{n,m,s}\sum_i [u_i, x](s)u^i(m)\alpha_{nr}^s l_k^{nm}\psi(\epsilon n). \end{aligned}$$

For the elements not in normal ordering we collect a commutator. We write $A_\epsilon = A_\epsilon^{(1)} + A_\epsilon^{(2)}$ and $B_\epsilon = B_\epsilon^{(1)} + B_\epsilon^{(2)}$ with $A_\epsilon^{(1)}$ respectively $B_\epsilon^{(1)}$ the above expressions, but now with the symbols for normal ordering. First we examine $A_\epsilon^{(2)}$. The commutator can be written as

$$[u_i(n), [u^i, x](s)] = [u_i, [u^i, x]](A_n A_s) - \gamma_{ns}(u_i|[u^i, x])c.$$

We have $(u_i|[u^i,x]) = ([u_i,u^i]|x)$. Hence, after summation with respect to i the second term vanishes due to Lemma 10.3 (3). Lemma 10.3 (5) yields $2\kappa\cdot x(A_n A_s) = 2\kappa\sum_v \alpha_{ns}^v x(v)$ for the first term. Carrying out the same for $B_\epsilon^{(2)}$, we get

$$A_\epsilon^{(2)} + B_\epsilon^{(2)} = 2\kappa\sum_v\Bigg(\sum_{s,m}\sum_{n>s}\alpha_{ns}^v\alpha_{mr}^s l_k^{nm}\psi(\epsilon n) - \sum_{n,m}\sum_{s>m}\alpha_{sm}^v\alpha_{nr}^s l_k^{nm}\psi(\epsilon n)\Bigg)x(v). \tag{10.76}$$

We have to keep in mind that neither sum makes sense alone when we put $\epsilon = 0$.

Before we continue with (10.76), we first show that $A_0^{(1)} + B_0^{(1)}$ vanishes. For this we change the variables in the summation for $B_\epsilon^{(1)}$ in the way that $s \to n \to m \to s$. By normal ordering $A_0^{(1)}$ and $B_0^{(1)}$ are well-defined operators in the sense that applied to a fixed $v \in V$ only a finite number of summands operate nontrivially. Hence, we ignore the ψ-factor.

Lemma 10.26. *Set*

$$F_{kr}^{sn} := \sum_m \alpha_{mr}^s l_k^{nm}, \tag{10.77}$$

then

$$F_{kr}^{sn} = \frac{1}{2\pi i}\int_{C_S} A_r(Q)\omega^s(Q)\omega^n(Q)e_k(Q) \tag{10.78}$$

and we have $F_{kr}^{sn} = F_{kr}^{ns}$. *This expression is symmetric in n and s.*

Proof. By definition

$$\begin{aligned}
F_{kr}^{sn} &= \sum_m \frac{1}{(2\pi i)^2}\iint_{C_S\,C_S}{}' A_m(Q)A_r(Q)\omega^s(Q)\omega^n(Q')\omega^m(Q')e_k(Q') \\
&= \frac{1}{(2\pi i)^2}\iint_{C_S\,C_S}{}' A_r(Q)\omega^s(Q)\omega^n(Q')e_k(Q')\Delta(Q,Q') \\
&= \frac{1}{2\pi i}\int_{C_S} A_r(Q)\omega^s(Q)\omega^n(Q)e_k(Q).
\end{aligned}$$

Hence, we obtain the claimed form which is obviously symmetric in n and s. □

Now $A_0^{(1)} + B_0^{(1)} = \sum_{n,s}\sum_i\left(:u_i(n)[u^i,x](s): + :[u_i,x](n)u^i(s):\right)F_{k,r}^{sn}$.

Lemma 10.27. $\sum_i\left(:u_i(n)[u^i,x](s): + :[u_i,x](n)u^i(s):\right) = 0$.

Proof. We calculate $\sum_i u_i(n)[u^i,x](s) = \sum_i u_i(n)\sum_j([u^i,x]|u_j)u^j(s) = -\sum_{i,j}u_i(n)(u^i|[u_j,x])u^j(s) = -\sum_j[u_j,x](n)u^j(s)$. □

Hence, $A_0^{(1)} + B_0^{(1)} = 0$.

Now we take up again expression (10.76) for $A_\epsilon^{(2)} + B_\epsilon^{(2)}$.

Lemma 10.28. *Given an $N \in \mathbb{Z}$, then the expression inside the v summation in* (10.76) *for $\lim_{\epsilon\to 0}$ is identical to the limit $\lim_{\epsilon\to 0}$ of*

$$E_\epsilon^{(N)} := \sum_{m,s}\sum_{n>N}\alpha_{ns}^v\alpha_{mr}^s l_k^{nm}\psi(\epsilon n) - \sum_{n,m}\sum_{s>N}\alpha_{sm}^v\alpha_{nr}^s l_k^{nm}\psi(\epsilon n). \tag{10.79}$$

Proof. After calculating the difference between (10.76) and (10.79) we obtain[1]

$$\sum_{m,s}\sum_{n=s+1}^{N}\alpha_{ns}^{v}\alpha_{mr}^{s}l_{k}^{nm}\psi(\epsilon n)-\sum_{n,m}\sum_{s=m+1}^{N}\alpha_{sm}^{v}\alpha_{nr}^{s}l_{k}^{nm}\psi(\epsilon n). \tag{10.80}$$

Note, that due to the almost-grading in every sum for fixed v, k, r, only a finite number of terms are involved. Hence, we can ignore $\psi(\epsilon n)$ and change the variables in the second sum as $(s \to n \to m \to s)$. By applying Lemma 10.26 the difference vanishes. □

This proof shows that the result does not depend on the chosen normal ordering. Again, the difference will consist of a finite number of terms which annihilate each other.

We examine

$$E_{\epsilon}^{(0)}=\sum_{s}\sum_{n>0}\alpha_{ns}^{v}F_{kr}^{sn}\psi(\epsilon n)-\sum_{n}\sum_{s>0}\alpha_{nr}^{s}F_{ks}^{vn}\psi(\epsilon n). \tag{10.81}$$

The second range of summation will be replaced as follows:
$(n, s>0)=(s, n>0)+(n>0, s\le 0)-(s>0, n\le 0)$.

We obtain

$$E_{\epsilon}^{(0)}=\sum_{n>0}\sum_{s}\left(\alpha_{ns}^{v}F_{kr}^{sn}-\alpha_{nr}^{s}F_{ks}^{vn}\right)\psi(\epsilon n)+\left(\sum_{\substack{n>0\\ s\le 0}}-\sum_{\substack{s>0\\ n\le 0}}\right)\alpha_{nr}^{s}F_{ks}^{vn}\psi(\epsilon n). \tag{10.82}$$

After summation over s in the first sum, and using the "delta distribution", we see that the sum vanishes. With the help of the integral representation of F_{ks}^{vn} given in Lemma 10.26 and $\alpha_{nr}^{s}=\frac{1}{2\pi\mathrm{i}}\int_{C_S}A_nA_r\omega^s$, the second sum can be rewritten as

$$\frac{1}{(2\pi\mathrm{i})^2}\iint_{C_S\,C_S}{}^{\prime} A_r(Q')e_k(Q)\omega^v(Q)\times\left(\sum_{\substack{n>0\\ s\le 0}}-\sum_{\substack{s>0\\ n\le 0}}\right)A_n(Q')A_s(Q)\omega^s(Q')\omega^n(Q)\psi(\epsilon n). \tag{10.83}$$

Next we use the following proposition.

Proposition 10.29. *For every $N\in\mathbb{Z}$ we have*

$$\left(\sum_{n>N}\sum_{s\le N}-\sum_{s>N}\sum_{n\le N}\right)A_n(Q')A_s(Q)\omega^s(Q')\omega^n(Q)=d'\Delta(Q',Q). \tag{10.84}$$

Here d' denotes differentiation with respect to the variable Q'.

1 If the upper bound is smaller than the lower, then for the sum it should be understood that we change the summation order and the sign of the expression. This rule should be used for all following summations.

Before supplying the proof we show Proposition 10.24. If we insert (10.84) into (10.83) we obtain

$$\begin{aligned} E_0^{(0)} &= \frac{1}{(2\pi \mathrm{i})^2} \iint_{C_S\, C_S}{}' A_r(Q')e_k(Q)\omega^v(Q)d'\Delta(Q',Q) \\ &= -\frac{1}{(2\pi \mathrm{i})^2} \iint_{C_S\, C_S}{}' d'A_r(Q')e_k(Q)\omega^v(Q)\Delta(Q',Q) \\ &= -\frac{1}{2\pi \mathrm{i}} \int_{C_S} dA_r(Q)e_k(Q)\omega^v(Q) = -K_{k,r}^v. \end{aligned} \tag{10.85}$$

If we collect all non-zero terms we obtain

$$\begin{aligned} 2\,[L_k, x(n)] &= \lim_{\epsilon\to 0}(A_\epsilon^{(2)} + B_\epsilon^{(2)}) - \lim_{\epsilon\to 0}(C_\epsilon + D_\epsilon) \\ &= -2\kappa \sum_v K_{kr}^v - -2c\kappa \sum_n K_{kr}^v. \end{aligned} \tag{10.86}$$

This gives the claim. □

Let $\alpha_{r,n}^t(\lambda)$ be the structure constants of the $\mathcal{A}$-module $\mathcal{F}^\lambda$, i.e.,

$$A_r \cdot f_n^\lambda = \sum_{t=r+n}^{r+n+S} \alpha_{r,n}^t(\lambda) f_t^\lambda. \tag{10.87}$$

To prove Proposition 10.29 we first show the following lemma.

Lemma 10.30. *Let $N \in \mathbb{Z}$, $\lambda \in 1/2\,\mathbb{Z}$, then*

$$\gamma_{rk} = \left(\sum_{n>N} \sum_{s\le N} - \sum_{s>N} \sum_{n\le N} \right) \alpha_{rn}^s(\lambda)\alpha_{ks}^n(\lambda). \tag{10.88}$$

Proof. In the two-point case it was a very fruitful idea of Bonora and colleagues [24] to use representations via semi-infinite wedge forms to prove this for $\lambda = 0$. Indeed, the technique works in the case $K = 1$ word for word and can also easily be extended to the multi-point case. The case of arbitrary λ also goes along the same lines. Semi-infinite wedge forms were introduced and discussed in Chapter 7. We consider forms of weight λ and the subspace generated from the vacuum vector $\Phi_{N+1} = f_{N+1}^\lambda \wedge f_{N+2}^\lambda \wedge \ldots$ by the action of the differential operator algebra $\widehat{\mathcal{D}^1}$. The element $A_i \in \mathcal{A}$ operates according to the Leibniz rule

$$A_i.\Phi_{N+1} = (A_i \cdot f_{N+1}^\lambda) \wedge f_{N+2}^\lambda \wedge \ldots + f_{N+1}^\lambda \wedge (A_i \cdot f_{N+2}^\lambda) \wedge f_{N+3}^\lambda \cdots + \ldots. \tag{10.89}$$

As shown in Section 7.2, this definition makes sense for $|i|$ large. For a critical strip of indices (in particular also for $A_0 = 1$), the action has to be modified[2]. We obtain a representation of a centrally extended algebra $\widehat{\mathcal{A}}'$ with a defining cocycle which is local.

2 In physicist's language: it has to be regularized.

Crucial is that the action of $\mathcal{A}$ comes from an action of $\mathcal{D}^1$. Hence, an action of a centrally extended $\widehat{\mathcal{D}^1}$ is given and the action $\widehat{\mathcal{A}}'$ is obtained by restriction. Consequently, the cocycle for $\mathcal{A}$ is an $\mathcal{L}$-invariant cocycle and by our classification results formulated in Theorem 6.38 it is a scalar multiple of $\gamma_{C_S}^{\mathcal{A}}$; see (6.134).

We obtain $[\widehat{A}_r, \widehat{A}_k].\Phi_{N+1} = \beta\gamma(A_r, A_k)\Phi_{N+1} = \beta\gamma_{rk}\Phi_{N+1}$. For r and k outside the critical strip for the indices, the actions of $\widehat{A}_r$ and $\widehat{A}_k$ coincide with the actions of the corresponding elements A_r and A_k given by (10.89). Outside the critical strip $[\widehat{A}_r, \widehat{A}_k].\Phi$ can be calculated as follows. We have only to take into account the possibilities that the element f_s^λ inside Φ_{N+1} reproduces itself. We have $A_k \cdot f_s^\lambda = \sum_n \alpha_{ks}^n(\lambda) f_n^\lambda$. This term only shows up if $s \geq N+1$. Only the terms with $n < N+1$ will not be annihilated by neighboring elements. To bring back the result by the operation of A_r to f_s^λ we obtain $\sum_{s>N}\sum_{n\leq N} \alpha_{rn}^s(\lambda)\alpha_{ks}^n(\lambda)$. If we apply the same for $-A_k \cdot A_r$ and change the variables we get

$$[\widehat{A}_r, \widehat{A}_k].\Phi_{N+1} = -\left(\sum_{n>N}\sum_{s\leq N} - \sum_{s>N}\sum_{n\leq N}\right)\alpha_{rn}^s(\lambda)\alpha_{ks}^n(\lambda)\Phi_{N+1} = \beta\gamma_{rk}\Phi_{N+1}.$$

To determine the constant β we calculate the expression for $r = i$ and $k = -i$ if $i \gg 0$. We have $A_i\Phi_{N+1} = 0$ and $[A_i, A_{-i}]\Phi_{N+1} = A_i(A_{-i}\Phi_{N+1})$. Furthermore, $A_k \cdot f_s^\lambda = f_{k+s}^\lambda + f_j^\lambda$-terms with j-indices larger than $k+s$. We collect for every s the factor 1 as long as $s - i \leq N$. Now $s \geq N+1$, and hence $N+1 \leq s \leq N+i$, yields exactly i terms. We obtain $[A_i, A_{-i}]\Phi_{N+1} = i \cdot \Phi_{N+1} = \beta \cdot \gamma(A_i, A_{-i})\Phi_{N+1}$. By calculating the residues we obtain $\gamma(A_i, A_{-i}) = -i$ and hence the claim. □

Now we prove Proposition 10.29. We have

$$dA_k(Q) = \sum_r \beta_{kr}\omega^r(Q), \quad \beta_{kr} = \frac{1}{2\pi \mathrm{i}}\int_{C_S} dA_k(Q)A_r(Q) = \gamma_{rk}, \tag{10.90}$$

and hence with Lemma 10.30

$$\begin{aligned} d'\Delta(Q', Q) &= \sum_k d'A_k(Q')\omega^k(Q) = \sum_{k,r}\gamma_{r,k}\omega^r(Q')\omega^k(Q) \\ &= \sum_{k,r}\left(\sum_{\substack{n>N\\ s\leq N}} - \sum_{\substack{s>N\\ n\leq N}}\right)\alpha_{rn}^s\alpha_{ks}^n\omega^k(Q)\omega^r(Q'). \end{aligned} \tag{10.91}$$

Here we set $\alpha_{ks}^n = \alpha_{ks}^n(0)$. We use $A_i\omega^j = \sum_r \alpha_{ir}^j\omega^r$ and obtain

$$= \left(\sum_{\substack{n>N\\ s\leq N}} - \sum_{\substack{s>N\\ n\leq N}}\right)A_n(Q')\omega^s(Q')A_s(Q)\omega^n(Q). \tag{10.92}$$

□

10.4.2 Proof of Proposition 10.10

Again we write for $\epsilon \neq 0$

$$[L_k(\epsilon), L_l] = \frac{1}{2} \sum_{n,m} \sum_i l_k^{nm} [:u_i(n)u^i(m):, L_l]\psi(\epsilon n). \tag{10.93}$$

As explained above, we can ignore the normal ordering under the commutators and rewrite (10.93) as

$$\frac{1}{2} \sum_{n,m} \sum_i l_k^{nm} \left(u_i(n)[u^i(m), L_l] + [u_i(n), L_l]u^i(m) \right) \psi(\epsilon n). \tag{10.94}$$

We use Proposition 10.24 to calculate the commutators and obtain

$$\frac{1}{2}(c+\kappa) \sum_{n,m,v} \sum_i l_k^{nm} \left(K_{l,m}^v u_i(n)u^i(v) + K_{l,n}^v u_i(v)u^i(m) \right) \psi(\epsilon n). \tag{10.95}$$

To separate this into two sums which will be well-defined for $\epsilon = 0$, we have to rewrite it in normal ordering. In this way, we collect commutators for all indices which are not yet in normal ordering. We evaluate these commutators with the help of Lemma 10.3 and rewrite the result as the sum $A_\epsilon + B_\epsilon + C_\epsilon + D_\epsilon$, where we define

$$A_\epsilon := \frac{1}{2}(c+\kappa) \sum_{n,m,v} \sum_i l_k^{nm} K_{l,m}^v :u_i(n)u^i(v): \psi(\epsilon n),$$

$$B_\epsilon := \frac{1}{2}(c+\kappa) \sum_{n,m,v} \sum_i l_k^{nm} K_{l,n}^v :u_i(v)u^i(m): \psi(\epsilon n),$$

$$C_\epsilon := -\frac{1}{2}c(c+\kappa) \dim \mathfrak{g} \sum_{v,m} \sum_{n>v} l_k^{nm} K_{l,m}^v \gamma_{nv} \psi(\epsilon n),$$

$$D_\epsilon := -\frac{1}{2}c(c+\kappa) \dim \mathfrak{g} \sum_{n,m} \sum_{v>m} l_k^{nm} K_{l,n}^v \gamma_{vm} \psi(\epsilon n).$$

Taking into account the range in which the coefficients l_k^{nm} and $K_{l,m}^v$ are different from zero and the normal ordering, we see that both A_0 and B_0 are well-defined operators. Hence, we ignore the factor $\psi(\epsilon n)$. If we change the variables in B_0 in the way ($v \to n \to m \to v$), we obtain

$$A_0 + B_0 = \frac{1}{2}(c+\kappa) \sum_{n,m,v} \sum_i (l_k^{nm} K_{lm}^v + l_k^{mv} K_{lm}^n) :u_i(n)u^i(v):. \tag{10.96}$$

The structure constants of the vector field algebra $\mathcal{L}$ calculate via the duality as

$$C_{kl}^s = \frac{1}{2\pi \mathrm{i}} \int_{C_S} ([e_k, e_l]) \cdot \Omega^s. \tag{10.97}$$

Lemma 10.31.

$$\sum_m (l_k^{nm} K_{lm}^v + l_k^{mv} K_{lm}^n) = -\frac{1}{2\pi \mathrm{i}} \int_{C_S} [e_k, e_l] \cdot \omega^n \omega^v = -\sum_s C_{kl}^s l_s^{nv}. \tag{10.98}$$

Proof. The right-hand side writes as

$$-\sum_s \frac{1}{(2\pi \mathrm{i})^2} \iint\limits_{C_S\, C_S}{}' [e_k, e_l](Q) \cdot \Omega^s(Q) \omega^n(Q') \omega^v(Q') e_s(Q').$$

After summation over s we obtain the "delta distribution" for the weight pair $(-1, 2)$. We integrate over Q' and get

$$-\frac{1}{2\pi \mathrm{i}} \int\limits_{C_S} [e_k, e_l](Q) \omega^n(Q) \omega^v(Q),$$

which is the expression in the middle. On the left-hand side for the first sum we get

$$\sum_m l_k^{nm} K_{lm}^v = \sum_m \frac{1}{(2\pi \mathrm{i})^2} \iint\limits_{C_S\, C_S}{}' \omega^n(Q) \omega^m(Q) e_k(Q) d' A_m(Q') e_l(Q') \omega^v(Q').$$

After applying $\sum_m d' A_m(Q') \omega^m(Q) = d' \Delta(Q', Q)$, we obtain by Q' integration $-\frac{1}{2\pi \mathrm{i}} \int_{C_S} \omega^n(Q) e_k(Q) d(e_l(Q) \omega^v(Q))$. For the second sum we get

$$-\frac{1}{2\pi \mathrm{i}} \int\limits_{C_S} \omega^v(Q) e_k(Q) d(e_l(Q) \omega^n(Q)) = \frac{1}{2\pi \mathrm{i}} \int\limits_{C_S} d(\omega^v(Q) e_k(Q)) e_l(Q) \omega^n(Q).$$

Together,

$$-\frac{1}{2\pi \mathrm{i}} \int\limits_{C_S} (\omega^n(Q) e_k(Q) d(e_l(Q) \omega^v(Q)) - \omega^n(Q) e_l(Q) d(e_k(Q) \omega^v(Q))).$$

If we replace every form by its local representing function, we obtain for the integrand

$$\omega^n(z) \omega^v(z) \left(e_k(z) \frac{de_l}{dz}(z) - e_l(z) \frac{de_k}{dz}(z) \right) = \omega^n(z) \omega^v(z) [e_k, e_l](z).$$

From which the claim follows. □

Hence,

$$\begin{aligned} A_0 + B_0 &= -\frac{1}{2}(c + \kappa) \sum_s \sum_{n,v} \sum_i C_{kl}^s l_s^{nv} :u_i(n) u^i(v): \\ &= -(c + \kappa) \sum_s C_{kl}^s L_s. \end{aligned}$$

It remains to examine $\alpha(k, l) := \lim_{\epsilon \to 0}(C_\epsilon + D_\epsilon)$. As L_k and L_l are well-defined elements of $gl(V)$, the scalar $\alpha(k, l)$ is well-defined too.

If $(c + \kappa) = 0$, then everything will be zero independent of ϵ and the general claim is true. If $(c+\kappa) \neq 0$, then we rescale everything by $\frac{-1}{c+\kappa}$. If we consider now $L_k^* = \frac{-1}{c+\kappa} L_k$, we see that they are well-defined operators inside $gl(V)$. Hence, they fulfill the Jacobi identity. Furthermore, the C_{kl}^s as structure constants of $\mathcal{L}$ also have to fulfill the Jacobi identity. Hence the $\alpha(k, l)$ has to be a Lie algebra two cocycle for $\mathcal{L}$. If we check the orders of the forms contributing to defining l_k^{nm}, $K_{l,m}^{v}$, and γ_{nv} at the points in I and O, and check when the result could have a residue there (we need poles both at at least one point in I and one point in O), we see that constants M_1, M_2, M_3 exist, such that[3]

$$\begin{aligned} K_{l,m}^{v} &\neq 0 \implies -M_1 \leq -v + l + m \leq 0, \\ l_k^{nm} &\neq 0 \implies -M_2 \leq k - (n+m) \leq 0, \\ \gamma_{nv} &\neq 0 \implies -M_3 \leq n + v \leq 0. \end{aligned} \tag{10.99}$$

To have a non-zero contribution, all 3 conditions have to be fulfilled. Hence, with $M_4 := M_1 + M_2 + M_3$ we get

$$\alpha(k,l) \neq 0 \implies -M_4 \leq k + l \leq 0. \tag{10.100}$$

In particular the cocycle is local. Now we use our classification result formulated in Theorem 6.41 and conclude that $\alpha(k,l)$ is a scalar multiple of the standard geometric cocycle $\gamma_{C_S,R}^{\mathcal{L}}$ for the vector field algebra with a suitable projective connection R. In particular, the fact that we get a representation of a centrally extended vector field algebra is now proven.

Now we deal with the question of determining the scalar. The cocycle $\alpha(k,l)$ is given by

$$-\frac{1}{2}c(c+\kappa)\dim\mathfrak{g}\sum_n\left(\sum_{m,s}\sum_{v<n} l_k^{nm} l_l^{vs}\gamma_{sm}\gamma_{nv} + \sum_{m,s}\sum_{v>m} l_k^{nm} l_l^{vs}\gamma_{sn}\gamma_{vm}\right). \tag{10.101}$$

We have already rescaled the situation and hence have to divide it by $\frac{1}{(c+\kappa)^2}$ to compare it with the cocycle for the vector field algebra. We will even ignore for the moment the remaining factor $\frac{-c\dim\mathfrak{g}}{c+\kappa}$ in front of the sum and will denote the obtained cocycle by $\tilde{\alpha}$. Recall that it is a local cocycle and, by (10.100), it is bounded by zero from above. The scalar can be determined by Theorem 6.41. In particular, it is enough to calculate $\tilde{\alpha}(k,-k)$. In fact, the knowledge of $\tilde{\alpha}(1,-1)$ and $\tilde{\alpha}(2,-2)$ will be sufficient. Because of the fact that it is bounded by zero, it follows that the projective connection will have at most poles of order two at the points in I. We will show that

$$\tilde{\alpha}(1,-1) = 0 \quad \text{and} \quad \tilde{\alpha}(2,-2) = -1. \tag{10.102}$$

Using Theorem 6.41, the scalar multiple with respect to the the standard cocycle is

$$\frac{c\dim\mathfrak{g}}{c+\kappa}. \tag{10.103}$$

Moreover, from $\alpha(1,-1) = 0$ it follows that the pole order of the projective connection at the points in I could be at most one.

The expression as sum of two sums for the cocycle is well-defined. But we cannot split it, as the individual sums over n have an infinite number of terms. We use again

3 For the N=2 point situation and generic n, m, v values, we have $M_1 = 3g$, $M_2 = g$ and $M_3 = 2g$.

the technique by which, after multiplying the expression by $\psi(\epsilon n)$, we are allowed to split it and finally consider the limit $\epsilon \to 0$. Let us denote the expression of the cocycle $\tilde{a}(k,-k)$ by E, the modified by E_ϵ, and the individual sums by $E_{1,\epsilon}$ and $E_{2,\epsilon}$, $E_\epsilon = E_{1,\epsilon} + E_{2,\epsilon}$.

The terms under the first sum of (10.101) are

$$\sum_{m,s}\sum_{v<n} l_k^{nm} l_{-k}^{v,s} \gamma_{sm}\gamma_{nv}. \tag{10.104}$$

These coefficients are given via integrals over a separating cocycle, respectively equivalently by calculating residues at I. If we check the orders of the involved basis elements at the points in I, we see that it is only possible to obtain a non-zero residue if the following system of inequalities is fulfilled:

$$k-(n+m)\le 0, \quad -k-(v+s)\le 0, \quad s+m\le 0, \quad n+v\le 0. \tag{10.105}$$

For fixed k and n the system has only one solution, which is given by

$$m=k-n, \quad s=-(k-n), \quad v=-n. \tag{10.106}$$

In fact, for the unique solution all inequalities above are equalities. To indicate the simple types of arguments let us calculate m. From the first equation we obtain $m \ge k-n$. Adding up all three other inequalities, we obtain $m \le k-n$, hence $m = k-n$. For the coefficients at these values we calculate $l_k^{n,k-n} = l_{-k}^{-n,-(k-n)} = 1$. The condition $v < n$ for the summation and $v = -n$ implies a summation range $n > 0$. Hence

$$E_{1,\epsilon} = \sum_{n>0} \gamma_{-(k-n),(k-n)} \cdot \gamma_{n,-n}\psi(\epsilon n). \tag{10.107}$$

In a similar way, for the nonvanishing of

$$\sum_{m,s}\sum_{v>m} l_k^{nm} l_{-k}^{v,s} \gamma_{sn}\gamma_{vm}, \tag{10.108}$$

we obtain the unique solution

$$m=k-n, \quad s=-n, \quad v=-(k-n). \tag{10.109}$$

Now the summation condition $v > m$ yields for the summation over n the condition $n > k$. Hence

$$E_{2,\epsilon} = \sum_{n>k} \gamma_{-n,n} \cdot \gamma_{-(k-n),k-n}\psi(\epsilon n). \tag{10.110}$$

As $\gamma_{-n,n} = -\gamma_{n,-n}$, there will be cancellations of terms. Without restrictions we assume $k > 0$ and get

$$E_\epsilon = E_{1,\epsilon} + E_{2,\epsilon} = \sum_{n=1}^{k} \gamma_{-(k-n),k-n} \cdot \gamma_{n,-n}\psi(\epsilon n). \tag{10.111}$$

This sum is a finite sum. Hence, if ϵ is small enough it will be constant and we get

$$\tilde{\alpha}(k,-k) = E_2 = \sum_{n=1}^{k} \gamma_{-(k-n),k-n} \cdot \gamma_{n,-n}. \tag{10.112}$$

We obtain the value $\gamma_{-l,l} = \frac{1}{2\pi i}\int_{C_S} A_{-l}d(A_l) = l$. Hence,

$$\begin{aligned} \tilde{\alpha}(1,-1) &= \gamma_{0,0}\gamma_{1,-1} = 0 \\ \tilde{\alpha}(2,-2) &= \gamma_{-1,1}\gamma_{1,-1} + \gamma_{0,0}\gamma_{2,-2}. \end{aligned} \tag{10.113}$$

From (6.141) we conclude[4]

$$\tilde{\alpha}(k,-k) = -\frac{1}{6}(k^3 - k). \tag{10.114}$$

Hence, the claim and Proposition 10.10 are demonstrated. □

Proof of Proposition 10.16. In the proof of Proposition 10.10 we only used the relation $[L_l, x(r)] = -(c+\kappa).x \otimes (e_l \,.\, A_r)$ of Proposition 10.8. Hence, we obtain for an operator E fulfilling the same relation $[L_k(\epsilon), E] = [L_k(\epsilon), L_l]$, and hence in the limit $\epsilon \to 0$: $[L_k, E] = [L_k, L_l]$. □

Remark 10.32. In [225], expressions for the cocycle in terms of structure constants of the algebra and the modules are also given. We will not need them and we only note them here for curiosity. Let

$$E_{kl}^{nv} = \frac{1}{2\pi i}\int_{C_S} [e_k, e_l] \cdot \omega^n \omega^v, \tag{10.115}$$

then

$$\begin{aligned} \chi_{kl} &= \psi_{kl} + \widehat{\chi}_{kl}, \quad \psi_{kl} = \sum_{s,v}\sum_{n=0}^{v+1} C_{kl}^{s} l_s^{nv} \gamma_{nv} = \sum_{v}\sum_{n=0}^{v+1} E_{kl}^{nv}\gamma_{nv}, \\ \widehat{\chi}_{kl} &= \left(\sum_{\substack{n>0\\ v\le 0}} - \sum_{\substack{n\le 0\\ v>0}}\right) K_{k,v}^{n} K_{l,n}^{v}. \end{aligned} \tag{10.116}$$

Using this expression, we see directly how the coboundary enters.

The expressions ψ_{kl} in (10.116) define a coboundary (in the sense of Lie algebra cohomology). The appearing γ_{nv} are due to the normal ordering of the operators, and in fact only the terms ψ_{kl} involve γ_{nv}.

We recall that a two-cocycle ψ is a coboundary for the Lie algebra $\mathcal{L}$ if there is a linear form $\phi : \mathcal{L} \to \mathbb{C}$ such that $\psi(e,f) = \phi([e,f])$. If we define ϕ by

$$\phi(e_s) := \sum_{v}\sum_{n=0}^{v+1} l_s^{nv}\gamma_{nv} \tag{10.117}$$

4 Of course we could also easily deduce this directly from the expression (10.112).

(this is a finite sum), then

$$\phi([e_k, e_l]) = \phi\Big(\sum_s C_{k,l}^s e_s\Big) = \psi_{kl}. \tag{10.118}$$

A different normal ordering would yield a different range for the summation in the definition of ϕ. Of course, we already know from the classification result that another normal ordering will produce an equivalent cocycle, as the local cocycle class is unique, up to rescaling.

10.4.3 The case $K > 1$

The proof presented in the previous section generalizes easily to $K > 1$. Where we had a summation over $n \in \mathbb{Z}$, we have to add another summation over the finite range $p = 1, \ldots, K$. Beside a notational blow-up nothing disturbs the chain of arguments. In the proofs presented above where we used the vanishing of certain coefficients such as l_k^{nm}, we already argued with the general situation. See for example (10.99). Also, the uniqueness results for local cocycles are completely the same in the $K > 1$ case. Correspondingly, Lemma 10.30 can be extended directly.

Lemma 10.33. *Let $N \in \mathbb{Z}$, $\lambda \in 1/2\,\mathbb{Z}$, then*

$$\gamma_{(r,u)(k,v)} = \Big(\sum_{n>N}\sum_{s\le N} - \sum_{s>N}\sum_{n\le N}\Big)\sum_{w,t} \alpha_{(r,u)(n,w)}^{(s,t)}(\lambda)\alpha_{(k,v)(s,t)}^{(n,w)}(\lambda). \tag{10.119}$$

Its proof via semi-infinite forms works in completely the same manner. To fix the scalar, we calculate $[A_{i,p}, A_{-i,p}].\Phi_{N+1}$. We have to pay attention to the fact that by the almost-graded structure $A_{k,s} f_{n,r}^{\lambda} = f_{k+n,r}^{\lambda}\delta_s^r + f\cdot$, terms of higher degree than $k+n$ appear. Furthermore, $\gamma(A_{i,p}, A_{-i,p}) = (-i)$, as there is only one residue and this is at the point P_p.

With this result we obtain for example the generalized result.

Proposition 10.34. *For every N we have*

$$\Big(\sum_{\substack{n>N\\ m\le N}} - \sum_{\substack{m>N\\ n\le N}}\Big)\sum_{p,s} A_{n,p}(Q')A_{m,s}(Q)\omega^{m,s}(Q')\omega^{n,p}(Q) = d'\Delta(Q',Q). \tag{10.120}$$

Here d' denotes differentiation with respect to the variable Q'.

11 Wess–Zumino–Novikov–Witten models and Knizhnik–Zamolodchikov connection

Wess–Zumino–Novikov–Witten models (WZNW) are important examples of models of two-dimensional conformal field theories and their quantized versions. They can roughly be described as follows. The gauge algebra of the theory is the affine algebra associated with a finite-dimensional gauge algebra (i.e., a simple finite-dimensional Lie algebra). The geometric data consists of a compact Riemann surface (with complex structure) of genus g and a finite number of marked points on this surface. Starting from representations of the gauge algebra, the space of conformal blocks can be defined. It depends on the geometric data. Varying the geometric data yields a bundle over the moduli space of the geometric data.

In [131] Knizhnik and Zamolodchikov considered the case of genus 0 (i.e., the Riemann sphere). There, changing the geometric data consists of moving the marked points on the sphere. The space of conformal blocks could be found completely inside the part of the representation associated with the finite-dimensional gauge algebra. On this space an important set of equations, the Knizhnik–Zamolodchikov (KZ) equations, were introduced. In a geometric setting, solutions are the flat sections of the bundle of conformal blocks over the moduli space with respect to the Knizhnik–Zamolodchikov connection.

For higher genus it is not possible to realize the space of conformal blocks inside the representation space associated with the finite-dimensional algebra. Different attacks on the generalization have been made. Some of them add additional structure to these representation spaces (e.g., twists, representations of the fundamental group...). A very incomplete list of references includes Bernard [13, 14], Felder, Wieczerkowski, Enriquez [58, 66, 67], Hitchin [105], and Ivanov [115].

An important approach, very much in the spirit of the original Knizhnik–Zamolodchikov approach, was made by Tsuchiya, Ueno, and Yamada [254]. The main point of their approach is that at the marked points, after choosing local coordinates, local constructions are made. In this setting, the well-developed theory of representations of the traditional affine Lie algebras (Kac-Moody affine type algebras) can be used. It appears a mixture between local and global objects, and considerable effort is necessary to extend the local constructions to global ones.

Oleg Sheinman and the author presented in [226] and [227] a different approach to the WZNW models which uses global objects. These objects are the Krichever–Novikov algebras and their representations which we study in this book. A crucial point is that a certain subspace of the Krichever–Novikov algebra of vector fields is identified with tangent directions on the moduli space of the geometric data. Conformal blocks can be defined. Finally, with the help of the global Sugawara construction given in Chapter 10, it is possible to define the higher genus multi-point Knizhnik–Zamolodchikov connection (see (11.82)). This connection is well-defined on the vector bundle of conformal blocks at least for those representations of the higher genus multi-point affine algebras which fulfill certain conditions. An important class of such representations consists of the Verma module and fermionic representations discussed in Chapter 9. The connection is projectively flat.

This global operator approach to WZNW models will be presented in this chapter. Complete proofs appeared in [226, 227]. See also [250] for an improved presentation and additional details.

I would like to point out that until now, the global construction works only over an open dense subset of the moduli space. This is in contrast to the approach of Tsuchiya, Ueno, and Yamada [254], which provides a valid theory for the compactified moduli space, and hence includes stable singular curves. In this way they were able to prove the Verlinde formula. For a pedagogical introduction to their approach see [257].

11.1 Moduli space of curves with marked points

In this chapter we will switch to the point of view that we consider compact Riemann surfaces Σ as smooth projective curves C over $\mathbb{C}$. For simplicity we call them just curves, but will also continue to use the symbol Σ for them. Without further explanation we will use the language of algebraic geometry here. Using the fact that all global holomorphic objects will be algebraic for compact Riemann surfaces, we will allow ourselves, if necessary, to switch between the two languages.

To compare it to other approaches we will consider our set A, where poles are allowed to be separated into

$$I := \{P_1, P_2, \dots, P_N\}, \quad O := \{P_\infty\}. \tag{11.1}$$

In particular, our usual K will be N, and our usual N will be $N + 1$. This change of notation also reflects the fact that the point P_∞ will play a role as a reference point.

Ignoring for the moment the necessary reference point P_∞, our principal moduli space object is the moduli space $\mathcal{M}_{g,N}$ of genus g curves with N marked points. A point $b \in \mathcal{M}_{g,N}$ is given as

$$b = [\Sigma, P_1, \dots, P_N], \tag{11.2}$$

where $[\dots]$ denotes equivalence with respect to algebraic isomorphisms $\phi : \Sigma \to \Sigma'$, with $\phi(P_i) = P_i'$. We will work over a generic open subset W of $\mathcal{M}_{g,N}$, and hence there will be a universal family of curves $\mathcal{U} \to W$. See (11.9) below for the discussion in the extended context.

For the approach presented here we will need an additional reference point, hence we should consider $\mathcal{M}_{g,N+1}$. Furthermore, we will need (at least) first order jets of coordinates at the points in I, and for technical intermediate reasons also some jets of coordinates at the point P_∞. Recall that a k-jet of coordinates at P_i is an equivalence class of coordinates where two coordinates z_i and z_i' are identified if they coincide up to order k in their power series expansion. This says $z_i'(z_i) = z_i + O(z_i^{k+1})$. Note that the zero order term of a coordinate is always fixed by the point itself, and consequently a 0-jet means that we can ignore the coordinates at the point P_i.

To allow this, we start rather generally and denote by $\mathcal{M}_{g,N+1}^{(k,p)}$ the moduli space of smooth projective curves of genus g (over $\mathbb{C}$) with $N+1$ ordered distinct marked points, fixed k-jets of local coordinates at the first N points, and a fixed p-jet of a local coordinate at the last point. The elements of $\mathcal{M}_{g,N+1}^{(k,p)}$ are given as

$$\tilde{b}^{(k,p)} = [\Sigma, P_1, \ldots, P_N, P_\infty, z_1^{(k)} \ldots, z_N^{(k)}, z_\infty^{(p)}], \tag{11.3}$$

where Σ is a smooth projective curve of genus g, P_i, $i = 1, \ldots, N, \infty$ are distinct points on Σ, z_i is a coordinate at P_i with $z_i(P_i) = 0$, and $z_i^{(l)}$ is a l-jet of z_i ($l \in \mathbb{N}_0$). Here [..] denotes an equivalence class of such tuples in the following sense. Two tuples representing $\tilde{b}^{(k,p)}$ and $\tilde{b}^{(k,p)\prime}$ are equivalent if an algebraic isomorphism $\phi : \Sigma \to \Sigma'$ with $\phi(P_i) = P_i'$ for $i = 1, \ldots, N, \infty$ exists, such that after identification via ϕ we have

$$z_i' = z_i + O(z_i^{k+1}), \quad i = 1, \ldots, N \quad \text{and} \quad z_\infty' = z_\infty + O(z_i^{p+1}). \tag{11.4}$$

For the following two special cases we introduce the notation

$$\mathcal{M}_{g,N+1} := \mathcal{M}_{g,N+1}^{(0,0)} \quad \text{and} \quad \mathcal{M}_{g,N+1}^{(1)} := \mathcal{M}_{g,N+1}^{(1,0)}. \tag{11.5}$$

Also we will use $\tilde{b}^{(1)} := \tilde{b}^{(1,0)}$. In the first case there will be no coordinates at all, in the second case only the first jets of coordinates appear at the points in I. These elements will be given by

$$\begin{aligned} \tilde{b} &= [\Sigma, P_1, \ldots, P_N, P_\infty] \quad \in \quad \mathcal{M}_{g,N+1} \\ \tilde{b}^{(1)} &= [\Sigma, P_1, \ldots, P_N, P_\infty, z_1^{(1)} \ldots, z_N^{(1)}] \quad \in \quad \mathcal{M}_{g,N+1}^{(1)}. \end{aligned} \tag{11.6}$$

By forgetting either coordinates or higher order jets we obtain natural projections

$$\nu^{k,p} : \mathcal{M}_{g,N+1}^{(k,p)} \to \mathcal{M}_{g,N+1}, \tag{11.7}$$

or more generally

$$\mathcal{M}_{g,N+1}^{(1,p)} \to \mathcal{M}_{g,N+1}, \quad \mathcal{M}_{g,N+1}^{(k,p)} \to \mathcal{M}_{g,N+1}^{(k',p')} \tag{11.8}$$

for any $k' \le k$ and $p' \le p$.

Here we deal with the situation in the neighborhood of a moduli point corresponding to a generic curve Σ with a generic marking $(P_1, P_2, \ldots, P_N, P_\infty)$. Let $\widetilde{W} \subseteq \mathcal{M}_{g,N+1}$ be an open subset around such a generic point $\tilde{b} = [\Sigma, P_1, P_2, \ldots, P_N, P_\infty]$. A generic curve of $g \ge 2$ with a generic marking doesn't admit nontrivial automorphisms, and we may assume that a universal family of curves with marked points over $\widetilde{W}$ exists. In particular, this means that there is a proper, flat family of smooth curves over $\widetilde{W}$

$$\pi : \mathcal{U} \to \widetilde{W}, \tag{11.9}$$

such that for the points $\tilde{b} = [\Sigma, P_1, P_2, \dots, P_N, P_\infty] \in \widetilde{W}$ we have $\pi^{-1}(\tilde{b}) = \Sigma$, and that the sections defined as

$$\sigma_i : \widetilde{W} \to \mathcal{U}, \quad \sigma_i(\tilde{b}) = P_i, \quad i = 1, \dots, N, \infty \tag{11.10}$$

will not meet and are algebraic.

The subset $\widetilde{W}$ can be pullbacked via $\nu^{k,p}$ to $\mathcal{M}^{(k,p)}_{g,N+1}$, and we obtain $\widetilde{W}^{(k,p)}$ as open subset. Now the sections (11.10) can additionally be complemented by choosing infinitesimal neighborhoods of the order under consideration. As cases of special importance we obtain the subsets

$$\widetilde{W}^{(1)} = \widetilde{W}^{(1,0)} \subseteq \mathcal{M}^{(1)}_{g,N+1}, \quad \widetilde{W}^{(1,1)} \subseteq \mathcal{M}^{(1,1)}_{g,N+1}. \tag{11.11}$$

Remark 11.1. There is another relation to be taken into account. If we "forget" the last point P_∞, the reference point, we obtain maps

$$\mathcal{M}_{g,N+1} \to \mathcal{M}_{g,N}, \quad \mathcal{M}^{(0,p)}_{g,N+1} \to \mathcal{M}_{g,N}, \quad \mathcal{M}^{(k,p)}_{g,N+1} \to \mathcal{M}^{(k)}_{g,N}. \tag{11.12}$$

Let us fix an algebraic section $\hat{\sigma}_\infty$ of the universal family of curves (without marking). In particular, for every curve a point is chosen in a manner depending algebraically on the moduli. (Recall, we are only dealing with the local and generic situation.) The subset

$$W' := \{\tilde{b} = [\Sigma, P_1, P_2, \dots, P_N, P_\infty] \mid P_\infty = \hat{\sigma}_\infty([\Sigma])\} \quad \subseteq \quad \widetilde{W} \tag{11.13}$$

can be identified with the open subset W of $\mathcal{M}_{g,N}$ via

$$\tilde{b} = [(\Sigma, P_1, P_2, \dots, P_N, \hat{\sigma}_\infty([\Sigma]))] \to b = [(\Sigma, P_1, P_2, \dots, P_N)]. \tag{11.14}$$

By genericity, the map is one-to-one.

By choosing not only a section $\hat{\sigma}_\infty$, but also a p-th order infinitesimal neighborhood of this section, we even get an identification of the open subset W of $\mathcal{M}_{g,N}$ with an analytic subset $W'^{,(p)}$ of $\widetilde{W}^{(0,p)}$ of $\mathcal{M}^{(0,p)}_{g,N+1}$. It is defined in a way similar to W'.

All constructions below for the $N + 1$ point situation will yield results for the N point situation. However, it is important to keep in mind that the results will generally depend on the chosen section $\hat{\sigma}_\infty$ yielding the reference points P_∞.

Remark 11.2. All of these considerations can be extended to the case where we allow infinite jets of local coordinates at P_∞. In this way we obtain the moduli space $\mathcal{M}^{(k,\infty)}_{g,N+1}$.

Next we consider the tangent space at the moduli spaces. Recall that our moduli point is generic, hence the moduli space there is smooth. The Kodaira–Spencer map for a versal family of complex analytic manifolds $Y \to B$ over the base B at the base point $b \in B$,

$$\mathrm{T}_b(B) \to \mathrm{H}^1(Y_b, T_{Y_b}) \tag{11.15}$$

is an isomorphism (e.g. [149]). Here, $\mathrm{T}_b(B)$ denotes the tangent space of B at the point b, Y_b is the fiber over b, and T_{Y_b} the (holomorphic) tangent sheaf of Y_b. The space $\mathrm{H}^1(Y_b, T_{Y_b})$ is also sometimes called the *Kuranishi tangent space*.

We are in the local generic situation where we have a universal family. Hence we can employ (11.15). Let Σ be the curve fixed by b, $\tilde{b}$, respectively $\tilde{b}^{(k,p)}$. We obtain

$$\mathrm{T}_{[\Sigma]}(\mathcal{M}_{g,0}) \cong \mathrm{H}^1(\Sigma, T_\Sigma). \tag{11.16}$$

Denote by S the divisor $S = \sum_{i=1}^N P_i$ on Σ. Then the Kodaira Spencer map (applied for the open subset of the moduli space which we are considering) gives

$$\mathrm{T}_{\tilde{b}}\mathcal{M}_{g,N+1} \cong \mathrm{H}^1(\Sigma, T_\Sigma(-S - P_\infty)), \tag{11.17}$$

$$\mathrm{T}_{\tilde{b}^{(1,p)}}\mathcal{M}^{(1,p)}_{g,N+1} \cong \mathrm{H}^1(\Sigma, T_\Sigma(-2S - (p+1)P_\infty)), \tag{11.18}$$

$$\mathrm{T}_{\tilde{b}^{(k,p)}}\mathcal{M}^{(k,p)}_{g,N+1} \cong \mathrm{H}^1(\Sigma, T_\Sigma(-(k+1)S - (p+1)P_\infty)). \tag{11.19}$$

Starting from the divisor

$$D_{k,p} := (k+1)S + (p+1)P_\infty, \quad k, p \in \mathbb{Z},\ k, p \geq -1. \tag{11.20}$$

With Serre duality[1] (E any vector bundle over Σ)

$$\mathrm{H}^1(\Sigma, E)^* \cong \mathrm{H}^0(\Sigma, E^* \otimes \mathcal{K}), \tag{11.21}$$

we obtain

$$\dim \mathrm{H}^1(\Sigma, T_\Sigma(-D_{k,p})) = \dim \mathrm{H}^0(\Sigma, \mathcal{K}^2(D_{k,p})). \tag{11.22}$$

For the term on the right we have

$$\deg(\mathcal{K}^2(D_{k,p})) = 4(g-1) + (k+1)N + (p+1). \tag{11.23}$$

(1) For $g \geq 2$ this expression is always $\geq 2g - 1$. Hence, by Riemann–Roch (see Section 4.1) we obtain

$$\dim \mathrm{H}^1(\Sigma, T_\Sigma(-D_{k,p})) = 3(g-1) + (k+1)N + (p+1). \tag{11.24}$$

(2) For $g = 1$ if $D_{k,p} \neq 0$, then this formula is also true. If $D_{k,p} = 0$, then we use $T_\Sigma = \mathcal{O}$, and hence $\dim \mathrm{H}^1(\Sigma, T_\Sigma) = 1$.

(3) For $g = 0$ the only information needed is the degree of the divisor and we obtain

$$\begin{aligned}\dim \mathrm{H}^1(\Sigma, T_\Sigma(-D_{k,p})) &= \max(0, \deg \mathcal{K}^2(D_{k,p}) + 1)\\ &= \max(0, -3 + (k+1)N + (p+1)).\end{aligned} \tag{11.25}$$

This allows us to give the dimension of the moduli spaces. The following cases will be of importance for us.

$$\dim_{[\Sigma]}(\mathcal{M}_{g,0}) = \begin{cases} 3g-3, & g \geq 2 \\ 1, & g = 1 \\ 0, & g = 0\,, \end{cases} \tag{11.26}$$

1 See e.g. [100, 203].

for $N \neq 0$

$$\dim_b(\mathcal{M}_{g,N}) = \begin{cases} 3g - 3 + N, & g \geq 1 \\ \max(0, N-3), & g = 0\,, \end{cases} \tag{11.27}$$

and correspondingly, for $\dim_{\tilde{b}}(\mathcal{M}_{g,N+1})$. Finally

$$\dim_{\tilde{b}^{(1,p)}}(\mathcal{M}^{(1,p)}_{g,N+1}) = \begin{cases} 3g - 2 + 2N + p, & g \geq 1 \\ \max(0, -2 + 2N + p), & g = 0. \end{cases} \tag{11.28}$$

Remark 11.3. After the choice of a section σ_∞, respectively of an infinitesimal neighborhood of the right order we obtain also the results

$$\begin{aligned} \mathrm{T}_b(\mathcal{M}_{g,N}) &\cong \mathrm{H}^1(\Sigma, T_\Sigma(-S)), \\ \mathrm{T}_{b^{(1)}}(\mathcal{M}^{(1)}_{g,N}) &\cong \mathrm{H}^1(\Sigma, T_\Sigma(-2S)). \end{aligned} \tag{11.29}$$

The first order vanishing condition at the points $P_1, \ldots, P_N$ comes from the fact that the vector fields which do not generate a nontrivial complex deformation of the curve should also not move the points to be a trivial deformation of the marked curve. The second order vanishing condition corresponds to the fact that it should additionally not change the first order infinitesimal neighborhood.

Note that for the Kodaira–Spencer mapping, a sheaf version [257, Corollary 1.2.5] also exists

$$T_W \cong R^1\pi_* T_{\mathcal{U}/W}(-S). \tag{11.30}$$

Remark 11.4. For genus 0 and 1 the situations are in a certain sense special. First we consider $g = 0$. There is only one isomorphy type. By an automorphism of $\mathbb{P}^1$ it is always possible to move three distinct points to the triple $(0, 1, \infty)$. If this is done there are no further automorphisms. Hence the moduli space $\mathcal{M}_{0,N}$ has a non-zero dimension exactly for $N \geq 4$. Its dimension is $\min(0, N-3)$ in accordance with the formula (11.27). It is quite useful to map the reference point P_∞ always to ∞; see Section 11.4.4. In this case one usually works with the configuration space

$$\widehat{W} := \{(P_1, P_2, \ldots, P_N) \mid P_i \in \mathbb{C},\ P_i \neq P_j,\ \text{for } i \neq j\}$$

of N points and studies the remaining invariance at the end.

For $g = 1$ the situation is similar. The moduli space of elliptic curves (i.e., Riemann surfaces of genus 1) is one-dimensional. For an elliptic curve E we always have the translations by a point of E as automorphisms. After fixing a point as the zero of the group law on E, respectively in the analytic picture $z \bmod L = 0$ (which might be chosen as the reference point P_∞), these automorphisms are not possible anymore. The only nontrivial automorphism which remains for the generic curve is the involution $x \to -x$. This means that with one point fixed there is no additional degree of freedom. Hence, the dimension of $\mathcal{M}_{g,N}$ equals N, which is in accordance with (11.27). Again, it is useful to work with the "configuration space" picture; see Section 11.4.5. Note that for higher genus in the generic situation the moduli space locally coincides with the "configuration space".

11.2 Tangent spaces of the moduli spaces and the Krichever–Novikov vector field algebra

Next we relate the tangent spaces considered above to a certain part of the Krichever-Novikov vector field algebra. This is done by showing that the cohomology spaces identified above can be identified with elements of the critical strips of the Krichever–Novikov vector field algebra. Hence the elements of the latter can be identified with tangent vectors to the moduli spaces.

To do this, we first have to recall the triangular decomposition from Section 3.5 of the vector field algebra $\mathcal{L}$ (and will do it for later use also already for $\mathcal{A}$ and hence for $\overline{\mathfrak{g}}$).

$$\begin{aligned} \mathcal{A} &= \mathcal{A}_+ \oplus \mathcal{A}_{(0)} \oplus \mathcal{A}_-, \\ \mathcal{L} &= \mathcal{L}_+ \oplus \mathcal{L}_{(0)} \oplus \mathcal{L}_-. \end{aligned} \tag{11.31}$$

These decompositions are defined with the help of the Krichever–Novikov basis elements.

Due to the almost-grading, the subspaces $\mathcal{A}_\pm$, and $\mathcal{L}_\pm$ are subalgebras, but the subspaces $\mathcal{A}_{(0)}$, and $\mathcal{L}_{(0)}$ in general are not. We use the term *critical strip* for them.

Note that $\mathcal{A}_+$, respectively $\mathcal{L}_+$, can be described as the algebra of functions (vector fields) having a zero of at least order one (two) at the points $P_i, i = 1, \ldots, N$. These algebras can be enlarged by adding all elements which are regular at all P_i's. This can be achieved by moving the set of basis elements $\{A_{0,i}, i = 1, \ldots, N\}$ (respectively $\{e_{0,i}, e_{-1,i}, i = 1, \ldots, N\}$) from the critical strip to these algebras. We denote the enlarged algebras by $\mathcal{A}_+^*$, respectively by $\mathcal{L}_+^*$.

On the other hand, $\mathcal{A}_-$ and $\mathcal{L}_-$ could also be enlarged such that they contain all elements which are regular at P_∞. We obtain $\mathcal{A}_-^*$ and $\mathcal{L}_-^*$ respectively.

In the same way, for every $p \in \mathbb{N}_0$ let $\mathcal{L}_-^{(p)}$ be the subalgebra of vector fields vanishing of order $\geq p+1$ at the point P_∞, and $\mathcal{A}_-^{(p)}$ the subalgebra of functions vanishing of order $\geq p$ at the point P_∞. We obtain decompositions

$$\begin{aligned} \mathcal{L} &= \mathcal{L}_+ \oplus \mathcal{L}_{(0)}^{(p)} \oplus \mathcal{L}_-^{(p)}, \quad && \text{for } p \geq 0, \\ \mathcal{A} &= \mathcal{A}_+ \oplus \mathcal{A}_{(0)}^{(p)} \oplus \mathcal{A}_-^{(p)}, \quad && \text{for } p \geq 1, \end{aligned} \tag{11.32}$$

with "critical strips" $\mathcal{L}_{(0)}^{(p)}$ and $\mathcal{A}_{(0)}^{(p)}$, which are only subspaces.

Of particular interest to us is $\mathcal{L}_{(0)}^{(1)}$, which we call *reduced critical strip*. For $g \geq 2$ its dimension is

$$\dim \mathcal{L}_{(0)}^{(1)} = N + N + (3g-3) + 1 + 1 = 2N + 3g - 1. \tag{11.33}$$

The first two terms correspond to $\mathcal{L}_0$ and $\mathcal{L}_{-1}$. The intermediate term comes from the vector fields in the basis which have poles at $P_i, i = 1, \ldots, N$ and P_∞. The "1 + 1" corresponds to the vector fields in the basis with exact order zero, respectively one, at P_∞.

Also a special role will be played by the *reduced regular subalgebras*

$$\mathcal{A}^r = \mathcal{A}_-^{(1)} \subset \mathcal{A}_-^*, \quad \mathcal{L}^r = \mathcal{L}_-^{(0)} \subset \mathcal{L}_-^*, \tag{11.34}$$

containing the function, respectively the vector fields vanishing at P_∞.

Recall that the almost-grading extends to the *higher genus current algebra* $\overline{\mathfrak{g}}$ by setting $\deg(x \otimes A_{n,p}) := n$, and we obtain obtain a triangular decomposition as above

$$\overline{\mathfrak{g}} = \overline{\mathfrak{g}}_+ \oplus \overline{\mathfrak{g}}_{(0)} \oplus \overline{\mathfrak{g}}_-, \quad \text{with} \quad \overline{\mathfrak{g}}_\beta = \mathfrak{g} \otimes \mathcal{A}_\beta, \quad \beta \in \{-, (0), +\}. \tag{11.35}$$

In particular, $\overline{\mathfrak{g}}_\pm$ are subalgebras. The corresponding is true for the enlarged subalgebras. Among them, $\overline{\mathfrak{g}}^r := \overline{\mathfrak{g}}_-^{(1)} = \mathfrak{g} \otimes \mathcal{A}_-^{(1)}$ is of special importance. It is called the *reduced regular subalgebra* of $\overline{\mathfrak{g}}$.

Remark 11.5. All these subalgebras can be considered subalgebras of the corresponding almost-graded central extensions $\widehat{\mathcal{A}}$, $\widehat{\mathcal{L}}$, and $\widehat{\mathfrak{g}}$ respectively. This is obvious, as the defining cocycles which are integrated over a separating cycle C_S can be calculated by calculating residues either at the points $P_1, \ldots, P_N$ or at P_∞. But the elements of the subalgebras are holomorphic at one of these sets. Also note that the finite-dimensional Lie algebra $\mathfrak{g}$ can naturally be considered as subalgebra of $\overline{\mathfrak{g}}$. It lies in the subspace $\overline{\mathfrak{g}}_0$; see Proposition 9.1.

Let Σ be the Riemann surface we are dealing with, and let U_∞ be a coordinate disc around P_∞, such that $P_1, \ldots, P_N \notin U_\infty$. Let $U_1 = \Sigma \setminus \{P_\infty\}$. Because U_1 and U_∞ are affine (respectively Stein) [100, p. 297], we get $\mathrm{H}^1(U_j, F) = 0$, $j = 1, \infty$ for every coherent sheaf F. Hence, the sheaf cohomology can be given as Cech cohomology with respect to the covering $\{U_1, U_\infty\}$. Set $U_\infty^* = U_1 \cap U_\infty = U_\infty \setminus \{P_\infty\}$. The Cech two-cocycles can be given by $s_{1,\infty} \in F(U_\infty^*)$ ($s_{0,0} = s_{\infty,\infty} = 0$), hence by arbitrary sections over the punctured coordinate disc U_∞^* We recall that it does not make any difference whether we calculate the Cech cohomology in the algebraic-geometric category or in the complex-analytic category.

Returning to the holomorphic (algebraic) tangent bundle T_Σ. For every element f of the Krichever–Novikov vector field algebra, its restriction to U_∞^* is holomorphic and indeed algebraic, as the poles (outside U_∞^*) are of finite order. Hence it defines an element of $\mathrm{H}^1(\Sigma, T_\Sigma)$. Note that it defines also an element of $\mathrm{H}^1(\Sigma, T_\Sigma(D))$, where D is any divisor supported outside U_∞^*. We introduce the map

$$\theta_D : \mathcal{L} \to \mathrm{H}^1(\Sigma, T_\Sigma(D)), \quad f \mapsto \theta_D(f) := [f_{|U_\infty^*}]. \tag{11.36}$$

If the divisor D is clear from the context we will suppress it in the notation. For us only the divisors $D_{k,p}$ are of importance, and we set $\theta_{k,p} = \theta_{D_{k,p}}$.

Theorem 11.6. *Let $g \geq 2$ and $k, p > -1$, then there is a surjective linear map from the Krichever–Novikov vector field algebra $\mathcal{L}$ to the cohomology space*

$$\theta = \theta_{k,p} : \mathcal{L} \to \mathrm{H}^1(\Sigma, T_\Sigma(-(k+1)S - (p+1)P_\infty), \tag{11.37}$$

such that θ restricted to the following subspace gives an isomorphism

$$\begin{aligned}\mathcal{L}_{k-1} \oplus \cdots \oplus \mathcal{L}_{(0)}^{(p)} &\cong \mathrm{H}^1(\Sigma, T_\Sigma(-(k+1)S - (p+1)P_\infty) \\ &\cong \mathrm{T}_{\tilde{b}^{(k,p)}} \mathcal{M}_{g,N+1}^{(1,p)}.\end{aligned} \tag{11.38}$$

Moreover,

$$\ker \theta_{k,p} = \bigoplus_{n \geq k} \mathcal{L}_n \oplus \mathcal{L}_{-}^{(p)}. \tag{11.39}$$

For $g = 1$ the same is true if at least k or p is ≥ 0. For $g = 0$ it is true except for some small values of N, k, or p.

Proof. The map $\theta_{k,p}$ is linear. The second map in (11.38) is the Kodaira–Spencer map (11.19). By (11.24) the cohomology spaces and the spaces on the left of (11.38) have the same dimension. We start by showing (11.39). We work in the algebraic-geometric setting. This means all our holomorphic functions on U_∞^* will have at most algebraic poles outside. First we assume $\theta_{k,p}(f) = 0$. Hence, we can write

$$f_{|U_\infty^*} = f_{1|U_\infty^*} - f_{0|U_\infty^*}$$

with f_0 defined and regular on U_∞ with $\mathrm{ord}_{P_\infty}(f_0) \geq p + 1$, and f_1 regular outside P_∞ and with order at least $k + 1$ at every point $P_i, i = 1, \ldots, N$. As f_0 is regular at P_∞, the vector field f_1 has to have the same singular part as f at P_∞. In particular it can be extended to a global meromorphic vector field. This implies $f_1 \in \mathcal{L}$. From $\mathrm{ord}_{P_i}(f_1) \geq k + 1$, $i = 1, \ldots, N$ it follows that $f_1 \in \bigoplus_{n \geq k} \mathcal{L}_n$. Now $f_{0|U_\infty^*} = (f_1 - f)_{|U_\infty^*}$. But $f_1 - f$ is a globally defined meromorphic vector field with poles at most at $\{P_1, \ldots, P_N, P_\infty\}$. Hence the same is true for the extension of f_0 (which we denote by the same symbol). In particular $f_0 \in \mathcal{L}$. Due to the regularity at P_∞ we get $f_0 \in \mathcal{L}_{-}^{(p)}$. This shows $\subseteq$.

Vice versa, for $f \in \bigoplus_{n \geq k} \mathcal{L}_n$ we set $f_1 = f_{|U_1}$, $f_0 = 0$, and for $f \in \mathcal{L}_{-}^{(p)}$ we set $f_1 = 0$, $f_0 = -f_{|U_\infty}$. We see that their cohomology classes will vanish. Hence, (11.39). From (11.39) it follows that $\theta_{k,p}$ is injective if restricted to the complementary space. From the equality of the dimension follows surjectivity and hence the first isomorphism in (11.38). □

This can be specialized to the following important cases:

$$\begin{aligned}\mathcal{L}_0 \oplus \mathcal{L}_{-1} \oplus \mathcal{L}_{(0)}^{(p)} &\cong \mathrm{H}^1(\Sigma, T_\Sigma(-2S - (p+1)P_\infty) \cong \mathrm{T}_{\tilde{b}^{(1,p)}} \mathcal{M}_{g,N+1}^{(1,p)}, \\ \mathcal{L}_{-1} \oplus \mathcal{L}_{(0)}^{(p)} &\cong \mathrm{H}^1(\Sigma, T_\Sigma(-S - (p+1)P_\infty)) \cong \mathrm{T}_{\tilde{b}^{(0,p)}} \mathcal{M}_{g,N+1}, \\ \mathcal{L}_{(0)}^{(p)} &\cong \mathrm{H}^1(\Sigma, T_\Sigma(-(p+1)P_\infty) \cong \mathrm{T}_{[\Sigma, P_\infty^{(p)}]} \mathcal{M}_{g,1}^{(p)}.\end{aligned} \tag{11.40}$$

For the infinite jets we obtain

$$\mathrm{T}_{\tilde{b}^{(1,\infty)}} \mathcal{M}_{g,N+1}^{(1,\infty)} = \lim_{p \to \infty} \mathrm{H}^1(\Sigma, T_\Sigma(-2S - pP_\infty)) \cong \mathcal{L}_{(0)} \oplus \mathcal{L}_{-}. \tag{11.41}$$

Remark 11.7. The spaces $\ker \theta_{k,p}$ are invariantly defined. They neither depend on the ordering of the points, nor on any special recipe for fixing the Krichever–Novikov basis elements. Because the vector fields in $\ker\theta \subset \mathcal{L}$ do not correspond to deformations in the moduli, we sometimes call these vector fields *vertical vector fields*. They should not be confused with the sections of the relative tangent sheaf of the universal family. The following should also be kept in mind. The definition of the critical strip, hence of the complementary subspace to $\ker\theta$, is only fixed by the order prescription for the basis. A different prescription (which involves changing the required orders) will yield a different identification of tangent vectors on the moduli space with vector fields in the fiber. The inverse of the map $\theta_{k,p}$ will be denoted by

$$\overline{\rho} : \mathrm{T}_{\tilde{b}^{(k,p)}}\mathcal{M}^{(1,p)}_{g,N+1} \to \mathcal{L}, \quad X \mapsto \overline{\rho}(X), \tag{11.42}$$

where strictly speaking the image should lie in the subspace $\mathcal{L}_{k-1} \oplus \cdots \oplus \mathcal{L}^{(p)}_{(0)}$. In fact, we could also describe $\overline{\rho}$ in a more natural manner as a composition of the canonical map to $\mathcal{L}/\ker\theta_{k,p}$ with a non-canonical identification with the critical strip.

$$\overline{\rho} : \mathrm{T}_{\tilde{b}^{(k,p)}}\mathcal{M}^{(k,p)}_{g,N+1} \to \mathcal{L}/\ker\theta_{k,p} \cong \mathcal{L}_{k-1} \oplus \cdots \oplus \mathcal{L}^{(p)}_{(0)}. \tag{11.43}$$

In this more natural description, $\overline{\rho}(X)$ is an element of the critical strip modulo vertical vector fields. In this description it does not depend on choices. For the equation (11.43) we could take any complement of $\ker\theta_{k,p}$ instead of the critical strip. The map $\overline{\rho}$ is the Kodaira–Spencer map. In Section 11.4.1 the map $\overline{\rho}$ will be described in more detail with respect to the variation of the complex structure.

Remark 11.8. We return to the modification in genus 0 and 1. In these cases there are global holomorphic vector fields. Hence the decomposition of the critical strip and its identification in (11.38) are not valid anymore. The spaces $\mathcal{L}_0 \oplus \mathcal{L}_{-1}$ and $\mathcal{L}^*_{(0)} \oplus \mathcal{L}^{(0)}_-$ have a nontrivial intersection. This is in complete conformity with the corrected dimensions of the moduli space.

11.3 Sheaf versions of the Krichever–Novikov type algebras

Let $\widetilde{W}$ be a generic open subset of $\mathcal{M}_{g,N+1}$ which we consider here. Let

$$\tilde{b} = [\Sigma, P_1, P_2, \ldots, P_N, P_\infty] \tag{11.44}$$

be a point of $\widetilde{W}$. For each $\tilde{b}$ we construct the Krichever–Novikov objects

$$\mathcal{A}_{\tilde{b}},\ \mathcal{L}_{\tilde{b}},\ \widehat{\mathcal{L}}_{\tilde{b}},\ \overline{\mathfrak{g}}_{\tilde{b}},\ \widehat{\mathfrak{g}}_{\tilde{b}},\ \mathcal{F}^{\lambda}_{\tilde{b}},\ \text{etc.}, \tag{11.45}$$

with respect to the splitting $I = \{P_1, P_2, \ldots, P_N\}$ and $O = \{P_\infty\}$.

There are sheaf versions of these objects

$$\mathcal{A}_{\widetilde{W}},\ \mathcal{L}_{\widetilde{W}},\ \widehat{\mathcal{L}}_{\widetilde{W}},\ \overline{\mathfrak{g}}_{\widetilde{W}},\ \widehat{\mathfrak{g}}_{\widetilde{W}},\ \mathcal{F}^{\lambda}_{\widetilde{W}}. \tag{11.46}$$

To this end we consider the universal family

$$\pi : \mathcal{U} \to \widetilde{W}. \tag{11.47}$$

By assigning to $\tilde{b}$ the point P_i we obtain sections

$$\sigma_i : \widetilde{W} \to \mathcal{U}, \quad \sigma_i(\tilde{b}) = P_i, \quad i = 1 \ldots, N, \infty. \tag{11.48}$$

Varying $\tilde{b}$ defines the divisor of sections

$$\widetilde{S} = \sum_{i=1}^{N} \sigma_i(\widetilde{W}) + \sigma_\infty(\widetilde{W}) \tag{11.49}$$

in the family $\mathcal{U}$.

Denote by $\mathcal{O}_\mathcal{U}$ the sheaf of regular functions on $\mathcal{U}$. As usual, let $\mathcal{O}_\mathcal{U}(k\widetilde{S})$, $k \in \mathbb{Z}$ be the sheaf of functions which have zeros of order at least $-k$ along the divisor $\widetilde{S}$. In particular, for $k \in \mathbb{N}$ this means that the functions have poles of order at most k at the divisor $\widetilde{S}$. By $\mathcal{O}_\mathcal{U}(*\widetilde{S})$ we understand the sheaf of functions which might have poles along the divisor $\widetilde{S}$. Also we set

$$\pi_* \mathcal{O}_\mathcal{U}(*\widetilde{S}) := \lim_{k\to\infty} \pi_*(\mathcal{O}_\mathcal{U}(k\widetilde{S})). \tag{11.50}$$

It is a locally free $\mathcal{O}_{\widetilde{W}}$-sheaf (with $\mathcal{O}_{\widetilde{W}}$ the structure sheaf of the space $\widetilde{W}$). Its vector space fiber over $\tilde{b} \in \widetilde{W}$ can be identified with

$$\mathrm{H}^0(\Sigma_{\tilde{b}}, \mathcal{O}_{\Sigma_{\tilde{b}}}(*\widetilde{S})) = \mathcal{A}_{\tilde{b}}, \quad \Sigma_{\tilde{b}} = \pi^{-1}(\tilde{b}).$$

The sheaf $\pi_* \mathcal{O}_\mathcal{U}(*\widetilde{S})$ can be made to a sheaf of commutative associative $\mathcal{O}_{\widetilde{W}}$-algebras by fiber-wise multiplication. We denote it as $\mathcal{A}_{\widetilde{W}}$. As a sheaf of abelian Lie algebras it can be centrally extended to the $\mathcal{O}_{\widetilde{W}}$-sheaf

$$\widehat{\mathcal{A}}_{\widetilde{W}} := \widehat{\pi_* \mathcal{O}_\mathcal{U}(*\widetilde{S})} := \pi_* \mathcal{O}_\mathcal{U}(*\widetilde{S}) \oplus \mathcal{O}_W \cdot t,$$

where t is the central element and the structure is defined as follows. The elements $f, g \in \mathcal{A}_{\widetilde{W}}(U)$ can be represented as functions on $\pi^{-1}(U)$, possibly with poles only along $\widetilde{S}$. Let γ be the cocycle $\gamma_S^\mathcal{A}$ (6.134). Here S in the index means that we integrate over a separating cycle C_S on $\Sigma_{\tilde{b}}$. As the topological type does not change, the cycle C_S will do for all $\Sigma_{\tilde{b}}$. The cocycle values can be obtained by calculating residues either along S or at the section σ_∞. By fiber-wise evaluation we obtain with $r, s \in \mathcal{O}_W(U)$

$$[f + r \cdot t, g + s \cdot t] := \gamma(f, g) \cdot t,$$

an element of $\widehat{\mathcal{A}}_W(U)$. Note that $\gamma(f, g) \in \mathcal{O}_W(U)$.

This construction can be extended to the affine algebra situation.

Definition 11.9. Given a finite-dimensional Lie algebra $\mathfrak{g}$, the *sheaf of the associated current algebra* $\overline{\mathfrak{g}}_{\widetilde{W}}$ and the *sheaf of the associated affine algebra* $\widehat{\mathfrak{g}}_{\widetilde{W}}$ are defined as

$$\overline{\mathfrak{g}}_{\widetilde{W}} := \mathcal{A}_{\widetilde{W}} \otimes \mathfrak{g}, \quad \widehat{\mathfrak{g}}_{\widetilde{W}} := \overline{\mathfrak{g}}_{\widetilde{W}} \oplus \mathcal{O}_{\widetilde{W}} \cdot t, \tag{11.51}$$

where the Lie structure is given by the naturally extended form of Section 9.2 (respectively without its central term for $\overline{\mathfrak{g}}_{\widetilde{W}}$).

Clearly, these are $\mathcal{O}_{\widetilde{W}}$-sheaves. Let $\tilde{b} \in \widetilde{W}$ and let $\mathcal{O}_{\widetilde{W},\tilde{b}}$ be the local ring at $\tilde{b}$ and $M_{\tilde{b}}$ its maximal ideal. Set $\mathbb{C}_{\tilde{b}} \cong \mathcal{O}_{\widetilde{W},\tilde{b}}/M_{\tilde{b}}$, then we obtain the following canonical isomorphisms:

$$\mathbb{C}_{\tilde{b}} \otimes \mathcal{A}_{\widetilde{W}} \cong \mathcal{A}_{\tilde{b}}, \quad \mathbb{C}_{\tilde{b}} \otimes \widehat{\mathcal{A}}_{\widetilde{W}} \cong \widehat{\mathcal{A}}_{\tilde{b}}, \quad \mathbb{C}_{\tilde{b}} \otimes \overline{\mathfrak{g}}_{\widetilde{W}} \cong \overline{\mathfrak{g}}_{\tilde{b}}, \quad \mathbb{C}_{\tilde{b}} \otimes \widehat{\mathfrak{g}}_{\widetilde{W}} \cong \widehat{\mathfrak{g}}_{\tilde{b}}.$$

Definition 11.10. A sheaf $\mathfrak{V}$ of $\mathcal{O}_{\widetilde{W}}$-modules is called a *sheaf of representations* for the affine algebra $\widehat{\mathfrak{g}}_{\widetilde{W}}$ if the $\mathfrak{V}(U)$ are modules over $\widehat{\mathfrak{g}}_{\widetilde{W}}(U)$ for all open subsets U of $\widetilde{W}$.

For a sheaf of representations $\mathfrak{V}$, we obtain that $\mathfrak{V}_{\tilde{b}}$ is a module over $\widehat{\mathfrak{g}}_{\tilde{b}}$ for every point in $\widetilde{W}$.

At certain places in the following section we have to deal with the Krichever–Novikov basis elements. They are fixed uniquely by the orders at the points $P_1, \ldots, P_N$ and P_∞ and by some normalizing conditions given in terms of the first order jet of the coordinates $z_1, z_2, \ldots, z_N$ at the points in I. More precisely, by Proposition 3.4 we know that for $z'_p = \alpha_p \cdot z_p + O(z_p^2)$ another coordinate at P_p, the Krichever–Novikov basis element of $\mathcal{F}^\lambda$ w.r.t. z_p (w.r.t. z'_p), are related as

$$f^{\lambda\prime}_{n,p} = (\alpha_p)^n f^\lambda_{n,p}, \quad \text{and} \quad f^{\lambda\prime}_{n,s} = f^\lambda_{n,s}, \quad s \neq p. \tag{11.52}$$

If we globalize this, then the α_p are nonvanishing local functions on $\mathcal{M}_{g,N+1}$. Hence, over $\widetilde{W} \subset \mathcal{M}_{g,N+1}$ the basis elements are not uniquely fixed. We already introduced the moduli space

$$\widetilde{W}^{(1)} \subset \mathcal{M}^{(1)}_{g,N+1} = \mathcal{M}^{(1,0)}_{g,N+1} \tag{11.53}$$

containing first order jets of coordinates at P_i. Let $\eta : \mathcal{M}^{(1)}_{g,N+1} \to \mathcal{M}_{g,N+1}$ be the surjective analytic map obtained by forgetting the coordinates. It is a surjective analytic map. Over $W^{(1)}$ also the basis elements $f^\lambda_{n,p}$ are well-defined. This was the reason for enlarging the moduli space. Due to the explicit description given in Chapter 5, the basis elements depend analytically on the moduli. As now the basis elements are given globally (at least as long as our moduli points correspond to generic situations), we obtain the following proposition.

Proposition 11.11. *The sheaves*

$$\mathcal{A}_{\widetilde{W}^{(1)}}, \ \mathcal{L}_{\widetilde{W}^{(1)}}, \ \widehat{\mathcal{L}}_{\widetilde{W}^{(1)}}, \ \overline{\mathfrak{g}}_{\widetilde{W}^{(1)}}, \ \widehat{\mathfrak{g}}_{\widetilde{W}^{(1)}}, \ \mathcal{F}^\lambda_{\widetilde{W}^{(1)}} \tag{11.54}$$

are free sheaves of $\mathcal{O}_{\widetilde{W}^{(1)}}$-modules of infinite rank.

Pulling back a sheaf of representation $\mathfrak{V}$ over W, we obtain a sheaf of representation $\mathfrak{V}^{(1)} = \eta^*\mathfrak{V}$ of $\widehat{\mathfrak{g}}_{W^{(1)}}$. More generally, we can define sheaves of representations over $W^{(1)}$ directly. In particular, for these sheaves of representations, operators depending on the Krichever–Novikov basis are well-defined.

With the help of the Krichever–Novikov basis elements, an almost-grading was also introduced for the algebraic objects. Hence, clearly the sheaves (11.46) carry an almost-grading too, and the homogeneous subspaces globalize immediately to the sheaf version. The definition of the homogeneous subspaces does not depend on a rescaling of the basis elements given by (11.52). Hence, already the sheaves (11.46) over $\widetilde{W}$ carry an almost-graded structure. This allows the definition of a sheaf of admissible representations to be a sheaf of representations either over $\widetilde{W}$ or over $\widetilde{W}^{(1)}$, where all fiber-wise representations are admissible. In addition, we will usually require (if not otherwise stated) the central element t to operate as $c \cdot id$, with c a function on W. This function is called the level function. Very often we will even assume c to be a constant c, which is just called the level of the representation.

Remark 11.12. If we fix a section σ_∞, then we can pullback our sheaves also to $W \subset \mathcal{M}_{g,N}$ and $W^{(1)} \subset \mathcal{M}_{g,N}^{(1)}$. For the dimension of the latter moduli space we get

$$\dim_{b^{(1)}}(\mathcal{M}_{g,N}^{(1)}) = \begin{cases} 3g-3+2N, & g \geq 2 \\ \max(0, 2N-3), & g = 0. \end{cases} \tag{11.55}$$

Example 11.13 (The Verma module sheaf). Due to the possibility of globalizing the grading to $\widehat{\mathfrak{g}}_{\widetilde{W}}$ it is possible to define the *Verma module sheaf*, both on $\widetilde{W}$ and on $\widetilde{W}^{(1)}$, in a straightforward manner by extending Definition 9.51

$$\widehat{V}_{(\chi,c),\widetilde{W}} := U(\widehat{\mathfrak{g}}_{\widetilde{W}}) \otimes_{U(\widehat{\mathfrak{b}}_{\widetilde{W}})} V. \tag{11.56}$$

The same works over $\widetilde{W}^{(1)}$. On first sight it looks as if the Verma module sheaf is only defined over $\widetilde{W}^{(1)}$ because the basis elements $A_{0,p}$ are involved. But as now $n = 0$, it follows from (11.52) that they are independent of the coordinate classes. The Verma module sheaves are sheaves of admissible representations. Note, it is even possible to vary the data (χ, c) over the moduli. See Section 9.9 for the notation. Recall that we constructed the Verma module starting from assigning finite-dimensional representations of the finite-dimensional Lie algebra $\mathfrak{g}$ to the points $P_1, P_2, \ldots, P_N$. This is exactly the situation studied in [254].

Example 11.14 (The fermionic Fock representation sheaves). Another example is given by the fermionic modules $\mathfrak{V}_{\widetilde{W}^{(1)}}$ as a representation sheaf. The basis fermions do not depend on moduli at all, only the Lie algebra action does via the structure constants. We trivialize the sheaf $\mathfrak{V}_{\widetilde{W}^{(1)}}$ using these bases and obtain the corresponding trivial vector bundle $V \times \widetilde{W}^{(1)}$, where V is a standard fermion space $\mathcal{H}_{F,\tau}^{(m)}$ as discussed in Section 9.10. Again it is an admissible representation sheaf.

Remark 11.15 (Sheaf version of the Sugawara construction). Recall the Sugawara construction from Chapter 10 which is well-defined for admissible representations of the affine Lie algebra $\widehat{\mathfrak{g}}$ associated with a finite dimensional reductive Lie algebra $\mathfrak{g}$. The construction again can be sheafified. Hence, let $\mathfrak{V}$ be an admissible representation sheaf of the sheaf $\widehat{\mathfrak{g}}_{\widetilde{W}^{(1)}}$. For simplicity, let $\mathfrak{g}$ be either abelian or simple[2]. Let 2κ be the eigenvalue of Casimir in the adjoint representation of the finite Lie algebra, and let c be the level of the representation under consideration. Assume that the level function c obeys $c + \kappa \neq 0$ for the moduli points we are considering. For every sheaf of admissible representations over $\widetilde{W}^{(1)}$ the Sugawara operators are well-defined. Note that the individual operators $u(n, p)$ depend on the coordinates. From (11.52) we know how the operators transform if we choose a different coordinate $z_p' = \alpha_p z_p + O(z_p^2)$. We obtain $(u(n, p))' = (\alpha_p)^n u(n, p)$. The factor will be canceled out by the contribution of $\omega^{n,p\prime} = (\alpha_p)^{-n}\omega^{n,p}$. Hence it follows from (10.19) for the Sugawara operators that $T(Q)' = T(Q)$, and consequently the Sugawara field is independently defined over $\widetilde{W}$. In contrast, the individual operators $L_{k,s}^*$ depend on the coordinates. They are only well-defined over $\widetilde{W}^{(1)}$. They transform as the vector fields $e_{k,s}$ do. If we assign to the vector field $e \in \mathcal{L}$ (which of course does not depend on the coordinates), the operator

$$T[e] := \frac{-1}{c+\kappa} \cdot \frac{1}{2\pi \mathrm{i}} \int_{C_\tau} T(Q)e(Q) \tag{11.57}$$

then it will also not depend on them. As this is the Sugawara representation, it is also defined over $\widetilde{W}$. We get $T[e_{n,p}] = L_{n,p}^*$.

Recall the definition of the *reduced regular subalgebras* $\mathcal{A}^r$, $\mathcal{L}^r$, and $\overline{\mathfrak{g}}^r = \mathfrak{g} \otimes \mathcal{A}^r$ as consisting of those elements vanishing at P_∞. Denote by $\overline{\mathfrak{g}}_{\widetilde{W}}^r$, $\mathcal{L}_{\widetilde{W}}^r$ etc., the corresponding sheaves.

Definition 11.16. Let $\mathfrak{V}_{\widetilde{W}}$ be a sheaf of (fiber-wise) representations of $\widehat{\mathfrak{g}}_{\widetilde{W}}$. The sheaf of *conformal blocks* (associated with the representation $\mathfrak{V}_{\widetilde{W}}$) is defined as the sheaf of coinvariants

$$C_{\widetilde{W}} := C_{\widetilde{W}}(\mathfrak{V}) := \mathfrak{V}_{\widetilde{W}} / \overline{\mathfrak{g}}_{\widetilde{W}}^r \mathfrak{V}_{\widetilde{W}}. \tag{11.58}$$

Here, $\overline{\mathfrak{g}}_{\widetilde{W}}^r \mathfrak{V}_{\widetilde{W}}$ should mean that we take the sheaf of point-wise vector spaces generated by these elements. Of course, we can also define conformal blocks for representation sheaves over $\widetilde{W}^{(1)}$ and even more generally. In the "opposite direction" we can, by fixing a section $\hat{\sigma}_\infty$, define conformal blocks for representation sheaves over W and $W^{(1)}$; see Remark 11.3. See also (9.107) for the point-wise definition.

Moreover, if for $p \in \mathbb{N}$ or $p = \infty$ we take $\widetilde{W}^{(1,p)} = (\nu^{(1,p)})^{-1}(\widetilde{W})$, then by pulling back the above sheaves over $\widetilde{W}$ via $\nu^{(1,p)}$, we obtain sheaves on the open subset $\widetilde{W}^{(1,p)}$ of $\mathcal{M}_{g,N+1}^{(p)}$. Starting from a fiber-wise representation $\mathfrak{V}_{\widetilde{W}^{(1,p)}}$, the sheaf of conformal blocks

2 See Section 10.2.1 for how to deal with the reductive case.

can be defined in the same way as (11.58) by

$$C_{\widetilde{W}^{(1,p)}} = \mathfrak{V}_{\widetilde{W}^{(1,p)}} / \overline{\mathfrak{g}}^{r}_{\widetilde{W}^{(1,p)}} \mathfrak{V}_{\widetilde{W}^{(1,p)}}. \tag{11.59}$$

Clearly, if $\mathfrak{V}_{\widetilde{W}}$ is already defined on $\widetilde{W}$, then $v_p^*(C_{\widetilde{W}}) = C_{\widetilde{W}^{(1,p)}}$.

For a general representation sheaf, the sheaf of conformal blocks will only be a sheaf. For the examples given by Verma modules (at least for simple Lie algebras $\mathfrak{g}$) and fermionic Fock modules (at least if $\mathfrak{g} = \mathfrak{gl}(n)$ and τ is the standard representation), the sheaf of conformal blocks will be locally free of finite rank on $\widetilde{W}^{(1)}$. For more information see [227] and [250].

11.4 The Knizhnik–Zamolodchikov connection

For the following let $\mathfrak{V}$ be a locally free representation sheaf of $\widehat{\mathfrak{g}}$ over $\widetilde{W}^{(1,1)} = (v^{(1,1)})^{-1}(\widetilde{W})$. Recall that this means that we take the first order jet of coordinates at $P_1, \dots, P_N$ and P_∞ into account. Also recall that if $\mathfrak{V}$ is a sheaf over $\widetilde{W}$, then its pull-back $(v^{(1,1)})^*(\mathfrak{V})$ will be a sheaf over $\widetilde{W}^{(1,1)}$. To avoid cumbersome notation we will use in this section instead of $\widetilde{W}^{(1,1)}$ just $\widetilde{W}$. We assume that for $\mathfrak{V}$ the fiber-wise representations are admissible. Furthermore, we assume that the sheaf of conformal blocks is locally free of finite rank. This is the case for the Verma module sheaf and the fermionic module sheaf in the cases discussed at the very end of the previous section. Following [226, 227] and the improved version in [250], we will introduce here for the sheaf of conformal blocks the Knizhnik–Zamolodchikov connection and show certain properties. To do this we need an additional assumption formulated later (11.81). Again this assumption is true for Verma module sheaves and fermionic module sheaves. It is not clear whether this assumption is really necessary, or what the necessary modification will be if it is not fulfilled.

Also recall that our constructions are done over an open subset of moduli space corresponding to generic configurations.

11.4.1 Variation of the complex structure

We will now examine how our objects behave under variation of the moduli point (moduli parameter) $\tau \in \mathcal{M}$. If X is an infinitesimal direction on the moduli space $\mathcal{M}^{(1,1)}_{g,N+1}$, then via the Kodaira–Spencer cocycle map ρ we can assign to it a local vector field $\rho(X)$ lying in $\widetilde{\mathcal{L}}_\tau$. The Lie algebra $\widetilde{\mathcal{L}}_\tau$ will be a certain completion of the algebra $\mathcal{L}_\tau$ and its definition will be explained below. The infinitesimal vector field X on the moduli space will also act on the Krichever–Novikov elements. The corresponding operator will be denoted by ∂_X. This action will be described next. If required, the reader can find more details in [227, 250].

Let $\mathcal{A}_\tau$ be the stalk of the sheaf $\mathcal{A}_{\overline{W}}$ at the moduli point τ. This means that elements are functions on Σ_τ. If we differentiate for example $g \in \mathcal{A}_\tau$ with respect to moduli parameters it will not be a Krichever–Novikov function on the same Σ_τ anymore, but can be given as infinite series with respect to expansion at the point P_∞. Hence, we have to consider completions of the algebras at P_∞ by allowing (infinite) Laurent series of Krichever–Novikov basis elements. These series are only defined locally in a neighborhood of P_∞ and might have algebraic poles there. The appearing sums are of the type

$$f = \sum_{n=-\infty}^{n<M} \sum_{p=1}^{N} \alpha_{n,p} f_{n,p}^{\lambda}. \tag{11.60}$$

In this way we obtain the algebra of local functions $\widetilde{\mathcal{A}}$, local vector fields $\widetilde{\mathfrak{L}}$, and local currents $\widetilde{\mathfrak{g}}$.

Remark 11.17. By the construction of the basis elements, the homogeneous spaces of degree m have dimension N. The order at the point P_∞ will be constant for all base elements $f_{m,p}^{\lambda}$, $p = 1, \ldots, N$. It will jump by N if we pass to the next degree. It is possible to change the basis inside every degree such that the orders at the point P_∞ differ exactly by 1. Of course, we lose the given prescription at I. In this way we obtain a basis $\hat{f}_m^{\lambda}$ which is now indexed by $\mathbb{Z}$. If we invert the index and make if necessary a shift such that

$$\operatorname{ord}_{P_\infty}(\hat{f}_m^{\lambda}) = m - \lambda, \tag{11.61}$$

then (11.60) can be written as

$$f = \sum_{m>M}^{\infty} \beta_m \hat{f}_m^{\lambda}, \tag{11.62}$$

which looks more similar to a Laurent series.

Obviously,

$$\mathcal{A} \subset \widetilde{\mathcal{A}}, \quad \mathfrak{L} \subset \widetilde{\mathfrak{L}}, \quad \overline{\mathfrak{g}} \subset \widetilde{\mathfrak{g}}. \tag{11.63}$$

Besides these algebras we also consider the corresponding regular and reduced regular algebras

$$\widetilde{\mathcal{A}}^{reg}, \quad \widetilde{\mathcal{A}}^{r}, \quad \widetilde{\mathfrak{L}}^{reg}, \quad \widetilde{\mathfrak{L}}^{r}, \quad \widetilde{\mathfrak{g}}^{reg}, \quad \widetilde{\mathfrak{g}}^{r}. \tag{11.64}$$

For later applications we note that in particular for $\mathcal{A}^r$ and $\mathfrak{g}^r$ (with $\lambda = 0$) and $\mathfrak{L}_-^{(1)}$ (with $\lambda = -1$), the expansion in (11.62) starts with $m = 1$. This is also true for $\widetilde{\mathcal{A}}^r$ and $\widetilde{\mathfrak{g}}^r$ and $\widetilde{\mathfrak{L}}_-^{(1)}$.

Our cocycles for the above algebras can obviously be extended as cocycles to the algebra of local elements, as they are calculated via residues at P_∞. Furthermore, they define sheaves $\widetilde{\mathcal{A}}_{\overline{W}}$, $\widetilde{\mathfrak{L}}_{\overline{W}}$, and $\widetilde{\mathfrak{g}}_{\overline{W}}$ over $\widetilde{W}$. Elements of these algebras or of the corresponding sheaves are called local elements.

To avoid cumbersome notations we will sometimes not mention the space if we talk about a sheaf. For example, $\mathcal{A}$ could mean one special copy of the algebra $\mathcal{A}$ or the sheaf over $\widetilde{W}$, etc.

Remark 11.18. In the following section we will need the action of $u(g)$ and $T[e]$ on a representation $\mathfrak{V}$, not only for honest Krichever–Novikov elements g and e, but for local elements. To make this well-defined we have to complement the representation space $\mathfrak{V}$ in negative degree direction (or equivalently in positive degree direction with respect to the P_∞ degree). This completion is denoted by $\overline{\mathfrak{V}}$. For those representations which we are considering here such a completion is possible. The above operators will then be well-defined and the important result from Proposition 10.14

$$[T[e], u(g)] = u(e\,.\,g), \tag{11.65}$$

stays valid for local objects. This comes from the almost-gradedness. Indeed, only a finite number of terms of the expansions (of the operator and the corresponding element in $\overline{\mathfrak{V}}$) contribute in the result to the component of a given degree. See [227] and [250] for more details. We will define the Knizhnik–Zamolodchikov connection later only for conformal blocks

$$C(\overline{\mathfrak{V}}) = \overline{\mathfrak{V}}/\overline{\mathfrak{g}}^r\overline{\mathfrak{V}} = \mathfrak{V}/\overline{\mathfrak{g}}^r\mathfrak{V} = C(\mathfrak{V}), \tag{11.66}$$

and low negative degrees will be truncated, see e.g. Lemma 11.19.

Lemma 11.19.

$$T[\mathcal{L}_-^{(1)}]\,.\,\mathfrak{V} \subseteq \overline{\mathfrak{g}}^r\mathfrak{V}, \quad T[\overline{\mathcal{L}}_-^{(1)}]\,.\,\overline{\mathfrak{V}} \subseteq \overline{\mathfrak{g}}^r\overline{\mathfrak{V}}. \tag{11.67}$$

Proof. Recall that we can re-express our homogeneous subspaces with respect to an adapted basis for P_∞; see Remark 11.17. We denote the corresponding elements in $\mathcal{L}$ by $\hat{e}_m$. The set $\{\hat{e}_m \mid m \geq 1\}$ is a basis of $\mathcal{L}_-^{(1)}$. In the sum for $T[e_k]\,.\,v$ up to a scaling factor we have the following terms

$$\sum_{m,n}\left(\frac{1}{2\pi \mathrm{i}}\int_{C_S} \hat{\omega}^n\hat{\omega}^m\hat{e}_k\right) :u_a(\hat{g}_m)u^a(\hat{g}_n):\ v. \tag{11.68}$$

Here we also have to invert the normal ordering prescription as our representation is done with the original grading. The integral can be determined by calculating the residue at P_∞. We have the orders

$$\operatorname{ord}_{P_\infty}(\hat{\omega}^m) = -m-1, \quad \operatorname{ord}_{P_\infty}(\hat{\omega}^n) = -n-1, \quad \operatorname{ord}_{P_\infty}(\hat{e}_k) = k+1.$$

Hence there is only a residue if $k \leq n+m$. Also note that $\hat{g}_m \in \mathcal{A}^r$ if $m \geq 1$. As $k \geq 1$, at least one n or m has to be ≥ 1. Hence if both are positive this term will clearly be an element of $\overline{\mathfrak{g}}^r\mathfrak{V}$. If one of them is ≤ 0, then the other is > 0 and according to normal ordering the other has to move to the left. Hence, again the term lies in $\overline{\mathfrak{g}}^r\mathfrak{V}$. From this the claim (11.67) follows. □

After these preliminaries we discuss the variation of the complex structure in more detail. Above we introduced U_∞^* and identified the vector field with tangent vectors on moduli via Cech cohomology. Choose a generic point in the moduli space with moduli parameters τ_0 in $\mathcal{M}$. In particular, Σ_{τ_0} has a fixed conformal structure representing the algebraic curve corresponding to the moduli parameters τ_0. For τ lying in a small enough neighborhood of τ_0, the conformal structure on Σ_τ can be obtained by deforming the conformal structure Σ_{τ_0} as described in the following section. Roughly speaking, we cut a coordinate patch at P_∞, deform the gluing function and re-glue it. This is exactly how the Kodaira–Spencer cocycle is constructed.

On the Riemann surface Σ_{τ_0}, we choose a local coordinate w at P_∞. Rigorously speaking, this amounts to passing (temporarily) to the moduli space $\widetilde{W}^{(\infty)}$. Let $U_\infty \subset \mathbb{C}$ be a unit disc with natural coordinate z. After identification of z with w, we can think of U_∞ as a subset of the coordinate chart at P_∞. In fact, we might even assume that the disc with radius 2 is still inside this chart. Let $v \in \mathcal{L}$ be a Krichever–Novikov vector field. By restriction, it defines a meromorphic vector field on U_∞. In turn, this vector field defines a family of local diffeomorphisms ϕ_t – the corresponding local flow. For t small enough they map an annulus $U_v \subset U_\infty$, which is bounded by the unit circle from the outside to a deformed annulus in the disc with radius 2. We take the set on the coordinate chart at P_∞, which (after the identification of z with w) is the interior complement to $\phi_t(U_v)$, cut it out of the curve and use ϕ_t to define a gluing of U_∞ to the rest of the curve along the subset U_v. In this way, for every t we obtain another conformal structure. Depending on the vector field (and corresponding diffeomorphisms) the equivalence class of the conformal structure will change or not. In any case we obtain via this process any conformal structure which is close to the given one in moduli space (see [102] for further information). Moreover, the deformation in any tangent direction in the moduli spaces $\widetilde{W}$, and $\widetilde{W}^{(1,\infty)}$ can be realized.

In abuse of notation we assign to every nearby τ such a diffeomorphism d_τ, $w = d_\tau(z)$, where d_τ is defined on the annulus U_τ; keeping in mind that τ uniquely defines neither d_τ nor U_τ. Here we use w for the new coordinate and z the standard coordinate.

Consider a family of local diffeomorphisms d_τ, where τ runs over a disc in the space of moduli parameters. Recall that for every τ the corresponding d_τ is nothing but the gluing function $w = d_\tau(z)$ for Σ_τ. From this point of view, the family d_τ is nothing but a family of (local) functions depending on parameters. To stress this interpretation, define the function $d(z,\tau) = d_\tau(z)$. Define for $X = \sum_i X_i \frac{\partial}{\partial \tau_i}$ the $\partial_X d_\tau$ as follows:

$$(\partial_X d_\tau)(z) = \sum_i X_i(\tau) \frac{\partial d(z,\tau)}{\partial \tau_i}. \tag{11.69}$$

Now we assign to a vector field X on $\widetilde{W}$ a local vector field on Σ_τ

$$\rho(X) := \rho_z(X) \frac{d}{dz} \quad \text{with} \quad \rho_z(X) := d_\tau^{-1}(\partial_X d_\tau(z)). \tag{11.70}$$

This is the *Kodaira–Spencer cocycle* of the corresponding deformation family.

Remark 11.20. The $\rho(X)$ is a priori a local vector field, i.e., $\rho(X) \in \widetilde{\mathcal{L}}$. By adding suitable coboundary terms (which corresponds to composing the diffeomorphism d_τ with a diffeomorphism of the disc - which again corresponds to taking a different local coordinate at P_∞) we could even obtain

$$\rho(X) = d_\tau^{-1} \cdot \partial_X d_\tau \in \mathcal{L}. \tag{11.71}$$

Furthermore, with (11.43) we get

$$\overline{\rho}(X) \in \mathcal{L}/\ker\theta_{k,p} \cong \overline{\mathcal{L}}, \tag{11.72}$$

where $\overline{\mathcal{L}}$ is a fixed chosen complement. For example, we can take the corresponding critical strip.

As intermediate data in the construction we had to choose a coordinate at P_∞. Hence, the element $\rho(X)$ will depend on it. It is only fixed up to the addition of elements of $\mathcal{L}_-^{(1)}$, as its elements correspond to the changes of the coordinate at P_∞ (but leaving its first order jet invariant as we work over $\widetilde{W} = \widetilde{W}^{(1,1)}$). We call every such element $\rho(X)$ a *pullback* of X. Note that this ambiguity comes from the diffeomorphism of the disc. If $\rho(X)$ and $\rho'(X)$ represents the same X, then

$$\rho(X) - \rho'(X) \in \widetilde{\mathcal{L}}_-^{(1)}. \tag{11.73}$$

Note that $\overline{\rho}(X) = \overline{\rho}'(X)$ after passing to the critical strip.

Given a vector field X on $\widetilde{W}$ by $\partial_X g_\tau$, we mean a full derivative in τ of $\tilde{g}_\tau(d_\tau^{-1}(w))$ along the vector field X. We interpret this as a differentiation (in τ) of a Krichever–Novikov function as a function in the variable z, taking account of the dependence of the local coordinate z on the moduli τ. After the substitution $w = d_\tau(z)$ we consider $\partial_X g_\tau$ an element of the sheaf $\tilde{\mathcal{A}}$. In formulas

$$\partial_X g_\tau(z) = \partial_X(\tilde{g}_\tau(d_\tau^{-1}(w)))_{|w=d_\tau(z)}. \tag{11.74}$$

The question is how $\partial_X g$ and $\rho(X).g$ are related. This is answered by the following proposition.

Proposition 11.21 ([227, Proposition 4.2], [250, Proposition 3.10]). *For every section g of the sheaf $\mathcal{A}$, and every vector field* [3] *X, let $g^X \in \tilde{\mathcal{A}}$ be given by the relation*

$$\partial_X g = -(\rho(X)).g + g^X. \tag{11.75}$$

Then $g^X \in \tilde{\mathcal{A}}^r$ for $g \in \mathcal{A}^r$.

Let $\mathcal{L}$ be the sheaf of Krichever–Novikov vector field algebras, and e be a meromorphic section of it. Its derivative with respect to vector fields X can be described in the usual way to give derivatives of operators

$$(\partial_X e).g = [\partial_X, e].g = \partial_X(e.g)) - e.(\partial_X g). \tag{11.76}$$

3 If we use vector field X, we always assume a local vector field on moduli.

This relation is equivalent to the Leibniz rule

$$(\partial_X e).g = \partial_X(e.g) - e.\partial_X g. \tag{11.77}$$

Proposition 11.22 ([227, Proposition 4.3], [250, Proposition 3.11]). *For every section e of the sheaf $\mathfrak{L}$, and every local vector field X let $e^X \in \tilde{\mathfrak{L}}$ be defined by the relation*

$$\partial_X e = -[\rho(X), e] + e^X. \tag{11.78}$$

Then $e^X \in \tilde{\mathfrak{L}}^r$ for $e \in \mathfrak{L}^r$.

For both proofs we refer to [227] or [250].

11.4.2 Defining the connection

Let $\mathfrak{V}$ be a locally free representation sheaf of the sheaf $\overline{\mathfrak{g}}_{\widetilde{W}}$, with the convention $\widetilde{W} = \widetilde{W}^{(1,1)}$ we use in this section. Recall that the sheaf of conformal blocks is given by

$$C(\mathfrak{V}) = \mathfrak{V}/\overline{\mathfrak{g}}^r_{\widetilde{W}}. \tag{11.79}$$

In the following section we will ignore the fact that we temporarily have to use the completion $\overline{\mathfrak{V}}$. By passing to the conformal blocks it will disappear again.

Consider a sheaf of operators on the local sections of the sheaf $\mathfrak{V}$. Assume B to be a local section of it. Then its derivative is defined as

$$\partial_X B\,.\,v = [\partial_X, B]\,.\,v = \partial_X(B\,.\,v) - B(\partial_X v)\,, \tag{11.80}$$

where v denotes a local section of the sheaf $\mathfrak{V}$. In this way, a differentiation ∂_X is defined on $\mathfrak{V}$. We have to deal with the following cases:

(1) $B = u(g)$ where $u \in \mathfrak{g}$ and $g \in \mathcal{A}$;

(2) $B = T[e] =: T(e)$, the Sugawara operator associated with $e \in \mathfrak{L}$, introduced in Chapter 10 and discussed already in Remark 11.15[4].

Recall that we extended the operation to local objects; see Remark 11.18. To avoid cumbersome notation we already used $\mathcal{A}$, $\mathfrak{L}$, etc., but meant of course the sheaves, and the elements are supposed to be sections of the sheaves. We will use this simplified notation also here.

We assume now that for the derivative of the operators $u(g)$ on $\mathfrak{V}$ we have

$$\partial_X u(g) = u(\partial_X g). \tag{11.81}$$

4 In this section we choose to denote the Sugawara operator by $T(e)$ to avoid confusion with the Lie bracket.

As shown in [227], this is fulfilled both for the fermionic representation and for the Verma module representations.

Given an X, we choose the local vector field $\rho(X)$ as described above and define the first order differential operator on local sections of $\mathfrak{V}$

$$\nabla_X = \partial_X + T(\rho(X)). \tag{11.82}$$

The operator ∇_X on $\mathfrak{V}$ will depend on the pullback of $\rho(X)$, i.e., on the coordinate at P_∞.

The following will be the key result.

Proposition 11.23 ([227, Proposition 4.4.]). *The operator ∇_X is well-defined on conformal blocks $C(\mathfrak{V})$ and does not depend on a choice of the pullback $\rho(X)$ of X.*

All relations in the following are with respect to conformal blocks.

Proof. Using (11.81) and (11.65) we obtain

$$\begin{aligned}[\nabla_X, u(g)] &= [\partial_X + T(\rho(X)), u(g)] = [\partial_X, u(g)] + [T(\rho(X)), u(g)]\\ &= u(\partial_X g) + u(\rho(X).g).\end{aligned}$$

Assume $g \in \mathcal{A}^r$. Then, by Proposition 11.21, we have

$$u(\partial_X g) = -u(\rho(X).g) + u(g^X),$$

where $g^X \in \mathcal{A}^r$. Hence, $[\nabla_X, u(g)] = u(g^X)$ and

$$[\nabla_X, u(\mathcal{A}^r)] \subseteq u(\mathcal{A}^r).$$

Hence $\overline{\mathfrak{g}}^r\mathfrak{V}$ is a ∇_X-invariant subspace and ∇_X is well-defined on $\mathfrak{V}/\overline{\mathfrak{g}}^r\mathfrak{V}$.

Two pullbacks of the same $\rho(X)$ differ by an element of $\widetilde{\mathfrak{L}}_-^{(1)}$; see (11.73). By Lemma 11.19 in passing to conformal blocks the dependency will vanish. □

Proposition 11.24 ([227, Proposition 4.6]). *Let $\mathfrak{V}$ be either an irreducible representation of $\overline{\mathfrak{g}}$ or a fermionic representation, then for every $X \in T\widetilde{W}$ we have*

$$\partial_X T(e) = T(\partial_X e) + \lambda \cdot id,$$

where $\lambda = \lambda(X, e) \in \mathbb{C}$.

Proof. By the fundamental relation (11.65) (see Proposition 10.14) for the Sugawara representation we have for every $e \in \mathfrak{L}$, $u \in \mathfrak{g}$, $g \in \mathcal{A}$ (and for the elements in $\widetilde{\mathfrak{L}}$ and $\widetilde{\mathcal{A}}$)

$$[T(e), u(g)] = u(e.g). \tag{11.83}$$

We take the derivative on both sides of the relation (11.83) along a vector field $X \in T\widetilde{W}^{(1)}$. By (11.81) we obtain

$$[\partial_X T(e), u(g)] + [T(e), \partial_X u(g)] = u((\partial_X e).g) + u(e.(\partial_X g)). \tag{11.84}$$

Again, by (11.83) and (11.81) the second terms on both sides of (11.84) are equal. Therefore,

$$[\partial_X T(e), u(g)] = u((\partial_X e).g).$$

Applying (11.83) once more, we replace the right-hand side of the latter relation by $[T(\partial_X e), u(g)]$ (making use of the fact that (11.83) also works for the local vector field $\partial_X e$). Therefore,

$$[\partial_X T(e) - T(\partial_X e), u(g)] = 0 \tag{11.85}$$

for every $X \in T\widetilde{W}^{(1)}$, $u \in \mathfrak{g}$, $g \in \mathcal{A}$. In case our representation is irreducible, the operator $\partial_X T(e) - T(\partial_X e)$ commutes with all operators of the representation. Hence, by Schur's lemma it is a scalar operator. This was the claim. For the fermionic representation the relation (11.85) will give that $v = (\partial_X T(e) - T(\partial_X e))v_{vac}$ will be also a vacuum vector. By the uniqueness of the vacuum vector (up to scalars) this shows the claim. □

Proposition 11.25 ([254, Lemma 1.3.8]).

$$\rho([X, Y]) = [\rho(X), \rho(Y)] + \partial_X \rho(Y) - \partial_Y \rho(X). \tag{11.86}$$

Theorem 11.26 ([227, Theorem 4.8.], [250, Theorem 3.14]). *The operator ∇_X is a projectively flat connection on the vector bundle of conformal blocks, i.e.,*

$$[\nabla_X, \nabla_Y] = \nabla_{[X,Y]} + \lambda(X, Y) \cdot id, \quad \lambda(X, Y) \in \mathbb{C}. \tag{11.87}$$

Proof. By the definition of the connection we have

$$\begin{aligned} [\nabla_X, \nabla_Y] &= [\partial_X + T(\rho(X)), \partial_Y + T(\rho(Y))] \\ &= [\partial_X, \partial_Y] + [\partial_X, T(\rho(Y))] - [\partial_Y, T(\rho(X))] \\ &\quad + [T(\rho(X)), T(\rho(Y))]. \end{aligned} \tag{11.88}$$

Since T is a projective representation of $\mathcal{L}$, and due to the relations

$$[\partial_X, T(\rho(Y))] = \partial_X T(\rho(Y)), \quad [\partial_Y, T(\rho(X))] = \partial_Y T(\rho(X)),$$

we rewrite (11.88) in the form

$$[\nabla_X, \nabla_Y] = \partial_{[X,Y]} + T\left(\partial_X \rho(Y) - \partial_Y \rho(X) + [\rho(X), \rho(Y)]\right) + \lambda(X, Y) \cdot id. \tag{11.89}$$

By Proposition 11.25, this reads as

$$\begin{aligned} [\nabla_X, \nabla_Y] &= \partial_{[X,Y]} + T\left(\rho([X, Y])\right) + \lambda(X, Y) \cdot id \\ &= \nabla_{[X,Y]} + \lambda(X, Y) \cdot id. \end{aligned} \tag{11.90}$$

□

11.4.3 Knizhnik–Zamolodchikov equations

Let $\mathfrak{V}$ be a sheaf of admissible representations of the affine algebra $\widehat{\mathfrak{g}}$ over $\widetilde{W}^{(1,1)}$, as introduced above. We assume the level c to be constant and obeying the condition $(c+\kappa) \neq 0$. Moreover, let all the additional assumptions be fulfilled which were needed in the previous section to define the Knizhnik–Zamolodchikov connection on conformal blocks.

Definition 11.27 ([226]). The Knizhnik–Zamolodchikov equations are the equations

$$\nabla_X \Phi = (\partial_X + T(\rho(X)))\Phi = 0, \qquad X \in \mathrm{H}^0(\widetilde{W}^{(1,1)}, \mathrm{T}\mathcal{M}^{(1,1)}_{g,N+1}) \tag{11.91}$$

respectively $X \in \mathrm{H}^0(\widetilde{W}, \mathrm{T}\mathcal{M}_{g,N+1})$, where Φ is a section of the sheaf of conformal blocks $\mathrm{C}(\mathfrak{V})$.

Hence, the Knizhnik–Zamolodchikov equations are the equations which have as solutions the horizontal sections of the connection ∇_X.

After fixing $\widehat{\sigma}_\infty$ and a first order infinitesimal neighborhood of it, we can again formulate this over the moduli space $\mathcal{M}^{(1)}_{g,N}$. This means that we do not consider the point P_∞ as a moving point, hence there is no tangent direction corresponding to this.

Denote the tangent vectors by X_k, $k = 1, \ldots, 3g-3+2N$ and set $e_k = \rho(X_k)$ for the corresponding element of the critical strip.

The elements of (the centrally extended) vector field algebra operate vertically on the fiber of the representation sheaf via the Sugawara representation. Recall the definition (11.57) of the operator $T[e]$ which is defined for every vector field $e \in \mathcal{L}$ and after passing to the conformal blocks also for local vector fields in $\widetilde{\mathcal{L}}$. We set $\partial_k := \partial_{X_k}$.

We define for sections Φ of $\mathfrak{V}$ and for every k the operator

$$\nabla_k \Phi := (\partial_k + T[e_k])\, \Phi. \tag{11.92}$$

Then the Knizhnik–Zamolodchikov equations read as

$$\nabla_k\, \Phi = 0\,, \qquad k = 1, \ldots, 3g-3+2N. \tag{11.93}$$

Using (10.21) and

$$l_k^{(n,p)(m,s)} := \frac{1}{2\pi \mathrm{i}} \int_{C_\tau} \omega^{n,p} \omega^{m,s} e_k, \tag{11.94}$$

we rewrite this as

$$\left(\partial_k - \frac{1}{c+\kappa} \sum_{\substack{n,m\\p,s}} l_k^{(n,p)(m,s)} \sum_a :u_a(n,p) u^a(m,s): \right) \Phi = 0\,, \quad k = 1, \ldots, 3g-3+2N. \tag{11.95}$$

Here the summation over a is a summation over a system of dual basis elements in $\mathfrak{g}$. Note that the coefficients $l_k^{(n,p)(m,s)}$ encode the geometric information about the complex structure and the position of the points.

Sometimes we consider sections which take their values in certain subspaces of the conformal blocks. For the Verma module sheaf, for example, the subsheaf consisting of the elements in $C(\mathfrak{V})$ of degree zero, or the subspace of the elements annihilated by $\widehat{\mathfrak{g}}_+$ (in its sheaf versions). Or we will consider induced actions on quotient sheaves in such cases when the operator ∇_k maps the subspaces which are factored out to themselves.

Keep in mind that (for example) it is not possible to talk about changing the complex structure and "fixing the points". Neither is it possible to move the points (in $\mathcal{M}^{(1)}_{g,N}$) and "fix the coordinates"). Nevertheless, if we have global coordinates, we have for every movement of the points also a definite change of coordinates. In other words, we are considering a special subspace of $W^{(1)} \subseteq \mathcal{M}^{(1)}_{g,N}$ isomorphic to $W \subseteq \mathcal{M}_{g,N}$. In this case it is possible to consider the sheaves of representation and the corresponding Sugawara operators which are defined over $\mathcal{M}^{(1)}_{g,N}$ as sheaves of representations over $\mathcal{M}_{g,N}$ and to drop the corresponding part of the equations. But let us stress the fact that this depends on the coordinate prescription given.

Such coordinates exist for $g = 0$ (the quasi-global coordinate z) and for $g = 1$ (the coordinate on the simply-connected covering $\mathbb{C}$). In Section 11.4.4 and Section 11.4.5 we deal with exactly this situation.

Remark 11.28. If we pass over to the Kodaira–Spencer class $\overline{\rho}(X)$, we obtain the Knizhnik–Zamolodchikov equations in terms of a fixed critical strip. By Theorem 11.6 the elements of the standard critical strip

$$\mathcal{L}_0 \oplus \mathcal{L}_{-1} \oplus L^*_{(0)}, \quad \text{respectively} \quad \mathcal{L}_{-1} \oplus L^*_{(0)}, \tag{11.96}$$

correspond to tangent vectors along the moduli space $\mathcal{M}^{(1)}_{g,N}$, or respectively along $\mathcal{M}_{g,N}$. For example, the N equations related to $e_{0,p}$, $p = 1, \dots, N$ are responsible for changing the coordinates, the N equations related to $e_{-1,p}$, $p = 1, \dots, N$. correspond to moving the points, and the other ones corresponding to the $3g-3$ elements $e_k \in \mathcal{L}^*_{(0)}$ (for $g \geq 2$) are responsible for changing the complex structure of the curve.

Without further assumptions on the representations under consideration, respectively additional conditions on the solutions of the Knizhnik–Zamolodchikov equations, the equations (with $\overline{\rho}(X)$) will depend on the critical strip chosen. One such assumption guaranteeing independence is that we require from the solutions Φ that $\widehat{\mathfrak{g}}_+\Phi = 0$. But this might be too restrictive for certain applications.

11.4.4 Example $g = 0$

In genus zero the only relevant variations are moving the points and the first order deformations of the coordinates. Let z_i, $i = 1, \dots, N$ be the N moving points and fix the reference point z_∞ to be ∞. In Section 5.1 the expressions for the λ-forms are given.

For the normalized basis elements we obtain for $i = 1, \dots, N$

$$f^{\lambda}_{n,i}(z) = (z - z_i)^{n-\lambda} \left(\prod_{j \neq i}^{N} \frac{z - z_j}{z_i - z_j} \right)^{n-\lambda+1} (dz)^{\lambda}. \tag{11.97}$$

Here we incorporated the normalizing constants

$$\alpha(i) := \prod_{\substack{l=1 \\ l \neq i}}^{N} (z_i - z_l)^{-1}, \quad i = 1, \dots, N, \tag{11.98}$$

already in the expression. The elements from the standard critical strip responsible for moving the points are

$$e_i(z) := e_{-1,i}(z) = \left(\alpha(i) \cdot \prod_{j \neq i} (z - z_j) \right) . \frac{d}{dz}. \tag{11.99}$$

Its Krichever–Novikov dual basis elements for the quadratic differentials are

$$\Omega^i(z) := \Omega^{-1,i} = \frac{\mathrm{d}z^2}{z - z_i}. \tag{11.100}$$

The vector field e_i evaluates to $\frac{d}{dz}$ at the point z_i and vanishes at all other points z_j, $j \neq i$. Therefore e_i corresponds to the basic direction ∂_i on the configuration space which is responsible for moving the point z_i.

Using the expression (11.94), the coefficients in (11.95) can be easily calculated. The following coefficients are non-zero ($k = 1, \dots, N$):

$$l_k^{(0,i)(0,i)} = \frac{\alpha(i)^{-1}\alpha(k)}{z_i - z_k}, \ (i \neq k), \qquad l_k^{(0,i)(0,k)} = l_k^{(0,k)(0,i)} = \frac{1}{z_k - z_i}, \ (i \neq k),$$
$$l_k^{(0,k)(0,k)} = \sum_{i \neq k} \frac{1}{z_k - z_i}, \qquad l_k^{(-1,k)(0,k)} = l_k^{(0,k)(-1,k)} = \alpha(k)^2.$$

Now we can write down the Knizhnik–Zamolodchikov equation using (11.95).

To show that these equations are equivalent to the classical equations in the situations considered, we will change the basis elements by adding vector fields from $\mathcal{L}_0$. We consider

$$e'_k := e_k + \sum_{i=1}^{N} \lambda_{ki} E_i, \quad \lambda_{ki} \in \mathbb{C}, \tag{11.101}$$

with $\{E_s \mid s = 1, \dots, N\}$ the Krichever–Novikov basis elements of degree zero, i.e,

$$E_i(z) := e_{0,i}(z) = (z - z_i) \prod_{s \neq i} \frac{(z - z_s)^2}{(z_i - z_s)^2} \frac{d}{dz}. \tag{11.102}$$

If we adjust the λ_{ki} accordingly and

(1) take the standard normal ordering

$$:x(n,i)y(m,j):= y(m,j)x(n,i), \quad \text{if} \quad n > m; \tag{11.103}$$

(2) consider the subspace of sections for which $\widehat{\mathfrak{g}}_+\psi = 0$;
(3) take the conformal block quotient $\mathfrak{V}/\overline{\mathfrak{g}}^r\mathfrak{V}$,

then in the new basis e_k' the coefficients yield

$$\Big(\partial_i - \frac{1}{c+\kappa}\sum_{j\neq i}\sum_a \frac{:u_a(0,i)u^a(0,j): + :u_a(0,j)u^a(0,i):}{z_i - z_j}\Big)\Phi = 0, \quad i = 1,\dots,N. \tag{11.104}$$

The above change of basis elements corresponds to the fact that we made the vector fields moving the points free from parts which would change the first jet of coordinates.

As $\overline{\mathfrak{g}}^r$ contains all degree ≤ -1 elements, everything can be realized in the degree zero part of the representation.

In the Verma module case where $\mathfrak{V}$ is the sheaf given by

$$\widehat{V}_{(\chi,c)} := U(\widehat{\mathfrak{g}}) \otimes_{U(\widehat{\mathfrak{b}})} V_\chi\,, \tag{11.105}$$

we take Φ from the $\mathfrak{g}$-module

$$V_\chi = V_{\chi_1} \otimes V_{\chi_2} \otimes \cdots \otimes V_{\chi_N} \tag{11.106}$$

which lies in degree zero. Furthermore, we take a self-dual basis u_a. Then the action of $u_a(0,i)$ can be described as

$$u_a(0,i)\,\Phi = t_i^a\,\Phi,$$

and furthermore

$$u_a(0,j)u_a(0,i)\,\Phi = t_j^a t_i^a\,\Phi = t_i^a t_j^a\,\Phi\,, \quad \text{for } j \neq i.$$

Here i says that the representation V_{χ_i} at the point P_i is considered. Hence,

$$\Big(\frac{\partial}{\partial z_i} - \frac{2}{c+\kappa}\sum_{j\neq i}\frac{t_i^a t_j^a}{z_i - z_j}\Big)\Phi = 0\,, \quad i = 1,\dots,N.$$

This is exactly the usual form of the rational Knizhnik–Zamolodchikov equation found in [131].

For details of the calculations see [227] or the book [250].

11.4.5 Example $g = 1$

In the genus one case we have, besides directions responsible for changing the coordinates and for moving the points, also one direction varying the complex moduli of the torus, respectively the elliptic curve. As above we ignore the deformations of the local coordinates. In Chapter 5 explicit expressions of the Krichever–Novikov basis elements were given with the help of the Weierstraß σ-function.

Let the moving points be represented by $z_1, \dots, z_N$ and the reference point represented by z_0. Then the moduli directions corresponding to moving the points are given for $k = 1, \dots, N$ by

$$e_k(z) = \prod_{s \neq k} \frac{\sigma(z - z_s)}{\sigma(z_k - z_s)} \cdot \frac{\sigma(z_k - z_0)^N}{\sigma(z - z_0)^N} \cdot \frac{\sigma(z + \sum_{s \neq k} z_s - N z_0)}{\sigma(\sum_s z_s - N z_0)} \frac{d}{dz}. \tag{11.107}$$

After adding certain vector fields of degree 0 to e_k (compensating for the effect of changing the local coordinates), we obtain Knizhnik–Zamolodchikov equations corresponding to moving the points ($k = 1, \dots, N$):

$$\begin{aligned} \partial_k \Phi &- \frac{1}{c + \kappa} \sum_{i \neq k} l_k^{(0,k)(0,i)} \left(:u(0,k)u(0,i): + :u(0,i)u(0,k):\right) \Phi \\ &- \frac{1}{c + \kappa} \sum_{i=1}^{N} \left(:u(0,i)u(-1,i): + :u(-1,i)u(0,i):\right) \Phi = 0. \end{aligned} \tag{11.108}$$

Here again we used the same restriction as in the genus zero case. Now the condition $\widehat{\mathfrak{g}}_+ \Phi$ is a really a restriction.

Remark 11.29. In the expression (11.108), elements of degree -1 will appear. This corresponds to the fact that the function $A_{-1,i}$ cannot be regular at z_0, as otherwise it would be a function on the torus which only has a pole of order 1 in contradiction to the residue theorem. Hence

$$\overline{\mathfrak{g}}_{-1} \cap \overline{\mathfrak{g}}^r = \{0\}.$$

This behavior is different to the $g = 0$ case, where everything after dividing out the subrepresentation of the reduced regular algebra takes place in degree zero.

Next we consider the Knizhnik–Zamolodchikov equation corresponding to the deformation of the complex structure (still under the restriction $\widehat{\mathfrak{g}}_+ \Phi = 0$). For this we take one element from the critical strip $\mathcal{L}^*_{(0)}$. In fact, each of the $e_{-2,i}$ would do. These elements have exactly poles of order 1 at z_i and at z_0. If we take one, the others will be linear combinations of this and vector fields which are regular at z_0. An equivalent but more symmetric choice is

$$e_0(z) = \sigma(z - w)^{N+1} \sigma(z - z_0)^{-1} \prod_{s=1}^{N} \sigma(z - z_s)^{-1} \frac{d}{dz},$$

where $w = (\sum_{i=0}^{N} z_i)/(N + 1)$. The vector field e_0 has simple poles at all the points $z_1, \dots, z_N, z_0$. If the points are in generic position then the point w cannot be equal to one of the z_i. Hence, there are no cancellations of poles. The element e_0 will have a nonvanishing residue at z_0. In particular, it cannot be a linear combination of the above vector fields $e_{-1,i}$ and vertical vector fields, as all of them have zero residue at z_0. Hence, it corresponds indeed to a deformation of the complex structure.

All the terms of the Knizhnik–Zamolodchikov equation which contain $u(n, j)$, $n < -1$ can be eliminated by factorization over $\overline{\mathfrak{g}}^r$ as in Section 11.4.4, and we obtain the equation corresponding to the deformation of the complex structure in the following form:

$$\begin{aligned}\partial_0\Phi - \frac{1}{c+\kappa}\sum_{i,j=1}^{N} l_0^{(0,i)(0,j)} \, :u(0,i)u(0,j):\,\Phi \\ - \frac{1}{c+\kappa}\sum_{i,j=1}^{N} l_0^{(0,i)(-1,j)}(:u(0,i)u(-1,j): + :u(-1,j)u(0,i):)\Phi \\ - \frac{1}{c+\kappa}\sum_{i=1}^{N} l_0^{(-1,i)(-1,i)} \, :u(-1,i)u(-1,i):\,\Phi &= 0.\end{aligned} \tag{11.109}$$

The appearing coefficients can be given in terms of values of the σ-function. Hence indirectly in terms of the points and the complex moduli τ. For details see [227] or the book [250].

12 Degenerations and deformations

In this chapter we will consider examples of families of Krichever–Novikov vector field and current algebras. These families are defined for Riemann surfaces of genus one. It will turn out to be more convenient to pass to the equivalent language of smooth projective curves over the complex numbers. Hence, we use examples of elliptic curves. The points where poles are allowed are special points. More precisely, they are two-torsion points. We will give complete expressions of the algebra structures. Hence, as a result aside we give examples of Krichever–Novikov type algebras by basis elements and structure equations. The basis elements and the structure constants will depend on the moduli (or configurations) of the geometric structure. Hence, we will obtain families of Lie algebras.

If the structure maximally degenerates the elliptic curve becomes a cuspidal cubic and the points move accordingly. It turns out that the structure of the vector field algebra degenerates to the Witt algebra. Interpreted differently, the Witt algebra will be deformed into these elliptic vector field algebras. There is a geometric reason behind this, which we will explain.

In joint work with Alice Fialowski [76–78] we had a closer look at these examples, in this case from the point of view that these families of algebras are deformations of the classical algebras. For example, the Witt algebra $\mathcal{W}$ will be a special element in the family. These families are locally nontrivial. This has to be seen in contrast to the statement that the Witt algebra cannot be deformed, or (what is the same) that it is rigid. The argument is that the second cohomology group $\mathrm{H}^2(\mathcal{W}, \mathcal{W})$ of the Witt algebra with values in the adjoint module is trivial. The later statement is true, as shown in [71, 73, 221]. However, the conclusion about rigidity has to be interpreted correctly, as we found nontrivial families. The point is that in the infinite-dimensional situation, the vanishing of $\mathrm{H}^2(\mathcal{W}, \mathcal{W})$ only tells us that the algebra is formally rigid. This does not imply that it is rigid in the geometric sense.

We will also consider here the case of current algebras, where the same phenomena appears. For $\mathfrak{g}$ a simple Lie algebra, the current algebra $\mathfrak{g} \otimes \mathbb{C}[z, z^{-1}]$ is rigid in the formal sense [87, 156]. Again, our examples produce interesting nontrivial deformation families.

We will not only consider cuspidal degenerations, but also nodal degenerations. In this case we obtain as degenerations interesting algebras associated to genus zero but now with 3 points where poles are allowed.

In this chapter we mainly point out the results and explain them. For the proofs and additional details we refer to the original literature [50, 76–78, 197, 208]. For some deformation families of Krichever–Novikov algebras over the moduli space see also Wagemann [261].

12.1 Deformations of Lie algebras

Before we deal with the examples of Krichever–Novikov type algebras, we recall the background on deformations of Lie algebras. The following section is essentially taken from the above cited joint work with Alice Fialowski. More details can be found there.

A Lie algebra[1] $\mathcal{L}$ is a (complex) vector space (also denoted by $\mathcal{L}$) endowed with the Lie product $[.,.]$. Equivalently, we describe it as $\mathcal{L}$ endowed with an anti-symmetric bilinear form

$$\mu_0 : \mathcal{L} \times \mathcal{L} \to \mathcal{L}, \quad \mu_0(x,y) = [x,y],$$

fulfilling the equivalent of the Jacobi identity ($\mu = \mu_0$)

$$\mu(\mu(x,y),z) + \mu(\mu(y,z),x) + \mu(\mu(z,x),y) = 0, \quad \forall x,y,z \in \mathcal{L}. \tag{12.1}$$

Consider on the same vector space $\mathcal{L}$ is modeled on, a family of Lie algebra structures

$$\mu_t = \mu_0 + t \cdot \phi_1 + t^2 \cdot \phi_2 + \cdots, \tag{12.2}$$

with bilinear maps $\phi_i : \mathcal{L} \times \mathcal{L} \to \mathcal{L}$ such that $\mathcal{L}_t := (\mathcal{L}, \mu_t)$ is a Lie algebra and $\mathcal{L}_0$ is the Lie algebra we started with. In particular, (12.1) has to be fulfilled for μ_t. The family $\{\mathcal{L}_t\}$ is a *deformation* of $\mathcal{L}_0$ [89, 181, 182].

Until now we have not specified the "parameter" t. Indeed, different choices are possible.

(1) The parameter t might be a variable which allows plugging in numbers $\alpha \in \mathbb{C}$. In this case, $\mathcal{L}_\alpha$ is a Lie algebra for every α for which the expression (12.2) is defined. The family can be considered as deformation over the affine line $\mathbb{C}[t]$ or over the convergent power series $\mathbb{C}\{\{t\}\}$. The deformation is called a *geometric* or an *analytic deformation* respectively.

(2) We consider t a formal variable and we allow an infinite number of terms in (12.2). It might be the case that μ_t does not exist if we plug in for t any other value different to 0. In this way we obtain deformations over the ring of formal power series $\mathbb{C}[[t]]$. The corresponding deformation is a *formal deformation.*

(3) The parameter t is considered an infinitesimal variable, i.e., we take $t^2 = 0$. We obtain *infinitesimal deformations* defined over the quotient $\mathbb{C}[X]/(X^2) = \mathbb{C}[[X]]/(X^2)$.

More general situations for the parameter space can be considered; see Section 12.2.

There is always the trivially deformed family given by $\mu_t = \mu_0$ for all values of t. Two families μ_t and μ_t' deforming the same μ_0 are *equivalent* if a linear automorphism (with the same vagueness regarding the meaning of t)

$$\psi_t = id + t \cdot \alpha_1 + t^2 \cdot \alpha_2 + \cdots \tag{12.3}$$

1 Here $\mathcal{L}$ is not necessarily the vector field algebra.

exists with $\alpha_i : \mathcal{L} \to \mathcal{L}$ linear maps, such that

$$\mu_t'(x, y) = \psi_t^{-1}(\mu_t(\psi_t(x), \psi_t(y))). \tag{12.4}$$

A Lie algebra $(\mathcal{L}, \mu_0)$ is called *rigid* if every deformation μ_t of μ_0 is locally equivalent to the trivial family. Intuitively, this means that $\mathcal{L}$ cannot be deformed.

The word "locally" in the definition of rigidity means that we only consider the situation for t "near 0". Of course, this depends on the category we consider. As on the formal and the infinitesimal level, only one closed point exists, i.e., the point 0 itself, every deformation over $\mathbb{C}[[t]]$ or $\mathbb{C}[X]/(X^2)$ is already local. This is different on the geometric and analytic level. Here it means that an (etale) open neighborhood U of 0 exists, such that the family restricted to it is equivalent to the trivial one. In particular, this implies $\mathcal{L}_\alpha \cong \mathcal{L}_0$ for all $\alpha \in U$.

An important question is to decide whether a given Lie algebra is rigid. Moreover, the question of rigidity will depend on the category we consider. Depending on the set-up, we will have to consider infinitesimal, formal, geometric, and analytic rigidity.

Deformation problems and moduli space problems are related to adapted cohomology theories. To a certain extent (in particular for the finite-dimensional case) this is also true for deformations of Lie algebras. In the finite-dimensional case, the relation is rather close. Our examples in the following sections will show that this relation is not so strong in the infinite-dimensional context. Differences appear if geometric or analytic deformations are considered.

For Lie algebra deformations, the relevant cohomology space is $\mathrm{H}^2(\mathcal{L}, \mathcal{L})$, the space of Lie algebra two-cohomology classes with values in the adjoint module $\mathcal{L}$. This means that $\mathcal{L}$ operates on itself by the Lie product. We introduced Lie algebra cohomology for arbitrary modules in Section 6.1. We specialize now the definition for the adjoint module.

An antisymmetric bilinear map $\phi : \mathcal{L} \times \mathcal{L} \to \mathcal{L}$ is a Lie algebra *two-cocycle* if $d_2\phi = 0$. For the adjoint module this means

$$\begin{aligned}\phi([x, y], z) + \phi([y, z], x) + \phi([z, x], y)&\\ - [x, \phi(y, z)] + [y, \phi(x, z)] - [z, \phi(x, y)] &= 0.\end{aligned} \tag{12.5}$$

The map ϕ will be a *coboundary*, if a linear map $\psi : \mathcal{L} \to \mathcal{L}$ with $\phi = d_1\psi$ exists. For the adjoint module this means

$$\phi(x, y) = (d_1\psi)(x, y) := \psi([x, y]) - [x, \psi(y)] + [y, \psi(x)]. \tag{12.6}$$

If we expand the Jacobi identity (12.1) for the family μ_t given by (12.2) in the variable t, we get the individual expressions

$$\begin{aligned}\mu_t(\mu_t(x, y), z) &= \mu_0(\mu_0(x, y), z) + t\,(\phi_1(\mu_0(x, y), z) + \mu_0(\phi_1(x, y), z)) + O(t^2)\\ &= [[x, y], z] + t\,(\phi_1([x, y], z) + [\phi_1(x, y), z]) + O(t^2).\end{aligned} \tag{12.7}$$

Here we identified μ_0 with the original Lie bracket in $\mathcal{L}_0$. Now we add up the cyclic permutation of (12.7). This sum has to be zero up to all orders in t. In zero order it is zero as we obtain as sum the Jacobi identity in $\mathcal{L}_0$. In order one it will be zero if and only if ϕ_1 is a two-cocycle with values in the adjoint module as defined by (12.5). If ϕ_1 vanishes identically, the same shows that the first nonvanishing ϕ_k will be a two-cocycle.

Assume now that μ_t and μ_t' are equivalent. If we plug in the expansions (12.2) and (12.3) into (12.4), use

$$\psi_t^{-1} = id - t\alpha_1 + O(t^2), \tag{12.8}$$

make an expansion in powers of t, we obtain for the term linear in t

$$\phi_1'(x, y) = \phi_1(x, y) + [x, \alpha_1(y)] - \alpha_1[x, y]. \tag{12.9}$$

Hence the two two-cocycles ϕ_1 and ϕ_1' will be cohomologous, i.e., their difference is a coboundary.

In the case where we consider infinitesimal deformations, i.e., $t^2 = 0$, there is no higher power of t present. The above shows that a two-cocycle ϕ defines via

$$\mu_t(x, y) = \mu_0(x, y) + t\phi(x, y) \tag{12.10}$$

an infinitesimal deformation. Two deformations will be equivalent if their cocycles are cohomologous.

If we have a more general deformation, e.g., a formal one, then by truncating the series (12.2) by setting $t^2 = 0$, we obtain an infinitesimal deformation. This deformation is called *differential of the deformation*. Hence, every deformation yields an infinitesimal deformation. This is not true in the opposite direction. Not every infinitesimal one can be lifted to every order. As far as formal deformations are concerned, there are obstructions which can be formulated inside the full cohomology. These are given in terms of the Massey Lie product in the third cohomology, see [68, 89].

The following are known results:

(1) the space $\mathrm{H}^2(\mathcal{L}, \mathcal{L})$ classifies infinitesimal deformations of $\mathcal{L}$ (this is simple and was demonstrated above);
(2) if $\dim \mathrm{H}^2(\mathcal{L}, \mathcal{L}) < \infty$, then all formal deformations of $\mathcal{L}$ up to equivalence can be realised in this vector space [75];
(3) if $\mathrm{H}^2(\mathcal{L}, \mathcal{L}) = 0$, then $\mathcal{L}$ is infinitesimally and formally rigid (this follows directly from (1) and (2));
(4) if $\dim \mathcal{L} < \infty$, then $\mathrm{H}^2(\mathcal{L}, \mathcal{L}) = 0$ implies that $\mathcal{L}$ is also rigid in the geometric and analytic sense [89, 181, 182].

The examples presented here will show that without the condition $\dim \mathcal{L} < \infty$, point (4) is not true anymore. Our locally nontrivial vector field families will contain the Witt algebra $\mathcal{W}$ as special element. But as is announced in [73] and shown in [71, 72, 221], we have $\mathrm{H}^2(\mathcal{W}, \mathcal{W}) = 0$. Hence $\mathcal{W}$ is formally rigid but not in the geometric or analytic sense. The same is true for the classical current algebras $\overline{\mathfrak{g}} = \mathfrak{g} \otimes \mathbb{C}[z^{-1}, z]$, with $\mathfrak{g}$ a finite-dimensional simple Lie algebra. Lecomte and Roger [156], and Garland [87] showed that $\overline{\mathfrak{g}}$ is formally rigid.

Needless to say, deformations of algebraic and geometric structure are of fundamental importance in mathematics and physics. Hence, such results showing fundamental differences between the finite-dimensional and infinite-dimensional, or rather the formal and non-formal situation, are important. For example, in the context of the Wess–Zumino–Novikov–Witten models examined in Chapter 11, such deformations show up both from the geometric and the algebraic point of view. Starting from the finite-dimensional symmetry algebra (i.e., the Lie algebra $\mathfrak{g}$), families of higher genus affine algebras and their representations are constructed, associated with the geometric data corresponding to the point in $\mathcal{M}_{g,N}$. The bundle of conformal blocks appears as a bundle of coinvariants. Now, clearly, the following question is fundamental. What happens if we approach the boundary of the moduli space? The boundary components correspond to curves with singularities. Resolving the singularities yields curves of lower genera. By geometric degeneration, we obtain families of (Lie) algebras containing a lower genus algebra (or sometimes a subalgebra of it), corresponding to a suitable collection of marked points, as special element. Or, reverting the perspective, we obtain a typical situation of the deformation of an algebra corresponding in some way to a lower genus situation, containing higher genus algebras as the other elements in the family. Such kinds of geometric degenerations are fundamental if one wants to prove Verlinde-type formula via factorization and normalization technique; see [254]. Further, see [274] and [275] for collections of results, applications and further topics.

Remark 12.1. We will not need it in the following section, but for completeness it should be mentioned that under the condition that $\dim \mathrm{H}^2(\mathcal{L}, \mathcal{L}) < \infty$ (without any further condition on $\mathcal{L}$ itself), a universal family of infinitesimal deformations exists [89]. In the formal setting it was shown by Fialowski and Fuks [75] that a versal formal family exists which can be described in $\mathrm{H}^2(\mathcal{L}, \mathcal{L})$ by a finite system of formal equations.

Remark 12.2. There is another approach to deformations of Lie algebras. Unfortunately, it works only in the finite dimension. In this case we can choose a basis and then the Lie algebra is given by the structure constants with respect to this basis. Of course, all isomorphy types of a fixed dimension n can be realized on a fixed vector space. The change of a basis corresponds to an action of $\mathrm{GL}(n)$ on the subset of structure constants $\mathbb{C}^{3n}$. If the orbit of the point realizing a Lie algebra $\mathcal{L}$ under the $\mathrm{GL}(n)$ action is Zariski dense in the variety of all structure constants, this algebra is *rigid in the orbit sense*. In fact, in the finite-dimensional situation this means that every "nearby" algebra is isomorphic to $\mathcal{L}$, hence it is rigid in any of the above senses. But this approach fails in infinite dimension.

12.2 Definition of a general deformation of a Lie algebra

Above we were not very precise about the character of the "parameter" t. A precise concept of deformations of Lie algebras is given by Fialowski and Fuks [74, 75]. We recall this definition for completeness here. Our different types of deformations are examples of it. The reader might skip this subsection, as for the deformation we consider the naive definition will be enough.

In the following section we will assume that A is a commutative algebra over $\mathbb{K}$ (where $\mathbb{K}$ is a field of characteristic zero) which admits an augmentation $\epsilon : A \to \mathbb{K}$. This means that ϵ is a $\mathbb{K}$-algebra homomorphism, e.g., $\epsilon(1_A) = 1$. The ideal $m_\epsilon := \operatorname{Ker} \epsilon$ is a maximal ideal of A. Vice versa, given a maximal ideal m of A with $A/m \cong \mathbb{K}$, the natural quotient map defines an augmentation.

If A is a finitely generated $\mathbb{K}$-algebra over an algebraically closed field $\mathbb{K}$, then $A/m \cong \mathbb{K}$ is true for every maximal ideal m. Hence, in this case every such A admits at least one augmentation and all maximal ideals come from augmentations.

Let us consider a Lie algebra $\mathcal{L}$ over the field $\mathbb{K}$, ϵ a fixed augmentation of A, and $m = \operatorname{Ker} \epsilon$ the associated maximal ideal.

Definition 12.3 ([75]). A *deformation* λ of $\mathcal{L}$ with base (A, m), or simply with base A, is a Lie A-algebra structure on the tensor product $A \otimes_{\mathbb{K}} \mathcal{L}$, with bracket $[.,.]_\lambda$ such that

$$\epsilon \otimes \mathrm{id} : A \otimes \mathcal{L} \to \mathbb{K} \otimes \mathcal{L} = \mathcal{L} \tag{12.11}$$

is a Lie algebra homomorphism.

This means that for all $a, b \in A$ and $x, y \in \mathcal{L}$,
(1) $[a \otimes x, b \otimes y]_\lambda = (ab \otimes \mathrm{id})[1 \otimes x, 1 \otimes y]_\lambda$;
(2) $[.,.]_\lambda$ is skew-symmetric and satisfies the Jacobi identity;
(3) $\epsilon \otimes \mathrm{id}([1 \otimes x, 1 \otimes y]_\lambda) = 1 \otimes [x, y]$.

By condition (1) to describe a deformation it is enough to give the elements $[1 \otimes x, 1 \otimes y]_\lambda$ for all $x, y \in \mathcal{L}$. If $B = \{z_i\}_{i \in J}$ is a basis of $\mathcal{L}$, it follows from condition (3) that the Lie product has the form

$$[1 \otimes x, 1 \otimes y]_\lambda = 1 \otimes [x, y] + \textstyle\sum_i' a_i \otimes z_i, \tag{12.12}$$

with $a_i = a_i(x, y) \in m$, $z_i \in B$. Here $\sum'$ denotes a finite sum. Clearly, condition (2) is an additional condition which has to be satisfied.

If we use $A = \mathbb{C}[t]$, we get exactly the notion of a one-parameter geometric deformation discussed above.

A deformation is called *trivial* if $A \otimes_{\mathbb{K}} \mathcal{L}$ carries the trivially extended Lie structure, i.e., (12.12) reads as $[1 \otimes x, 1 \otimes y]_\lambda = 1 \otimes [x, y]$.

Two deformations of a Lie algebra $\mathcal{L}$ with the same base A are called *equivalent* if a Lie algebra isomorphism exists between the two copies of $A \otimes \mathcal{L}$, with the two Lie algebra structures compatible with $\epsilon \otimes \mathrm{id}$.

Formal deformations are defined in a similar way. Let A be a complete local algebra over $\mathbb{K}$, so $A = \varprojlim_{n\to\infty}(A/m^n)$, where m is the maximal ideal of A. Furthermore, we assume that $A/m \cong \mathbb{K}$, and $\dim(m^k/m^{k+1}) < \infty$ for all k.

Definition 12.4. A *formal deformation* of $\mathcal{L}$ with base A is a Lie algebra structure on the completed tensor product $A\hat{\otimes}\mathcal{L} = \varprojlim_{n\to\infty}((A/m^n)\otimes\mathcal{L})$ such that

$$\epsilon\hat{\otimes}\mathrm{id} : A\hat{\otimes}\mathcal{L} \to \mathbb{K}\otimes\mathcal{L} = \mathcal{L} \tag{12.13}$$

is a Lie algebra homomorphism.

If $A = \mathbb{C}[[t]]$, then a formal deformation of $\mathcal{L}$ with base A is the same as a formal one-parameter deformation discussed above. There is an analogous definition for equivalence of deformations parameterized by a complete local algebra.

12.3 The geometric families in the case of the torus

12.3.1 Complex tori

Let $\tau \in \mathbb{C}$ with $\operatorname{Im}\tau > 0$ and L be the lattice

$$L = \langle 1, \tau\rangle_{\mathbb{Z}} := \{m + n\cdot\tau \mid m, n \in \mathbb{Z}\} \subset \mathbb{C}. \tag{12.14}$$

The complex one-dimensional torus is the quotient $T = \mathbb{C}/L$. It carries a natural structure of a complex manifold coming from the structure of $\mathbb{C}$. It will be a compact Riemann surface of genus 1.

The field of meromorphic functions on T is generated by the doubly-periodic Weierstraß $\wp$ function and its derivative $\wp'$. Recall that

$$\wp(z,\tau) := \frac{1}{z^2} + \sum_{\omega\in\Gamma}{}' \left(\frac{1}{(z-\omega)^2} - \frac{1}{\omega^2}\right). \tag{12.15}$$

The $'$ at the sum denotes that we leave out 0 from the summation. This series converges for all $z \notin \Gamma$ to a holomorphic function. It is an even function which has a pole of second order at the lattice points. It is doubly periodic. If we differentiate we get

$$\wp'(z,\tau) = -\sum_{\omega\in\Gamma} \frac{2}{(z-\omega)^3}. \tag{12.16}$$

The function $\wp'$ is obviously doubly periodic too and has poles of order 3 at the lattice points. Both functions fulfill the differential equation

$$(\wp')^2 = 4(\wp - e_1)(\wp - e_2)(\wp - e_3) = 4\wp^3 - g_2\wp - g_3, \tag{12.17}$$

with

$$\Delta := g_2{}^3 - 27g_3{}^2 = 16(e_1 - e_2)^2(e_1 - e_3)^2(e_2 - e_3)^2 \neq 0. \tag{12.18}$$

Furthermore,

$$g_2 = -4(e_1e_2 + e_1e_3 + e_2e_3), \quad g_3 = 4(e_1e_2e_3). \tag{12.19}$$

The numbers e_i are pairwise distinct, can be given as

$$\wp(\frac{1}{2}, \tau) = e_1, \quad \wp(\frac{\tau}{2}, \tau) = e_2, \quad \wp(\frac{\tau+1}{2}, \tau) = e_3, \tag{12.20}$$

and fulfill

$$e_1 + e_2 + e_3 = 0. \tag{12.21}$$

Note that the "constants" $\Delta, e_1, e_2, e_3, g_2, g_3$ depend also on the lattice parameter τ.

The points[2]

$$\overline{1/2}, \quad \overline{\tau/2}, \quad \overline{(\tau+1)/2} \tag{12.22}$$

are the nontrivial two-torsion points. A point $\overline{z} \in T/L$ is called a two-torsion point if $2\overline{z} = \overline{0}$. Of course, $\overline{0}$ itself is also a two-torsion point. It is called the trivial two-torsion point.

As $\wp'$ is odd and doubly-periodic, it has to have zeros of order one at the nontrivial two-torsion points (12.22) and all their translates under the lattice L.

For the Krichever–Novikov type vector field algebra we have to fix two points where poles are allowed. To obtain a very symmetric situation we choose as points

$$\overline{0}, \quad \overline{1/2}. \tag{12.23}$$

12.3.2 The family of elliptic curves

One-dimensional tori correspond to elliptic curves. As we have degenerations in mind, it is more convenient to use this purely algebraic-geometric picture. The map

$$T \to \mathbb{P}^2(\mathbb{C}), \quad z \bmod L \mapsto \begin{cases} (\wp(z) : \wp'(z) : 1), & z \notin L \\ (0 : 1 : 0), & z \in L, \end{cases} \tag{12.24}$$

realizes T as a complex-algebraic smooth curve in the projective plane. As its genus is one it is an elliptic curve. The affine coordinates are $X = \wp(z,\tau)$ and $Y = \wp'(z,\tau)$. From (12.17) it follows that the affine part of the curve can be given by the smooth cubic curve defined by

$$Y^2 = 4(X - e_1)(X - e_2)(X - e_3) = 4X^3 - g_2X - g_3 =: f(X). \tag{12.25}$$

The point at infinity on the curve is the point $\infty = (0 : 1 : 0)$. It is a non-singular point.

2 Here $\bar{z}$ does not denote conjugation, but taking the residue class modulo the lattice.

The points (12.23) where poles are allowed are mapped under (12.24) to

$$\infty \cong (0:1:0), \quad (e_1, 0) \cong (e_1 : 0 : 1). \tag{12.26}$$

By choosing a different nontrivial two-torsion point we would only obtain an isomorphic situation. Recall that the elliptic curves can be given in the projective plane by

$$Y^2Z = 4X^3 - g_2XZ^2 - g_3Z^3, \quad g_2, g_3 \in \mathbb{C}, \quad \text{with } \Delta := g_2{}^3 - 27g_3{}^2 \neq 0. \tag{12.27}$$

The condition $\Delta \neq 0$ ensures that the curve will be non-singular.

A more convenient description instead of (12.27) is given by

$$Y^2Z = 4(X - e_1Z)(X - e_2Z)(X - e_3Z), \tag{12.28}$$

with

$$e_1 + e_2 + e_3 = 0, \quad \text{and} \quad \Delta = 16(e_1 - e_2)^2(e_1 - e_3)^2(e_2 - e_3)^2 \neq 0. \tag{12.29}$$

These presentations are related via

$$g_2 = -4(e_1e_2 + e_1e_3 + e_2e_3), \quad g_3 = 4(e_1e_2e_3). \tag{12.30}$$

The coefficients g_2, g_3, or respectively e_1, e_2, e_3, are algebraic parameters. They can be varied to obtain all elliptic curves as long as

$$\Delta = \Delta(g_1, g_2) = \Delta(e_1, e_2, e_3) \neq 0. \tag{12.31}$$

If $\Delta = 0$, the cubic curve will have singularity. From (12.29) follows that this is the case if and only if two of the e_i and e_j, $i \neq j$ coincide.

In the following we will need the *elliptic modular parameter*

$$j = 1728 \, \frac{g_2^3}{\Delta} \tag{12.32}$$

classifying all elliptic curves up to isomorphy.

Next we want to realize these families. We set as base variety

$$B := \{(e_1, e_2, e_3) \in \mathbb{C}^3 \mid e_1 + e_2 + e_3 = 0, \quad e_i \neq e_j \text{ for } i \neq j\}. \tag{12.33}$$

In the product $B \times \mathbb{P}^2$ we consider the family of elliptic curves $\mathcal{E}$ over B defined via (12.28). As the curves are given via the coefficients of the defining polynomial, the family can be extended to

$$\widehat{B} := \{e_1, e_2, e_3) \in \mathbb{C}^3 \mid e_1 + e_2 + e_3 = 0\}. \tag{12.34}$$

As new fibers above $\widehat{B} \backslash B$, singular cubic curves appear. By assigning the marked points (individually) to every point in the base $\widehat{B}$, we obtain two sections of the family. The sections are

$$\sigma_1(e_1, e_2, e_3) = \infty, \quad \sigma_2(e_1, e_2, e_3) = (e_1 : 0 : 1). \tag{12.35}$$

Over the points in the base corresponding to smooth curves the two sections will not meet.

Resolving the one linear relation in $\widehat{B}$ via $e_3 = -(e_1 + e_2)$, we obtain a family over $\mathbb{C}^2$. We consider the complex lines in $\mathbb{C}^2$

$$\begin{aligned} D_s &:= \{(e_1, e_2) \in \mathbb{C}^2 \mid e_2 = s \cdot e_1\}, \quad s \in \mathbb{C}, \\ D_\infty &:= \{(0, e_2) \in \mathbb{C}^2\}, \end{aligned} \tag{12.36}$$

and set

$$D_s^* = D_s \setminus \{(0,0)\} \tag{12.37}$$

for the punctured line.

By excluding all possibilities when two of the e_i coincides we obtain

$$B \cong \mathbb{C}^2 \setminus (D_1 \cup D_{-1/2} \cup D_{-2}). \tag{12.38}$$

As we have to check what happens to our sections (12.35), we have to consider these exceptional lines in detail. Above D_1^* we have $e_1 = e_2 \neq e_3$, above $D_{-1/2}^*$ we have $e_2 = e_3 \neq e_1$, and above D_{-2}^* we have $e_1 = e_3 \neq e_2$. In all these cases we obtain the nodal cubic. The nodal cubic E_N can be given as

$$Y^2 Z = 4(X - eZ)^2 (X + 2eZ), \tag{12.39}$$

where e denotes the value of the coinciding $e_i = e_j$ ($-2e$ is then necessarily the remaining one). The singular point is the point $(e : 0 : 1)$. It is a node. It is up to isomorphy the only singular cubic which is stable in the sense of Mumford-Deligne.

Above the unique common intersection point $(0,0)$ of all D_s there is the cuspidal cubic E_C

$$Y^2 Z = 4X^3. \tag{12.40}$$

The singular point is $(0 : 0 : 1)$. The curve is not stable in the sense of Mumford-Deligne. In both cases the complex projective line is the desingularisation. We will return to this in Section 12.6. In all cases (non-singular or singular), the point $\infty = (0 : 1 : 0)$, which is our image of the section σ_1, lies on the curves. It is the only intersection with the line at infinity, and is a non-singular point. If we keep in mind that there are poles at ∞, we will lose nothing by passing to an affine chart.

For the curves above the points in D_s^* we calculate $e_2 = se_1$ and $e_3 = -(1+s)e_1$ (or respectively $e_3 = -e_2$ if $s = \infty$). Due to homogeneity, the modular parameter j for the curves above D_s^* will be constant along the line. In particular, the curves in the family lying above D_s^* will be isomorphic. By a direct calculation we obtain from (12.32)

$$j(s) = 1728 \, \frac{4(1 + s + s^2)^3}{(1-s)^2 (2+s)^2 (1+2s)^2}, \qquad j(\infty) = 1728. \tag{12.41}$$

12.4 Basis for the meromorphic forms

Even if we are only interested in the Krichever–Novikov like basis for our situation, it is more conceptional to consider $\mathcal{F}^\lambda$ for all $\lambda \in \mathbb{Z}$. As we are in the genus one case, the canonical bundle is trivial, hence if we have a basis $\{A_n \mid n \in \mathbb{Z}\}$ for $\lambda = 0$, we obtain a basis for $\mathcal{F}^\lambda$ by taking the elements $A_{n-\lambda}(dz)^\lambda$ up to an index shift.

The field of meromorphic functions on the torus corresponds to the field of rational functions on the curve E given by

$$\mathbb{C}(X)[Y]/(Y^2 - f(X)), \quad f(T) = 4(T - e_1)(T - e_2)(T - e_3). \tag{12.42}$$

Every meromorphic function on the torus can be described as a rational function (i.e., as a function which is a quotient of polynomials) in X and Y. In fact, (12.42) shows that it can even be given as a rational function in X plus Y times another rational function in X. We will use the terms meromorphic and rational interchangeably.

Proposition 12.5 ([208]). *A basis of the vector space of meromorphic (rational) functions on the elliptic curve E holomorphic outside the points ∞ and $(e_1, 0)$ is given by*

$$A_{2k} = (X - e_1)^k, \quad A_{2k+1} = \frac{1}{2} Y \cdot (X - e_1)^{k-1}, \quad k \in \mathbb{Z}. \tag{12.43}$$

Using the facts given about the zero and pole order of $\wp$ and $\wp'$, we have with $\gamma \in \mathbb{C}$, $a \neq e_1, e_2, e_3$ the following results for the orders:

$$\begin{aligned} \operatorname{ord}_\infty(X - \gamma) &= -2; \quad \operatorname{ord}_\infty(Y) = -3; \\ \operatorname{ord}_{(e_i,0)}(X - e_i) &= 2; \quad \operatorname{ord}_{(e_i,0)}(Y) = 1; \\ \operatorname{ord}_{(a,b)}(X - a) &= 1; \quad \operatorname{ord}_{(a,-b)}(X - a) = 1. \end{aligned} \tag{12.44}$$

This shows for the divisors of the basis elements

$$\begin{aligned} (A_{2k}) &= -2k[\infty] + 2k[(e_1, 0)], \\ (A_{2k+1}) &= -(2k+1)[\infty] + (2k-1)[(e_1, 0)] + [(e_2, 0)] + [(e_3, 0)]. \end{aligned} \tag{12.45}$$

Next we have to express the analytic differential dz as

$$dz = \frac{dX}{Y}. \tag{12.46}$$

To see that it is indeed globally holomorphic, we observe that $(X - \gamma)$ for $\gamma \notin \{e_1, e_2, e_3\}$ is a uniformizing variable at $(\gamma, \pm\sqrt{f(\gamma)})$ and Y does not vanish there. For $\gamma = e_1, e_2$ or e_3 the function $(X - \gamma)$ vanishes of second order at $(e_i, 0)$, and hence compensates for the pole of $1/Y$ at this point.

For the vector fields we take as basis

$$V_m = A_{m-1} Y \frac{d}{dX}, \quad m \in \mathbb{Z}. \tag{12.47}$$

We introduce a degree in the modules by setting

$$\deg(V_m) = \deg(A_m) := m, \tag{12.48}$$

to yield an almost-grading.

Remark 12.6. The reader should not be puzzled by the fact that we did the shift in a different direction than in other places in the book. However, if we check the divisor (12.45) of the elements A_n, we see that the almost-grading is given with respect to the degree at ∞ (or in the analytic picture with respect to $\bar{0}$). Moreover, with respect to the standard degree n, we took $-n$, hence the inverted behavior. As a consequence, our almost-grading will have lower order terms instead of higher order terms. It would have been a simple task to rewrite everything, but I decided not to do this as we will refer for certain results to other articles with the convention used here.

Using Proposition 12.5 we obtain the next one.

Proposition 12.7. *The elements ($k \in \mathbb{Z}$)*

$$\begin{aligned} V_{2k} &:= \frac{1}{2} f(X)(X - e_1)^{k-2} \frac{d}{dX}, \\ V_{2k+1} &:= (X - e_1)^k Y \frac{d}{dX}, \end{aligned} \tag{12.49}$$

constitute a basis of the vector field algebra $\mathcal{L}$.

12.5 Families of algebras

We consider the algebras of Krichever–Novikov type corresponding to the elliptic curve and possible poles at $\bar{z} = \bar{0}$ and $\bar{z} = \overline{1/2}$ (respectively in the algebraic-geometric picture, at the points ∞ and $(e_1, 0)$).

We will consider the algebra structure above the moduli points corresponding to smooth curves, e.g., elliptic curves. It will turn out that they also define Lie algebras over the points in the base corresponding to the singular curves. We will study these algebras and give an explanation for this in Section 12.6.

12.5.1 Function algebras

The function algebra $\mathcal{A}$ has as basis the elements (12.43). One easily calculates its structure equations

$$A_n \cdot A_m = \begin{cases} A_{n+m}, & n \text{ or } m \text{ even}, \\ A_{n+m} + 3e_1 A_{n+m-2} & \\ \quad +(e_1 - e_2)(2e_1 + e_2) A_{n+m-4}, & n \text{ and } m \text{ odd}. \end{cases} \tag{12.50}$$

If we let e_1 and e_2 (and hence also e_3) go to zero, we obtain the classical function algebra, i.e., the algebra of Laurent polynomials, as degeneration. Obviously, it is an almost-graded associative algebra.

12.5.2 Vector field algebras

Proposition 12.8. *With respect to the basis* (12.49), *the vector field algebra given by the parameters e_1 and e_2 has the following structure equation.*

$$[V_n, V_m] = \begin{cases} (m-n)V_{n+m}, & n, m \text{ odd}, \\ (m-n)(V_{n+m} + 3e_1 V_{n+m-2} \\ \quad +(e_1-e_2)(e_1-e_3)V_{n+m-4}), & n, m \text{ even}, \\ (m-n)V_{n+m} + (m-n-1)3e_1 V_{n+m-2} \\ \quad +(m-n-2)(e_1-e_2)(e_1-e_3)V_{n+m-4}, & n \text{ odd}, m \text{ even}. \end{cases} \tag{12.51}$$

Furthermore, with the degree definition (12.48), *the algebra is an almost-graded Lie algebras.*

Proof. This is formulated in [76]. It is shown and discussed in [208]. Alternatively, it can be obtained by transferring the analytic picture from Deck and Ruffing [50, 195, 197] into the algebraic-geometric picture. □

In fact, these relations define Lie algebras for every pair $(e_1, e_2) \in \mathbb{C}^2$. Hence, we obtain a two-dimensional family of Lie algebras. We denote by $\mathcal{L}^{(e_1,e_2)}$ the Lie algebra corresponding to (e_1, e_2). Obviously, $\mathcal{L}^{(0,0)} \cong \mathcal{W}$, where $\mathcal{W}$ is the Witt algebra

Remark 12.9. This two-dimensional family could also be written with the parameters p and q instead of $3e_1$ and $(e_1 - e_2)(e_1 - e_3)$. In this form, it was algebraically found by Deck [50]. Guerrini [94, 95] related it later to deformations of the Witt algebra over certain spaces of polynomials. See also [25].

Remark 12.10. We had chosen one special two-torsion point $1/2 \bmod L$, respectively e_1. From the purely algebraic geometric point of view, any of the nontrivial two-torsion points $1/2, \tau/2$, or $(1+\tau)/2$ will define isomorphic families, see e.g., Bremner [25]. In [208] also the case of three points by considering four-torsion points was considered. In this respect see also [49, 50, 195, 197].

Proposition 12.11 ([76, Proposition 5.1]). *For $(e_1, e_2) \neq (0,0)$, the algebras $\mathcal{L}^{(e_1,e_2)}$ are not isomorphic to the Witt algebra $\mathcal{W}$, but $\mathcal{L}^{(0,0)} \cong \mathcal{W}$.*

If we restrict our two-dimensional family to one of the lines D_s, then we obtain a one-dimensional family. The following results are taken from [76]. First consider $s \neq \infty$, then $e_2 = se_1$ and we calculate $(e_1 - e_2)(e_1 - e_3) = e_1^2(1-s)(2+s)$. Rewriting (12.51) for these curves yields

$$[V_n, V_m] = \begin{cases} (m-n)V_{n+m}, & n, m \text{ odd}, \\ (m-n)(V_{n+m} + 3e_1 V_{n+m-2} & \\ \qquad + e_1^2(1-s)(2+s)V_{n+m-4}), & n, m \text{ even}, \\ (m-n)V_{n+m} + (m-n-1)3e_1 V_{n+m-2} & \\ \qquad + (m-n-2)e_1^2(1-s)(2+s)V_{n+m-4}, & n \text{ odd}, m \text{ even}. \end{cases} \tag{12.52}$$

For D_∞ we have $e_3 = -e_2$ and $e_1 = 0$, and hence

$$[V_n, V_m] = \begin{cases} (m-n)V_{n+m}, & n, m \text{ odd}, \\ (m-n)(V_{n+m} - e_2^2 V_{n+m-4}), & n, m \text{ even}, \\ (m-n)V_{n+m} - (m-n-2)e_2^2 V_{n+m-4}, & n \text{ odd}, m \text{ even}. \end{cases} \tag{12.53}$$

Proposition 12.12. *For a fixed s in all cases the algebras will be isomorphic above every point in D_s, as long as we are not above* $(0, 0)$.

Proof. In the first case, $s \neq \infty$ and $e_1 \neq 0$ a rescaling $V_n^* = (\sqrt{e_1})^{-n} V_n$ of generators will always yield the same algebraic structure equation with $e_1 = 1$. For $s = \infty$, a rescaling with $(\sqrt{e_2})^{-n} V_n$ will do the same (for $e_2 \neq 0$). □

Hence, as long as $e_1 \neq 0$ the algebras over two points in D_s are pairwise isomorphic but not isomorphic to the algebra over 0, which is the Witt algebra. Using the result $\mathrm{H}^2(\mathcal{W}, \mathcal{W}) = \{0\}$ of [71, 73, 221] we obtain the following.

Theorem 12.13 (Fialowski and Schlichenmaier, [76]). *Despite its infinitesimal and formal rigidity, the Witt algebra $\mathcal{W}$ admits deformations $\mathcal{L}_t$ over the affine line with $\mathcal{L}_0 \cong \mathcal{W}$ which, restricted to every (Zariski or analytic) neighborhood of $t = 0$, are nontrivial.*

Remark 12.14. Mathematicians and physicists working in contractions of algebraic structures sometimes call such families over D_s *jump deformations*, as the isomorphy type stays the same and then jumps. We would like to point out that the theorem above has nothing to do with the fact that it is a jump deformation. Clearly the two-dimensional family (12.51) is not a jump deformation. Also, one-dimensional deformations exist as subfamilies which are not jump deformations [78]. We take, for example, the smooth rational curve given by

$$C := \{(e_1, e_2) \in B \mid e_2 = 2e_1^2,\ e_1 \in \mathbb{C}\}. \tag{12.54}$$

The rational parameter along the curve will be e_1 and the curve passes through $(0, 0)$. For every line D_s there will be just one other point of intersection with C. Its parameter value is given by $e_1 = 1/2s$. Hence, for e_1 small, the curve will not meet the exceptional lines D_s, $s = 1, -1/2, -2$ a second time. The curves corresponding to small $e_1 \neq 0$ values will be non-singular cubics, i.e., elliptic curves. If we evaluate the modular function j (given by (12.32)) along the curve C, we obtain

$$j(e_1) = 1728 \frac{(1 + 2e_1 + 4e_1^2)^3}{(1 - 2e_1)^2(1 + e_1)^2(1 + 4e_1)^2}. \tag{12.55}$$

This value will not be constant along C. Furthermore, for small e_1, the values will be different. This implies that the elliptic curves will be pairwise non-isomorphic, and the vector field algebras along this curve in the neighborhood of 0 will also be pairwise non-isomorphic. As limit, again the Witt algebra shows up.

Remark 12.15. Of course, we can also consider lines in the base manifold not passing through $(0,0)$ and take the subfamily of the two-dimensional family. These lines will meet all exceptional lines once. In this way we obtain a deformation family with special elements given by the algebras lying over the points of the exceptional lines. We will consider these algebras also in Section 12.7.

Remark 12.16. The cuspidal cubic E_C is a non-stable cubic curve in the sense of Mumford. Hence, from the point of view of the moduli space of curves, it should not be taken into account for its compactification. Clearly, forgetting the algebras our families over D_s give examples of families which show that the cuspidal cubic is a "bad curve". In these families, the curves over the non-special point are isomorphic, hence the sequence in the moduli space is a constant sequence. Nevertheless, it has as "limit point" the cuspidal cubic, which corresponds to a point different from this constant.

Using the cocycle (6.88) defining central extensions, in the families (12.51), (12.52), and (12.53) a central term can be easily incorporated. With respect to the flat coordinate $z - a$, we can take the projective connection $R \equiv 0$. The integral along a separating cocycle C_S is obtained by taking the residue at $\overline{z} = 0$. In this way we obtain geometric families of deformations for the Virasoro algebra. They are locally nontrivial, despite the fact that the Virasoro algebra is formally rigid.

12.5.3 The current algebra

Let $\mathfrak{g}$ be a simple finite-dimensional Lie algebra (similar results are true for general Lie algebras), and $\mathcal{A}$ the algebra of meromorphic functions corresponding to the geometric situation discussed above. The current algebra is defined as $\mathfrak{g} \otimes A$. With respect to the basis (12.43), generators are given as tensor product with the Lie algebra elements. Using (12.50) we deduce immediately

$$[x \otimes A_n, y \otimes A_m] = \begin{cases} [x,y] \otimes A_{n+m}, & n \text{ or } m \text{ even}, \\ [x,y] \otimes A_{n+m} + 3e_1 [x,y] \otimes A_{n+m-2} & \\ +(e_1 - e_2)(2e_1 + e_2)[x,y] \otimes A_{n+m-4}, & n \text{ and } m \text{ odd}. \end{cases} \tag{12.56}$$

If we let e_1 and e_2 (and hence also e_3) go to zero, we obtain the classical current algebra as degeneration. Again it can be shown that the family, even if restricted on D_s, is locally nontrivial; see [77]. Recall that by the results of Lecomte and Roger [156] and Garland [87], the current algebra is formally rigid if $\mathfrak{g}$ is simple. But our families show that it is neither geometrically nor analytically rigid.

Also in this case we can construct families of centrally extended algebras by considering the cocycle (6.116). In this way we obtain nontrivial deformation families for the formally rigid classical affine algebras of Kac-Moody type. The cocycle (6.116) is

$$\gamma(x \otimes A_n, y \otimes A_m) = p(e_1, e_2) \cdot \beta(x, y) \cdot \frac{1}{2\pi \mathrm{i}} \int_{C_S} A_n dA_m. \tag{12.57}$$

Here $p(e_1, e_2)$ is an arbitrary polynomial in the variables e_1 and e_2, and β the Cartan–Killing form. The integral can be calculated [77, Theorem 4.6] as

$$\frac{1}{2\pi \mathrm{i}} \int_{C_S} A_n dA_m = \begin{cases} -n\delta_m^{-n}, & n, m \text{ even}, \\ -n\delta_m^{-n} + 3e_1(-n+1)\delta_m^{-n+2} + \\ +(e_1 - e_2)(2e_1 + e_2)(-n+2)\delta_m^{-n+4}, & n, m \text{ odd}. \\ 0, & \text{otherwise}. \end{cases}$$

12.6 The geometric background of the degenerated cases

Here we want to explain why, for example, we obtain the Witt algebra as a degeneration of the families (12.52) and (12.53) for $(e_1, e_2) = (0, 0)$. Moreover, we want to study what happens over the exceptional lines D_1, D_{-2}, $D_{-1/2}$. Recall that our cubic curve given by (12.28) will become singular if at least two of the e_i coincide. Depending on whether two or all three coincide, we obtain the nodal cubic (12.39) or the cuspidal cubic (12.40); see the discussion in Section 12.3.2. In the cuspidal case, the singularity will be the point on the curve with the affine coordinates $(0, 0)$. In the nodal case, the singularity will be the point $(e, 0)$, where e is the coinciding value. By (12.35) we always have that one of the marked points stays the point ∞ on the curve, the other will be the point with the affine coordinate $(e_1, 0)$. Hence, over $D^*_{-1/2}$ the second point will stay non-singular, and for D^*_1 and D^*_{-2} it will move to the singularity.

The basis elements given in Section 12.4 are expressed as rational expressions in the affine coordinate functions X and Y. Hence they make complete sense in the degenerate case.

For singular curves there is a well-defined algebraic-geometric technique to desingularize them. In our case we obtain algebraic maps

$$\psi_N : \mathbb{P}^1 \to E_N, \qquad \psi_C : \mathbb{P}^1 \to E_C, \tag{12.58}$$

which are surjective and $1:1$ outside of the singular point. For the nodal cubic there are 2 points lying above the singularity, for the cuspidal cubic there is one point lying above the singularity. Via these maps our basis elements and derivation can be pulled back.

For a complete analysis we have to give explicit expressions for these maps. The points of $\mathbb{P}^1$ are given by homogeneous coordinates $(t : s)$. We define the maps by

$$\psi_N(t : s) = (\, t^2 s - 2es^3 \;:\; 2t(t^2 - 3es) \;:\; s^3\,), \tag{12.59}$$

$$\psi_C(t : s) = (\, t^2 s \;:\; 2t^3 \;:\; s^3\,). \tag{12.60}$$

The maps are given by homogeneous polynomials. Hence, they are obviously algebraic maps. Under these maps the point $\infty = (1 : 0)$ on $\mathbb{P}^1$ corresponds to (and only to) the point ∞ on both curves E_N and E_C. Again, it is enough to consider the affine part (i.e., we are setting $s = 1$) and we use the same symbols for the affine maps. A direct calculation shows:

(1) the maps are surjective;
(2) the map ψ_C is $1 : 1$;
(3) the map ψ_N is not $1 : 1$, only at the points $t = \sqrt{3e}$ and $t = -\sqrt{3e}$. Both points project onto the singular point $(e, 0)$. The point $(-2e, 0)$ corresponds to $t = 0$.

We set $a = \sqrt{3e}$ to simplify the following formulas.

With the help of these maps, the functions on the singular cubic can be pulled back to $\mathbb{P}^1$. Outside the singular points the orders stay the same. For the power of the differential we obtain as pullback

$$\psi_N^*\left(\frac{dX}{Y}\right)^\lambda = \frac{1}{(t^2 - a^2)^\lambda}(dt)^\lambda, \tag{12.61}$$

$$\psi_C^*\left(\frac{dX}{Y}\right)^\lambda = \frac{1}{t^{2\lambda}}(dt)^\lambda. \tag{12.62}$$

As for our vector fields $\lambda = -1$, we get additional zeros at the points lying above the singular point. In the nodal case these are the points $t = a$ and $t = -a$ and the additional zero is of order one. For the cuspidal case it is the point $t = 0$ and the order is two. The explicit expressions of the pulled back elements can be easily given but are not needed here. They can be found in [208].

Now we are able to understand the situation. In case our degeneration is maximal, meaning we have a cusp, then the singular point is a marked point. As there are poles allowed, the additional zero of order two does not disturb the fact that we get the classical situation.

In the nodal cubic case we have two subcases. First, if the singular point becomes a marked point. In this case we have on $\mathbb{P}^1$ two points for which poles are allowed. This yields the three-point Krichever–Novikov situation. In case the marked point is different to the singular point, we have to take into account that the elements coming from the pullback need to have the same value on these two points. Also we need to incorporate the pullback of the derivation. In total we will obtain subalgebras of the classical algebras.

As these algebras give important examples of infinite-dimensional Lie algebras we will be studying them in the next section in more detail.

12.7 Algebras appearing in the degenerate cases

In this section we identify which algebras appear as degenerations in the two-dimensional family (12.51). This means the algebras lying over the exceptional lines D_s^* for $s = 1, -2$ or $-1/2$, and above the point $(0, 0)$.

12.7.1 Witt algebra case

In the maximally degenerate case (i.e., lying above $(0, 0)$), the poles which might appear at the points which correspond to the point $(e_1, 0)$ move to the point zero if pulled back to $\mathbb{P}^1$. The pullback of the differential $\frac{dX}{Y}$ will only increase the zero order there by two. But as there are in any case poles allowed, we obtain as pullback the "classical" objects

$$A_n = z^n, \qquad V_n = l_n = z^{n+1}\frac{d}{dz}. \tag{12.63}$$

Consequently, we obtain also the classical algebras, meaning the associative algebra of Laurent polynomials, the Witt algebra, and the classical affine Lie algebra with the structures ($n, m \in \mathbb{Z}$, $x, y \in \mathfrak{g}$)

$$\begin{aligned} A_n \cdot A_m &= A_{n+m}, \\ [V_n, V_m] &= (m-n)V_{n+m}, \\ [x \otimes A_n, y \otimes A_m] &= [x, y] \otimes A_{n+m}. \end{aligned} \tag{12.64}$$

12.7.2 The genus zero and three-point situation

For $s = 1$ or $s = -2$ above the exceptional lines D_s^* (with the origin removed) we have $e = e_1$. This means that the point where poles are allowed will correspond to two points after the pullback to $\mathbb{P}^1$. We obtain objects on $\mathbb{P}^1$ which, besides the pole at ∞, might have poles at two other points. The pullback of the derivation $\frac{dX}{Y}$ will only increase the order at the points where poles are allowed anyway.

Hence, we obtain the three-point Krichever–Novikov type algebras of genus zero. We consider the Riemann sphere $S^2 = \mathbb{P}^1$ and a set A consisting of 3 points. Given any triple of 3 points, an analytic automorphism of $\mathbb{P}^1$ mapping this triple to $\{a, -a, \infty\}$ always exists, with $a \neq 0$. In fact $a = 1$ would suffice. Without restriction we can take

$$I := \{\infty\}, \quad O := \{a, -a\}.$$

Due to the symmetry of the situation, it is more convenient to take a symmetrized basis of $\mathcal{A}$ (with $k \in \mathbb{Z}$)

$$\begin{aligned} A_{2k} &:= (z-a)^k(z+a)^k, \\ A_{2k+1} &:= z(z-a)^k(z+a)^k, \end{aligned} \tag{12.65}$$

and for $\mathcal{L}$ (with $k \in \mathbb{Z}$)

$$
\begin{aligned}
V_{2k} &:= z(z-a)^k(z+a)^k \frac{d}{dz}, \\
V_{2k+1} &:= (z-a)^{k+1}(z+a)^{k+1} \frac{d}{dz}.
\end{aligned}
\tag{12.66}
$$

Again we inverted the grading.

By direct calculations we obtain for the algebras the following structures.

The function algebra

$$
A_n \cdot A_m = \begin{cases} A_{n+m}, & n \text{ or } m \text{ even}, \\ A_{n+m} + a^2 \otimes A_{n+m-2}, & n \text{ and } m \text{ odd}. \end{cases}
\tag{12.67}
$$

The current algebra

$$
[A_n, y \otimes A_m] = \begin{cases} [x,y] \otimes A_{n+m}, & n \text{ or } m \text{ even}, \\ [x,y] \otimes A_{n+m} + a^2[x,y] \otimes A_{n+m-2}, & n \text{ and } m \text{ odd}. \end{cases}
\tag{12.68}
$$

The vector field algebra

$$
[V_n, V_m] = \begin{cases} (m-n)V_{n+m}, & n, m \text{ odd}, \\ (m-n)(V_{n+m} + a^2 V_{n+m-2}), & n, m \text{ even}, \\ (m-n)V_{n+m} + (m-n-1)a^2 V_{n+m-2}, & n \text{ odd}, \ m \text{ even}. \end{cases}
\tag{12.69}
$$

We get exactly the structure for the algebras (12.50), (12.51), (12.52), (12.53), (12.56) if specialized over D_1^* and D_{-2}^*. In this case our $a = \sqrt{3e_1}$.

Now it turns out to be advantageous that we kept a as a parameter and did not fix it to 1. The families above are again families which have as special elements for $a = 0$ the classical algebras. The geometric context is clear, the two points a and $-a$ move together.

The affine algebra

The affine algebra is the almost-graded central extension $\widehat{\mathfrak{g}}_{\alpha,S}$ of the current algebra. It is obtained via the cocycle

$$
\gamma(x \otimes A_n, y \otimes A_m) = \alpha(x,y) \cdot \frac{1}{2\pi \mathrm{i}} \int_{C_S} A_n dA_m.
\tag{12.70}
$$

The integral in the cocycle calculates as (see [77, A.13 and A.14])

$$\frac{1}{2\pi i}\int_{C_S} A_n dA_m = \begin{cases} -n\delta_m^{-n}, & n, m \text{ even}, \\ 0, & n, m \text{ different parity}, \\ -n\delta_m^{-n} + a^2(-n+1)\delta_m^{-n+2}, & n, m \text{ odd}. \end{cases} \tag{12.71}$$

Example 12.17 (Three-point $\mathfrak{sl}(2,\mathbb{C})$-current algebra for genus 0). Of course, given a simple Lie algebra $\mathfrak{g}$ with generators and structure equations, the relations above can be written in these terms. An important example is $\mathfrak{sl}(2,C)$ with the standard generators

$$h := \begin{pmatrix} 1 & 0 \\ 0 & -1 \end{pmatrix}, \quad e := \begin{pmatrix} 0 & 1 \\ 0 & 0 \end{pmatrix}, \quad f := \begin{pmatrix} 0 & 0 \\ 1 & 0 \end{pmatrix}.$$

For illustration we give the results in their terms. We set $e_n := e \otimes A_n$, $n \in \mathbb{Z}$, and in the same way f_n and h_n. Recall that $\beta(x,y) = \operatorname{tr}(x \cdot y)$. We calculate

$$[e_n, f_m] = \begin{cases} h_{n+m}, & n \text{ or } m \text{ even}, \\ h_{n+m} + a^2 h_{n+m-2}, & n \text{ and } m \text{ odd}, \end{cases} \tag{12.72}$$

$$[h_n, e_m] = \begin{cases} 2e_{n+m}, & n \text{ or } m \text{ even}, \\ 2e_{n+m} + 2a^2 e_{n+m-2}, & n \text{ and } m \text{ odd}, \end{cases} \tag{12.73}$$

$$[h_n, f_m] = \begin{cases} -2f_{n+m}, & n \text{ or } m \text{ even}, \\ -2f_{n+m} - 2a^2 f_{n+m-2}, & n \text{ and } m \text{ odd}. \end{cases} \tag{12.74}$$

For the central extension we obtain

$$[e_n, f_m] = \begin{cases} h_{n+m} - n\delta_m^{-n}, & n \text{ or } m \text{ even}, \\ h_{n+m} + a^2 h_{n+m-2} - n\delta_m^{-n} - a^2(n-1)\delta_m^{-n+2}, & n \text{ and } m \text{ odd}. \end{cases} \tag{12.75}$$

For the other commutators we do not have contributions from the center.

12.7.3 Subalgebras of the classical algebras

If $e_1 \neq e_2 = e_3$, then the point of a possible pole will remain non-singular. This appears if we approach a point of $D^*_{-1/2}$ or if we stay on this line. It is necessary that the pullbacks of the functions have the same value at the points a and $-a$. Furthermore, as far as the vector fields are concerned, the pullback of the derivation $\frac{dX}{Y}$ will pick up a zero of order one at both points a and $-a$. As no additional poles will appear, besides possibly at 0 and ∞, we will obtain subalgebras of the classical algebra. They will be of independent interest too.

12.7.3.1 Function and current algebras

The set of elements

$$A_n := \begin{cases} z^n, & n \text{ even}, \\ z^n - a^2 z^{n-2} = z^{n-2}(z^2 - a^2), & n \text{ odd}, \end{cases} \tag{12.76}$$

for $n \in \mathbb{Z}$ is a basis of the subalgebra of meromorphic functions on $\mathbb{P}^1$ which are holomorphic outside 0 and ∞ and have the same value at a and $-a$. The additional factor $z^2 - a^2$ in the odd generators takes care that $A(a) = A(-a)$ in all cases.

We obtain as structure of the function algebra

$$A_n \cdot A_m = \begin{cases} A_{n+m}, & \text{for } n \text{ or } m \text{ even}, \\ A_{n+m} - 2a^2\, A_{n+m-2} & \\ \qquad +a^4\, A_{n+m-4}, & \text{for } n \text{ and } m \text{ both odd}. \end{cases} \tag{12.77}$$

This induces directly for the current algebra

$$[x \otimes A_n, y \otimes A_m] = \begin{cases} [x, y]\, A_{n+m}, & \text{for } n \text{ or } m \text{ even}, \\ [x, y]\, A_{n+m} - 2a^2 [x, y]\, A_{n+m-2} & \\ \qquad +a^4 [x, y]\, A_{n+m-4}, & \text{for } n \text{ and } m \text{ both odd}. \end{cases} \tag{12.78}$$

12.7.3.2 Vector field algebra

We consider the subalgebra of the Witt algebra defined by the basis elements

$$\begin{aligned} V_{2k} &= z^{2k-3}(z^2 - a^2)^2 \frac{d}{dz} = l_{2k} - 2a^2 l_{2k-2} + a^4 l_{2k-4}, \\ V_{2k+1} &= z^{2k}(z^2 - a^2)\frac{d}{dz} = l_{2k+1} - a^2 l_{2k-1}. \end{aligned} \tag{12.79}$$

Here to distinguish we use $l_n = z^{n+1}\frac{d}{dz}$ for the standard generators of Witt algebra.

One calculates directly

$$[V_n, V_m] = \begin{cases} (m-n)V_{n+m}, & n, m \text{ odd}, \\ (m-n)(V_{n+m} - 2a^2 V_{n+m-2} + a^4 V_{n+m-4}), & n, m \text{ even}, \\ (m-n)V_{n+m} + (m-n-1)(-2a^2)V_{n+m-2} & \\ \qquad +(m-n-2)a^4 V_{n+m-4}, & n \text{ odd}, \ m \text{ even}. \end{cases} \tag{12.80}$$

This subalgebra can be described as the subalgebra of meromorphic vector fields vanishing at a and $-a$, with possible poles at 0 and ∞, and such that in the representation of $V(z) = f(z)(z^2 - a^2)\frac{d}{dz}$ the function f fulfills $f(a) = f(-a)$.

In accordance with the geometric picture, the obtained algebra is isomorphic to the vector field algebras lying above $D^*_{-1/2}$. As in this case the singular point is $e = -(1/2)e_1$, we have $\alpha = \sqrt{-3/2e_1}$.

13 Lax operator algebras

Recently, a new class of global current type algebras has appeared, the *Lax operator algebras*. As the name indicates, they are related to integrable systems [248, 249]. The algebras were introduced and studied by Krichever [138], Krichever and Sheinman [148], and Sheinman [251]. Compared to the Krichever–Novikov current type algebras, additional singularities are allowed. They will play a special role. The points where these singularities are allowed are called *weak singular points*, and the singularities are called *weak singularities*. The construction works only for the matrix algebras $\mathfrak{gl}(n)$, $\mathfrak{sl}(n)$, $\mathfrak{so}(n)$, $\mathfrak{sp}(2n)$, and G_2 at the moment. For those algebras they generalize the Krichever–Novikov type algebras. It is expected that a definition for the other exceptional cases will be found too. The maximal pole order at the weak singular points will be one or two depending on the Lie algebra.

A very important fact is that it is possible for these Lax operator algebras to introduce an almost-grading and to classify associated almost-graded central extensions.

In this chapter I will report on these algebras and the corresponding results on their almost-grading and central extensions. From the applications there is again a need to classify such almost-graded central extensions. The author obtained this jointly with Sheinman in [228] in the two-point case, and in [224] for the multi-point case. For the Lax operator algebras associated to the simple algebras $\mathfrak{sl}(n)$, $\mathfrak{so}(n)$, $\mathfrak{sp}(n)$ (and now also for G_2, see [251]) it will be unique (meaning: given a splitting of A there is an almost-grading, and with respect to this there is up to equivalence and rescaling only one nontrivial almost-graded central extension). For $\mathfrak{gl}(n)$ we obtain two independent local cocycle classes if we assume $\mathcal{L}$-invariance on the abelian part.

For the proofs I refer to the original articles of Krichever, Sheinman and Schlichenmaier, written either alone or in changing collaborations [138, 148, 224, 228, 251]. See in particular also the book by Sheinman [250]. This book also deals with the relation to integrable systems; see also [248, 249].

13.1 Lax operator algebras

Let $\mathfrak{g}$ be one of the classical matrix algebras $\mathfrak{gl}(n)$, $\mathfrak{sl}(n)$, $\mathfrak{so}(n)$, $\mathfrak{sp}(2n)$, or $\mathfrak{s}(n)$, where the latter denotes the algebra of scalar matrices[1]. Our algebras will consist of certain $\mathfrak{g}$-valued meromorphic functions, forms, etc, defined on Riemann surfaces with additional structures (marked points, vectors associated to these points, ...).

The general set-up is similar to the general geometric situation considered in this monograph. Let Σ be a compact Riemann surface of genus g (g arbitrary), and A a

1 For the recently considered case G_2, we refer to [251].

finite subset of points in Σ divided into two non-empty disjoint subsets

$$I := \{P_1, P_2, \dots, P_K\}, \quad O := \{Q_1, Q_2, \dots, Q_M\}, \tag{13.1}$$

with $\#A = K + M$. The points in I are the in-points, the points in O the out-points.

To define Lax operator algebras we have to fix some additional data. Fix $R \in \mathbb{N}_0$ and a collection of points

$$W := \{\gamma_s \in \Sigma \setminus A \mid s = 1, \dots, R\}. \tag{13.2}$$

We assign to every point γ_s a vector $\alpha_s \in \mathbb{C}^n$ (respectively from $\mathbb{C}^{2n}$ for $\mathfrak{sp}(2n)$). The system

$$\mathcal{T} := \{(\gamma_s, \alpha_s) \in \Sigma \times \mathbb{C}^n \mid s = 1, \dots, R\} \tag{13.3}$$

is called *Tyurin data*. We allow also $R = 0$. In this case, the Tyurin data will be empty.

Remark 13.1. For $K = n \cdot g$ and for generic values of (γ_s, α_s) with $\alpha_s \neq 0$, the tuples of pairs $(\gamma_s, [\alpha_s])$ with $[\alpha_s] \in \mathbb{P}^{n-1}(\mathbb{C})$ parameterize framed semi-stable rank n and degree $n \cdot g$ holomorphic vector bundles as shown by Tyurin [256]. Hence, the name Tyurin data. In Section 13.2 we will give more details on this relation.

We fix local coordinates z_l, $l = 1, \dots, K$ centered at the points $P_l \in I$ and w_s centered at γ_s, $s = 1, \dots, R$. In fact, nothing will depend on the choice of w_s. This is essentially also true for z_l. Only its first jet will be used to normalize certain basis elements uniquely.

We consider $\mathfrak{g}$-valued meromorphic functions[2]

$$L : \Sigma \to \mathfrak{g}, \tag{13.4}$$

which are holomorphic outside $W \cup A$, have at most poles of order one (or respectively of order two for $\mathfrak{sp}(2n)$) at the points in W, and fulfill certain conditions at W depending on $\mathcal{T}$ and $\mathfrak{g}$. These conditions will be described in the following section. The singularities at W are called *weak singularities*. These objects were introduced by Krichever [138] for $\mathfrak{gl}(n)$ in the context of Lax operators for algebraic curves, and further generalized (including the above matrix groups) by Krichever and Sheinman in [148].

For $\mathfrak{gl}(n)$ the conditions are as follows. For $s = 1, \dots, R$ we require that $\beta_s \in \mathbb{C}^n$ and $\kappa_s \in \mathbb{C}$ exist such that the function L has the following expansion at $\gamma_s \in W$

$$L(w_s) = \frac{L_{s,-1}}{w_s} + L_{s,0} + \sum_{k>0} L_{s,k} w_s^k, \tag{13.5}$$

with

$$L_{s,-1} = \alpha_s {}^t\beta_s, \quad \operatorname{tr}(L_{s,-1}) = {}^t\beta_s \alpha_s = 0, \quad L_{s,0}\, \alpha_s = \kappa_s \alpha_s. \tag{13.6}$$

2 The interpretation as functions is not completely accurate; see Section 13.3.2.

In particular, if $L_{s,-1}$ is nonvanishing then it is a rank 1 matrix, and if $\alpha_s \neq 0$ then it is an eigenvector of $L_{s,0}$. The requirements (13.6) are independent of the chosen coordinates w_s.

We denote this set (in fact it will be an algebra) by $\overline{\mathfrak{gl}}(n)$. Of course, it will depend on the Riemann surface Σ, the finite set of points A, and the Tyurin data T.

Note that if one of the $\alpha_s = 0$, then the conditions at the point γ_s correspond to the fact that L has to be holomorphic there. We can erase the point from the Tyurin data. Also, if $\alpha_s \neq 0$ and $\lambda \in \mathbb{C}$, $\lambda \neq 0$, then α and $\lambda\alpha$ induce the same conditions at the point γ_s. Hence, only the projective vector $[\alpha_s] \in \mathbb{P}^{n-1}(\mathbb{C})$ plays a role.

The splitting of $\mathfrak{gl}(n) = \mathfrak{s}(n) \oplus \mathfrak{sl}(n)$ given by

$$X \mapsto \left(\frac{\operatorname{tr}(X)}{n} I_n\,,\ X - \frac{\operatorname{tr}(X)}{n} I_n \right), \tag{13.7}$$

where I_n is the $n \times n$-unit matrix, induces a corresponding splitting of $\overline{\mathfrak{gl}}(n)$:

$$\overline{\mathfrak{gl}}(n) = \overline{\mathfrak{s}}(n) \oplus \overline{\mathfrak{sl}}(n). \tag{13.8}$$

For $\overline{\mathfrak{sl}}(n)$ the only additional condition is that in (13.5) all matrices $L_{s,k}$ are trace-less. The conditions (13.6) remain unchanged.

For $\overline{\mathfrak{s}}(n)$ all matrices in (13.5) are scalar matrices. This implies that the corresponding $L_{s,-1}$ vanish. In particular, the elements of $\overline{\mathfrak{s}}(n)$ are holomorphic at W. Also $L_{s,0}$, as a scalar matrix, has every α_s as eigenvector. This means that besides the holomorphicity there are no further conditions. And we get $\overline{\mathfrak{s}}(n) \cong \mathcal{A}$, where $\mathcal{A}$ is the algebra of meromorphic functions on Σ holomorphic outside of A. This is the (multi-point) Krichever–Novikov type function algebra introduced in Section 2.3.

In the case of $\mathfrak{so}(n)$, we require that all $L_{s,k}$ in (13.5) are skew-symmetric. In particular, they are trace-less. Furthermore, the set-up has to be slightly modified. First, only those Tyurin parameters α_s are allowed which satisfy ${}^t\alpha_s\alpha_s = 0$. Then the first requirement in (13.6) is changed to obtain

$$L_{s,-1} = \alpha_s\,{}^t\beta_s - \beta_s\,{}^t\alpha_s, \quad \operatorname{tr}(L_{s,-1}) = {}^t\beta_s\alpha_s = 0, \quad L_{s,0}\,\alpha_s = \kappa_s\alpha_s. \tag{13.9}$$

For $\mathfrak{sp}(2n)$ we consider a symplectic form $\hat{\sigma}$ for $\mathbb{C}^{2n}$, given by a non-degenerate skew-symmetric matrix σ. The Lie algebra $\mathfrak{sp}(2n)$ is the Lie algebra of matrices X such that ${}^tX\sigma + \sigma X = 0$. The condition $\operatorname{tr}(X) = 0$ will be automatic. At the weak singularities we have the expansion

$$L(z_s) = \frac{L_{s,-2}}{w_s^2} + \frac{L_{s,-1}}{w_s} + L_{s,0} + L_{s,1}w_s + \sum_{k>1} L_{s,k}w_s^k. \tag{13.10}$$

The condition (13.6) is modified as follows: $\beta_s \in \mathbb{C}^{2n}$, $\nu_s, \kappa_s \in \mathbb{C}$ exists such that

$$\begin{gathered} L_{s,-2} = \nu_s\alpha_s\,{}^t\alpha_s\sigma, \quad L_{s,-1} = (\alpha_s\,{}^t\beta_s + \beta_s\,{}^t\alpha_s)\sigma, \\ {}^t\beta_s\sigma\alpha_s = 0, \quad L_{s,0}\,\alpha_s = \kappa_s\alpha_s. \end{gathered} \tag{13.11}$$

Moreover, we require

$$ {}^t\alpha_s \sigma L_{s,1} \alpha_s = 0. \tag{13.12}$$

Again, under the point-wise matrix commutator the set of such maps constitutes a Lie algebra.

For G_2 the pole order at the weak singular points is again two. The description corresponding to (13.10) is much more involved. For this reason we refer to [251].

The following theorem is due to Krichever and Sheinman [148]. The proof presented there for the two-point case applies for the multi-point case as well.

Theorem 13.2. *Let $\overline{\mathfrak{g}}$ be the space of Lax operators associated with $\mathfrak{g}$, one of the above introduced finite-dimensional classical Lie algebras. Then $\overline{\mathfrak{g}}$ is a Lie algebra under the point-wise matrix commutator. For $\overline{\mathfrak{g}} = \overline{\mathfrak{gl}}(n)$ it is an associative algebra under point-wise matrix multiplication.*

(Warning: In the whole chapter $\overline{\mathfrak{g}}$ does not denote the Krichever–Novikov current algebras but the Lax operator algebras.) To give some kind of feeling of the type of arguments used, we will reproduce the proof for $\mathfrak{gl}(n)$ here.

Proof (Only $\mathfrak{gl}(n)$). We will show that the product of two Lax operators for the algebra $\mathfrak{gl}(n)$ is again a Lax operator. This means that the equations (13.6) are fulfilled for their product. Hence, $\overline{\mathfrak{gl}}(n)$ will also be closed under commutator. We start with two elements L' and L'' with corresponding expansions (13.5) and examine their product $L = L'L''$. The singularities at the points in A are not bounded. Hence, they will not create problems and we need only consider the weak singular points. For this, we have to consider each point γ_s (with local coordinate w_s) of the weak singularities with $\alpha_s \neq 0$ separately. Taking into account only those parts which might contribute, we obtain for L

$$ L = \frac{L'_{s,-1}L''_{s,-1}}{w_s^2} + \frac{L'_{s,-1}L''_{s,0} + L'_{s,0}L''_{s,-1}}{w_s^1} + (L'_{s,-1}L''_{s,1} + L'_{s,0}L''_{s,0} + L'_{s,1}L''_{s,-1}) + O(w_s^1). \tag{13.13}$$

By expanding the first numerator, we get

$$ L'_{s,-1}L''_{s,-1} = \alpha_s {}^t\beta'_s \alpha_s {}^t\beta''_s = 0, \tag{13.14}$$

as ${}^t\beta'_s\alpha_s = 0$ by (13.6). Hence, there is no pole of order two appearing.

Next, we consider the expression which comes with pole order one.

$$ L_{s,-1} = L'_{s,-1}L''_{s,0} + L'_{s,0}L''_{s,-1} = \alpha_s {}^t\beta'_s L''_{s,0} + L'_{s,0}\alpha_s {}^t\beta''_s. \tag{13.15}$$

As by the conditions $L'_{s,0}\alpha_s = \kappa'_s\alpha_s$ we can write

$$ L_{s,-1} = \alpha_\sigma {}^t\beta_s, \quad \text{with} \quad {}^t\beta_s = {}^t\beta'_s L''_{s,0} + \kappa'_s {}^t\beta''_s. \tag{13.16}$$

For the trace condition we obtain

$$ \operatorname{tr}(L_{s,-1}) = ({}^t\beta'_s L''_{s,0} + \kappa'_s {}^t\beta''_s)\alpha_s = \kappa''_s {}^t\beta'_s\alpha_s + \kappa'_s {}^t\beta''_s\alpha_s = 0. \tag{13.17}$$

Hence, we have the required form.

Finally, we have to verify that α_s is an eigenvector of $L_{s,0}$. First, we note that $L''_{s,-1}\alpha_s = 0$ and $L'_{s,0}L''_{s,0}\alpha_s = \kappa'_s\kappa''_s\alpha_s$. Also

$$L'_{s,-1}L''_{s,1}\alpha_s = \alpha_s({}^t\beta'_s L''_{s,1}\alpha_s) = ({}^t\beta'_s L''_{s,1}\alpha_s)\alpha_s. \tag{13.18}$$

Hence, indeed α_s is an eigenvector with eigenvalue ${}^t\beta'^t_s L''_{s,1}\alpha_s + \kappa'_s\kappa''_s$. This shows the claim that $L \in \overline{\mathfrak{gl}}(n)$.

In a similar manner, the other cases are treated (but now only for the commutators). The original proofs (involving partly tedious calculations) can be found in [148, 228]. See also the book [250] for a rather detailed presentation of these calculations. □

Definition 13.3. The Lie algebra $\overline{\mathfrak{g}} = \overline{\mathfrak{g}}(\mathcal{T}, A)$ introduced above is called *Lax operator algebra* associated with the finite-dimensional Lie algebra $\mathfrak{g}$ (with respect to the Tyurin data $\mathcal{T}$ and set of points A).

For simplicity of notation we will just use $\overline{\mathfrak{g}}$.

The next step is to introduce an almost-graded structure for these Lax operator algebras induced by splitting the set $A = I \cup O$. This was done for the two-point case in [148] and for the multi-point case in [224]. The statement follows.

Theorem 13.4. *Induced by the splitting $A = I \cup O$, the (multi-point) Lax operator algebra $\overline{\mathfrak{g}}$ becomes a (strongly) almost-graded Lie algebra*

$$\begin{gathered} \overline{\mathfrak{g}} = \bigoplus_{m\in\mathbb{Z}} \overline{\mathfrak{g}}_m, \quad \dim \overline{\mathfrak{g}}_m = K \cdot \dim \mathfrak{g} \\ [\overline{\mathfrak{g}}_m, \overline{\mathfrak{g}}_n] \subseteq \bigoplus_{h=n+m}^{n+m+S} \overline{\mathfrak{g}}_h, \end{gathered} \tag{13.19}$$

with a constant S independent of n and m.

Recall that the usual Krichever–Novikov current type algebras are tensor products of the finite-dimensional Lie algebra $\mathfrak{g}$ with the almost-graded function algebra $\mathcal{A}$. Hence, an almost-grading can be given via the almost-grading of $\mathcal{A}$, i.e., $\overline{\mathfrak{g}}_m = \mathfrak{g} \otimes \mathcal{A}_m$ and Theorem 13.4 is obvious.

In the case that $\mathcal{T}$ is not trivial there is no such tensor product and we have to approach the statement directly. The technique of fixing $\overline{\mathfrak{g}}_m$ is similar to the technique developed in Chapter 3 and Chapter 4. We fix basis elements of $\overline{\mathfrak{g}}$ by prescribing the order at the points in I and corresponding orders at O to make it unique. Important in this context is that at each weak singular point the degree of freedom due to a possible pole is compensated by the additional conditions there. For example, in the relation (13.6) we have n degrees of freedom for the polar part $L_{s,-1}$ by the choice for β. The trace condition reduces by 1 the degree of freedom. For $L_{s,0}$, the fact that α_s is an eigenvector with arbitrary eigenvalue κ_s reduces by $n-1$ the degree for freedom there. Hence, in total the additional degree of the poles is compensated. For existence and uniqueness, the Riemann–Roch theorem will be used heavily. In this way we will single out for each $m \in \mathbb{Z}$ a subspace $\overline{\mathfrak{g}}_m$ of $\overline{\mathfrak{g}}$, called *(quasi-)homogeneous* subspace of degree m. The degree is essentially related to the order of the elements of $\overline{\mathfrak{g}}$ at the points in I.

Proposition 13.5 ([224, Proposition 3.3.]). *Let X be an element of $\mathfrak{g}$. For each (m, s), $m \in \mathbb{Z}$, and $s = 1, \ldots, K$ there is a unique element $X_{m,s}$ in $\overline{\mathfrak{g}}_m$ such that locally in the neighborhood of the point $P_p \in I$ we have*

$$X_{m,s|}(z_p) = Xz_p^m \cdot \delta_s^p + O(z_p^{m+1}), \quad \forall p = 1, \ldots, K. \tag{13.20}$$

Proposition 13.6 ([224, Proposition 3.4.]). *Let $\{X^u \mid u = 1, \ldots, \dim \mathfrak{g}\}$ be a basis of the finite dimensional Lie algebra $\mathfrak{g}$. Then*

$$\mathcal{B}_m := \{X^u_{m,p},\ u = 1, \ldots, \dim \mathfrak{g},\ p = 1, \ldots, K\} \tag{13.21}$$

is a basis of $\overline{\mathfrak{g}}_m$, and $\mathcal{B} = \cup_{m \in \mathbb{Z}} \mathcal{B}_m$ is a basis of $\overline{\mathfrak{g}}$.

We introduce the associated filtration

$$\overline{\mathfrak{g}}_{(k)} := \bigoplus_{m \geq k} \overline{\mathfrak{g}}_m, \quad \overline{\mathfrak{g}}_{(k)} \subseteq \overline{\mathfrak{g}}_{(k')}, \quad k \geq k'. \tag{13.22}$$

It can be given independent of any choice by

$$\overline{\mathfrak{g}}_{(m)} := \{L \in \overline{\mathfrak{g}} \mid \operatorname{ord}_{P_s}(L) \geq m,\ s = 1, \ldots, K\}. \tag{13.23}$$

Note that the elements L are meromorphic maps from Σ to $\mathfrak{g}$, hence it makes sense to talk about the orders of the component functions with respect to a basis. The minimum of these orders is meant in (13.23). That these two definitions yield the same filtration is the statement of [224, Proposition 3.6].

Proposition 13.7 ([224, Proposition 3.7.]). *Let $X_{k,s}$ and $Y_{m,p}$ be the elements in $\overline{\mathfrak{g}}_k$ and $\overline{\mathfrak{g}}_m$ corresponding to $X, Y \in \mathfrak{g}$ respectively, then*

$$[X_{k,s}, Y_{m,p}] = [X, Y]_{k+m,s}\delta_s^p + L, \tag{13.24}$$

with $[X, Y]$ the bracket in $\mathfrak{g}$ and $L \in \overline{\mathfrak{g}}_{(k+m+1)}$.

13.2 The geometric meaning of the Tyurin parameters

It will not be needed in the following section, but it might be interesting for the reader to see the geometric relevance of the Tyurin parameters in relation to the moduli space of bundles. The reader not interested in this connection might directly jump to the next section.

As usual, let Σ be a compact Riemann surface (or, in the language of algebraic geometry, a projective smooth curve over $\mathbb{C}$) of genus g. Fix a number $n \in \mathbb{N}$. Given a rank n holomorphic (respectively algebraic) vector bundle E, its determinant $\det E$ is defined as $\det E = \wedge^n E$. The degree $\deg E$ of the bundle E is defined as $\deg(\det(E))$. Recall that for a line bundle M over a Riemann surface, the degree of M can be determined by taking a global meromorphic section of M and counting the number of zeros minus the number of poles of this section.

For vector bundles over compact Riemann surfaces we have the Hirzebruch–Riemann–Roch formula

$$\dim \mathrm{H}^0(\Sigma, E) - \dim \mathrm{H}^1(\Sigma, E) = \deg E - \mathrm{rk}(E)(g-1). \tag{13.25}$$

If one wants to construct a moduli space for vector bundles, one has to restrict the set of vector bundles to the subset of stable, or more general semi-stable bundles.

Definition 13.8. A bundle E over a projective smooth curve is called *stable*, if for all sub-bundles $F \neq E$ one has

$$\frac{\deg F}{\mathrm{rk}\, F} < \frac{\deg E}{\mathrm{rk}\, E}. \tag{13.26}$$

The bundle E is called *semi-stable* if in (13.26) the strict inequality $<$ is replaced by $\le$.

In the following section we consider bundles E which are of rank n and degree $n \cdot g$. If we evaluate (13.25) for such bundles, on the left-hand side we obtain the value n. For a generic semi-stable bundle E one has $\dim \mathrm{H}^1(\Sigma, E) = 0$, hence we obtain

$$\dim \mathrm{H}^0(\Sigma, E) = n. \tag{13.27}$$

If we choose a basis $S := \{s_1, s_2, \ldots, s_n\}$ of the space of global holomorphic sections of E, their exterior power is a global holomorphic section

$$s_1 \wedge s_2 \wedge \cdots \wedge s_n \ \in \ \mathrm{H}^0(\Sigma, \det E). \tag{13.28}$$

Denote by E_P the fiber of E over the point P. Then the zeros of the above section are exactly the points $P \in \Sigma$ for which the set of section S fails to be a basis of E_P. As $\deg E = ng$ (counted with multiplicities) exactly ng such points exist. For a generic choice of the set of section, all zeros will be simple zeros. Hence, we will obtain ng such points. They correspond exactly to the weak singularities W appearing in the definition of the Lax operator algebras. Accordingly, we denote the zero points by γ_s, $s = 1, \ldots, ng$.

Furthermore, as the zero is of order one at such a γ_s, the sections evaluated at γ_s span a $(n-1)$-dimensional subspace of E_{γ_s}. We have relations

$$\sum_{i=1}^{n} \alpha_{s,i} s_i(\gamma_s) = 0, \qquad s = 1, \ldots, ng. \tag{13.29}$$

In this way, to every γ_s a vector $\alpha_s \in \mathbb{C}^n$, $\alpha_s \neq 0$ can be assigned. This vector is unique up to multiplication by a non-zero scalar, hence unique as element $[\alpha_s] \in \mathbb{P}^{n-1}(\mathbb{C})$. Again these vectors are exactly the vectors used in the definition of the Lax operator algebra.

As already remarked above, the conditions (13.6) are independent of a rescaling. Hence only the projective class $[\alpha_s]$ matters.

Obviously everything depends on the set S of basis elements. The choice of such a basis is called a *framing* of the bundle E. As described above, the space of Tyurin parameters parameterizes an open dense subset of semi-stable framed vector bundles of rank n and degree ng. Note also that given such a bundle E with fixed set S of basis elements of $\mathrm{H}^0(\Sigma, E)$ it can be trivialized over $\Sigma \setminus W$.

Associated with these moduli spaces, integrable hierarchies of Lax equations can be constructed. See [138] and [249] for results in these directions.

13.3 Module structure of Lax operator algebras

13.3.1 Structure over $\mathcal{A}$

The space $\overline{\mathfrak{g}}$ is an $\mathcal{A}$-module with respect to the point-wise multiplication. Obviously, the relations (13.5), (13.6), (13.9), (13.11), are not disturbed. In particular, the order of the poles at the weak singularities is not increased. Moreover, see the proposition below.

Proposition 13.9 ([224]).
(a) *The Lax operator algebra $\overline{\mathfrak{g}}$ is an almost-graded module over $\mathcal{A}$, i.e., a constant S (not depending on k and m) exists, such that*

$$\mathcal{A}_k \cdot \overline{\mathfrak{g}}_m \subseteq \bigoplus_{h=k+m}^{k+m+S_4} \overline{\mathfrak{g}}_h. \tag{13.30}$$

(b) *For $X \in \mathfrak{g}$*

$$A_{m,s} \cdot X_{n,p} = X_{m+n,s}\,\delta_p^s + L, \quad L \in \overline{\mathfrak{g}}_{(m+n+1)}. \tag{13.31}$$

13.3.2 Structure over $\mathcal{L}$

Next we introduce an action of $\mathcal{L}$ on $\overline{\mathfrak{g}}$. We cannot just take the derivative of the elements of $\overline{\mathfrak{g}}$ considered as functions. A pole of order one at a weak singularity will become a pole of order two, etc. Hence, $\overline{\mathfrak{g}}$ is not closed under the naive action of $\mathcal{L}$ on $\overline{\mathfrak{g}}$. In this sense our interpretation (13.4) is misleading. The interpretation in Section 13.2 says that we should better consider the Lax operators as operators acting on sections of a vector bundle. Differentiation of sections of nontrivial vector bundles can only be done after the choice of a connection. Such a connection for the vector bundle will introduce a connection on the endomorphisms.

Along the lines of [138, 139, 148], with the modification made in [228], the connection

$$\nabla^{(\omega)} = d + [\omega, .] \tag{13.32}$$

is given by a connection form ω, which is a $\mathfrak{g}$-valued meromorphic 1-form, holomorphic outside I, O, and W, and has a certain prescribed behavior at the points in W. For $\gamma_s \in W$ with $\alpha_s = 0$ the requirement is that ω is also regular there. For the points γ_s with $\alpha_s \neq 0$ it is required that it has an expansion of the form

$$\omega(z_s) = \left(\frac{\omega_{s,-1}}{z_s} + \omega_{s,0} + \omega_{s,1} z_s + \sum_{k>1} \omega_{s,k} z_s^k \right) dz_s. \tag{13.33}$$

For $\mathfrak{gl}(n)$: $\tilde{\beta}_s \in \mathbb{C}^n$ and $\tilde{\kappa}_s \in \mathbb{C}$ exist, such that

$$\omega_{s,-1} = \alpha_s {}^t\tilde{\beta}_s, \quad \omega_{s,0}\, \alpha_s = \tilde{\kappa}_s \alpha_s, \quad \operatorname{tr}(\omega_{s,-1}) = {}^t\tilde{\beta}_s \alpha_s = 1. \tag{13.34}$$

For $\mathfrak{so}(n)$: $\tilde{\beta}_s \in \mathbb{C}^n$ and $\tilde{\kappa}_s \in \mathbb{C}$ exist, such that

$$\omega_{s,-1} = \alpha_s {}^t\tilde{\beta}_s - \tilde{\beta}_s {}^t\alpha_s, \quad \omega_{s,0}\, \alpha_s = \tilde{\kappa}_s \alpha_s, \quad {}^t\tilde{\beta}_s \alpha_s = 1. \tag{13.35}$$

For $\mathfrak{sp}(2n)$: $\tilde{\beta}_s \in \mathbb{C}^{2n}$, $\tilde{\kappa}_s \in \mathbb{C}$ exist, such that

$$\omega_{s,-1} = (\alpha_s {}^t\tilde{\beta}_s + \tilde{\beta}_s {}^t\alpha_s)\sigma, \quad \omega_{s,0}\, \alpha_s = \tilde{\kappa}_s \alpha_s, \quad {}^t\alpha_s \sigma \omega_{s,1} \alpha_s = 0, \quad {}^t\tilde{\beta}_s \sigma \alpha_s = 1. \tag{13.36}$$

Such connection forms exist again via Riemann–Roch type arguments. We are even allowed (and will do so) to demand that ω is holomorphic in I. Note also that if all $\alpha_s = 0$ we could take $\omega = 0$.

Let $e \in \mathcal{L}$ be a vector field. In a local coordinate z, the connection form and the vector field are represented as $\omega = \tilde{\omega} dz$ and $e = \tilde{e}\frac{d}{dz}$ with a local function $\tilde{e}$ and a local matrix valued function $\tilde{\omega}$. The covariant derivative in the direction of e is given by

$$\nabla_e^{(\omega)} = dz(e)\frac{d}{dz} + [\omega(e), .\,] = e\,. + [\,\tilde{\omega}\tilde{e}\,, .\,] = \tilde{e} \cdot \left(\frac{d}{dz} + [\,\tilde{\omega}\,, .\,] \right). \tag{13.37}$$

Here the first term ($e.$) corresponds to taking the usual derivative of functions in each matrix element separately, whereas $\tilde{e}\cdot$ means multiplication with the local function $\tilde{e}$.

Using the last description we obtain for $L \in \overline{\mathfrak{g}}$, $g \in \mathcal{A}$, $e, f \in \mathcal{L}$

$$\nabla_e^{(\omega)}(g \cdot L) = (e\,.\,g) \cdot L + g \cdot \nabla_e^{(\omega)} L, \quad \nabla_{g\cdot e}^{(\omega)} L = g \cdot \nabla_e^{(\omega)} L, \tag{13.38}$$

and

$$\nabla_{[e,f]}^{(\omega)} = [\nabla_e^{(\omega)}, \nabla_f^{(\omega)}]. \tag{13.39}$$

In [228] it is shown that indeed in all cases for $\mathfrak{g}$

$$\nabla_{[e,f]}^{(\omega)} \overline{\mathfrak{g}} \subseteq \overline{\mathfrak{g}}. \tag{13.40}$$

The calculation can also be found in [250].

Proposition 13.10.

(a) $\nabla_e^{(\omega)}$ *acts as a derivation on the Lie algebra* $\overline{\mathfrak{g}}$, *i.e.,*

$$\nabla_e^{(\omega)}[L, L'] = [\nabla_e^{(\omega)} L, L'] + [L, \nabla_e^{(\omega)} L']; \tag{13.41}$$

(b) *the covariant derivative makes* $\overline{\mathfrak{g}}$ *a Lie module over* $\mathcal{L}$;

(c) *the decomposition* $\overline{\mathfrak{gl}}(n) = \overline{\mathfrak{s}}(n) \oplus \overline{\mathfrak{sl}}(n)$ *is a decomposition into* $\mathcal{L}$*-modules, i.e.,*

$$\nabla_e^{(\omega)} : \overline{\mathfrak{s}}(n) \to \overline{\mathfrak{s}}(n), \quad \nabla_e^{(\omega)} : \overline{\mathfrak{sl}}(n) \to \overline{\mathfrak{sl}}(n). \tag{13.42}$$

Moreover, the $\mathcal{L}$*-module* $\overline{\mathfrak{s}}(n)$ *is equivalent to the* $\mathcal{L}$*-module* $\mathcal{A}$.

The proof presented for the two-point case in [228] works in general.

Furthermore, see the following proposition.

Proposition 13.11 ([224, Proposition 4.3]).
(a) $\overline{\mathfrak{g}}$ *is an almost-graded* $\mathcal{L}$*-module;*
(b) *for the corresponding* $\mathcal{L}$*-action we have*

$$\nabla_{e_{k,s}}^{(\omega)} X_{m,r} = m \cdot X_{k+m,s}\, \delta_s^r + L, \quad L \in \overline{\mathfrak{g}}_{(k+m+1)}. \tag{13.43}$$

13.3.3 Structure over $\mathcal{D}^1$ and the algebra $\mathcal{D}^1_{\mathfrak{g}}$

We defined the Lie algebra $\mathcal{D}^1$ of meromorphic differential operators on Σ of degree ≤ 1 holomorphic outside $I \cup O$ as a semi-direct sum of $\mathcal{A}$ and $\mathcal{L}$, with the commutator between them given by the action of $\mathcal{L}$ on A. The introduced module structures of $\mathcal{A}$ and $\mathcal{L}$ extend to $\mathcal{D}^1$.

Proposition 13.12. *The Lax operator algebras* $\overline{\mathfrak{g}}$ *are (strongly) almost-graded Lie modules over* $\mathcal{D}^1$ *via*

$$e\,.\,L := \nabla_e^{(\omega)} L, \quad h\,.\,L := h \cdot L. \tag{13.44}$$

Proof. As $\overline{\mathfrak{g}}$ is an almost-graded $\mathcal{A}$- and $\mathcal{L}$-module, it is enough to show that for $e \in \mathcal{L}, h \in \mathcal{A}, L \in \overline{\mathfrak{g}}$ the relation

$$[e, h] = e\,.\,h. \tag{13.45}$$

is respected. Using (13.37) we calculate

$$\begin{aligned} e\,.\,(h\,.\,L) - h\,.\,(e\,.\,L) &= \nabla_e^{(\omega)}(hL) - h\nabla_e^{(\omega)}(L) = \tilde{e}\left(\frac{d(hL)}{dz} + [\tilde{\omega}, hL]\right) - h\tilde{e}\left(\frac{dL}{dz} + [\tilde{\omega}, L]\right) \\ &= \left(\tilde{e}\frac{dh}{dz}\right) L = (e\,.\,h)L = [e, h]\,.\,L. \end{aligned}$$

□

The Lax operator algebra $\overline{\mathfrak{g}}$ is a module over the Lie algebra $\mathcal{L}$ which acts on $\overline{\mathfrak{g}}$ by derivations. Proposition 13.10 says that this action of $\mathcal{L}$ on $\overline{\mathfrak{g}}$ is an action by derivations. Hence, as above we can consider the semi-direct sum $\mathcal{D}^1_{\mathfrak{g}} = \overline{\mathfrak{g}} \oplus \mathcal{L}$ with Lie product given by

$$[e, L] := e\,.\,L = \nabla_e^{(\omega)} L, \tag{13.46}$$

for the mixed pairs. See Section 9.7 for the corresponding construction for the usual Krichever–Novikov algebras of current type.

13.4 Almost-graded central extensions of Lax operator algebras

As explained in Chapter 6, equivalence classes of central extensions of $\overline{\mathfrak{g}}$ will be given by $\mathrm{H}^2(\overline{\mathfrak{g}}, \mathbb{C})$. Explicit central extensions will be given by Lie algebra two-cocycles. Furthermore, the almost-grading of $\mathfrak{g}$ can be extended to the central extension if and only if the defining cocycle is a local cocycle in the sense introduced there.

In the case of the usual Krichever–Novikov type current algebra $\overline{\mathfrak{sl}}(n)$ with respect to the Cartan–Killing form $\alpha(x, y) = \mathrm{tr}(xy)$, $x, y \in \mathfrak{sl}(n)$, if we write $L = x \otimes f$ and $L' = y \otimes g$, then the cocycle can be expressed as

$$\gamma(L, L') = \alpha(x, y)\frac{1}{2\pi\mathrm{i}}\int_{C_S} f dg = \frac{1}{2\pi\mathrm{i}}\int_{C_S} \mathrm{tr}(L \cdot dL'). \tag{13.47}$$

In this way we (re-)expressed the cocycle without using the tensor product. Nevertheless this expression cannot yet be used for the Lax operator algebra, as for dL' the order at the weak singular points might increase. But if we replace the d by the covariant derivative $\Delta^{(\omega)}$, then the order will not increase.

Now we are able to introduce geometric 2-cocycles. Let ω be a connection form as introduced above. Furthermore, let C be a (not necessarily connected) differentiable cycle on Σ not meeting the sets $A = I \cup O$ and W. We define the following bilinear forms on $\overline{\mathfrak{g}}$:

$$\gamma_{1,\omega,C}(L, L') = \frac{1}{2\pi\mathrm{i}}\int_C \mathrm{tr}(L \cdot \nabla^{(\omega)} L'), \quad L, L' \in \overline{\mathfrak{g}}, \tag{13.48}$$

and

$$\gamma_{2,\omega,C}(L, L') = \frac{1}{2\pi\mathrm{i}}\int_C \mathrm{tr}(L) \cdot \mathrm{tr}(\nabla^{(\omega)} L'), \quad L, L' \in \overline{\mathfrak{g}}. \tag{13.49}$$

Proposition 13.13. *The bilinear forms $\gamma_{1,\omega,C}$ and $\gamma_{2,\omega,C}$ are cocycles.*

Proposition 13.14.
(a) *The cocycle $\gamma_{2,\omega,C}$ does not depend on the choice of the connection form ω.*
(b) *The cohomology class $[\gamma_{1,\omega,C}]$ does not depend on the choice of the connection form ω. More precisely*

$$\gamma_{1,\omega,C}(L, L') - \gamma_{1,\omega',C}(L, L') = \frac{1}{2\pi\mathrm{i}}\int_C \mathrm{tr}((\omega - \omega')[L, L']). \tag{13.50}$$

Their proofs remain the same as in [228] for the two-point situation.

For $\gamma_{2,\omega,C}$ we will drop ω in the notation. Note that $\gamma_{2,C}$ vanishes on $\overline{\mathfrak{g}}$ for $\mathfrak{g} = \mathfrak{sl}(n), \mathfrak{so}(n), \mathfrak{sp}(2n)$. But it does not vanish on $\overline{\mathfrak{s}}(n)$, hence not on $\overline{\mathfrak{gl}}(n)$.

As explained in Section 13.3.2, after fixing a connection form ω', the vector field algebra $\mathcal{L}$ operates on $\overline{\mathfrak{g}}$ via the covariant derivative $e \mapsto \nabla_e^{(\omega')}$.

Definition 13.15. A cocycle γ for $\overline{\mathfrak{g}}$ is called *$\mathcal{L}$-invariant* (with respect to ω') if

$$\gamma(\nabla_e^{(\omega')} L, L') + \gamma(L, \nabla_e^{(\omega')} L') = 0, \quad \forall e \in \mathcal{L}, \quad \forall L, L' \in \overline{\mathfrak{g}}. \tag{13.51}$$

Proposition 13.16.
(a) *The cocycle* $\gamma_{2,C}$ *is* $\mathcal{L}$*-invariant.*
(b) *If* $\omega = \omega'$*, then the cocycle* $\gamma_{1,\omega,C}$ *is* $\mathcal{L}$*-invariant.*

The proof is the same as presented in [228] for the two-point case.

We will call a cohomology class $\mathcal{L}$*-invariant* if it has a representing cocycle which is $\mathcal{L}$-invariant. Clearly, the $\mathcal{L}$-invariant classes constitute a subspace of $\mathrm{H}^2(\overline{\mathfrak{g}}, \mathbb{C})$, which we denote by $\mathrm{H}^2_{\mathcal{L}}(\overline{\mathfrak{g}}, \mathbb{C})$.

In the following let $\omega = \omega'$. The property of $\mathcal{L}$-invariance of a cocycle has a deeper meaning. Above we introduced the algebra $\mathcal{D}^1_{\mathfrak{g}}$. The Lax operator algebra $\overline{\mathfrak{g}}$ is a subalgebra of $\mathcal{D}^1_{\mathfrak{g}}$. Given a 2-cocycle γ for $\overline{\mathfrak{g}}$, we might extend it to $\mathcal{D}^1_{\mathfrak{g}}$ as a bilinear form by setting ($L, L' \in \overline{\mathfrak{g}}$, $e, f \in \mathcal{L}$)

$$\tilde{\gamma}(L, L') = \gamma(L, L'), \quad \tilde{\gamma}(e, L) = \tilde{\gamma}(L, e) = 0, \quad \tilde{\gamma}(e, f) = 0. \tag{13.52}$$

The following proposition is shown in completely the same way as Proposition 9.35

Proposition 13.17. *The extended bilinear form* $\tilde{\gamma}$ *is a cocycle for* $\mathcal{D}^1_{\mathfrak{g}}$ *if and only if* γ *is* $\mathcal{L}$*-invariant.*

As our Lax operator algebras $\overline{\mathfrak{g}}$ are almost-graded, we have the notions of bounded and local cocycles; see Definition 6.32. Recall that we need the cocycle to be local if we want to extend the almost-grading to the central extension by assigning a degree to the central element. We call a cohomology class *local*, respectively *bounded*, if it contains a local (or bounded) representing cocycle. Again, not every representing cocycle of a local (or bounded) class will be local (or bounded). The set of bounded cohomology classes is a subspace of $\mathrm{H}^2(\overline{\mathfrak{g}}, \mathbb{C})$ which we denote by $\mathrm{H}^2_b(\overline{\mathfrak{g}}, \mathbb{C})$. It contains the subspace of local cohomology classes denoted by $\mathrm{H}^2_{loc}(\overline{\mathfrak{g}}, \mathbb{C})$. This space classifies the almost-graded central extensions of $\overline{\mathfrak{g}}$ up to equivalence. Both spaces admit subspaces consisting of those cohomology classes admitting a representing cocycle which is both bounded (or local) and $\mathcal{L}$-invariant. The subspaces are denoted by $\mathrm{H}^2_{b,\mathcal{L}}(\overline{\mathfrak{g}}, \mathbb{C})$ and $\mathrm{H}^2_{loc,\mathcal{L}}(\overline{\mathfrak{g}}, \mathbb{C})$ respectively.

If we consider our geometric cocycles $\gamma_{2,C}$ and $\gamma_{1,\omega,C}$, obtained by integrating over an arbitrary cycle, then they will neither be bounded nor local, nor will they define a bounded or local cohomology class.

As we did for the Krichever–Novikov type algebras, we will consider special integration paths. Let C_i be positively oriented (deformed) circles around the points P_i in I, $i = 1, \dots, K$ and C_j^* positively oriented ones around the points Q_j in O, $j = 1, \dots, M$. The cocycle values of γ if integrated over such cycles can be calculated via residues, e.g.,

$$\gamma_{1,\omega,C_i}(L, L') = \operatorname{res}_{P_i}(\operatorname{tr}(L \cdot \nabla^{(\omega)} L')), \quad i = 1, \dots, N. \tag{13.53}$$

Proposition 13.18.
(a) *The 1-form* $\operatorname{tr}(L \cdot \nabla^{(\omega)} L')$ *has no poles outside* $A = I \cup O$.
(b) *The 1-form* $\operatorname{tr}(L) \cdot \operatorname{tr}(dL')$ *has no poles outside* $A = I \cup O$.

Proof. For (a) see [148]. For (b) see [228]. □

For our cocycles (13.48), (13.49), we integrate the forms given in Proposition 13.18 over closed curves C. As by the proposition the integrands do not have poles in W, the integrals will yield the same results if $[C] = [C']$ in $\mathrm{H}(\Sigma \setminus A, \mathbb{Z})$. Let C_S be a separating cycle. Again, we write in this sense for every separating cycle as in the usual Krichever–Novikov situation

$$[C_S] = \sum_{i=1}^{K} [C_i] = -\sum_{j=1}^{M} [C_j^*]. \tag{13.54}$$

In particular, the cocycle values obtained by integrating over a C_S can be obtained by calculating residues either over the points in I or the points in O.

Theorem 13.19 ([224, Theorem 5.8]).
(a) *For $i = 1, \dots, K$ the cocycles γ_{1,ω,C_i} and γ_{2,C_i}, with C_i a circle around P_i, will be bounded from above by zero and $\mathfrak{L}$-invariant.*
(b) *For $j = 1, \dots, M$ the cocycles γ_{1,ω,C_j^*} and γ_{2,C_j^*}, with C_j^* a circle around Q_j, will be bounded from below and $\mathfrak{L}$-invariant.*
(c) *The cocycles γ_{1,ω,C_S} and γ_{2,C_S}, with C_S a separating cycle, will be local, bounded by zero from above , and $\mathfrak{L}$-invariant.*
(d) *In cases (a) and (c) the upper bound will be zero.*

We have similar classification and uniqueness results as for the usual Krichever–Novikov current case.

Theorem 13.20 ([224, Theorem 6.4]).
(a) *If $\mathfrak{g}$ is simple (i.e., $\mathfrak{g} = \mathfrak{sl}(n), \mathfrak{so}(n), \mathfrak{sp}(2n)$), then the space of bounded cohomology classes is K-dimensional. If we fix any connection form ω, then this space has as basis the classes of γ_{1,ω,C_i}, $i = 1, \dots, K$. Every $\mathfrak{L}$-invariant (with respect to the connection ω) bounded cocycle is a linear combination of the γ_{1,ω,C_i}.*
(b) *For $\overline{\mathfrak{g}} = \overline{\mathfrak{gl}}(n)$, the space of bounded cohomology classes which are $\mathfrak{L}$-invariant having been restricted to the scalar subalgebra is $2K$-dimensional. If we fix any connection form ω, then the space has as basis the classes of the cocycles γ_{1,ω,C_i} and γ_{2,C_i}, $i = 1, \dots, K$. Every $\mathfrak{L}$-invariant bounded cocycle is a linear combination of the γ_{1,ω,C_i} and γ_{2,C_i}.*

We will make some remarks on the proof below.

Corollary 13.21. *Let $\mathfrak{g}$ be a simple classical Lie algebra and $\overline{\mathfrak{g}}$ the associated Lax operator algebra. Let ω be a fixed connection form. Then in each $[\gamma] \in \mathrm{H}_b(\overline{\mathfrak{g}}, \mathbb{C})$ a unique representative γ' exists which is bounded and $\mathfrak{L}$-invariant (with respect to ω). Moreover, $\gamma' = \sum_{i=1}^{N} a_i \gamma_{1,\omega,C_i}$, with $a_i \in \mathbb{C}$.*

After these results, which are valid for bounded cocycles, the corresponding classification theorem for local cocycles follows by using this theorem with respect to the grading and the inverted grading obtained by switching the point set I and O. The arguments are similar to the arguments used in Section 6.8.

Theorem 13.22 ([224, Theorem 6.7]).
(a) *If $\mathfrak{g}$ is simple (i.e., $\mathfrak{g} = \mathfrak{sl}(n), \mathfrak{so}(n), \mathfrak{sp}(2n)$), then the space of local cohomology classes is one-dimensional. If we fix any connection form ω, then this space will be generated by the class of γ_{1,ω,C_S}. Every $\mathcal{L}$-invariant (with respect to the connection ω) local cocycle is a scalar multiple of γ_{1,ω,C_S}.*
(b) *For $\overline{\mathfrak{g}} = \overline{\mathfrak{gl}}(n)$. the space of local cohomology classes which are $\mathcal{L}$-invariant having been restricted to the scalar subalgebra is two-dimensional. If we fix any connection form ω, then the space will be generated by the classes of the cocycles γ_{1,ω,C_S} and γ_{2,C_S}. Every $\mathcal{L}$-invariant local cocycle is a linear combination of γ_{1,ω,C_S} and γ_{2,C_S}.*

As in the bounded case, we obtain also for the local case the following corollary.

Corollary 13.23. *Let $\mathfrak{g}$ be a simple classical Lie algebra and $\overline{\mathfrak{g}}$ the associated Lax operator algebra. Let ω be a fixed connection form. Then in each $[\gamma] \in \mathrm{H}_{loc}(\overline{\mathfrak{g}}, \mathbb{C})$ a unique representative γ' exists which is local and $\mathcal{L}$-invariant (with respect to ω). Moreover, $\gamma' = a\gamma_{1,\omega}$, with $a \in \mathbb{C}$.*

In fact, the proof of Theorem 13.20 does not make any reference to the special structure of the individual simple Lie algebra. Beside general results about the structure of simple Lie algebras (e.g. the Chevalley–Serre relations), the only thing which is crucially needed is the almost-graded structure (see below) of $\overline{\mathfrak{g}}$ and its almost-graded module structure over $\mathcal{L}$. Hence, the statement of Theorem 13.20 and of course also of Theorem 13.22, including their corollaries, will be true for all other simple matrix Lie algebras $\mathfrak{g}$, not only for $\mathfrak{sl}(n), \mathfrak{so}(n), \mathfrak{sp}(2n)$ if one is able to establish these properties. For example, in [251] Sheinman has established them for the exceptional Lie algebra G_2.

Recall that the classification of almost-graded central extensions is obtained by the classification of local cocycles. Hence, from Theorem 13.22 we conclude for example the following theorem.

Theorem 13.24. *For the Lax operator algebra $\overline{\mathfrak{g}}$ with $\mathfrak{g}$ simple (e.g., $\mathfrak{g} = \mathfrak{sl}(n), \mathfrak{so}(n), \mathfrak{sp}(2n), G_2$) endowed with an almost-grading induced by splitting $A = I \cup O$, up to equivalence and rescaling of the central element exactly one nontrivial almost-graded central extension exists.*

Remark 13.25 (On the proof of Theorem 13.20). The complete proof can be found in [224]. See also [228] for the two-point case. Of course, in the two-point case every bounded cocycle will be a local cocycle and the proof simplifies accordingly. In the following section we indicate a few steps of the proof to give some hint about its spirit.
(1) We start with a bounded, $\mathcal{L}$-invariant cocycle γ and use the almost-graded $\mathcal{L}$-module structure to show that everything can be reduced to level zero. It is exactly the same technique as used in Section 6.8. Recall that if L_m and L'_k are homoge-

neous elements of degree m and k respectively, then the level of the pair (L_m, L'_k) is the sum of their degrees $m + k$. If the level $l = m + k \neq 0$, the cocycle values for pairs of elements of level l can be linearly expressed by values of the cocycle evaluated for pairs at higher level with universal coefficients only depending on the algebra $\overline{\mathfrak{g}}$. By boundedness, the cocycle values are zero for high enough level. Hence, they will vanish for all levels $l > 0$. Moreover, by recursion their values at level $l < 0$ are fixed by knowing the values at level zero.

(2) Let $X \in \mathfrak{g}$, then we denote by $X_{n,s}$, $s = 1, \dots, K$ the element in $\overline{\mathfrak{g}}_n$ with leading term Xz_s^n at P_s and higher orders at the other points in I. By Proposition 13.5 such a unique element always exists. Let $\{L^1, L^2, \dots, L^{\dim \mathfrak{g}}\}$ be a basis of $\mathfrak{g}$. By the almost-graded structure we show that at level zero the cocycle is uniquely fixed by its values

$$\gamma(L^u_{1,s}, L^v_{-1,s}) \quad \text{for} \quad u, v = 1, \dots, \dim \mathfrak{g},\ s = 1, \dots, K. \tag{13.55}$$

The other cocycle values at level zero are given by

$$\begin{aligned} \gamma(L^u_{n,s}, L^v_{-n,r}) &= 0 \quad \text{if } s \neq r, \quad \gamma(L^u_{0,s}, L^v_{0,s}) = 0 \\ \gamma(L^u_{n,r}, L^v_{-n,s}) &= n \cdot \gamma(L^u_{1,r}, L^v_{-1,r})\delta_r^s . \end{aligned} \tag{13.56}$$

Next we define for $s = 1, \dots, K$ the maps

$$\psi_{\gamma,s} : \mathfrak{g} \times \mathfrak{g} \to \mathbb{C} \quad \psi_{\gamma,s}(X, Y) := \gamma(X_{1,s}, Y_{-1,s}). \tag{13.57}$$

We show that $\psi_{\gamma,s}$ is a symmetric invariant bilinear form. As the cocycle γ is fixed by the values $\gamma(L^u_{1,s}, L^v_{-1,s})$, $s = 1, \dots, K$ and they are fixed by the bilinear maps $\psi_{\gamma,s}$, we obtain that we have at most as many different $\mathcal{L}$-invariant bounded cocycles as we have possible combinations with these bilinear forms. Recall that for simple Lie algebras $\mathfrak{g}$ there is up to a rescaling only the Cartan–Killing form which is symmetric and invariant. In fact, in our case we obtain that the cocycle γ will be a linear combination of the cocycles γ_{1,ω,C_i} (and γ_{2,C_i} for $\overline{\mathfrak{gl}(n)}$).

(3) The abelian part of the algebra is now covered, as we put $\mathcal{L}$-invariance into the requirement. For the simple part, we have to show that in every cohomology class there is an $\mathcal{L}$-invariant cocycle. To show this we consider the Chevalley generators and the Chevalley-Serre relations of the finite-dimensional simple Lie algebra $\mathfrak{g}$. We use the almost-graded structure inside $\overline{\mathfrak{g}}$ and the boundedness from above of the cocycle. By Proposition 13.7, the Chevalley-Serre relations are valid in $\overline{\mathfrak{g}}$ modulo higher order terms.
We make appropriate cohomological changes and end up with the fact that the cohomologous cocycle is bounded from above by zero and is fixed by its value at K special pairs of elements in $\overline{\mathfrak{g}}$. More precisely, it is given by $\gamma(H^\alpha_{1,s}, H^\alpha_{-1,s})$ for one fixed simple root α. Here H^α is the corresponding element from the Cartan subalgebra. We use the internal structure of $\mathfrak{g}$ related to the root system of $\mathfrak{g}$ (and of course the almost-graded structure). Hence the space is at most K-dimensional. But the γ_{1,ω,C_i} are K bounded and $\mathcal{L}$-invariant cocycles which are not coboundaries

and we get the result. As a side effect we obtain that in every bounded cohomology class (for $\mathfrak{g}$ simple) there is a unique $\mathcal{L}$-invariant representing cocycle. In the very classical case $\mathfrak{g} \otimes \mathbb{C}[z, z^{-1}]$, the algebra will be graded. In this case the line of arguments becomes simpler and is similar to those of Garland [87].

14 Some related developments

In this chapter we will mention some related further developments and related subjects. Mainly this is done by giving references.

14.1 Vertex algebras

Occasionally in this book we also treated the field theoretical background of the Krichever–Novikov type algebras. For example, in Section 8.2 we described the mathematical notion of an operator field. Such fields also showed up in the context of the Sugawara operator in Chapter 10. Of course, the WZNW models are related to operator fields in Conformal Field Theory.

From the definitions given it would not be a long way to discuss operator product expansions also in the Krichever–Novikov setting. Krichever and Novikov already did some steps in this direction [142, 144].

In this book fermionic representations mainly appeared. In physics bosonic representations are also needed. From the physicists' point of view, vertex operators give a "boson-fermion correspondence". In fact, we have seen in Section 7.4 an example of such a correspondence for the classical infinite dimensional Heisenberg algebra.

For the mathematical background of vertex algebras in the classical genus zero setting see [82, 118, 119, 123]. We will not recall their definition here. Let me only say, that there is a *state-field correspondence* fulfilling certain axioms.

It has to be pointed out that vertex algebras do not only play a role in field theory. They were also crucial in understanding the *Monster and Moonshine* phenomena, which refers to the fact that the dimensions of the irreducible representations of the largest sporadic finite group, the monster group, show up in the q-expansion coefficients of the elliptic modular function j. This was first seen experimentally and later explained with the help of representations of a certain vertex algebra which was related to the monster. The j-function appears as graded dimension of a representation of this vertex algebra. The details can be found in [82]. Also, with the help of vertex algebras, representations of Kac-Moody algebras can be constructed.

To construct vertex algebras in higher genus there are different strategies. One is by some kind of semi-local approach, very much in the spirit of Tsuchiya, Ueno, and Yamada [254]. An example is given by Zhu [271]. Another direction is based on an operadic approach. See for example Huang and Lepowsky [107–110]. Also, there is a sheaf theoretical approach due to Frenkel and Ben-Zvi [80, 81].

A physicist's approach via Krichever–Novikov objects in the context of explicit types of field theories and their special properties is given by Bonora and collaborators [22, 198].

A mathematical treatment which stays very close to the axiomatic treatment in genus zero is given by Linde [161, 162]. With the help of normal ordering and delta-distribution introduced here and some more objects which he introduces, he is able to give a possible definition for vertex algebras for higher genus. Strictly speaking, he does it only for the two-point case, but his objects, as they are formulated in terms of the Krichever–Novikov basis, should extend to the multi-point situation too. The details are not yet complete. Finally, starting from the fermionic representation of the Heisenberg algebra which we discussed in Chapter 7, he constructed a vertex algebra.

14.2 Other geometric algebras

Lie superalgebra and Jordan superalgebras of Krichever–Novikov type already appeared in the main text. Lie superalgebras were discussed in detail. See also Schlichenmaier [222], Kreusch [137] for more information on the supercase, Leidwanger, and Morier-Genoud [158, 159] for the Jordan algebra case.

We considered geometric vector fields and differential operators. But Donin and Khesin [56]also showed that pseudo-differential symbols could be treated via some Krichever–Novikov like objects.

The path from the classical picture as taken by Krichever–Novikov goes from genus zero to higher genus. There is another path in which, by considering instead of the classical algebra of Laurent polynomial $C[z^{-1}, z]$, the algebra generated by several "Laurent variables". This generalization is quite natural in the context of current algebras, as the finite-dimensional Lie algebra can be tensorized by an arbitrary commutative algebra. The algebras obtained in this way are called *toroidal algebras*. They correspond to increasing the dimension instead of the genus. A short collection of references and names is given by Bermann, Billig, Buelk, Cox, Futorny, Hu, Jurisich, Kashuba, Penkov, Szmigielski, Xia, Yokunuma, [10–12, 16, 32, 42, 43, 51, 59, 86, 175, 266].

14.3 Discretized and q-deformed Krichever–Novikov type algebras

In the definition of the vector field Krichever–Novikov type algebra, the differentiation can be replaced by a difference operator, assuming that we have a geometric situation appropriate to this discretization. See for example Meiler and Ruffing [170, 196].

Such discretizations are also related to q-deformed versions of the Krichever–Novikov vector field algebra - again for special geometric cases. In some sense the structure equations are deformed by expressions depending on a formal parameter q. One does not obtain Lie algebras anymore, but q-Lie algebras. The q-Witt and

q-Virasoro algebras are of certain importance in the context of integrable systems. One might guess that the same will be the case for the q-deformed Krichever–Novikov vector field algebra. See for example Kuang [152], and Oh and Singh [187].

14.4 Genus zero multi-point algebras – integrable systems

Witt and Virasoro algebra in genus zero with two points where poles are allowed are mathematically highly interesting objects which have, for example, a nontrivial representation theory. If we remain on the Riemann sphere but allow more than two poles, we obtain an even more demanding mathematical theory. The related systems are important. For example, the classical Knizhnik-Zamolodchikov models are of this type. We discussed them here in Chapter 11; see for example [131]. Typically, integrable systems show up.

For the genus zero multi-point situation quite a number of publications exist. Some references are [3, 26, 28, 33, 41, 45, 46, 76, 77, 205, 217]. For genus one, the complex torus case, there is [25, 29, 50, 76, 77, 197, 205].

From the point of view of symmetries of integrable systems, the concept of *automorphic Lie algebras* shows up. It was developed by, for example, Lombardo, Mikailov, and Sanders in [163–165]. Invariant objects under finite subgroups of $PGL(2, \mathbb{C})$, the symmetry group of the Riemann sphere, are studied. Of course, there are relations to the $g = 0$, multi-point Krichever–Novikov type algebras. Chopp [38] obtained some results for the genus one multi-point setting.

Another line of research in this context is given by Terwilliger and collaborators [9, 103, 112–114, 183, 184].

The genus zero Krichever–Novikov algebras turn out to be related to algebras appearing in statistical mechanics. For example, the *Onsager algebra* appears as subalgebra of the 3-point $g = 0$ Krichever–Novikov algebra.

14.5 Related works in theoretical physics

The global operator approach to Conformal Field Theory (CFT) from the mathematical point of view was initiated by Krichever and Novikov in 1987 [140–144]. They did this for the two-point case. Sheinman joined in by investigating the affine algebras and their representations [233–236, 240, 241]. In 1989, for the multi-point case, the corresponding objects were developed by the author in [204–207]. In this setting, the current book presents the mathematical knowledge in the field of Krichever–Novikov type algebras.

In the context of CFT and string theory, this approach has been taken up by quite a number of physicists. In their articles the applications were developed further and different aspects treated. Some of them had their appearance also in the main text, see for example Bonora in the context of the Sugawara construction. In the following, we give a list of references and some names of physicists who worked in the field. Neither one is intended to be complete.

List of references

[1–3, 8, 17–24, 30, 31, 35–37, 40, 54, 55, 60, 106, 132–136, 150–155, 157, 166–169, 171, 185, 186, 188–190, 192, 198, 200, 201, 229–231, 264, 266, 268–270, 272].

List of names

Alberty, Ali, Anzaldo-Meneses, Benhamou, Bonora, Bryant, Bregola, Ceresole, Cheng, Cotta-Ramusino, Chao, Dick, Fairlie, Fletcher, Guo, Huang, Kashaev, Konisi, Koibuchi, Kuang, Kubo, Kumar, Kunimasa, Lee, Lhallabi, Lugo, Maharana, Matone, Martellini, Mesincescu, Mintchev, Nepomechie, Nuyts, Osipov, Ojima, Paul, Platten, Rinaldi, Russo, Saito, Semikhatov, Sengupta, Takahas, Taormina, Toppan, van Baal, Viswanatha, Wang, Wu, Xu, Zachos, Zha, Zhao, Zhang, Zucchini.

To close I would like to point out that at the very beginning, the Krichever–Novikov type algebras were mainly used in the context of CFT. Nowadays they are also quite useful in other applications (integrable systems, deformations and many more). Also, from the pure mathematical point of view they are important. For example, they provide a lot of infinite-dimensional Lie algebras which are still controllable by their geometric nature.

Bibliography

[1] J. Alberty, A. Taormina, and P. van Baal, *Relating Kac-Moody, Virasoro and Krichever-Novikov algebras*. Comm. Math. Phys. **120**(2), (1988), 249–260.

[2] A. Ali, A. Kumar, J. Maharana, and G. Sengupta, *Algebraic structure of toplogical superconformal field theory on Riemann surfaces*. Comm. Math. Phys. **148**, (1992), 117–137.

[3] A. Anzaldo-Meneses, *Krichever-Novikov algebras on Riemann surfaces of genus zero and one with N punctures*. J. Math. Phys. **33**(12), (1992), 4155–4163.

[4] E. Arbarello, C. deConcini, V. Kac, C. Procesi, *Moduli spaces of Curves and representation theory*, Comm. Math. Phys. **117** (1988), 1–36.

[5] M. F. Atiyah, *Riemann surfaces and spin structures*, Ann. Scient. de L'E.N.S., **4** (1), (1971), 47–62.

[6] K. Bardakçi and M. B. Halpern, *New dual quark models*, Phys. Rev. D. **3** (1971), 2493.

[7] A. A. Belavin, A. M. Polyakov, A. B. Zamolodchikov, *Infinite conformal symmetry in two-dimensional quantum field theory.* Nucl. Phys. B. **241**, (1984), 333–380.

[8] N. Benhamou and T. Lhallabi, *$N = 2$ supercovariant operators and Krichever-Novikov algebra*. J. Math. Phys. **41**(1), (2000), 468–479.

[9] G. Benkart and P. Terwilliger, *The universal central extension of the three-point $\mathfrak{sl}_2$ loop algebra*. Proc. Amer. Math. Soc. **135**(6), (2007), 1659–1668.

[10] S. Berman and Y. Billig, *Irreducible representations for toroidal Lie algebras*. J. Algebra **221**(1), (1999), 188–231.

[11] S. Berman, Y. Billig, and J. Szmigielski, *Vertex operator algebras and the representation theory of toroidal algebras*. In: *Recent developments in infinite-dimensional Lie algebras and conformal field theory (Charlottesville, VA, 2000), Contemp. Math.* Providence, RI: Amer. Math. Soc., **297**, (2002) 1–26.

[12] S. Berman and B. Cox, *Enveloping algebras and representations of toroidal Lie algebras*. Pacific J. Math. **165**(2), (1994), 239–267.

[13] D. Bernard, *On the Wess–Zumino–Witten models on the torus*, Nucl. Phys. B. **303** (1988), 77–93.

[14] D. Bernard, *On the Wess–Zumino–Witten models on Riemann surfaces*. Nucl. Phys. **309** (1988), 145–174.

[15] I. N. Bernstein, I. M. Gelfand, S. I. Gelfand, *Differential Operators on the Base Affine Space and a Study of g-modules*, (in) Lie groups and their representations (ed. I.M. Gelfand), Adam Hilger Ltd., 1975.

[16] Y. Billig, *Representations of toroidal Lie algebras and Lie algebras of vector fields*. Resenhas **6**(2-3), (2004), 111–119.

[17] L. Bonora, *Global operator formalism on Riemann surfaces*. In: *String theory (Rome, 1988)*, Sympos. Math., XXXIII. London: Academic Press, 1–15, 1990.

[18] L. Bonora, M. Bregola, P. Cotta-Ramusino, M. Martellini, *Virasoro-type algebras and BRST operators on Riemann surfaces*, Phys. Lett. B, **205**(1), (1988), 53–56.

[19] L. Bonora, M. Matone, and M. Rinaldi, *Relation between representations of KN and Virasoro algebras*. Phys. Lett. B., **216**(3-4), (1989), 313–319.

[20] L. Bonora, M. Matone, and F. Toppan, *$b - c$ system approach to minimal models. The genus-zero case*, Phys. Lett. B., **224**(1-2), (1989), 115–120.

[21] L. Bonora, M. Martellini, M. Rinaldi, J. Russo, *Neveu–Schwarz- and Ramond-type Superalgebras on genus g Riemann surfaces*, Phys. Lett. B., **206**(3), (1988), 444–450.

[22] L. Bonora, A. Lugo, M. Matone, J. Russo, *A global operator formalism on higher genus Riemann surfaces: $b - c$ systems*, Comm. Math. Phys. **123** (1989), 329–352.

[23] L. Bonora, M. Matone, F. Toppan, K. Wu, *Real weight $b-c$ systems and conformal field theories in higher genus*, Nuclear Phys. B., **334**(3), (1990), 717–744.

[24] L. Bonora, M. Rinaldi, J. Russo, K. Wu, *The Sugawara construction on genus g Riemann surfaces*, Phys. Lett. B., **208** (1988), 440-446.

[25] M. R. Bremner, *On a Lie algebra of vector fields on a complex torus*, J. Math. Phys. **31** (1990), 2033–2034.

[26] M. R. Bremner, *Structure of the Lie algebra of polynomial vector fields on the Riemann sphere with three punctures*, J. Math. Phys. 32 (1991), 1607–1608.

[27] M. Bremner, *Universal central extensions of elliptic affine Lie algebras*, J. Math. Phys. **35**(12), (1994), 6685–6692.

[28] M. R. Bremner, *Four-point affine Lie algebras,* Proc. Amer. Math. Soc., **123** (1995), 1981–1989.

[29] A. Bruder, M. Meiler, and A. Ruffing, *Topological symmetries in Krichever-Novikov algebras.* Nonlinear Phenom. Complex Syst. **9**(3), (2006), 288–293.

[30] P. Bryant, *Graded Riemann surfaces and Krichever-Novikov algebras.* In: *Proceedings of the Winter School on Geometry and Physics (Srní, 1989).* 15–24: 1990a.

[31] P. Bryant, *Graded Riemann surfaces and Krichever-Novikov algebras.* Lett. Math. Phys. 19(2), 97–108: 1990.

[32] S. Buelk, B. Cox, and E. Jurisich, *A Wakimoto type realization of toroidal $\mathfrak{sl}_{n+1}$.* Algebra Colloq. **19**(Special Issue No. 1), (2012), 841–866.

[33] A. Bueno, B. Cox, and V. Futorny, *Free field realizations of the elliptic affine Lie algebra* $\mathfrak{sl}(2, \boldsymbol{R}) \oplus (\Omega_R/dR)$. J. Geom. Phys. **59**(9), (2009), 1258–1270.

[34] V.M. Bukhshtaber and D. V. Leĭkin, *Functional equations that define multiplication in the continuous Krichever-Novikov basis.* Uspekhi Mat. Nauk **61**(1(367)), (2006), 171–172.

[35] A. Ceresole and C. S. Huang, *Extended KN algebras and extended conformal field theories over higher genus Riemann surfaces.* Phys. Lett. B., **247**(2-3), (1990), 331–336.

[36] Q. H. Cheng, *Sewing construction in KN operator formalism of conformal field theory.* Comm. Theoret. Phys. ,**18**(4), (1992), 435–442.

[37] Q. H. Cheng, *Krichever-Novikov operator formalism and sewing conformal fields.* Internat. J. Theoret. Phys., **32**(4), (1993), 529–540.

[38] M. Chopp, *Lie-admissible structures on Witt type algebras and automorphic algebras*, Ph.D. thesis, University of Luxembourg and University of Metz, 2011.

[39] A. Connes, *Noncommutative differential geometry. Chapter I: The Chern character in K homology; Chapter II: de Rham homology and noncommutative algebra.* Publ. Math. IHES no. 62 (1985), 41–144.

[40] P. Cotta-Ramusino, M. Martellini, and M. Mintchev, *Deformations of complex structures and representations of Krichever-Novikov algebras.* Phys. Lett. B., **215**(2), (1988), 331–337.

[41] B. Cox, *Realizations of the four point affine Lie algebra $\mathfrak{sl}(2, R) \oplus (\Omega_R/dR)$.* Pacific J. Math., **234**(2), (2008), 261–289.

[42] B. Cox and V. Futorny, *DJKM algebras I: their universal central extension.* Proc. Amer. Math. Soc., **139**(10), (2011), 3451–3460.

[43] B. Cox and V. Futorny, *Borel subalgebras and categories of highest weight modules for toroidal Lie algebras.* J. Algebra, **236**(1), (2001), 1–28.

[44] B. Cox, V. Futorny, and J. A. Tirao, *DJKM algebras and non-classical orthogonal polynomials.* J. Differential Equations, **255**(9), (2013), 2846–2870.

[45] B. Cox, X. Guo, R. Lu, and K. Zhao, n-Point Virasoro algebras and their modules of densities, Comm. Contemp. Math. **16**(3) (2014) 1350047, (27 pages), DOI: 10.1142/D0219199713500478.

[46] B. Cox, and E. Jurisich, Realizations of the three point Lie algebra $\mathfrak{sl}(2, \mathcal{R}) \oplus (\Omega_{\mathcal{R}}/d\mathcal{R})$, arXiv:1303.6973, to appear in Pac. J. of Math.

[47] R. Dashen and Y. Frishman, *Thirring model with $\mathrm{U}(n)$ symmetry: scale invariant only for fixed values of a coupling constant.* Phys. Lett. B., **46** (1973), 439–442.

[48] E. Date, M. Jimbo, T. Miwa, and M. Kashiwara, *Transformation groups for soliton equations*, Publ. RIMS, Kyoto 394 (1982).
[49] T. Deck, *Stringtheorie und Krichever–Novikov–Algebren*, Diplomarbeit (in German), Universität Karlsruhe. 1990.
[50] T. Deck, *Deformations from Virasoro to Krichever-Novikov algebras*, Phys. Lett. B., **251** (1990) 535–540.
[51] I. Dimitrov, V. Futorny, and I. Penkov, *A reduction theorem for highest weight modules over toroidal Lie algebras*. Comm. Math. Phys., **250**(1), (2004), 47–63.
[52] E. D'Hoker, I. Krichever, and D. H. Phong, *Seiberg-Witten theory, symplectic forms, and Hamiltonian theory of solitons*. In: *Superstring theory*, Vol. 1 of *Adv. Lect. Math. (ALM)*. Int. Press, Somerville, MA, pp. 124–177: 2008.
[53] R. Dick, *Krichever-Novikov-like bases on punctured Riemann surfaces*, Lett. Math. Phys., **18** (1989), 255–265
[54] R. Dick, *Holomorphic differentials on punctured Riemann surfaces*. In: *Differential geometric methods in theoretical physics (Davis, CA, 1988)*, Vol. 245 of *NATO Adv. Sci. Inst. Ser. B Phys.* New York: Plenum, pp. 475–483: 1990.
[55] R. Dick, *Topological aspects of chiral fields in two dimensions and superstring vertices*. Fortschr. Phys., **40**(6), (1992), 519–591.
[56] D. Donin and B. Khesin, *Pseudodifferential symbols on Riemann surfaces and Krichever-Novikov algebras*. Comm. Math. Phys., **272**(2), (2007), 507–527.
[57] T. Eguchi and H. Ooguri, *Conformal and current algebras on general Riemann surface*, Nuclear Phys. B., **283** (1987), 308.
[58] B. Enriquez and G. Felder, *Solutions of the KZB equations in genus* ≥ 1. math.QA/9912198.
[59] S. Eswara Rao and R. V. Moody, *Vertex representations for* n*-toroidal Lie algebras and a generalization of the Virasoro algebra*. Comm. Math. Phys., **159**(2), (1994), 239–264.
[60] D B. Fairlie, P. Fletcher, and J. Nuyts, *Eigenvalue equations for Krichever-Novikov algebras*. J. Math. Phys., **30**(5), (1989), 957–964.
[61] H. M. Farkas and I. Kra, *Riemann surfaces*, Graduate Texts in Mathematics. 71., Springer, Berlin New York, 1992.
[62] J. D. Fay, *Theta functions on Riemann surfaces*, Lecture Notes in Math., 352, Springer, Berlin New York, 1973.
[63] B. L. Feigin, *Cohomology of groups and current algebras.*, Usp. Mat. Nauk, **35** (1980), 224–226.
[64] B. L. Feigin, *Conformal field theory and cohomologies of the Lie algebra of holomorphic vector fields on a complex curve.*, Proceedings ICM, Kyoto, Japan (1990), pp. 71–85.
[65] B. L. Feigin, D. B. Fuchs, Cohomologies of Lie Groups and Lie Algebras. In: Onishchik, A.L., and Vinberg, E.B. (eds) *Lie groups and Lie algebras II*. Encyclopaedia of Math. Sciences, Vol. 21, pp. 125–215, Springer, Berlin New York, 2000.
[66] G. Felder, *The KZB equations on Riemann surfaces*, Quantum symmetries/ Symetries quantiques. Proceedings of the Les Houches summer school (Les Houches, France, August 1–September 8, 1995) (A. Connes et. al., ed.), North-Holland, Amsterdam, 1998, pp. 687–725, hep-th/9609153.
[67] G. Felder, C. Wieczerkowski, *Conformal blocks on elliptic curves and the Knizhnik-Zamolodchikov-Bernard equation*. Commun. Math. Phys. **176** (1996), 133–161.
[68] A. Fialowski, Deformations of the Lie algebra of vector fields on the line. Uspkhi Mat. Nauk, **38** (1983) 201–202; English translation: Russian Math. Surveys, **38**(1), (1983), 185–186.
[69] A. Fialowski, Deformations of Lie algebras. Math. USSR Sbornik **55**, (1986) 467–472.
[70] A. Fialowski, An example of formal deformations of Lie algebras. In: *Proceedings of NATO Conference on Deformation Theory of Algebras and Applications, Il Ciocco, Italy, 1986*, pp. 375–401, Kluwer, Dordrecht, 1988.

[71] A. Fialowski, *Formal rigidity of the Witt and Virasoro Algebra*, J. Math. Phys., **53**(7), 073501, 5 p. (2012), arXiv:1202.3132.

[72] A. Fialowski, *Unpublished notes*. 1989.

[73] A. Fialowski, *Deformations of some infinite dimensional Lie algebras*. J. Math. Phys., **31**(6), (1990), 1340–1343.

[74] A. Fialowski and D. Fuchs, Singular deformations of Lie algebras. Example: Deformations of the Lie algebra L_1. In: *Topics in singularity theory, V.I. Arnold's 60th anniversary collection*, Transl. Ser. 2, Am. Math. Soc., **180**(34), (1997), 77–92.

[75] A. Fialowski and D. Fuchs, *Construction of Miniversal Deformations of Lie Algebras*. J. Funct. Anal. ,**161** (1999) 76–110.

[76] A. Fialowski and M. Schlichenmaier, *Global deformations of the Witt algebra of Krichever-Novikov type*, Comm. Contemp. Math., **5**(6), (2003), 921–946.

[77] A. Fialowski and M. Schlichenmaier, *Global geometric deformations of current algebras as Krichever-Novikov type algebras*, Comm. Math. Phys., **260** (2005), 579 –612.

[78] A. Fialowski and M. Schlichenmaier, *Global Geometric Deformations of the Virasoro algebra, current and affine algebras by Krichever-Novikov type algebras*, Int. Jour. Theor. Phys., **46**(11), (2007), 2708–2724.

[79] O. Forster, *Lectures on Riemann Surfaces*, Springer, Berlin New York, 1981.

[80] E. Frenkel, *Vertex algebras and algebraic curves*. Astérisque (276), 299–339. Séminaire Bourbaki, Vol. 1999/2000: 2002.

[81] E. Frenkel and D. Ben-Zvi, Vertex algebras and algebraic curves, Vol. 88 of *Mathematical Surveys and Monographs*. Providence, RI: American Mathematical Society, 2001.

[82] I. Frenkel, J. Lepowsky, and A. Meurman, *Vertex operator algebras and the Monster*, Vol. 134 of *Pure and Applied Mathematics*. Academic Press Inc., Boston, MA, 1988.

[83] I. Frenkel and M. Zhu, *Vertex algebras associated to modified regular representations of the Virasoro algebra*. Adv. Math. **229**(6), (2012), 3468–3507.

[84] J. Fuchs and C. Schweigert, *Symmetries, Lie algebras and representations. A graduate course for physicists*. Cambridge Monographs on Mathematical Physics. Cambridge University Press, 1997.

[85] D. B. Fuks, *Cohomology of infinite-dimensional Lie algebras*. Consultants Bureau, Plenum 1986

[86] V. Futorny and I. Kashuba, *Verma type modules for toroidal Lie algebras*. Comm. Algebra, **27**(8), (1999), 3979–3991.

[87] H. Garland, *The arithmetic theory of loop groups.*, Publ. Math. Inst. Hautes Etud. Sci. **52** (1980), 5–136.

[88] H. Garland, *Dedekind's η-function and the cohomology of infinite dimensional Lie algebras*, Proc. Nat. Acad. Sci. USA, **72** (1975) 2493–2495.

[89] M. Gerstenhaber, *On the Deformation of Rings and Algebras I,II,III*, Ann. Math. **79** (1964) 59–10 (1964); **84** (1966) 1–19; **88** (1968) 1–34.

[90] L. V. Goncharova, *Cohomology of Lie algebras of formal vector fields on the line*, Funct. Anal. Appl. 7, 2 (1973), 6–14

[91] A. Z. Górski and J. Szmigielski, *Representations of the Heisenberg algebra by difference operators*. Acta Phys. Polon. B., **31**(4), (2000), 789–799.

[92] P. Griffiths and J. Harris, *Principles of algebraic geometry*, John Wiley, New York, 1978.

[93] A. Grothendieck and J. A. Dieudonné, *Eléments de géométrie algébrique I*, Springer, Berlin New York, 1971.

[94] L. Guerrini, Formal and analytic deformations of the Witt algebra. Lett. Math. Phys., **46** (1998) 121–129.

[95] L. Guerrini, Formal and analytic rigidity of the Witt algebra. Rev. Math. Phys., **11** (1999) 303–32.

[96] L. Guieu, C. Roger, *L'algèbre et le groupe de Virasoro*. Les publications CRM, Montreal 2007.

[97] R. C. Gunning, *Lectures on Riemann surfaces*, Princeton Math. Notes, N.J. 1966.

[98] H. Guo, J. Na, J. Shen, S. Wang, and Q. Yu, *The algebras of meromorphic vector fields and their realisation on the spaces of meromorphic λ-differentials on Riemann surfaces*. J. Phys. A, Math. Gen., **23**(4), (1990), 379–384.

[99] A. Haddi, *Homologie des algèbres de Lie étendues à une algèbre commutative.*, Commun. Algebra **20**(4), (1992), 1145–1166.

[100] R. Hartshorne, *Algebraic Geometry*, Springer, Berlin New York, 1977.

[101] N. S. Hawley and M. Schiffer, *Half-order differentials on Riemann surfaces*, Acta Math., **115** (1966), 199–236.

[102] J. Harris and I. Morrison, *Moduli of Curves*. Springer, Berlin New York, 1998.

[103] B. Hartwig and P. Terwilliger, *The tetrahedron algebra, the Onsager algebra, and the $\mathfrak{sl}_2$ loop algebra*. J. Algebra, **308**(2), (2007), 840–863.

[104] P. Hilton and U. Stammbach, *A course in homological algebra*, Springer, Berlin New York, 1971

[105] N. Hitchin, *Flat connections and geometric quantization*. Commun. Math. Phys. **131** (1990), 347–380.

[106] C S. Huang and Z. Y. Zhao, *Krichever-Novikov algebras and string theories on group manifolds*. Phys. Lett. B., **220**(1-2), (1989), 87–92.

[107] Y.-Z. Huang, *Operadic formulation of topological vertex algebras and Gerstenhaber or Batalin-Vilkovisky algebras*. Comm. Math. Phys., **164**(1), (1994), 105–144.

[108] Y.-Z. Huang, *Meromorphic open-string vertex algebras*. J. Math. Phys., **54**(5), (2013) 051702, 33.

[109] Y.-Z. Huang and L. Kong, *Open-string vertex algebras, tensor categories and operads*. Comm. Math. Phys., **250**(3), (2004), 433–471.

[110] Y.-Z. Huang and J. Lepowsky, *On the $\mathcal{D}$-module and formal-variable approaches to vertex algebras*. In: *Topics in geometry*, Vol. 20 of *Progr. Nonlinear Differential Equations Appl.* Boston, MA: Birkhäuser Boston, pp. 175–202, 1996.

[111] A. Hurwitz and R. Courant, *Allgemeine Funktionentheorie und elliptische Funktionen*, Springer, Berlin New York, 1964.

[112] T. Ito and P. Terwilliger, *The q-tetrahedron algebra and its finite dimensional irreducible modules*. Comm. Algebra, **35**(11), (2007), 3415–3439.

[113] T. Ito and P. Terwilliger, *Tridiagonal pairs and the quantum affine algebra $U_q(\widehat{\mathrm{sl}}_2)$*. Ramanujan J., **13**(1-3), (2007), 39–62.

[114] T. Ito and P. Terwilliger, *Finite-dimensional irreducible modules for the three-point $\mathfrak{sl}_2$ loop algebra*. Comm. Algebra, **36**(12), (2008), 4557–4598.

[115] D. Ivanov, *Knizhnik-Zamolodchikov-Bernard equations on Riemann surfaces*, Int. J. Mod. Phys. A., **10** (1995), 2507–2536.

[116] D. A. Jordan, *On the ideals of a Lie algebra of derivations*, J. London Math. Soc (2), **33** (1986), 33–39.

[117] V. G. Kac, *Simple irreducible graded Lie algebras of finite growth. (Russian).*, Izv. Akad. Nauk SSSR Ser. Mat. **32** (1968), 1323–1367.

[118] V. G. Kac, *Infinite dimensional Lie algebras*. Cambridge Univ. Press, Cambridge, 1990.

[119] V. Kac, *Vertex algebras for beginners*, Vol. 10 of *University Lecture Series*. Providence, RI: American Mathematical Society, second edition: 1998.

[120] V. G. Kac, Petersen, D. H., *Spin and wedge representations of infinite-dimensional Lie algebras and groups*. Proc. Nat. Acad. Sci. USA 78 (1981) 3308–3312.

[121] V. Kac and A. Radul, *Representation theory of the vertex algebra $W_{1+\infty}$*. Transform. Groups, **1**(1-2), (1996), 41–70.

[122] V. Kac and A. Raina, *Highest weight representations of infinite dimensional Lie algebras*, World Scientific, 1987.

[123] V. Kac, A. Raina, and N. Rozhkovskaya, *Highest weight representations of infinite dimensional Lie algebras, 2. Edition*, World Scientific, 2013.

[124] V. G. Kac and M. Wakimoto, *Modular and conformal invariance constraints in representation theory of affine algebras.*, Adv. in Math., **70** (1988), 156–234.

[125] N. Kawamato, Y. Namikawa, A. Tsuchia, and Y. Yamada, *Geometric realizations of Conformal Field Theory on Riemann surfaces*, Commun. Math. Phys., **116**(1988), 247–308.

[126] N. Kawazumi, *On the complex analytic Gel'fand-Fuks cohomology of open Riemann surfaces*, Ann. de l'Institut Fourier, 40(3), (1993), 655–712.

[127] C. Kassel, *Kähler differentials and coverings of complex simple Lie algebras extended over a commutative algebra*, J. Pure Appl. Algebra, **34** (1984), 265–275.

[128] C. Kassel and J.-L. Loday, *Extensions centrales d'algébres de Lie.*, Ann. Inst. Fourier **32**(4), (1982), 119–142.

[129] N. Kawazumi, *On the complex analytic Gel'fand-Fuks cohomology of open Riemann surfaces*, Ann. Inst. Fourier **43**(3), (1993), 655–712.

[130] S. Klimek and A. Lesniewski, *Global Laurent expansions on Riemann surfaces.* Commun. Math. Phys., **126** (1989), 421–431.

[131] V. G. Knizhnik and A. B. Zamolodchikov, *Current algebra and Wess–Zumino model in two dimensions.* Nucl. Phys. B., **247** (1984), 83–103.

[132] R. M. Kashaev and A. A. Osipov, *The gauge fixing extension of the Krichever-Novikov algebra in the closed string theory.* Phys. Lett. B., **222**(3-4), (1989), 391–394.

[133] G. Konisi, T. Saito, and W. Takahasi, *Super Schwarzian connections in Krichever-Novikov superalgebras.* Progr. Theoret. Phys., **86**(1), (1991), 229–242.

[134] H. Koibuchi, *Virasoro algebra in the KN algebra: bosonic string with fermionic ghosts on Riemann surfaces.* Modern Phys. Lett. A., **6**(31), (1991), 2861–2870.

[135] G. Konisi, T. Saito, and W. Takahasi, *KN algebra derived from Virasoro algebra with vertex operators.* Progr. Theoret. Phys., **82**(1), (1989), 162–170.

[136] G. Konisi, T. Saito, and W. Takahasi, *KN superalgebras with vertex operators.* Progr. Theoret. Phys., **82**(4), (1989), 813–828.

[137] M. Kreusch, *Extensions of superalgebras of Krichever-Novikov type.* Lett. Math. Phys., **103**(11), (2013), 1171–1189.

[138] I. M. Krichever, *Vector bundles and Lax equations on algebraic curves.* Comm. Math. Phys., **229**, (2002) 229–269.

[139] I. M. Krichever, *Isomonodromy equations on algebraic curves, canonical transformations and Whitham equations.* Mosc. Math. J. **2**(4), (2002) 717–752, 806.

[140] I. M. Krichever and S. P. Novikov, *Algebras of Virasoro type, Riemann surfaces and structures of the theory of solitons.* Funktional Anal. i. Prilozhen., **21**(2), (1987), 46–63.

[141] I. M. Krichever and S. P. Novikov, *Virasoro type algebras, Riemann surfaces and strings in Minkowski space.* Funktional Anal. i. Prilozhen., 21(4), (1987), 47–61.

[142] I. M. Krichever and S. P. Novikov, *Algebras of Virasoro type, energy-momentum tensors and decompositions of operators on Riemann surfaces.* Funktional Anal. i. Prilozhen., **23**(1), (1989), 46–63.

[143] I. M. Krichever and S. P. Novikov, *Virasoro-Gel′fand-Fuks type algebras, Riemann surfaces, operator's theory of closed strings.* J. Geom. Phys., **5**(4), (1989), 631–661.

[144] I. M. Krichever and S. P. Novikov, *Riemann surfaces, operator fields, strings. Analogues of the Fourier-Laurent bases.* In: *Physics and mathematics of strings.* World Sci. Publ., Teaneck, NJ, pp. 356–388, 1990.

[145] I. M. Krichever and S. P. Novikov, *Holomorphic bundles and commuting difference operators. Two-point constructions.* Uspekhi Mat. Nauk, **55**(3 (333)), 181–182, 2000.

[146] I. M. Krichever and S. P. Novikov, *Holomorphic bundles and difference scalar operators: single-point constructions*. Uspekhi Mat. Nauk, **55**(1(331)), (2000), 187–188.
[147] I. M. Krichever and S. P. Novikov, *A two-dimensionalized Toda chain, commuting difference operators, and holomorphic vector bundles*. Uspekhi Mat. Nauk. **58**(3(351)), (2003), 51–88.
[148] I. M. Krichever and O. K. Sheinman, *Lax operator algebras*, Funct. Anal. Appl., **41**(4), (2007), 284–294.
[149] K. Kodaira, *Complex manifolds and deformation of complex structures*, Springer, Berlin New York, 1986.
[150] L. M. Kuang, $N = 4$ *KN superalgebra on higher genus Riemann surfaces*. Acta Sci. Natur. Univ. Norm. Hunan., **17**(4), (1994), 24–28, 35.
[151] L. Kuang, *Primary superfield representation of* $N = 2$ *Krichever-Novikov superalgebra*. Acta Sci. Natur. Univ. Norm. Hunan., **18**(4), (1995), 17–21.
[152] L. M. Kuang, *The* q*-deformed Krichever-Novikov algebra and Ward* q*-identities for correlators on higher genus Riemann surfaces*. J. Phys. A., **26**(20), (1993), 5435–5441.
[153] R. Kubo, S. Ojima, and S. K. Paul, *Relations between Virasoro algebra and Krichever-Novikov algebra on a torus*. Modern Phys. Lett. A., **4**(15), (1989), 1423–1430.
[154] A. Kumar, J. Maharana, and G. Sengupta, *Krichever-Novikov formulation of topological conformal field theory*. Lett. Math. Phys., **23**(4), (1991), 271–277.
[155] T. Kunimasa and T. Saito, *Schwarzian derivative and connection in Krichever Novikov algebra*. Progr. Theoret. Phys., **83**(3), (1990), 385–389.
[156] P. Lecomte and C. Roger, *Rigidity of Current Lie Algebras of Complex Simple Type*, J. London Math. Soc., **37**(2), (1988), 232–240.
[157] T. Lee and K. S. Viswanathan, *Krichever-Novikov bases for the bosonic string*. Internat. J. Modern Phys. A., **4**(17), (1989), 4469–4474.
[158] S. Leidwanger and S. Morier-Genoud, *Superalgebras associated to Riemann surfaces: Jordan algebras of Krichever-Novikov type*. Int. Math. Res. Not. IMRN **19**, (2012), 4449–4474.
[159] S. Leidwanger and S. Morier-Genoud, *Universal enveloping algebras of Lie antialgebras*, Algebras and Representation Theory, **15**(1), (2012), 1–27.
[160] Wang-Lai Li, *2-cocycles on the algebra of differential operators*, J. Algebra, 122 (1989) 64–80.
[161] K. J. Linde, *Global vertex algebras on Riemann surfaces*. München, Dissertation, 2004.
[162] K. J. Linde, *Towards vertex algebras of Krichever–Novikov type, Part I*. arXiv: math/0305428, 2003.
[163] S. Lombardo and A. V. Mikhailov, *Reduction groups and automorphic Lie algebras*. Comm. Math. Phys., **258**(1), (2005), 179–202.
[164] S. Lombardo and A. V. Mikhailov, *Reductions of integrable equations and automorphic Lie algebras*. In: *SPT 2004—Symmetry and perturbation theory*. World Sci. Publ., Hackensack, NJ, 2005.
[165] S. Lombardo and J. A. Sanders, *On the classification of automorphic Lie algebras*. Comm. Math. Phys., **299**(3), (2010), 793–824.
[166] A. R. Lugo and J. Russo, *Multiloop (super) string amplitudes and other applications to CFT of the operator formalism on arbitrary genus Riemann surfaces*. In: *Functional integration, geometry and strings (Karpacz, 1989)*, Vol. 13 of *Progr. Phys.* Basel, Birkhäuser, 1989.
[167] J. Maharana, S. K. Paul, and G. Sengupta, *Krichever-Novikov global operator formalism: NSR superstring in curved background*. Comm. Math. Phys., **139**(3), (1991), 527–549.
[168] J. Maharana, S. K. Paul, and G. Sengupta, *Superstring in curved space: application of Krichever-Novikov algebra*. In: *High energy physics and cosmology (Trieste, 1990)*, Vol. 7 of *ICTP Ser. Theoret. Phys.* World Sci. Publ., River Edge, NJ, 1991.
[169] M. Matone, *Conformal field theory, real weight differentials and KDV equation in higher genus*, (in) Global geometry and mathematical physics, Lect. 2nd Sess. CIME, Montecatini Terme/Italy 1988, Lect. Notes Math. 1451, 163–175 Springer, Berlin New York, 1990.

[170] M. Meiler and A. Ruffing, *Constructing similarity solutions to discrete basic diffusion equations.* Adv. Dyn. Syst. Appl., **3**(1), (2008), 41–51.

[171] L. Mesincescu, R. I. Nepomechie, and C. K. Zachos, *(Super)conformal algebra on the (super)torus*, UMTG-144, ANL-Hep-Pr-88–33.

[172] D. V. Millionshchikov, *Krichever-Novikov algebras and the cohomology of the algebra of meromorphic vector fields.* In: *Solitons, geometry, and topology: on the crossroad*, Vol. 179 of *Amer. Math. Soc. Transl. Ser. 2.* Providence, RI: Amer. Math. Soc., pp. 101–108, 1997.

[173] Montserrat Teixidor i Bigas, *The divisor of curves with a vanishing theta-null*, Comp. Math., **66** (1988), 15–22.

[174] R. V. Moody, *Euclidean Lie algebras*, Canad. J. Math., **21** (1969), 1432 –1454.

[175] R. V. Moody, S. E. Rao, and T. Yokonuma, *Toroidal Lie algebras and vertex representations.* Geom. Dedicata, **35**(1-3), (1990), 283–307.

[176] D. Mumford, *Theta characteristics of an algebraic curve*, Ann. Scient. de l'E.N.S., **4**(2), (1971), 181–192.

[177] D. Mumford, *Tata lectures on Theta I, II*, Birkhäuser, Basel 1983 (1984).

[178] K.-H. Neeb and F. Wagemann, *The universal central extension of the holomorphic current algebra.* Manuscripta Math., **112**(4), (2003), 441–458.

[179] K.-H. Neeb and F. Wagemann, *Addendum to: "The universal central extension of the holomorphic current algebra" [Manuscripta Math. **112**(4) (2003), 441–458].* Manuscripta Math. **116**(2), (2005). 245–247.

[180] K.-H. Neeb and F. Wagemann, *The second cohomology of current algebras of general Lie algebras.* Canad. J. Math., **60**(4), (2008), 892–922.

[181] A. Nijenhuis, Richardson, R., *Cohomology and Deformations of Algebraic Structures.* Bull. Amer. Math. Soc., **70** (1964) 406–411.

[182] A. Nijenhuis and R. Richardson, *Cohomology and Deformations in Graded Lie Algebras.* Bull. Amer. Math. Soc., **72** (1966) 1–29.

[183] K. Nomura and P. Terwilliger, *The structure of a tridiagonal pair.* Linear Algebra Appl., **429**(7), (2008), 1647–1662.

[184] K. Nomura and P. Terwilliger, *Towards a classification of the tridiagonal pairs.* Linear Algebra Appl. **429**(2-3), (2008), 503–518.

[185] J. Nuyts, *Some properties of the Virasoro and of some Krichever-Novikov algebras.* In: *Mathematical physics (Islamabad, 1989).* World Sci. Publ., River Edge, NJ, pp. 317–325, 1990.

[186] J. Nuyts and I. Platten, *Presentation of the constant r-term Krichever-Novikov-type algebras.* J. Math. Phys., **31**(3), (1990), 527–532.

[187] C. H. Oh and K. Singh, *q-deformation of the Krichever-Novikov algebra.* Lett. Math. Phys., **42**(2), (1997), 115–122.

[188] S. Ojima, *A note on Krichever-Novikov-Kac-Moody algebra on Riemann surfaces.* Modern Phys. Lett. A., **5**(27), (1990), 2215–2221.

[189] A. A. Osipov, *Extensions of the Krichever-Novikov superalgebras in the Ramond and Neveu-Schwarz closed superstring theory.* Phys. Lett. B., **225**(1-2), (1989), 83–86.

[190] A. A. Osipov, *The gauge fixing extensions of the Krichever-Novikov superalgebras in the Ramond and Neveu-Schwarz closed superstring theory.* In: *Solitons and applications (Dubna, 1989).* World Sci. Publ., River Edge, NJ, pp. 152–158, 1990.

[191] V. Ovsienko, *Lie antialgebras: prémices*, J. of Algebra, **325**(1), (2011), 216–247.

[192] S. K. Paul, *Krichever-Novikov global operator formalism: NSR superstring in curved background.* In: *Modern quantum field theory (Bombay, 1990).* World Sci. Publ., River Edge, NJ, pp. 209–224. 1991.

[193] A. O. Radul, *A central extension of the Lie algebra of differential operators on the circle and W-algebras*, JETP Lett. **50**(8), (1990), 371–373, translated from Pis'ma Zh. Éksp. Teor. Fiz. **50**(8), (1989), 341–343.

[194] V. N. Reshetnikov, On the cohomology of the Lie algebra of vector fields on a circle. Usp. Mat. Nauk **26**, (1971), 231–232.

[195] A. Ruffing, *Prozessinduzierte symmetrische Basiskonstruktion verallgemeinerter Krichever-Novikov-Algebren für bosonische Stringtheorien auf maximal vierfach punktierten* $\mathbb{C}$*-Tori*, Diplomarbeit Universität Karlsruhe 1991.

[196] A. Ruffing, *On discretized generalizations of Krichever-Novikov algebras*. Open Syst. Inf. Dyn., **7**(3), (2000), 237–254.

[197] A. Ruffing, T. Deck, and M. Schlichenmaier, *String Branchings on complex tori and algebraic representations of generalized Krichever-Novikov algebras*, Lett. Math. Phys., **26** (1992), 23–32.

[198] J. Russo, *Bosonization and fermion vertices on an arbitrary genus Riemann surface by using a global operator formalism*. Modern Phys. Lett. A., **4**(24), (1989), 2349–2362.

[199] V. A. Sadov, *Bases on Multipunctured Riemann Surfaces and Interacting String Amplitudes*. Comm. Math. Phys., **136** (1991), 585–597.

[200] T. Saito, *Schwarzian connections in the Krichever-Novikov algebra*. Progr. Theoret. Phys., **85**(4), (1991), 743–750.

[201] T. Saito, *Schwarzian derivative in the Krichever-Novikov algebra*. Progr. Theoret. Phys., **85**(4), (1991), 705–709.

[202] L. J. Santharoubane, *The second cohomology group for Kac-Moody Lie algebras and Kähler differentials.*, J. Algebra **125**(1), (1989), 13–26.

[203] M. Schlichenmaier, *An Introduction to Riemann Surfaces, Algebraic Curves and Moduli Spaces, 2nd enlarged edition*, Springer, Berlin New York, 2007.

[204] M. Schlichenmaier, *Krichever-Novikov algebras for more than two points*, Lett. Math. Phys., **19** (1990), 151–165.

[205] M. Schlichenmaier, *Krichever-Novikov algebras for more than two points: explicit generators*, Lett. Math. Phys., **19**(1990), 327–336.

[206] M. Schlichenmaier, *Central extensions and semi-infinite wedge representations of Krichever-Novikov algebras for more than two points*, Lett. Math. Phys., **20**(1990), 33–46.

[207] M. Schlichenmaier,*Verallgemeinerte Krichever - Novikov Algebren und deren Darstellungen*. Ph.D. thesis, Universität Mannheim, 1990.

[208] M. Schlichenmaier, *Degenerations of generalized Krichever-Novikov algebras on tori*, Journal of Mathematical Physics, **34** (1993), 3809–3824.

[209] M. Schlichenmaier, *Differential operator algebras on compact Riemann surfaces*, in: Generalized Symmetries in Physics, Clausthal 1993 (eds. H.-D. Doebner, V.K. Dobrev, A.G. Ushveridze), World Scientific, Singapore London, 1994.

[210] M. Schlichenmaier, *Zwei Anwendungen algebraisch-geometrischer Methoden in der theoretischen Physik: Berezin-Toeplitz-Quantisierung und globale Algebren der zweidimensionalen konformen Feldtheorie*", Habilitation Thesis, University of Mannheim, 1996.

[211] M. Schlichenmaier, *Sugawara operators for higher genus Riemann surfaces*, Rep. on Math. Phys., **43**(1999), 323–339,

[212] M. Schlichenmaier, *Elements of a Global Operator Approach to Wess–Zumino–Novikov–Witten Models*, in "Lie theory and its applications in physics III" (July 1999, Clausthal), (eds. H.D. Doebner, V.K.Dobrev, and J. Hilgert), pp. 204–220, World Scientific, 2000, arXiv:math/0001040.

[213] M. Schlichenmaier, $W_{1+\infty}$ *algebra*, in Encyclopedia of Mathematics, Suppl. II (ed. M. Hazewinkel), pp. 486–487, Kluwer Academic Publishers, 2000.

[214] M. Schlichenmaier, *Local cocycles and central extensions for multi-point algebras of Krichever-Novikov type*, J. reine angew. Math., **559**(2003), 53–94.

[215] M. Schlichenmaier, *Higher genus affine algebras of Krichever-Novikov type*. Moscow Math. J., **3**(4), (2003), 1395–1427.

[216] M. Schlichenmaier, *Deformations of the Virasoro algebra of Krichever-Novikov type*. J. Geom. Symmetry Phys., **5**, (2006), 95–105.

[217] M. Schlichenmaier, *Higher genus affine Lie algebras of Krichever-Novikov type* Proceedings of the International Conference on Difference Equations, Special Functions, and Applications, Munich, World-Scientific, pp. 589–599, 2007.

[218] M. Schlichenmaier, *A global operator approach to Wess–Zumino–Novikov–Witten models*. In: *XXVI Workshop on Geometrical Methods in Physics*, Vol. 956 of *AIP Conf. Proc.* Melville, NY: Amer. Inst. Phys., pp. 107–119, 2007.

[219] M. Schlichenmaier, *Deformations of the Witt, Virasoro, and Current Algebra*, (in) Silvestrov, et. al., Generalized Lie theory in Mathematics, Physics and Beyond, pp. 219–234, Springer, Berlin New York, 2009

[220] M. Schlichenmaier, *Almost-graded central extensions of Lax operator algebras*, Banach Center Publ., **93**(2011), 129–144.

[221] M. Schlichenmaier, *An elementary proof of the vanishing of the second cohomology of the Witt and Virasoro algebra with values in the adjoint module*, arXiv:1111.6624v1, 2011, DOI: 10.1515/forum-2011-0143, February 2012.

[222] M. Schlichenmaier, *Lie superalgebras of Krichever-Novikov type and their central extensions*. Anal. Math. Phys., **3**(3), (2013), 235–261, arXiv:1301.0484.

[223] M. Schlichenmaier, *From the Virasoro algebra to Krichever-Novikov algebras and beyond*. (in) Vasiliev, A.(ed), Harmonic and Complex Analysis and its Applications, pp. 325–358, Springer, Berlin New York, 2014.

[224] M. Schlichenmaier, *Multipoint Lax operator algebras. Almost-graded structure and central extensions*, arXiv:1304.3902, (to appear in Math. Sbornik).

[225] M. Schlichenmaier, O. K. Sheinman, *Sugawara construction and Casimir operators for Krichever-Novikov algebras*. Jour. of Math. Science **92** (1998), 3807–3834, q-alg/9512016.

[226] M. Schlichenmaier, O. K. Sheinman, *Wess–Zumino–Witten–Novikov theory, Knizhnik-Zamolodchikov equations, and Krichever-Novikov algebras, I.*. Russian Math. Surv. (Uspekhi Math. Nauk.) **54** (1999), 213–250, math.QA/9812083.

[227] M. Schlichenmaier and O. K. Sheinman, *Knizhnik-Zamolodchikov equations for positive genus and Krichever-Novikov algebras*, Russian Math. Surv. **59**(4), (2004), 737–770,

[228] M. Schlichenmaier and O.K. Sheinman, *Central extensions of Lax operator algebras*, Usp.Math. Nauk. **382**(4), 2008, 187 –228, (engl. in Russ. Math. Surveys **63**(4), (2008), 727–766.

[229] A. M. Semikhatov, *A global operator construction of the bc-system on Riemann surfaces in terms of a bosonic conformal theory*. Phys. Lett. B., **212**(3), (1988), 357–361.

[230] A. M. Semikhatov, *Global operator formalism in the theory of bosonic strings on Riemann surfaces*. Trudy Fiz. Inst. Lebedev., **201** (1990), 180–215.

[231] A. M. Semikhatov, *Sewing $\beta\gamma$ theories on Riemann surfaces in the global operator formalism*. Phys. Lett. B., **236**(4), (1990), 433–442.

[232] L. O. Sheĭnman, *Sigma functions and Krichever-Novikov bases*. Uspekhi Mat. Nauk, **61**(6(372)), (2006), 189–190.

[233] O. K. Sheinman, *Elliptic affine Lie algebras*. Funktional Anal. i Prilozhen., **24**(3), (1992), 210–219.

[234] O. K. Sheinman, *Highest Weight modules over certain quasi-graded Lie algebras on elliptic curves*, Funktional Anal. i Prilozhen., **26**(3), (1992), 203–208.

[235] O. K. Sheinman, *Affine Lie algebras on Riemann surfaces*. Funktional Anal. i Prilozhen., **27**(4), (1993), 54–62.

[236] O. K. Sheinman, *Modules with highest weight for affine Lie algebras on Riemann surfaces*, Funktional Anal. i Prilozhen., **29**(1), (1995), 44–55-

[237] O. K. Sheinman, *Krichever-Novikov algebras and CCC-groups.* Uspekhi Mat. Nauk, **50**(5(305)), (1995), 253–254.

[238] O. K. Sheinman, *Representations of Krichever-Novikov algebras.* In: *Topics in topology and mathematical physics*, Vol. 170 of *Amer. Math. Soc. Transl. Ser. 2.* Providence, RI: Amer. Math. Soc., pp. 185–197, 1995.

[239] O. K. Sheinman, *Orbits and representations of Krichever-Novikov affine-type algebras.* J. Math. Sci., **82**(6), (1996), 3834–3843.

[240] O. K. Sheinman, *The fermion model of representations of affine Krichever-Novikov algebras.* Funktional Anal. i Prilozhen., **35**(3), (2001), 209–219.

[241] O. K. Sheinman, *Second order casimirs for the affine Krichever-Novikov algebras* $\widehat{\mathfrak{gl}}_{g,2}$ *and* $\widehat{\mathfrak{sl}}_{g,2}$, Moscow Math. J., **1** (2001), 605–628.

[242] O. K. Sheinman, *Krichever-Novikov algebras and self-duality equations on Riemann surfaces.* Uspekhi Mat. Nauk, **56**(1(337)), (2001), 185–186.

[243] O. K. Sheinman, *Second-order Casimir operators of the affine Krichever-Novikov algebras* $\widehat{\mathfrak{g}^1}$ *and* $\widehat{\mathfrak{sl}_{g,2}}$. Uspekhi Mat. Nauk, **56**(5(341)), (2001), 189–190.

[244] O. K. Sheinman, *Second-order Casimirs for the affine Krichever-Novikov algebras* $\widehat{\mathfrak{gl}}_{g,2}$ *and* $\widehat{\mathfrak{sl}}_{g,2}$. In: *Fundamental mathematics today (Russian).* Nezavis. Mosk. Univ., Moscow, pp. 372–404, 2003.

[245] O. K. Sheinman, *Krichever-Novikov algebras, their representations and applications.* In: Geometry, Topology and Mathematical Physics. S. P. Novikov's Seminar 2002–2003, V. M.Buchstaber, I. M.Krichever, eds., AMS Translations, Ser. 2, **212** (2004), 297–316, math.RT/0304020.

[246] O. K. Sheinman, *Highest-weight representations of Krichever-Novikov algebras and integrable systems.* Uspekhi Mat. Nauk, **60**(2(362)), (2005), 177–178.

[247] O. K. Sheinman, *Krichever-Novikov algebras and their representations.* In: *Noncommutative geometry and representation theory in mathematical physics*, Vol. 391 of *Contemp. Math.* Providence, RI: Amer. Math. Soc., pp. 313–321, 2005.

[248] O. K. Sheinman, *Lax operator algebras and integrable hierarchies.* Tr. Mat. Inst. Steklova 263 (Geometriya, Topologiya i Matematicheskaya Fizika. I), 216–226, 2008.

[249] O. K. Sheinman, *Lax operator algebras and Hamiltonian integrable hierarchies*, Russ. Math. Surv. **66** (2011), 145–171.

[250] O. K. Sheinman, *Current Algebras on Riemann surfaces*, Expositions in Mathematics, Vol. 58, De Gruyter, 2012.

[251] O. K. Sheinman, *Lax operator algebras of type* G_2, arXiv:1304.2510v1 (in print).

[252] S. Skryabin, *Degree one cohomology for the Lie algebras of derivations*, Lobachevskii Journal of Mathematics, **14** (2004), 69–107.

[253] H. Sugawara, *A field theory of currents.* Phys. Rev., **170** (1968), 1659–1662.

[254] K. Ueno A. Tsuchiya, and Y. Yamada, *Conformal field theory on universal family of stable curves with gauge symmetries.* Adv. Stud. Pure Math.. **19** (1989), 459–566.

[255] T. Tsujishita, On the continuous cohomology of the Lie algebra of vector fields. Proc. Japan Acad. **53** (1977) 134–138.

[256] A.N. Tyurin, *Classification of vector bundles on an algebraic curve of an arbitrary genus.* Soviet Izvestia, Ser. Mat., **29**(3), (1965), 657–688.

[257] K. Ueno, *Conformal field theory with gauge symmetry*, Fields Institute Monographs, AMS, 2008; see also Geometry and Physics, Proceed. Aarhus conference 1995 (Andersen J.E. et. al., ed.), pp. 603–745, Marcel Dekker, 1997.

[258] F. Wagemann, Some remarks on the cohomology of Krichever-Novikov algebras, Lett. Math. Phys., **47** (1999), 173–177, Erratum: Lett. Math. Phys. **52** (2000), 349

[259] F. Wagemann, *Differential graded cohomology and Lie algebras of holomorphic vector fields*. Comm. Math. Phys. **208**(2), (1999), 521–540; *Erratum*, Comm. Math. Phys., **220**(2), (2001), 453–454.

[260] F. Wagemann, *Explicit formulae for cocycles of holomorphic vector fields with values in λ densities*. J. Lie Theory, **11**(1), (2001), 173–184.

[261] F. Wagemann, *Deformations of Lie algebras of vector fields arising from families of schemes*. J. Geom. Phys., **58**(2), (2008), 165–178.

[262] F. Wagemann, *More remarks on the cohomology of Krichever-Novikov algebras*. Afr. Mat., **23**(1), (2012), 41–52.

[263] M. Wakimoto, *Fock representations of affine Lie algebras $A_1^{(1)}$*. Comm. Math. Phys., **104** (1986), 605–609.

[264] A. Wang and K. Xu, *Superconformal algebra of meromorphic vector fields with multipoles on a super Riemann sphere*, J. Math. Phys., **31**(8), (1990) 2040–2041.

[265] L.-M. Xia and N.-H. Hu, *Irreducible representations for Virasoro-toroidal Lie algebras*. J. Pure Appl. Algebra, **194**(1-2), (2004), 213–237.

[266] B. Q. Xu, D. Y. Xu, and C. G. Chao, *The KN algebras and the geometrical quantization of heterotic strings*. Internat. J. Theoret. Phys., **32**(11), (1993), 2161–2172.

[267] S. Wang and K. Xu, *Superconformal algebra of meromorphic vector fields with multipoles on a super Riemann sphere*, J. Math. Phys., **31**(8), (1990), 2040–2041.

[268] K. W. Xu and H. Y. Guo, *BRST quantization and quasifree field theory for the Krichever-Novikov string*. Acta Phys. Sinica, **38**(12), (1989), 2008–2018.

[269] C. Zachos, *Fermionic center in the superconformal algebra on the supertorus*, Nucl. Phys. B., (Proc. Suppl.) **11** (1989), 414–424.

[270] D. H. Zhang, *The Krichever-Novikov operator formalism of the superstring*. Phys. Lett. B., **216**(1-2), (1989), 107–111.

[271] Y. Zhu, *Global vertex operators on Riemann surfaces*. Comm. Math. Phys. **165**(3), (1994), 485–531.

[272] R. Zucchini, *A Krichever-Novikov formulation of classical W-algebras on Riemann surfaces*. Phys. Lett. B., **323**(3-4), (1994), 322–329.

[273] P. Zusmanovich, *The second homology group of current Lie algebras.*, Kassel, Christian (ed.) et al., K-theory., S.M.F., Astérisque. **226**, (1994), 435–452.

[274] *Deformations and Contractions in Mathematics and Physics*, Oberwolfach Reports, EMS Publishing House, **3**(1), (2006), 119–186.

[275] *Deformation Methods in Mathematics and Physics*, Oberwolfach Reports, EMS Publishing House **7**(3), (2010), 2503–2560.

Index

De Gruyter Studies in Mathematics

Volume 59
Jan de Vries
Topological Dynamical Systems, 2014
ISBN 978-3-11-034073-0, e-ISBN 978-3-11-034240-6, Set-ISBN 978-3-11-034241-3

Volume 58
Lubomir Banas, Zdzislaw Brzezniak, Mikhail Neklyudov, Andreas Prohl
Stochastic Ferromagnetism: Analysis and Numerics, 2014
ISBN 978-3-11-030699-6, e-ISBN 978-3-11-030710-8, Set-ISBN 978-3-11-030711-5

Volume 57
Dmitrii S. Silvestrov
American-Type Options: Stochastic Approximation Methods, Volume 2, 2014
ISBN 978-3-11-032968-1, e-ISBN 978-3-11-032984-1, Set-ISBN 978-3-11-032985-8

Volume 56
Dmitrii S. Silvestrov
American-Type Options: Stochastic Approximation Methods, Volume 1, 2013
ISBN 978-3-11-032967-4, e-ISBN 978-3-11-032982-7, Set-ISBN 978-3-11-032983-4

Volume 55
Lucio Boccardo, Gisella Croce
Elliptic Partial Differential Equations, 2013
ISBN 978-3-11-031540-0, e-ISBN 978-3-11-031542-4, Set-ISBN 978-3-11-031543-1

Volume 54
Yasushi Ishikawa
Stochastic Calculus of Variations for Jump Processes, 2013
ISBN 978-3-11-028180-4, e-ISBN 978-3-11-028200-9, Set-ISBN 978-3-11-028201-6

Volume 53
Martin Schlichenmaier
Krichever–Novikov Type Algebras: Theory and Applications, 2014
ISBN 978-3-11-026517-0, e-ISBN 978-3-11-027964-1, Set-ISBN 978-3-11-028025-8

Bei Fragen zur Produktsicherheit wenden Sie sich bitte an:
If you have any questions regarding product safety,
please contact:

Walter de Gruyter GmbH
Genthiner Straße 13
10785 Berlin
productsafety@degruyterbrill.com